AF602092

Operator Theory: Advances and Applications
Vol. 135

Toeplitz Matrices and Singular Integral Equations

The Bernd Silbermann Anniversary Volume

Albrecht Böttcher
Israel Gohberg
Peter Junghanns
Editors

Springer Basel AG

Editors:

Albrecht Böttcher
Faculty of Mathematics
Technical University Chemnitz
09107 Chemnitz
Germany
e-mail: aboettch@mathematik.tu-chemnitz.de

Peter Junghanns
Faculty of Mathematics
Technical University Chemnitz
09107 Chemnitz
Germany
e-mail: peter.junghanns@mathematik.tu-chemnitz.d

Israel Gohberg
School of Mathematical Sciences
Raymond and Beverly Sackler
Faculty of Exact Sciences
Tel Aviv University
Ramat Aviv 69978
Israel
e-mail: gohberg@math.tau.ac.il

2000 Mathematics Subject Classification 47-06; 47A56, 47B35, 47B48, 47N20, 47N70

A CIP catalogue record for this book is available from the
Library of Congress, Washington D.C., USA

Deutsche Bibliothek Cataloging-in-Publication Data

Toeplitz matrices and singular integral equations : the Bernd Silbermann
anniversary volume / Albrecht Böttcher ... ed. – Basel ; Boston ; Berlin :
Birkhäuser, 2002
(Operator theory ; Vol. 135)
ISBN 978-3-7643-6877-7 ISBN 978-3-0348-8199-9 (eBook)
DOI 10.1007/978-3-0348-8199-9

Originally published by Birkhäuser Verlag Basel, Switzerland in 2002

Printed on acid-free paper produced from chlorine-free pulp. TCF ∞
Cover design: Heinz Hiltbrunner, Basel

ISBN 978-3-7643-6877-7

www.birkhauser-scien

9 8 7 6 5 4 3 2 1

Contents

Editorial Preface

This volume is dedicated to Bernd Silbermann on the occasion of his sixtieth birthday. It consists of selected papers devoted to the inexhaustible and ever-young fields of Toeplitz matrices and singular integral equations, and thus to areas Bernd Silbermann has been enriching by fundamental contributions for the last three decades.

Most authors of this volume participated in the conference organized and sponsored by the Department of Mathematics (under the deans Dieter Happel and Jürgen vom Scheidt) of the Chemnitz University of Technology in honor of Bernd Silbermann in Pobershau, April 8–12, 2001. The majority of the papers presented here are based on the talks given on that conference. We thank all contributors for their enthusiasm when preparing the articles for this volume.

Bernd Silbermann

Operator Theory:
Advances and Applications, Vol. 135, 1–12

Essay on Bernd Silbermann

Albrecht Böttcher

Bernd Silbermann was born on 6 April 1941 in Langhennersdorf, a village in Saxony. His parents were farmers. To this day, he is proud of his ability to drive a tractor. He went to school in Langhennersdorf from 1947 to 1955, and in the subsequent two years he apprenticed to a grocer (and really sold fish). From 1958 to 1962 he attended a school in Chemnitz, and from 1962 to 1967, in the heyday of Soviet mathematics, he was a student of mathematics at the Lomonosov University in Moscow. His lecturers included such eminent mathematicians as P.S. Alexandrov, A.G. Kurosh, and G.E. Shilov. His diploma paper was supervised by E.A. Gorin and A.Ya. Helemskii and was devoted to the structure of radicals in certain normed rings. Since 1966 he has been married to Ludmilla Pavlovna; their children Sergej and Katja were born in 1971 and 1974. In 1967, Silbermann moved back to Chemnitz and started working under the supervision of Siegfried Prössdorf.

1967–1969. First beat of the drum

Following the advice of Prössdorf, he embarked on singular integral, Toeplitz and Wiener-Hopf operators whose symbols have zeros. From the work of Chebotarev, Cherskii, Dybin, Haikin, Karapetyants, and Prössdorf, to mention only a few principal figures, it was known that if the symbol is in C^∞ and has only a finite number of zeros of integral orders, then the corresponding Toeplitz operator is not normally solvable on the usual Banach spaces, but it is Fredholm on certain Frechet spaces of test functions and distributions. The big question of those days was whether the converse is true. Silbermann tackled this question and showed by an ingenious proof that the answer is yes: if a Toeplitz operator with a C^∞ symbol is Fredholm on the spaces of test functions or distributions mentioned, then the symbol has at most a finite number of zeros of integral orders. This result was a beat of the drum. It advanced him immediately to the first row of researchers in the field and it is up to the present the brightest star in the vault of Toeplitz operators whose symbols have zeros. The result is proved in Silbermann's paper "On singular integral operators in spaces of infinitely differentiable and generalized functions" (in Russian), which appeared in the then famous journal "Matematicheskie Issledovania" in 1971. This paper is Silbermann's actual opus 1 (another paper appeared earlier but was written later), and it has become one of his most frequently cited papers.

1970–1977. The years with Prössdorf

Within a short time, Silbermann grew from a student of Prössdorf's to a co-worker of equal rank. Together they made numerous significant contributions to the theory of singular integral operators that are not normally solvable, and to projection methods for such operators. The other co-workers of Silbermann during that period were Uwe Köhler, Christian Meyer, Johannes Steinmüller, Karla Rost, and Johannes Elschner. In 1970, Silbermann defended his Dr. rer. nat. thesis, and only four years later, in 1974, he completed his habilitation thesis. He was then 33 years old.

One cannot remember those years without mentioning Israel Gohberg. He visited Chemnitz in the late 60s and it was on his suggestion that Prössdorf and Silbermann started working on projection methods for singular integral operators whose symbols have zeros. Moreover, in 1974, Georg Heinig came to Chemnitz. He had then just accomplished his PhD under Gohberg's supervision and brought a good deal of Gohberg's spirit to Chemnitz.

In 1975, Prössdorf left Chemnitz and went to Berlin. The large amount of mathematics developed by Prössdorf and Silbermann during those years resulted in their book "Projektionsverfahren und die näherungsweise Lösung singulärer Gleichungen", which appeared in 1977. In 1976, Silbermann was appointed professor (Dozent) in Chemnitz. He then had one book and more than 20 published papers. In those times, this was an outstanding balance.

1978–1979. I entered the scene

Now it is time to introduce myself. I became a mathematics student in Chemnitz in 1975. I attended Silbermann's lecture course Analysis I–III and was fascinated by his charismatic teaching ability. In 1977, a couple of weeks before Easter, I turned to him with the request for a research problem. I remember perfectly telling him that I intended to bridge the time until Easter by doing some research. He gave me a problem and added on his turn that he would be impressed if I solved it by Christmas. The problem was as follows. Prössdorf had shown that if a function on the unit circle is Hölder continuous with the exponent α, then the partial sums of the Fourier series converge to the function in the βth Hölder norm provided $\beta < \alpha$. Silbermann asked me to check whether such a result is also true locally, that is, whether the partial sums of the Fourier series of a function that is Hölder continuous with the exponent α on some arc I converge in the βth Hölder norm ($\beta < \alpha$) to the function on every arc J properly contained in I. I solved the problem (in the negative), but this was in 1978, many months after the Christmas of 1977.

Some time in the second half of the 70s, Silbermann learned of the formula

$$(PaP)^{-1}P = Pa^{-1}P - Pa^{-1}Q(Qa^{-1}Q)^{-1}Qa^{-1}P$$

from Anatoli Kozak of Rostov-on-Don. He fell in ardent love with this formula and was soon able to do real wonders with it. One of those miracles was a new proof of the (strong) Szegő limit theorem on Toeplitz determinants. This proof allowed him to extend the theorem to exciting classes of symbols.

In 1979, Silbermann was appointed full professor and I wrote my diploma paper under his supervision. Its topic was the generalization of Silbermann's fresh results on Toeplitz determinants to the block Toeplitz case. During that time, I learned that Silbermann is not only an excellent teacher, but also an extraordinarily pleasant person, an enthusiastic researcher, and a stirring co-worker. Within a few months, we elaborated a pretty nice new approach to Toeplitz determinants and completed two papers on the topic. The mathematics we learned and developed in connection with Toeplitz determinants would become of deciding importance for our forthcoming work. Toeplitz determinants have never left us and have been an everlasting source of our inspiration for more than twenty years now.

1980–1981. The great breakthrough

One of the by-products of our research into Toeplitz determinants was a beautiful result on the finite section method for Toeplitz operators. In 1977, Naum Krupnik and Igor Verbitsky proved that if a is a piecewise smooth function with a single jump, then the finite section method is applicable to the Toeplitz operator $T(a)$ on ℓ^p if and only if $T(a)$ is invertible on both ℓ^p and ℓ^q, where $1/p + 1/q = 1$. In 1979, Silbermann and I developed a separation technique to extend this result to piecewise smooth functions with an arbitrary finite number of jumps. I was happy and satisfied, but Silbermann went further.

He was anxious for symbols with countably many jumps. It was clear that the treatment of this case required the replacement of our separation technique by some kind of a localization technique. A 1976 paper by Harold Widom contained the wonderful formula

$$T_n(a)T_n(b) = T_n(ab) - P_n H(a)H(\widetilde{b})P_n - W_n H(\widetilde{a})H(b)W_n$$

for the product of two $n \times n$ Toeplitz matrices. Here P_n is the projection onto the first n coordinates and W_n is P_n followed by reversal of the coordinates. The products of the Hankel operators, $H(a)H(\widetilde{b})$ and $H(\widetilde{a})H(b)$, were known to be compact when a and b have no common discontinuities. With this formula in mind, Silbermann understood that localization would work perfectly provided one could find an algebra of sequences of increasing matrices that contained the set of all sequences of the form

$$\{\{P_n K P_n + W_n L W_n + C_n\}_{n=1}^{\infty} : K \text{ and } L \text{ compact}, \ \|C_n\| \to 0\}$$

as an ideal. What a great and daring idea! I am still admiring it today, more than 20 years after its birth. Silbermann worked out the idea and extended the ℓ^p result on the finite section method mentioned above to symbols with countably many jumps. He published his theory in the paper "Lokale Theorie des Reduktionsverfahrens für Toeplitzoperatoren" in the Math. Nachrichten in 1981.

That short paper was a breakthrough. It laid the foundation for a new level of the application of Banach algebra techniques to numerical analysis and thus for an approach that has led to plenty of impressive results during the last twenty years.

1982–1989. Years of work and harvest

In the early 80s, Peter Junghanns and Steffen Roch entered the arena. Armed with his local principle, thrilling enthusiasm, and a wealth of ideas, Silbermann produced 2 books together with me and about 40 papers together with two, one, or none of Junghanns, Roch, and me during these eight years. As I cannot embark on all aspects of Silbermann's work of that period here, I will focus my attention on some selected topics.

Fisher-Hartwig. The Szegő limit theorem describes the aymptotic behavior of Toeplitz determinants whose symbols are smooth, have no zeros, and have winding number zero. In 1968, Michael Fisher and Robert Hartwig raised a conjecture on the asymptotic behavior of Toeplitz determinants generated by symbols with jumps and zeros. Roughly speaking, Szegő says that, after appropriate normalization,

$$\det T_n(a) = E(a)(1 + o(1)) \text{ as } n \to \infty$$

with some nonzero constant $E(a)$, while Fisher and Hartwig conjectured that, again after normalization,

$$\det T_n(a) = n^{\sum(\alpha_j^2 - \beta_j^2)} \widetilde{E}(a)(1 + o(1)) \text{ as } n \to \infty,$$

where $\widetilde{E}(a)$ is some nonzero constant (possibly different from $E(a)$) and α_j and β_j are complex numbers that measure the character of the zeros and jumps of the symbol a, respectively.

This conjecture has attracted many people over the years. In the 70s, Harold Widom and Estelle Basor proved important special cases of the conjecture: for example, Widom showed that it is true if $\beta_j = 0$ for all j, while Basor confirmed it in the cases where either $\operatorname{Re}\beta_j = 0$ for all j or $\alpha_j = 0$ and $|\operatorname{Re}\beta_j| < 1/2$ for all j. Finding a proof of the Fisher-Hartwig conjecture was also one of the favorite dreams of Silbermann. In 1980, he proved it under the assumption that $\alpha_j + \beta_j = 0$ for all j or $\alpha_j - \beta_j = 0$ for all j. Also in 1980, Silbermann and I pointed out that the conjecture is in general not true if $\alpha_j + \beta_j$ and $\alpha_j - \beta_j$ are nonzero integers for all j, and we proved a revised version of the conjecture in this case. The Fisher-Hartwig conjecture took much of our time and was the reason for many of our sleepless nights in the early 80s. However, all's well that ends well. In 1984, Silbermann and I were able to prove the conjecture provided $|\operatorname{Re}\alpha_j| < 1/2$ and $|\operatorname{Re}\beta_j| < 1/2$ for all j. This restriction to the α_j's and β_j's is a natural constraint (notice that this restriction prevents $\alpha_j \pm \beta_j$ from becoming nonzero integers). Up to the present we are proud of this result, which was communicated by Mark Krein to the Journal of Functional Analysis. In 1985, we were also able essentially to weaken the restriction to moduli less than 1/2 in the case of only a single singularity. Wonderful final stroke!? No, Torsten Ehrhardt, who will make his debut later, will set us right.

C*-algebras. When exploiting Silbermann's localization techniques, we soon realized that the prevailing Banach algebras are or can be replaced by C^*-algebras in many interesting situations. As C^*-algebras enjoy a lot of nice properties that are not shared by general Banach algebras, it was possible to sharpen various known results of numerical analysis significantly, to give extremely lucid proofs of several profound theorems, and to open the door to a wealth of new insights. Here is the beginning of the story.

Originally, I investigated the Fredholm properties of quarter-plane Toeplitz operators with discontinuous symbols. Let $\mathcal{A}$ and $\mathcal{B}$ be the Banach algebras generated by all one-dimensional Toeplitz operators on ℓ^2 with continuous and piecewise continuous symbols, respectively, and let $\mathcal{K}$ denote the compact operators. So-called bilocalization reduces invertibility in $\mathcal{B} \otimes \mathcal{B}/\mathcal{K} \otimes \mathcal{K}$ (which was what I was interested in) to invertibility in $\mathcal{B} \otimes \mathcal{B}/\mathcal{K} \otimes \mathcal{B}$. In 1977, Roland Duduchava realized that the latter algebra has the algebra $\mathcal{A} \otimes \mathbf{C}/\mathcal{K} \otimes \mathcal{B} \cong \mathcal{A}/\mathcal{K} \cong C(\mathbf{T})$ as a central subalgebra. He localized over this central subalgebra, arrived at local representatives of the form

$$T(b) \otimes T(c) + I \otimes T(d),$$

and was able to come up with these rather complicated operators. In 1981, in an attempt to extend Duduchava's approach from ℓ^2 to ℓ^p, I understood that localization over the much larger central subalgebra $\mathcal{B} \otimes \mathbf{C}/\mathcal{K} \otimes \mathcal{B} \cong \mathcal{B}/\mathcal{K} \cong C(\mathbf{T} \times [0,1])$ results in local representatives of the pretty nice form

$$I \otimes T(d),$$

which simplified Duduchava's ℓ^2 theory significantly and, moreover, also yielded the desired ℓ^p versions of his results.

In 1982, when discussing this new approach with Silbermann, we began to feel that an analogous localization should also be applicable to the finite section method for quarter-plane Toeplitz operators. The difference was that now $\mathcal{A}, \mathcal{B}, \mathcal{K}$, which are algebras of operators, must be replaced by certain algebras $\mathbf{A}, \mathbf{B}, \mathbf{K}$ of sequences of increasing matrices. A key ingredient to the localization sketched in the previous paragraph was the identification $\mathcal{B}/\mathcal{K} \cong C(\mathbf{T} \times [0,1])$. This is a 1969 result by Gohberg and Krupnik, who first observed that $\mathcal{B}/\mathcal{K}$ is a commutative C^*-algebra, then proved that the maximal ideal space can be identified with the cylinder $\mathbf{T} \times [0,1]$, and finally invoked the Gelfand-Naimark theorem to obtain that $\mathcal{B}/\mathcal{K}$ and $C(\mathbf{T} \times [0,1])$ are isometrically isomorphic. Silbermann and I did nothing but the same for $\mathbf{B}/\mathbf{K}$: we showed that it is a commutative C^*-algebra, identified the cylinder $\mathbf{T} \times [0,1]$ as the maximal ideal space, and so arrived at the conclusion that $\mathbf{B}/\mathbf{K}$ is isometrically isomorphic to $C(\mathbf{T} \times [0,1])$. To the best of my knowledge, this was the first time that C^*-algebras were deliberately used in connection with a question of numerical analysis.

Idempotents. Localization in Toeplitz algebras generated by piecewise continuous symbols leads to singly generated algebras. But now consider the Banach algebra $\mathcal{C}$ generated by all singular integral operators $aP+b(I-P)$ with piecewise continuous coefficients a, b. Suppose the underlying space is $L^p(\Gamma)$ over a closed sufficiently nice curve Γ. Localization is, in a sense, equivalent to "freezing" the coefficients at the points of Γ. Clearly, P remains P after localization. But the coefficient a can be replaced by $a(t-0)(1-\chi_t)+a(t+0)\chi_t$ at $t \in \Gamma$, where χ_t is 1 in some right half-neighborhood of t and 0 elsewhere. Thus, after localization we arrive at the algebra generated by P and χ_t. Since $P^2 = P$ and $\chi_t^2 = \chi_t$, this is an algebra that is generated by two idempotents (or, in more loose language, by two projections). If $p = 2$, then all algebras involved are C^*-algebras and P and χ_t are selfadjoint. This allows us to employ the two projections theorem by Halmos, which, via the correspondence

$$P \mapsto \begin{pmatrix} 1 & 0 \\ 0 & 0 \end{pmatrix}, \quad \chi_t \mapsto \begin{pmatrix} x & \sqrt{x(1-x)} \\ \sqrt{x(1-x)} & 1-x \end{pmatrix},$$

establishes an isometric isomorphism of the local algebra at $t \in \Gamma$ onto a C^*-subalgebra of $C^{2\times 2}([0,1])$, the 2×2 continuous matrix functions on $[0,1]$, and eventually gives the well-known symbol map of $\mathcal{C}$ into $C^{2\times 2}(\Gamma \times [0,1])$. This nifty way of deriving the symbol calculus for singular integral operators is due to Ronald Douglas (1978).

In 1987, Roch and Silbermann performed the brilliant feat of extending this approach from C^*-algebras to general Banach algebras (and, in particular, from $p = 2$ to all $p \in (1, \infty)$). They proved that everything works well if the points 0 and 1 are cluster points of a certain set X, which, roughly speaking, is the spectrum of the "Toeplitz operator" $P\chi_t P$. In Halmos's setting, X is $[0,1]$. In all other situations we have been confronted with, X is a circular arc, or a horn, or a logarithmic double-spiral, or something like this between 0 and 1, and hence the Roch-Silbermann condition is always satisfied. Roch and Silbermann's theorem on Banach algebras generated by two idempotents has stimulated much subsequent research into the topic (for example by Gohberg, Krupnik, Spitkovsky, Karlovich, myself, Finck, Ehrhardt, Samoilenko) and is now one of the pillars of the modern edifice of singular integral equations.

Numerical analysis. Silbermann's breakthrough with the finite section method resulted in a genuine renaissance of several approximation methods for singular integral equations. In 1984, Peter Junghanns presented a splendid dissertation on polynomial collocation methods for singular integral equations. A little bit later, splines and wavelets became the fashion. A Galerkin method for the approximate solution of the equation $Ax = y$ is the passage from this equation to n equations $(Ax^{(n)}, \psi_j) = (y, \psi_j)$ with certain test functions (polynomials, splines, wavelets) $\psi_1, \ldots, \psi_n$. People replaced the integral hidden in the scalar product $(\cdot,\cdot)$ by a quadrature formula $(\cdot,\cdot)_Q$ and called this a *qualocation* method. Or they first discretized the integral operator A to an operator A_D and then applied qualocation to

A_D, that is, they considered the equations $(A_D x^{(n)}, \psi_j)_Q = (y, \psi_j)_Q$. This method was named *quadrocation.* All these strategies have led to fascinating mathematics. The heros in this realm included Hagen, Roch, Junghanns, Prössdorf, Elschner, Rathsfeld, Schmidt, Mastroianni, Monegato, Vainikko, Chandler, Graham, Arnold, Wendland, Saranen, Sloan, Dahmen, Schneider, and, of course, Silbermann.

Books. In the 80s, Silbermann and I wrote two books: "Invertibility and Asymptotics of Toeplitz Matrices" (Akademie-Verlag 1983) and "Analysis of Toeplitz Operators" (Akademie-Verlag 1989 and Springer-Verlag 1990). The second book has now become a standard reference. So let me tell a story about the first.

In contrast to me, Silbermann is a very reasonable person. He is reasonable in all respects and able to bravely resist even the toughest temptations. Here is a convincing example. Some day in the late 80s, he received a letter from Reinhard Höppner, who was then the mathematics reader and editor of the Akademie-Verlag (and is now the Minister President of the Land Sachsen-Anhalt). Höppner wrote that the sale of the book has tended to zero, that the book will therefore be made to waste paper, but that the authors have the right to buy as many copies as they want, at a price between 2 and 5 marks per copy (I don't remember the exact price). The other day, Silbermann radiantly came to me and enthusiastically told me that he had made use of the opportunity and had ordered 5 copies! The same day I asked Höppner to send 25 more copies to me.

In 1990, Roch and Silbermann's report "Algebras of convolution operators and their image in the Calkin algebra" was published by the Karl Weierstrass Institute for Mathematics in Berlin. This is not a book in the true sense of the word (although it has an ISSN number), but it is a text of 156 pages that has found a prominent place on the bookshelves of many colleagues. At least my personal copy is totally worn out.

Finally, in the last years of the 80s, Prössdorf and Silbermann wrote their capital monograph "Numerical Analysis for Integral and Related Operator Equations" (Akademie-Verlag 1991 and OT 52 of the Birkhäuser Verlag). I myself was an immediate witness to their enormous efforts to complete this volume. They had to organize and systematize a gigantic amount of material, they were forced to close many subtle gaps, and in addition, Silbermann was obsessed by the ambition to include several topics on which work was still in progress. The sole existence of this opus provides us with an idea of the stamina, strength, and greatness of people of the calibre of Prössdorf and Silbermann.

1990 up to the present. Hard and golden years

The political events in Germany changed our life dramatically. All (Eastern) university teachers who had been appointed before 1990 were dismissed from their positions and had to apply anew. Silbermann survived this 2 or 3 year procedure thanks to his integrity and indisputably high scientific reputation. It was nevertheless a hard time that has left its imprint on everyone who was concerned.

In addition, Silbermann had been afflicted by cancer since the middle of the 80s and had to undergo three serious operations and countless medical treatments over a period of more than ten years. He had to suffer beyond words but never gave up the hopes for successful treatment and never forgot his care and responsibility for his family and his students and co-workers. For example, he held his regular lecture courses at the university when still undergoing chemotherapy! I had experienced all the ups and downs of the mathematics and university business together with Silbermann and had come to know his inexhaustible optimism, his remarkable sense of responsibility, and his ability of overcoming all obstacles in some manner. But all this was nothing in comparison with those cruel buffets of fate. This time Silbermann showed us what it means to fight as a real man. His unbelievable struggle against the disease was successful. Now his state of health is normal and his way of passionately teaching, glowingly doing research, and circumspectly managing several things makes me sometimes forget that he had to endure those hard ten years. My admiration for Silbermann, and also for Gohberg, who shared a similar fate, has grown by dimensions. Their greatness as mathematicians has always been beyond all question, but the will-power they showed to defeat this fate is beyond my imagination.

Oberwolfach. Before 1990, Silbermann was a guest of several renowned research groups of the Soviet Union, in Kishinev, Moscow, St. Petersburg, and Rostov-on-Don. He was an invited speaker of conferences in Poland, Hungary, Bulgaria, Romania, and in the Georgian Republic. He also quite regularly participated in the annual workshops organized by Vlastimil Pták in the Czech and Slovak Republics. He was not allowed to accept any of the many invitations he received from colleagues in the Western countries.

His first conference outside the Eastern area was the memorable Oberwolfach meeting "Wiener-Hopf Problems, Toeplitz Operators, and Applications" in December 1989. This meeting, organized by Israel Gohberg, Rien Kaashoek, and Ehrhard Meister, was a remarkable event. The participants included H. Bart, M. Costabel, R. Duduchava, H. Dym, J. Elschner, I. Feldman, L. Frank, R. Gorenflo, B. Gramsch, G. Heinig, G. Litvinchuk, R. Mennicken, S. Prössdorf, A. dos Santos, F. van Schagen, F.-O. Speck, E. Stephan, F. Teixera, H. Upmeier, N. Vasilevski, H. Widom, L. von Wolfersdorf, and me. Even today, both Silbermann and I are moved when remembering the irretrievable atmosphere of this conference and the talks with the people we had known from their work for a long time but met only during those days for the first time.

The scope of this essay does not allow me to list the many universities Silbermann visited and the numerous conferences he attended during the ten years after that Oberwolfach meeting. He is especially enthusiastic about his visits to Hermann Brunner in St. John's, Newfoundland, in 1994 and to Ian Sloan in Sydney, Australia, in 1999. He also benefited much from a half-year stay in Portugal in 2000. To give a fourth and final example, I want to mention that he was one of the Toeplitz lecturers at the Tel Aviv University in 1995.

Torsten Ehrhardt. This young man completed the school in 1990 and studied mathematics in Chemnitz from 1990 to 1994. His mathematical gift exceeds that of all other students of Silbermann's. In the early 90s, still being a student, Ehrhardt got involved in research by Harm Bart and Silbermann into logarithmic residues in Banach algebras, which was the starting point for a fruitful collaboration that is lasting up to the present.

Ehrhardt was the right man to attack the Fisher-Hartwig conjecture again. In the case of a single singularity, Silbermann and I had proved the conjecture under the assumption $\operatorname{Re}\alpha \geq 0$ and $|\operatorname{Re}\beta| < \operatorname{Re}\alpha + 1$. In 1996, Ehrhardt and Silbermann removed this restriction completely and proved Fisher-Hartwig under the sole condition $\operatorname{Re}\alpha > -1/2$. For more than one singularity, the Fisher-Hartwig conjecture was known not to be true in general. In 1991, Estelle Basor and Craig Tracy reformulated and generalized the Fisher-Hartwig conjecture to what is now the Basor-Tracy conjecture. In 1997, Ehrhardt defended an outstanding dissertation, in which he proved the Basor-Tracy conjecture in all the cases in which it coincides with the Fisher-Hartwig conjecture. Equivalently, he proved the Fisher-Hartwig conjecture in all the cases where it is surmised to be true. In addition, Ehrhardt has remarkable papers on the Szegő-Widom limit theorem for block Toeplitz determinants, papers with Estelle Basor on Toeplitz plus Hankel operators, a gigantic paper with Ilya Spitkovsky in the St. Petersburg Math. Journal on the factorization of piecewise constant matrix functions and the monodromy of Fuchsian differential equations ... In short, Torsten Ehrhardt is a stroke of luck for Silbermann's research group.

Singular values. Silbermann's scientific outcome of the 90s includes 3 books and more than 50 papers. The number of his co-authors has grown from 22 in 1989 to 37 in 2000, and his (and my) Erdős number went down to 3. In the order of their appearance, the 15 new co-authors are: Ilya Spitkovsky, Dietmar Berthold, Wolfram Hoppe, Viktor Didenko, Tilo Finck, Yuri Karlovich, Harm Bart, Torsten Ehrhardt, Harold Widom, Israel Gohberg, Sergei Grudsky, Pedro Santos, Volodya Rabinovich, Anatoli Kozak, and Roland Duduchava. Whatever proper subset of the 50 papers I would mention honorably here, I would definitely leave out essential pieces of Silbermann's work in the 90s. I therefore concentrate upon a single item: singular values.

The singular values $s_1(T_n(a)) \leq s_2(T_n(a)) \leq \cdots \leq s_n(T_n(a))$ of the $n \times n$ Toeplitz matrix $T_n(a)$ with the symbol a are the square roots of the eigenvalues of $T_n^*(a)T_n(a)$. Throughout what follows, I assume that a is a piecewise continuous function. In 1989, Harold Widom showed that the set of the singular values of $T_n(a)$ converges in the Hausdorff metric to some set one can completely identify. In 1992, Silbermann extended this result to locally normal symbols. A function is called locally normal if, for each point $t \in \mathbf{T}$, localization (=freezing) of the function at t yields a function whose range lies on a straight line (for piecewise continuous functions, localization gives two points at jumps and one point at the points of continuity). The result appeared as the paper "On the limiting set of singular values of Toeplitz matrices" in LAA 182, 35–43 (1993). The theorem proved in that paper was new and interesting. However, the real importance of that paper is the C^*-algebra techniques used by Silbermann in the course of the proof. As this is not the right place to describe these techniques, I confine myself to citing three other results that were subsequently obtained on the basis of these techniques.

First, it was just that paper by Silbermann which showed me the right way how to sharpen the classical result

$$\limsup_{n\to\infty} \|T_n^{-1}(a)\| < \infty \quad \Longleftrightarrow \quad \|T^{-1}(a)\| < \infty,$$

which was established by Gohberg and Feldman for continuous symbols a and extended by Gohberg to piecewise continuous symbols a, to the equality

$$\lim_{n\to\infty} \|T_n^{-1}(a)\| = \|T^{-1}(a)\|.$$

This equality is in turn the key to an alternative proof of Landau, Reichel, and Trefethen's theorem, according to which the pseudospectra of $T_n(a)$ converge to the pseudospectrum of $T(a)$ in the Hausdorff metric (which is not true for the usual spectra!). In 1994, I published the above equality and their consequences for pseudospectra, and since then I have observed with pleasure a steadily increasing interest in C^*-algebras in the pseudospectra community. In comparison herewith, the acceptance of C^*-algebra techniques by the numerical analysis community must be called hesitant.

Secondly, Gohberg and Feldman's result cited in the previous paragraph can be restated in the form

$$\liminf_{n\to\infty} s_1(T_n(a)) > 0 \quad \Longleftrightarrow \quad T(a) \text{ is invertible.}$$

Notice that a Toeplitz operator is invertible if and only if it is Fredholm of index zero. In the middle of the 90s, Roch and Silbermann generalized this result to the following theorem. If $T(a)$ is not Fredholm, then

$$\lim_{n\to\infty} s_k(T_n(a)) = 0 \text{ for each } k,$$

while if $T(a)$ is Fredholm of index k, then

$$\lim_{n\to\infty} s_{|k|}(T_n(a)) = 0 \quad \text{and} \quad \liminf_{n\to\infty} s_{|k|+1}(T_n(a)) > 0.$$

Thus, the number of the singular values of $T_n(a)$ that go to zero is equal to the modulus of the Fredholm index of $T(a)$. This so-called splitting theorem is certainly not the deepest and most important theorem by Roch and Silbermann, but it is probably the most beautiful. An illustration is on the next page.

Finally, in the early 90s, Georg Heinig and Frank Hellinger studied the problem of whether the Moore-Penrose inverses $T_n^+(a)$ of $T_n(a)$ converge strongly to the Moore-Penrose inverse $T^+(a)$ of $T(a)$. This problem is equivalent to the following: if $T(a)$ is normally solvable but not invertible, is there a sequence $\{B_n\}$

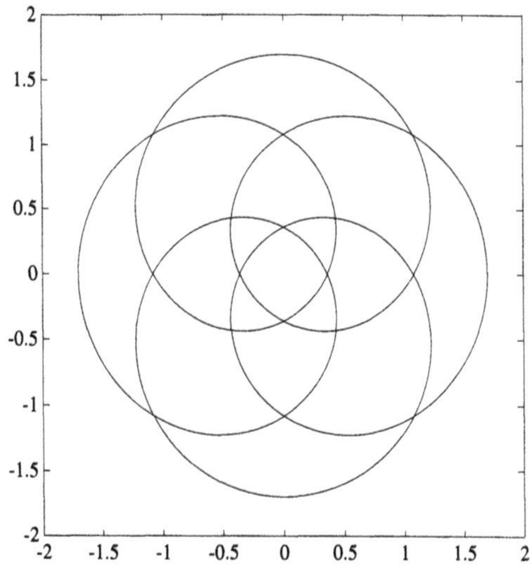

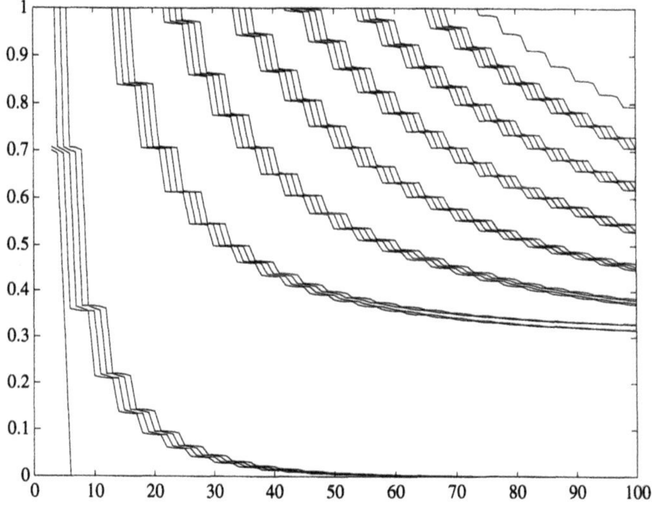

The top figure shows the range of the symbol $a(t) = 0.7\,t + t^5$ for $t \in \mathbf{T}$. The winding number of a about the point 0.01 is 5, and hence $T(a - 0.01)$ has index -5. The bottom figure plots the singular values $s_j(T_n(a - 0.01))$ for $3 \le n \le 100$ and $1 \le j \le \min(n, 30)$. As predicted by Roch and Silbermann, the 5 lowest singular values go to zero, while the remaining singular values stay away from zero.

of $n \times n$ matrices B_n such that

$$T_n(a)B_nT_n(a) - T_n(a) = 0, \quad B_nT_n(a)B_n - B_n = 0,$$
$$(T_n(a)B_n)^* - T_n(a)B_n = 0, \quad (B_nT_n(a))^* - B_nT_n(a) = 0,$$

and $B_n \to T^+(a)$ strongly? Heinig and Hellinger showed that this happens in rare cases only, that is, generically the answer to the question is NO. However, in 1995, Silbermann ingeniously replaced the above question by the following: if $T(a)$ is normally solvable but not invertible, is there a sequence $\{B_n\}$ of $n \times n$ matrices B_n such that

$$\|T_n(a)B_nT_n(a) - T_n(a)\| \to 0, \quad \|B_nT_n(a)B_n - B_n\| \to 0,$$
$$\|(T_n(a)B_n)^* - T_n(a)B_n\| \to 0, \quad \|(B_nT_n(a))^* - B_nT_n(a)\| \to 0,$$

and $B_n \to T^+(a)$ strongly? And Silbermann showed that the answer is always YES!

Books. With my good German bent for completeness, I want to add that in the 90s Silbermann wrote the two books "Spectral Theory of Approximation Methods for Convolution Equations" (Birkhäuser Verlag 1995, OT 74) and "C^*-Algebras and Numerical Analysis" (Marcel Dekker 2001, Monographs and Textbooks in Pure and Applied Mathematics 236) with Hagen and Roch and the book "Introduction to Large Truncated Toeplitz Matrices" (Springer-Verlag 1999, universitext) with me. The Hagen-Roch-Silbermann books are mathematical high technology. In addition, their C^*-algebras book is a methodological and aesthetical masterpiece. Although I am not supposed to praise the book co-authored by myself, I proudly state that I find it charming and that I love it. The two pictures on the previous page (done by Harald Heidler) are from that universitext.

I hope the above suffices to provide at least a modest idea of Silbermann as an exceptional mathematician and as a wonderful man. This essay is very personal, and others would probably write it differently and better. However, my entire career is interlaced with Silbermann in a way that makes it difficult to throw any subjective views overboard. In April 2001, we celebrated his 60th birthday. I wish him stable health and never-ending productivity, I wish him and his wife, Ludmilla Pavlovna, a good and exciting life, and I wish that he can take me, his family, his friends and colleagues under his wings for many more years.

Albrecht Böttcher
Fakultät für Mathematik
TU Chemnitz
D-09107 Chemnitz, Germany
e-mail: `aboettch@mathematik.tu-chemnitz.de`

Operator Theory:
Advances and Applications, Vol. 135, 13–23

Publications of Bernd Silbermann

Dissertation

B. SILBERMANN, *Einige Fragen zur Theorie singulärer Integraloperatoren nicht normalen Typs.* TU Chemnitz, 22 December 1970.

Habilitation

B. SILBERMANN, *Singuläre Integraloperatoren vom nicht normalen Typ mit unstetigen Koeffizienten.* TU Chemnitz, 12 July 1974.

Books

1. S. PRÖSSDORF AND B. SILBERMANN, *Projektionsverfahren und die näherungsweise Lösung singulärer Gleichungen.* Teubner-Texte zur Mathematik, Teubner, Leipzig 1977.
2. A. BÖTTCHER AND B. SILBERMANN, *Invertibility and Asymptotics of Toeplitz Matrices.* Mathematical Research, Vol. 17, Akademie-Verlag, Berlin 1983.
3. A. BÖTTCHER AND B. SILBERMANN, *Analysis of Toeplitz Operators.* Akademie-Verlag, Berlin 1989 and Springer-Verlag, Berlin, Heidelberg, New York 1990.
4. S. PRÖSSDORF AND B. SILBERMANN, *Numerical Analysis for Integral and Related Operator Equations.* Akademie-Verlag, Berlin 1991 and Operator Theory: Advances and Applications, Vol. 52, Birkhäuser Verlag, Basel 1991.
5. R. HAGEN, S. ROCH AND B. SILBERMANN, *Spectral Theory of Approximation Methods for Convolution Equations.* Operator Theory: Advances and Applications, Vol. 74, Birkhäuser Verlag, Basel 1995.
6. A. BÖTTCHER AND B. SILBERMANN, *Introduction to Large Truncated Toeplitz Matrices.* Universitext, Springer-Verlag, New York 1999.
7. R. HAGEN, S. ROCH AND B. SILBERMANN, *C^*-Algebras and Numerical Analysis.* Monographs and Textbooks in Pure and Applied Mathematics, Vol. 236, Marcel Dekker, New York 2001.

Papers

1. B. SILBERMANN, *Über eine Klasse von singulären Integralgleichungen, deren Symbol nicht mehr als eine endliche Anzahl von Nullstellen ganzzahliger und gebrochener Ordnung aufweist.* Math. Nachr. **47**, 245–260 (1970).
2. B. SILBERMANN, *On singular integral operators in spaces of infinitely differentiable and generalized functions.* (Russian) Matem. Issled. **6**, 168–179 (1971).

3. B. Silbermann, *Über einen Zugang zu einer Klasse eindimensionaler singulärer Integraloperatoren nicht normalen Typs.* Wiss. Z. TH Karl-Marx-Stadt **13**, 135–142 (1971).

4. B. Silbermann, *Über eine Klasse einseitig invertierbarer singulärer Integraloperatoren nicht normalen Typs in gewissen Paaren von Banach-Räumen I.* Math. Nachr. **51**, 327–342 (1971).

5. B. Silbermann, *Über eine Klasse einseitig invertierbarer singulärer Integraloperatoren nicht normalen Typs in gewissen Paaren von Banach-Räumen II.* Math. Nachr. **52**, 297–313 (1972).

6. S. Prössdorf and B. Silbermann, *Über die normale Auflösbarkeit des singulären Integraloperators vom nicht normalen Typ.* Math. Nachr. **55**, 73–88 (1973).

7. S. Prössdorf and B. Silbermann, *Über die normale Auflösbarkeit von Systemen singulärer Integralgleichungen vom nicht normalen Typ.* Math. Nachr. **56**, 131–144 (1973).

8. U. Köhler and B. Silbermann, *Einige Ergebnisse über Φ_+-Operatoren in lokalkonvexen, topologischen Vektorräumen.* Math. Nachr. **56**, 145–153 (1973).

9. U. Köhler and B. Silbermann, *Über algebraische Eigenschaften einer Klasse von Operatorenmatrizen und eine Anwendung auf singuläre Integraloperatoren.* Math. Nachr. **57**, 245–258 (1973).

10. B. Silbermann, *Zur Theorie eindimensionaler singulärer Integraloperatoren nicht normalen Typs mit stückweise stetigen Koeffizienten.* Math. Nachr. **57**, 371–384 (1973).

11. B. Silbermann, *Über paarige Operatoren nicht normalen Typs.* Math. Nachr. **60**, 79–95 (1974).

12. B. Silbermann, *Über die einseitige Invertierbarkeit gewisser Klassen paariger Operatoren nicht normalen Typs.* Wiss. Z. TH Karl-Marx-Stadt **16**, 277–381 (1974).

13. S. Prössdorf and B. Silbermann, *Ein Projektionsverfahren zur Lösung singulärer Gleichungen vom nicht normalen Typ.* Wiss. Z. TH Karl-Marx-Stadt **16**, 367–376 (1974).

14. S. Prössdorf and B. Silbermann, *Ein Projektionsverfahren zur Lösung abstrakter singulärer Gleichungen vom nicht normalen Typ und einige seiner Anwendungen.* Math. Nachr. **61**, 133–155 (1974).

15. B. Silbermann, *Ein Projektionsverfahren für schwach ausgeartete singuläre Integralgleichungen.* Z. Angew. Math. Mech. **55**, 525–527 (1975).

16. S. Prössdorf and B. Silbermann, *Verallgemeinerte Projektionsverfahren zur Lösung singulärer Gleichungen vom nicht normalen Typ.* Math. Nachr. **68**, 7–28 (1975).

17. B. SILBERMANN, *Über eine Klasse singulärer Integraloperatoren nicht normalen Typs mit stückweise stetigen Koeffizienten.* In: 5. Tagung über Probleme und Methoden der Mathematischen Physik (TH Karl-Marx-Stadt, 1975), pp. 305–308, TH Karl-Marx-Stadt 1975.
18. S. PRÖSSDORF AND B. SILBERMANN, *The convergence of the reduction and the collocation methods for systems of singular integral equations.* (Russian) Dokl. Akad. Nauk SSSR **226**, 516–519 (1976).
19. B. SILBERMANN, *Ein Projektionsverfahren für einen diskreten Wiener-Hopfschen Operator, dessen Koeffizientensymbole Nullstellen nicht ganzzahliger Ordnung besitzen.* Math. Nachr. **74**, 191–199 (1976).
20. S. PRÖSSDORF AND B. SILBERMANN, *General theorems on the convergence of projection methods for operator equations in Banach spaces.* (Russian) Dokl. Akad. Nauk SSSR **230**, 527–529 (1976).
21. S. PRÖSSDORF AND B. SILBERMANN, *Einige allgemeine Sätze zur Theorie der Projektionsverfahren für lineare Operatorgleichungen in Banachräumen.* Math. Nachr. **75**, 61–72 (1976).
22. S. PRÖSSDORF AND B. SILBERMANN, *Projektionsverfahren zur Lösung von Systemen singulärer Gleichungen vom nicht normalen Typ.* Rev. Roumaine Math. Pures Appl. **22**, 965–991 (1977).
23. CH. MEYER AND B. SILBERMANN, *Die Indexformel für eine Klasse von ausgearteten singulären Integraloperatoren mit Carlemanscher Verschiebung.* Demonstratio Math. **10**, 155–167 (1977).
24. S. PRÖSSDORF AND B. SILBERMANN, *Gestörte Projektionsverfahren und einige ihrer Anwendungen.* In: Theory of Nonlinear Operators (Berlin, 1977), pp. 229–237, Abh. Akad. Wiss. DDR, Abt. Math. Naturwiss. Tech., No. 6, Akademie-Verlag, Berlin 1978.
25. B. SILBERMANN, *Zur Berechnung von Toeplitz-Determinanten, die durch eine Klasse im wesentlichen beschränkter Funktionen erzeugt werden.* Wiss. Z. TH Karl-Marx-Stadt **20**, 683–687 (1978).
26. K. ROST AND B. SILBERMANN, *Das Reduktionsverfahren für eine Klasse ausgearteter Integrodifferenzengleichungen.* Wiss. Z. TH Karl-Marx-Stadt **20**, 689–691 (1978).
27. J. ELSCHNER AND B. SILBERMANN, *Eine Klasse entarteter gewöhnlicher Differentialgleichungen und das Kollokationsverfahren zu ihrer Lösung.* Czechoslovak Math. J. **29**, 551–563 (1979).
28. B. SILBERMANN, *Some remarks on the asymptotic behavior of Toeplitz determinants.* Applicable Anal. **11**, 185–197 (1980).
29. A. BÖTTCHER AND B. SILBERMANN, *Notes on the asymptotic behavior of block Toeplitz matrices and determinants.* Math. Nachr. **98**, 183–210 (1980).
30. A. BÖTTCHER AND B. SILBERMANN, *The asymptotic behavior of Toeplitz determinants for generating functions with zeros of integral orders.* Math. Nachr. **102**, 79–105 (1981).

31. P. Junghanns and B. Silbermann, *Zur Theorie der Näherungsverfahren für singuläre Integralgleichungen auf Intervallen.* Math. Nachr. **103**, 199–244 (1981).

32. B. Silbermann, *Ausgeartete paarige Integrodifferenzengleichungen und ein Projektionsverfahren zu ihrer Lösung.* Wiss. Z. TH Karl-Marx-Stadt **23**, 465–470 (1981).

33. B. Silbermann, *Das asymptotische Verhalten von Toeplitzdeterminanten für einige Klassen von Erzeugerfunktionen.* In: Nonlinear Analysis (Berlin, 1979), pp. 267–272, Abh. Akad. Wiss. DDR, Abt. Math. Naturwiss. Tech., 1981, No. 2, Akademie-Verlag, Berlin 1981.

34. B. Silbermann, *The strong Szegő limit theorem for a class of singular generating functions I.* Demonstratio Math. **14**, 647–667 (1981).

35. B. Silbermann, *Lokale Theorie des Reduktionsverfahrens für Toeplitzoperatoren.* Math. Nachr. **104**, 137–146 (1981).

36. A. Böttcher and B. Silbermann, *Über das Reduktionsverfahren für diskrete Wiener-Hopf-Gleichungen mit unstetigem Symbol.* Z. Anal. Anwendungen **1:2**, 1-5 (1982).

37. S. Roch and B. Silbermann, *Das Reduktionsverfahren für Potenzen von Toeplitzoperatoren mit unstetigem Symbol.* Wiss. Z. TH Karl-Marx-Stadt **24**, 289–294 (1982).

38. B. Silbermann, *Lokale Theorie des Reduktionsverfahrens für singuläre Integralgleichungen.* Z. Anal. Anwendungen **1:6**, 45–56 (1982).

39. B. Silbermann, *Notes on the asymptotic behavior of Toeplitz matrices and determinants.* In: Functions, Series, Operators, Vols. I, II (Budapest, 1980), pp. 1063–1073, Colloq. Math. Soc. János Bolyai, 35, North-Holland, Amsterdam 1983.

40. A. Böttcher and B. Silbermann, *The finite section method for Toeplitz operators on the quarter-plane with piecewise continuous symbols.* Math. Nachr. **110**, 279–291 (1983).

41. A. Böttcher and B. Silbermann, *Wiener-Hopf determinants with symbols having zeros of analytic type.* In: Seminar Analysis 1982/83, pp. 224–243, Akad. Wiss. DDR, Berlin 1983.

42. A. Böttcher and B. Silbermann, *Toeplitz determinants generated by symbols with one singularity of Fisher-Hartwig type.* Wiss. Z. TH Karl-Marx-Stadt **26**, 186–188 (1984).

43. P. Junghanns and B. Silbermann, *Numerical analysis for one-dimensional Cauchy-type singular integral equations.* In: Problems and Methods of Mathematical Physics (Karl-Marx-Stadt, 1983), pp. 122–129, Teubner-Texte Math., 63, Teubner, Leipzig 1984.

44. P. Junghanns and B. Silbermann, *Local theory of the collocation method for the approximate solution of singular integral equations I.* Integral Equations Operator Theory **7**, 791–807 (1984).

45. G. Heinig and B. Silbermann, *Factorization of matrix functions in algebras of bounded functions.* In: Spectral Theory of Linear Operators and Related Topics (Timişoara/Herculane, 1983), pp. 157–177, Oper. Theory Adv. Appl., 14, Birkhäuser, Basel 1984.

46. A. Böttcher and B. Silbermann, *Toeplitz determinants with symbols from the Fisher-Hartwig class.* (Russian) Dokl. Akad. Nauk SSSR **278**, 13-16 (1984).

47. B. Silbermann, *Harmonic approximation of Toeplitz operators and index formulas.* Integral Equations Operator Theory **8**, 842–853 (1985).

48. R. Hagen and B. Silbermann, *A finite element collocation method for bisingular integral equations.* Applicable Anal. **19**, 117–135 (1985).

49. S. Roch and B. Silbermann, *Toeplitz-like operators, quasicommutator ideals, numerical analysis I.* Math. Nachr. **120**, 141–173 (1985).

50. A. Böttcher and B. Silbermann, *Toeplitz matrices and determinants with Fisher-Hartwig symbols.* J. Funct. Anal. **63**, 178–214 (1985).

51. A. Böttcher, S. Roch and B. Silbermann, *Local constructions and Banach algebras associated with Toeplitz operators on H^p.* In: Seminar Analysis 1985/86, pp. 23–30, Akad. Wiss. DDR, Berlin 1986.

52. B. Silbermann, *Numerical analysis for Wiener-Hopf integral equations in spaces of measurable functions.* In: Seminar Analysis 1985/86, pp. 187–203, Akad. Wiss. DDR, Berlin 1986.

53. B. Silbermann, *Asymptotics for Toeplitz operators with piecewise quasicontinuous symbols and related questions.* Math. Nachr. **125**, 179–190 (1986).

54. B. Silbermann, *Local objects in the theory of Toeplitz operators.* Integral Equations Operator Theory **9**, 706–738 (1986).

55. A. Böttcher and B. Silbermann, *Toeplitz operators and determinants generated by symbols with one Fisher-Hartwig singularity.* Math. Nachr. **127**, 95–123 (1986).

56. A. Böttcher and B. Silbermann, *Local spectra of approximate identities, cluster sets, and Toeplitz operators.* Wiss. Z. TH Karl-Marx-Stadt **28**, 175–180 (1986).

57. R. Hagen and B. Silbermann, *Local theory of the collocation method for the approximate solution of singular integral equations II.* In: Seminar analysis 1986/87, pp. 41–56, Akad. Wiss. DDR, Berlin 1987.

58. S. Roch and B. Silbermann, *Finite sections of singular integral operators with Carleman shift.* In: Seminar Analysis 1986/87, pp. 149–180, Akad. Wiss. DDR, Berlin 1987.

59. B. Silbermann, *The C^*-algebra generated by Toeplitz and Hankel operators with piecewise quasicontinuous symbols.* Integral Equations Operator Theory **10**, 730–738 (1987).

60. S. Roch and B. Silbermann, *Toeplitz-like operators, quasicommutator ideals, numerical analysis II.* Math. Nachr. **134**, 245–255 (1987).

61. A. Böttcher and B. Silbermann, *Toeplitz operators in l^p spaces, with symbols from $C + H^\infty$.* (Russian) Zap. Nauchn. Sem. LOMI **157**, 124–128 (1987); Engl. transl. in J. Soviet Math. **44**, 834–836 (1989).

62. R. Hagen and B. Silbermann, *On the stability of the qualocation method.* In: Seminar Analysis 1987/88, pp. 43–52, Akademie-Verlag, Berlin 1988.

63. P. Junghanns and B. Silbermann, *Numerical analysis of the quadrature method for solving linear and nonlinear singular integral equations.* Wissensch. Schriftenreihe der TU Karl-Marx-Stadt 10/1988 (1988).

64. S. Roch and B. Silbermann, *Algebras generated by idempotents and the symbol calculus for singular integral operators.* Integral Equations Operator Theory **11**, 385–419 (1988).

65. S. Roch and B. Silbermann, *A symbol calculus for finite sections of singular integral operators with shift and piecewise continuous coefficients.* J. Funct. Anal. **78**, 365–389 (1988).

66. A. Böttcher, N. Krupnik and B. Silbermann, *A general look at local principles with special emphasis on the norm computation aspect.* Integral Equations Operator Theory **11**, 455–479 (1988).

67. H. Schulze and B. Silbermann, *One-dimensional singular integral operators on Hölder-Zygmund spaces.* In: Seminar Analysis 1988/89, pp. 129–139, Akad. Wiss. DDR, Berlin 1989.

68. B. Silbermann, *Approximation of solutions of periodic pseudodifferential equations and optimal error estimates.* In: Seminar Analysis 1988/89, pp. 141–152, Akad. Wiss. DDR, Berlin 1989.

69. S. Roch and B. Silbermann, *A symbol calculus for the algebra generated by shift operators.* Z. Anal. Anwendungen **8:4**, 293–306 (1989).

70. A. Böttcher and B. Silbermann, *Asymptotics of Toeplitz and Wiener-Hopf operators.* In: Proc. 9th Conf. on Problems and Methods in Math. Physics (Karl-Marx-Stadt, 1988), pp. 27–35, Teubner-Texte Math., 111, Teubner, Leipzig 1989.

71. S. Roch and B. Silbermann, *The Calkin image of algebras of singular integral operators.* Integral Equations Operator Theory **12**, 855–897 (1989).

72. S. Roch and B. Silbermann, *Nonstrongly converging approximation methods.* Demonstratio Math. **22**, 651–676 (1989).

73. R. Hagen and B. Silbermann, *A Banach algebra approach to the stability of projection methods for singular integral equations.* Math. Nachr. **140**, 285–297 (1989).

74. N. Krupnik and B. Silbermann, *The structure of some Banach algebras fulfilling a standard identity.* Math. Nachr. **142**, 175–180 (1989).

75. A. Böttcher, B. Silbermann and I. Spitkovsky, *Toeplitz operators with piecewise quasisectorial symbols.* Bull. London Math. Soc. **22**, 281–286 (1990).
76. S. Roch and B. Silbermann, *On algebras with standard identities.* Linear Algebra Appl. **137/138**, 239–247 (1990).
77. S. Roch and B. Silbermann, *Algebras of convolution operators and their image in the Calkin algebra.* Report MATH 90-05, 160 pages, Akad. Wiss. DDR, Karl-Weierstrass-Institut f. Mathematik, Berlin 1990.
78. B. Silbermann, *Symbol construction and numerical analysis.* In: Integral Equations and Inverse Problems (Varna, 1989), pp. 241–252, Pitman Res. Notes Math. Ser., 235, Longman Sci. Tech., Harlow 1991.
79. A. Böttcher, S. Roch, B. Silbermann and I. Spitkovsky, *A Gohberg-Krupnik-Sarason symbol calculus for algebras of Toeplitz, Hankel, Cauchy, and Carleman operators.* In: Topics in Operator Theory: Ernst D. Hellinger Memorial Volume, pp. 189–234, Oper. Theory Adv. Appl., 48, Birkhäuser, Basel 1990.
80. S. Roch and B. Silbermann, *Representations of noncommutative Banach algebras by continuous functions.* Algebra i Analiz **3**, 171–185 (1991) and St. Petersburg Math. J. **3**, 865–879 (1992).
81. D. Berthold, W. Hoppe and B. Silbermann, *The numerical solution of the generalized airfoil equation.* J. Integral Equations Appl. **4**, 309–336 (1992).
82. S. Roch and B. Silbermann, *The structure of algebras of singular integral operators.* J. Integral Equations Appl. **4**, 421–442 (1992).
83. D. Berthold, W. Hoppe and B. Silbermann, *A fast algorithm for solving the generalized airfoil equation.* In: Orthogonal Polynomials and Numerical Methods, J. Comput. Appl. Math. **43**, 185–219 (1992).
84. V.D. Didenko and B. Silbermann, *Symbols of some operator sequences and quadrature methods for solving singular integral equations with conjugation.* (Russian) Funktsional. Anal. i Prilozhen. **26**, 67–70 (1992); English transl. in Funct. Anal. Appl. **26**, 285–287 (1992).
85. B. Silbermann, *Asymptotische Invertierung von Faltungsoperatoren.* In: Jahrbuch Überblicke Mathematik, 1993, pp. 73–95, Vieweg, Braunschweig 1993.
86. B. Silbermann, *On the limiting set of singular values of Toeplitz matrices.* Linear Algebra Appl. **182**, 35–43 (1993).
87. V.D. Didenko and B. Silbermann, *On the stability of some operator sequences and the approximate solution of singular integral equations with conjugation.* Integral Equations Operator Theory **16**, 224–243 (1993).
88. T. Finck, S. Roch and B. Silbermann, *Two projection theorems and symbol calculus for operators with massive local spectra.* Math. Nachr. **162**, 167–185 (1993).

89. A. Böttcher and B. Silbermann, *Operator-valued Szegő-Widom limit theorems.* In: Toeplitz Operators and Related Topics (Santa Cruz, CA, 1992), pp. 33–53, Oper. Theory Adv. Appl., 71, Birkhäuser, Basel 1994.
90. A. Böttcher, Yu.I. Karlovich and B. Silbermann, *Singular integral equations with PQC coefficients and freely transformed argument.* Math. Nachr. **166**, 113–133 (1994).
91. R. Hagen, S. Roch and B. Silbermann, *Stability of spline approximation methods for multidimensional pseudodifferential operators.* Integral Equations Operator Theory **19**, 25–64 (1994).
92. H. Bart, T. Ehrhardt and B. Silbermann, *Zero sums of idempotents in Banach algebras.* Integral Equations Operator Theory **19**, 125–134 (1994).
93. H. Bart, T. Ehrhardt and B. Silbermann, *Logarithmic residues in Banach algebras.* Integral Equations Operator Theory **19**, 135–152 (1994).
94. A. Böttcher, B. Silbermann and H. Widom, *A continuous analogue of the Fisher-Hartwig formula for piecewise continuous symbols.* J. Funct. Anal. **122**, 222–246 (1994).
95. A. Böttcher, B. Silbermann and H. Widom, *Determinants of truncated Wiener-Hopf operators with Hilbert-Schmidt kernels and piecewise continuous symbols.* Arch. Math. (Basel) **63**, 60–71 (1994).
96. S. Roch and B. Silbermann, *Limiting sets of eigenvalues and singular values of Toeplitz matrices.* Asymptotic Anal. **8**, 293–309 (1994).
97. D. Berthold and B. Silbermann, *Corrected collocation methods for periodic pseudodifferential equations.* Numer. Math. **70**, 397–425 (1995).
98. V.D. Didenko, S. Roch and B. Silbermann, *Approximation methods for singular integral equations with conjugation on curves with corners.* SIAM J. Numer. Anal. **32**, 1910–1939 (1995).
99. B. Silbermann, *Toeplitz-like operators and their finite sections.* In: Recent Developments in Operator Theory and Its Applications (Winnipeg, MB, 1994), pp. 386–398, Oper. Theory Adv. Appl., 87, Birkhäuser, Basel 1996.
100. A. Böttcher, I. Gohberg, Yu.I. Karlovich, N. Krupnik, S. Roch, B. Silbermann and I. Spitkovsky, *Banach algebras generated by N idempotents and applications.* In: Singular Integral Operators and Related Topics (Tel Aviv, 1995), pp. 19–54, Oper. Theory Adv. Appl., 90, Birkhäuser, Basel 1996.
101. T. Ehrhardt, S. Roch and B. Silbermann, *Symbol calculus for singular integrals with operator-valued PQC-coefficients.* In: Singular Integral Operators and Related Topics (Tel Aviv, 1995), pp. 182–203, Oper. Theory Adv. Appl., 90, Birkhäuser, Basel 1996.
102. T. Ehrhardt, S. Roch and B. Silbermann, *Finite section method for singular integrals with operator-valued PQC-coefficients.* In: Singular Integral Operators and Related Topics (Tel Aviv, 1995), pp. 204–243, Oper. Theory Adv. Appl., 90, Birkhäuser, Basel 1996.

103. B. Silbermann, *Asymptotic invertibility of Toeplitz operators.* In: Singular Integral Operators and Related Topics (Tel Aviv, 1995), pp. 295–303, Oper. Theory Adv. Appl., 90, Birkhäuser, Basel 1996.

104. A. Böttcher and B. Silbermann, *Infinite Toeplitz and Hankel matrices with operator-valued entries.* SIAM J. Math. Anal. **27**, 805–822 (1996).

105. N. Krupnik, S. Roch and B. Silbermann, *On C^*-algebras generated by idempotents.* J. Funct. Anal. **137**, 303–319 (1996).

106. S. Roch and B. Silbermann, *Asymptotic Moore-Penrose invertibility of singular integral operators.* Integral Equations Operator Theory **26**, 81–101 (1996).

107. S. Roch and B. Silbermann, *C^*-algebra techniques in numerical analysis.* J. Operator Theory **35**, 241–280 (1996).

108. D. Berthold and B. Silbermann, *The fast solution of periodic pseudo-differential equations.* Appl. Anal. **63**, 3–23 (1996).

109. D. Berthold and B. Silbermann, *Fast algorithms for the solution of pseudodifferential equations.* In: Boundary Element Topics (Stuttgart, 1995), pp. 291–316, Springer, Berlin 1997.

110. A. Böttcher, S. Grudsky and B. Silbermann, *Norms of inverses, spectra, and pseudospectra of large truncated Wiener-Hopf operators and Toeplitz matrices.* New York J. Math. **3**, 1–31 (1997).

111. B. Silbermann, *Asymptotic Moore-Penrose inversion of Toeplitz operators.* Linear Algebra Appl. **256**, 219–234 (1997).

112. S. Roch, P. Santos and B. Silbermann, *Finite section method in some algebras of multiplication and convolution operators and a flip.* Z. Anal. Anwendungen **16:3**, 575–606 (1997).

113. T. Ehrhardt and B. Silbermann, *Toeplitz determinants with one Fisher-Hartwig singularity.* J. Funct. Anal. **148**, 229–256 (1997).

114. H. Bart, T. Ehrhardt and B. Silbermann, *Logarithmic residues, generalized idempotents, and sums of idempotents in Banach algebras.* Integral Equations Operator Theory **29**, 155–186 (1997).

115. V.D. Didenko and B. Silbermann, *Extension of C^*-algebras and Moore-Penrose stability of sequences of additive operators.* Linear Algebra Appl. **275/276**, 121–140 (1998).

116. S. Roch and B. Silbermann, *A note on singular values of Cauchy-Toeplitz matrices.* Linear Algebra Appl. **275/276**, 531–536 (1998).

117. S. Roch and B. Silbermann, *Index calculus for approximation methods and singular value decomposition.* J. Math. Anal. Appl. **225**, 401–426 (1998).

118. V.S. Rabinovich, S. Roch and B. Silbermann, *Fredholm theory and finite section method for band-dominated operators.* Integral Equations Operator Theory **30**, 452–495 (1998).

119. V.D. Didenko and B. Silbermann, *Stability of approximation methods on locally non-equidistant meshes for singular integral equations.* J. Integral Equations Appl. **11**, 317–349 (1999).

120. A. Böttcher, S. Grudsky, A. Kozak and B. Silbermann, *Norms of large Toeplitz band matrices.* SIAM J. Matrix Anal. Appl. **21**, 547–561 (1999).

121. S. Roch and B. Silbermann, *Continuity of generalized inverses in Banach algebras.* Studia Math. **136**, 197–227 (1999).

122. V.D. Didenko and B. Silbermann, *On real and complex spectra in some real C^*-algebras and applications.* Z. Anal. Anwendungen **18:3**, 669–686 (1999).

123. A. Böttcher, S. Grudsky, A. Kozak and B. Silbermann, *Convergence speed estimates for the norms of the inverses of large truncated Toeplitz matrices.* Calcolo **36**, 103–122 (1999).

124. T. Ehrhardt and B. Silbermann, *Approximate identities and stability of discrete convolution operators with flip.* In: The Maz'ya Anniversary Collection, Vol. 2 (Rostock, 1998), pp. 103–132, Oper. Theory Adv. Appl., 110, Birkhäuser, Basel 1999.

125. T. Finck, S. Roch and B. Silbermann, *Banach algebras generated by two idempotents and one flip.* Math. Nachr. **216**, 73–94 (2000).

126. P. Santos and B. Silbermann, *Galerkin method for Wiener-Hopf operators with piecewise continuous symbol.* Integral Equations Operator Theory **38**, 66–80 (2000).

127. R. Duduchava and B. Silbermann, *Boundary value problems in domains with peaks.* Mem. Differential Equations Math. Phys. **21**, 1–122 (2000).

128. P. Junghanns and B. Silbermann, *Numerical analysis for one-dimensional Cauchy singular integral equations.* J. Comput. Appl. Math. **125**, 395–421 (2000).

129. P. Junghanns, B. Silbermann and S. Roch, *Collocation methods for systems of Cauchy singular integral equations on an interval.* Comp. Technologies **6**, 88-126 (2001).

130. B. Silbermann, *Obituary: Professor Dr. rer. nat. habil. Siegfried Prössdorf. January 2, 1939–July 19, 1998.* In: Problems and Methods in Mathematical Physics (Chemnitz, 1999), pp. 1–3, Oper. Theory Adv. Appl., 121, Birkhäuser, Basel 2001.

131. H. Bart, T. Ehrhardt and B. Silbermann, *Sums of idempotents and logarithmic residues in matrix algebras.* In: Operator Theory and Analysis (Amsterdam, 1997), pp. 139–168, Oper. Theory Adv. Appl., 122, Birkhäuser, Basel 2001.

132. V.D. Didenko and B. Silbermann, *On the approximate solution of some two-dimensional singular integral equations.* Math. Methods Appl. Sci. **24**, 1125–1138 (2001).

133. S.M. Grudsky and B. Silbermann, *Approximate identities, almost-periodic functions and Toeplitz operators.* Acta Appl. Math. **65**, 237–271 (2001).

134. V.S. Rabinovich, S. Roch and B. Silbermann, *Algebras of approximation sequences: finite sections of band-dominated operators.* Acta Appl. Math. **65**, 315–332 (2001).

135. V.S. Rabinovich, S. Roch and B. Silbermann, *Band-dominated operators with operator-valued coefficients, their Fredholm properties and finite sections.* Integral Equations Operator Theory **40**, 342–381 (2001).

Dissertations directed by Bernd Silbermann

1. Christian Meyer, *Über einige Klassen von singulären Operatoren und ihre Beziehungen zu singulären Integralgleichungen.* TU Chemnitz, 1 July 1975.
2. Johannes Steinmüller, *Ein Projektionsverfahren zur Lösung einer Klasse singulärer Integralgleichungen, deren Symbol Nullstellen nichtganzzahliger Ordnung besitzt.* TU Chemnitz, 1 February 1978.
3. Karla Rost, *Invertierung einiger Klassen von Matrizen im Zusammenhang mit der näherungsweisen Lösung von Operatorgleichungen.* This dissertation was supervised by both Bernd Silbermann and Georg Heinig. TU Chemnitz, 5 February 1981.
4. Peter Junghanns, *Polynomiale Näherungsverfahren für singuläre Integralgleichungen auf beschränkten Intervallen.* TU Chemnitz, 24 February 1984.
5. Steffen Roch, *Lokale Theorie des Reduktionsverfahrens für singuläre Integraloperatoren mit Carlemanschen Verschiebungen.* TU Chemnitz, 1 July 1988.
6. Wolfram Hoppe, *Stabilität von Splineapproximationsverfahren für singuläre Integralgleichungen auf kompakten, glatten Mannigfaltigkeiten ohne Rand.* TU Chemnitz, 28 February 1994.
7. Torsten Ehrhardt, *Toeplitz determinants with several Fisher-Hartwig singularities.* TU Chemnitz, 29 September 1997.
8. Pedro Alexandre Simões dos Santos, *Approximation methods for convolution operators on the real line.* TU Chemnitz, 22 May 1998.
9. Tilo Finck, *Splineapproximationsverfahren für singuläre Integralgleichungen auf ebenen, glattberandeten Gebieten.* TU Chemnitz, 18 December 1998.
10. Marko Lindner, *Approximation methods and stability for multidimensional integral operators on spaces of essentially bounded functions.* TU Chemnitz, 2002.

Operator Theory:
Advances and Applications, Vol. 135, 25–37

Toeplitz Corona Problem for Algebras of Almost Periodic Functions

J.A. Ball, L. Rodman and I.M. Spitkovsky

Dedicated to Professor Bernd Silbermann on the occasion of his sixtieth birthday

Abstract. A version of the Toeplitz corona problem is solved in the class of Hermite subalgebras of Wiener algebras of almost periodic functions. A linear fractional formula for solutions of the suboptimal problem is obtained. The proofs are based on algebraic properties of Hermitian rings, almost periodic factorization, and Krein space techniques.

1. Introduction

We start with the basic definitions. A continuous complex-valued function is called *almost periodic* if it is a uniform (on the real line $\mathbb{R}$) limit of almost periodic polynomials, i.e. functions of the form

$$p(x) = \sum_{k=1}^{n} a_k e_{\lambda_k}, \quad x \in \mathbb{R}, \tag{1.1}$$

where $a_k \in \mathbb{C}$, $\lambda_k \in \mathbb{R}$, and e_λ is an abbreviated notation for the function $e_\lambda(x) = e^{i\lambda x}$. Denote by (AP) the class of almost periodic functions. Every almost periodic function is bounded; denote $\|f\|_\infty = \sup_{x\in\mathbb{R}}|f(x)|$, $f \in (AP)$. The class (AP) is a unital Banach algebra with respect to pointwise addition and multiplication and with respect to the norm $\|.\|_\infty$. See, e.g., [6, 7, 11] for more background information on almost periodic functions.

For every $f \in (AP)$ there exists the limit

$$\lim_{T\to\infty} \frac{1}{2T} \int_{-T}^{T} f(t)\, dt,$$

called the *mean* of f, and denoted $\mathbf{M}\{f\}$. Define $f_\lambda = \mathbf{M}\{e_{-\lambda} f\}, \lambda \in \mathbb{R}$. The set of $\lambda \in \mathbb{R}$ for which $f_\lambda \neq 0$ is at most countable and is called the *Fourier spectrum*,

---

All authors were partially supported by NSF grants.

denoted $\sigma(f)$, of $f \in (AP)$. A (formal) *Fourier series* is associated with every $f \in (AP)$:

$$f(x) \sim \sum_{\lambda \in \sigma(f)} f_\lambda e^{i\lambda x}.$$

The set (APW) of functions $f \in (AP)$ for which the Fourier series converges absolutely is also a Banach algebra, if supplied with a norm $\|f\|_W = \sum_{\lambda\in\sigma(f)} |f_\lambda|$. For any nonempty subset $\Lambda \subseteq \mathbb{R}$, let

$$(AP)_\Lambda = \{f \in (AP) \colon \sigma(f) \subseteq \Lambda\} \text{ and } (APW)_\Lambda = \{f \in (APW) \colon \sigma(f) \subseteq \Lambda\}.$$

If $\Lambda = \Sigma$ is an additive subgroup of $\mathbb{R}$ then $(AP)_\Sigma$ ($(APW)_\Sigma$) is a closed unital subalgebra of (AP) (resp., (APW)).

If X is a set (ring, Banach algebra, etc.) we denote by $X^{p\times q}$ the set (ring if $p = q$, Banach algebra if $p = q$, etc.) of $p \times q$ matrices with entries in X. Thus, $(AP)^{p\times q}$ is the Banach space (algebra if $p = q$) of $p \times q$ almost periodic matrix functions F with the norm $\|F\|_\infty = \sup_{t\in\mathbb{R}} \|F(t)\|$, where $\|X\|$ is the maximal singular value of the matrix X.

In this paper we study the following version of the corona problem for almost periodic functions (here and elsewhere we use the notation $\mathbb{R}_\pm = \{x\in\mathbb{R} \colon \pm x \geq 0\}$).

Problem 1.1. *Suppose that we are given* $A \in (APW)^{p\times m}_{\mathbb{R}_+}$, $B \in (APW)^{p\times p}_{\mathbb{R}_+}$, *and a positive number* γ. *For any additive subgroup* Λ' *of* $\mathbb{R}$ *that contains* $\sigma(A) \cup \sigma(B)$, *find criteria for existence of an* $F \in (APW)^{m\times p}_{\mathbb{R}_+\cap\Lambda'}$ *such that* $\|F\|_\infty \leq \gamma$ *and* $AF = B$.

If such an F *exists, describe all of them.*

The classical version of the corona theorem calls for the existence criterion in Problem 1.1 to be given in terms of existence of a number $\delta > 0$ so that $A(z)A(z)^* - \delta^2 B(z)B(z)^* \geq 0$ for all z in the upper half plane. The Toeplitz version of the corona theorem works instead with the seemingly stronger hypothesis that there exists a $\delta > 0$ so that $T_A T_A^* - \delta^2 T_B T_B^* \geq 0$, where T_A and T_B are the Toeplitz operators associated with A and B respectively. In the classical H^∞ setting, this Toeplitz-operator hypothesis is equivalent to the positivity of the kernel function $K(z,w) = \frac{A(z)A(w)^* - B(z)B(w)^*}{z - \overline{w}}$ over the upper half plane. The accomplishment of the paper [16] (see also [19]) was to streamline the proof of the classical corona theorem (for the classical H^∞-case) by using the Toeplitz corona theorem as a stepping stone. We note that the classical corona theorem for (AP) was solved by Arens and Singer [1], see also [21]. The interest in this problem was revived recently due to its newly discovered connection with the factorization problem via the so-called Portuguese transformation ([14], see also [6]). In [5], the corona problem was considered for almost periodic functions in several variables whose Bohr-Fourier spectra are contained in additive semi-groups of $[0,\infty)^n$.

A Toeplitz corona theorem for $(APW)_{\mathbb{R}_+}$ with $B = I$ (the constant identity matrix) was obtained in [4] by using techniques of factorization and a Grassmannian approach. The recent preprint [3] obtains the Toeplitz existence criterion for a general setup of H^∞ functions on (the dual space) of ordered discrete groups.

The present paper may be viewed as a continuation of [4]. Here we obtain the Toeplitz corona theorem for $(APW)_{\mathbb{R}_+}$ (as in Problem 1.1) for the case when A and B are almost periodic left coprime matrix polynomials. In particular, we give a linear fractional formula for solutions F in the suboptimal case. We note that the technique of [3] allows one to obtain the existence criteria for the solution of Problem 1.1, however with F guaranteed to be only in the suitable H^∞ space.

Our main results are Theorem 4.1 and Corollary 4.3, stated and proved in Section 4. In the next section we present some algebraic background. The main result here is that the (non-closed) algebra of almost periodic polynomials is Hermite. In Section 3 we quote a factorization theorem needed for the proof of Theorem 4.1.

2. Hermite rings and related properties

Throughout this section $\mathcal{R}$ is a commutative ring with unity e. An ordered n-tuple $(a_1, \dots, a_n)$ of elements of $\mathcal{R}$ is called *unimodular*, if there exist $b_1, \dots, b_n \in \mathcal{R}$ such that $a_1 b_1 + \cdots + a_n b_n = e$. A commutative ring with unity $\mathcal{R}$ is called *Hermite* if every unimodular row can be complemented. In other words, given elements $a_1, \dots, a_m \in \mathcal{R}$ that generate $\mathcal{R}$ (as an ideal), there exist $m \times m$ matrices F and G with entries in $\mathcal{R}$ such that $a_1, \dots, a_m$ form the first row of F and $FG = I$ (equivalently, $GF = I$). See [20] for some theory of Hermite rings.

An important application of Hermite rings $\mathcal{R}$ without divisors of zero has to do with coprime factorizations. Let $\mathcal{R}$ be a unital commutative ring without divisors of zero, and let $\mathcal{F}$ be its field of fractions. Two matrices G_1 and G_2 with entries in $\mathcal{R}$ are called *right coprime* if they have the same number of columns and there exist matrices X_1 and X_2 with entries in $\mathcal{R}$ of suitable sizes such that $X_1 G_1 + X_2 G_2 = I$. Dually, matrices G_1 and G_2 with entries in $\mathcal{R}$ are called *left coprime* if they have same number of rows and there exist Y_1 and Y_2 with entries in $\mathcal{R}$ such that $G_1 Y_1 + G_2 Y_2 = I$. Now let G be a rectangular matrix with entries in $\mathcal{F}$. A *right coprime factorization* of G is, by definition, a representation of the form $G = NM^{-1}$, where N and M are matrices with entries in $\mathcal{R}$, M is of square size with non-zero determinant, and N and M are right coprime. A *left coprime factorization* of G is a representation of the form $G = Q^{-1}P$, where P and Q are matrices with entries in $\mathcal{R}$, Q is of square size with non-zero determinant, and P and Q are left coprime.

In general, the existence of a coprime factorization from one side does not imply existence of coprime factorization from the other side. However, for Hermitian rings the situation is different.

Proposition 2.1. *Assume $\mathcal{R}$ is a Hermite ring without divisors of zero. If G is a matrix with entries in the field of fractions of $\mathcal{R}$, and G admits a left (resp., right) coprime factorization, then G admits also a right (resp., left) coprime factorization.*

For a proof of Proposition 2.1 see, e.g., [20].

Let $\mathcal{R}$ be a commutative ring with identity. All $\mathcal{R}$-modules will be considered right $\mathcal{R}$-modules. An $\mathcal{R}$-module K will be called *free* if there exists a finite set $k_1, \ldots, k_s \in K$ (called the *generators* of K) such that every element $x \in K$ admits a unique representation in the form $x = \sum_{j=1}^{s} k_j \alpha_j$ where $\alpha_j \in \mathcal{R}$. It is well known that the cardinality of a set of generators of a free $\mathcal{R}$-module is uniquely determined by the module (see, e.g., p. 455 in [2]). An $\mathcal{R}$-module K is called *projective* if K is finitely generated and is a direct summand in a free module. Clearly, every free module is projective. The ring $\mathcal{R}$ is called *projective-free* if every projective $\mathcal{R}$-module is free.

The following lemma is known. We present a proof for the reader's convenience; more definitive information about connections between various versions of the projective-free property and the Hermite property is found in [13], for example.

Lemma 2.2. *The properties* (i) *and* (ii) *below are equivalent for a commutative ring $\mathcal{R}$ with identity, and each of them implies the property* (iii)*:*

(i) *The ring $\mathcal{R}$ is projective-free.*
(ii) *For every idempotent $P \in \mathcal{R}^{n\times n}$ there exists an invertible matrix V in $\mathcal{R}^{n\times n}$ such that $V^{-1}PV = \begin{bmatrix} I_k & 0 \\ 0 & 0_{n-k} \end{bmatrix}$ for some k.*
(iii) *$\mathcal{R}$ is Hermite.*

Proof. (i) $\Rightarrow$ (ii). Let $P \in \mathcal{R}^{n\times n}$ be an idempotent. Let K be the submodule of $\mathcal{R}^{n\times 1}$ generated by the columns of P. Then $\mathcal{R}^{n\times 1} = K \oplus L$, where L is the submodule generated by the columns of $I - P$. Hence K and L are projective $\mathcal{R}$-modules. By (i), both K and L are free. Let $k_1, \ldots, k_s$ be generators in K, and let $\ell_1, \ldots, \ell_t$ be generators in L. Clearly, $k_1, \ldots, k_s, \ell_1, \ldots, \ell_t$ are generators of the free $\mathcal{R}$-module $\mathcal{R}^{n\times 1}$. By the uniqueness of the cardinality of the set of generators of $\mathcal{R}^{n\times 1}$ we must have $s + t = n$. Let V be the $n \times n$ matrix having $k_1, \ldots, k_s, \ell_1, \ldots, \ell_t$ as its columns. Then V is invertible, and

$$PV = V \begin{bmatrix} I_s & 0 \\ 0 & 0_{n-s} \end{bmatrix}.$$

(ii) $\Rightarrow$ (i). Let K be a direct summand in a free $\mathcal{R}$-module. We identify this free $\mathcal{R}$-module with $\mathcal{R}^{n\times 1}$ for some n. Thus, $\mathcal{R}^{n\times 1} = K \oplus L$ for some $\mathcal{R}$-module L. Let $P \in \mathcal{R}^{n\times n}$ be the projection on K along L. Now use (ii) to deduce that $V^{-1}(K)$ is a free module for some invertible $V \in \mathcal{R}^{n\times n}$. But then K itself is free.

(ii) ⇒ (iii). Let $a_1, \ldots, a_m \in \mathcal{R}$ generate $\mathcal{R}$. Then $\sum_{j=1}^m a_j b_j = 1$ for some $b_j \in \mathcal{R}$. Let

$$P = \begin{bmatrix} b_1 \\ b_2 \\ \vdots \\ b_m \end{bmatrix} \cdot \begin{bmatrix} a_1 & \ldots & a_m \end{bmatrix}.$$

Clearly, P is an idempotent. By (ii), there is an invertible $V \in \mathcal{R}^{m \times m}$ such that $V^{-1}PV = \begin{bmatrix} I_k & 0 \\ 0 & 0_{m-k} \end{bmatrix}$. (In fact $k = 1$, but we will not use this.) It follows that

$$\begin{bmatrix} a_1 & \ldots & a_m \end{bmatrix} \cdot V = \begin{bmatrix} x & 0 & \ldots & 0 \end{bmatrix}$$

for some invertible $x \in \mathcal{R}$. Clearly, the unimodular row $[a_1, \ldots, a_m]$ can be complemented. □

Theorem 2.3. *Let Σ be a finitely generated additive subgroup of $\mathbb{R}$. Then the (non-closed) algebra $(APP)_{\Sigma \cap \mathbb{R}_+}$ of almost periodic polynomials (1.1) with $0 \le \lambda_k \in \Sigma$ is projective-free.*

For the proof we need an algebraic lemma.

Lemma 2.4. *Given real numbers $\lambda_1, \lambda_2, \ldots, \lambda_n$, consider the algebra $\mathcal{P}(\lambda_1, \ldots, \lambda_n)$ generated (as a complex vector space) by monomials of the form $x_1^{a_1} x_2^{a_2} \cdots x_n^{a_n}$, where $x_1, \ldots, x_n$ are commuting independent variables, and the exponents a_j are integers subject to the inequality $a_1\lambda_1 + a_2\lambda_2 + \cdots + a_n\lambda_n \ge 0$, with multiplication of the monomials defined in the standard way. Then the algebra $\mathcal{P}(\lambda_1, \ldots, \lambda_n)$ is projective-free.*

Lemma 2.4 is a particular case of Gubeladze's theorem [8], [18].

Lemma 2.5. *Let Σ be the additive subgroup of $\mathbb{R}$ generated by nonzero real numbers $r_1, \ldots, r_k$. Then there exist $s_1, \ldots, s_\ell \in \Sigma$ such that $s_1, \ldots, s_\ell$ are linearly independent over the field of rationals $\mathbb{Q}$ and every r_j is a linear combination of $s_1, \ldots, s_\ell$ with integer coefficients.*

Proof. Using induction on k, we assume that there exist $s'_1, \ldots, s'_p$ that satisfy the requirements of the lemma with respect to $r_1, \ldots, r_{k-1}$ and to the subgroup Σ' generated by $r_1, \ldots, r_{k-1}$. If r_k is $\mathbb{Q}$-linearly independent of $s'_1, \ldots, s'_p$, we are done: take $\ell = p + 1$, $s_j = s'_j$ $(j = 1, \ldots, p)$, $s_{p+1} = r_k$. Otherwise,

$$mr_k + \sum_{j=1}^p m_j s'_j = 0,$$

where the integers $m, m_1, \ldots, m_p$ are not all zero; moreover, we may assume that the greatest common divisor of $m, m_1, \ldots, m_p$ is equal to 1. Since $\mathbb{Z}$, being a principal ideal domain, is an Hermite ring, there is a $(p + 1) \times (p + 1)$ matrix

V with integer entries and determinant ± 1 having the first row $[m\ m_1\ \dots\ m_p]$. Define $s_1, \dots, s_p \in \Sigma$ by

$$\begin{bmatrix} 0 \\ s_1 \\ s_2 \\ \vdots \\ s_p \end{bmatrix} = V \begin{bmatrix} r_k \\ s'_1 \\ \vdots \\ s'_p \end{bmatrix}; \quad \text{then} \quad \begin{bmatrix} r_k \\ s'_1 \\ \vdots \\ s'_p \end{bmatrix} = V^{-1} \begin{bmatrix} 0 \\ s_1 \\ s_2 \\ \vdots \\ s_p \end{bmatrix}.$$

So indeed every r_j $(j = 1, \dots, k)$ is a linear combination of $s_1, \dots, s_p$ with integer coefficients. It is easy to see that $s_1, \dots, s_p$ are $\mathbb{Q}$-linearly independent (because $s'_1, \dots, s'_p$ are $\mathbb{Q}$-linearly independent). □

Proof of Theorem 2.3. By Lemma 2.5, there exist linearly independent (over $\mathbb{Q}$) generators $s_1, \dots, s_\ell$ for Σ. Clearly, every element $\mu \in \Sigma$ can be uniquely represented in the form $\mu = \sum_{j=1}^{\ell} q_j s_j$, where q_j are integers. Now the rings $(APP)_{\Sigma\cap\mathbb{R}_+}$ and $\mathcal{P}(s_1, \dots, s_\ell)$ are isomorphic via the correspondence

$$e_{q_1 s_1 + \cdots + q_\ell s_\ell} \in (APP)_{\Sigma\cap\mathbb{R}_+} \leftrightarrow x_1^{q_1} x_2^{q_2} \cdots x_\ell^{q_\ell} \in \mathcal{P}(s_1, \dots, s_\ell), \quad q_j \in \mathbb{Z}.$$

It remains to appeal to Lemma 2.4. □

Corollary 2.6. *Let Σ be a subgroup of $\mathbb{R}$. Then the algebra $(APP)_{\Sigma\cap\mathbb{R}_+}$ of almost periodic polynomials with Fourier spectrum in $\Sigma \cap \mathbb{R}_+$ is Hermite.*

Proof. Let $a_1, \dots, a_n, b_1, \dots, b_n \in (APP)_{\Sigma\cap\mathbb{R}_+}$ be such that $a_1 b_1 + \cdots + a_n b_n = e$. Let $\Lambda \subseteq \Sigma$ be the subgroup generated by the finite set $\left(\cup_{j=1}^n \sigma(a_j)\right) \cup \left(\cup_{j=1}^n \sigma(b_j)\right)$. By Lemma 2.2 and Theorem 2.3, $(APP)_{\Lambda\cap\mathbb{R}_+}$ is Hermite. □

We do not know whether the algebra $(APW)_{\mathbb{R}_+\cap\Sigma}$, Σ a subgroup of $\mathbb{R}$, is Hermite.

3. AP factorizations of matrix functions

Fix an additive subgroup Σ of the real line. We say that a *canonical AP_Σ factorization* of an $n \times n$ AP matrix function G is a representation

$$G(x) = G^+(x) G^-(x), \tag{3.1}$$

where $(G^+)^{\pm 1} \in (AP)^{n\times n}_{\Sigma\cap\mathbb{R}_+}$ and $(G^-)^{\pm 1} \in (AP)^{n\times n}_{\Sigma\cap\mathbb{R}_-}$. If (APW) is used in place of (AP) in the above definition, then we obtain a canonical APW_Σ factorization.

Such factorizations, as well as more general noncanonical almost periodic factorizations, were introduced in [10] and extensively studied later on, see the book [6] and references therein for the case $\Sigma = \mathbb{R}$; see also [15, 4] for arbitrary subgroups $\Sigma \subseteq \mathbb{R}$.

To formulate a criterion for canonical factorization, let us recall that the formula

$$\langle f, g \rangle = \mathbf{M}\{f g^*\}, \quad f, g \in (AP) \tag{3.2}$$

defines an inner product on AP. The completion of (AP) with respect to this inner product is called the *Besicovitch space* and is denoted by (B). Thus (B) is a Hilbert space. For a nonempty set $\Lambda \subseteq \mathbb{R}$ we denote by Π_Λ the projection

$$\Pi_\Lambda f = \sum_{\lambda \in \Lambda \cap \sigma(f)} f_\lambda e_\lambda, \qquad f \in (APW). \tag{3.3}$$

The projection (3.3) extends by continuity to the orthogonal projection (also denoted Π_Λ) on (B). We denote by $(B)_\Lambda$ the range of Π_Λ, or, equivalently, the completion of $(AP)_\Lambda$ with respect to the inner product (3.2). The vector-valued Besicovitch space $(B)^{n\times 1}$ consists of $n \times 1$ columns with components in (B), with the standard Hilbert space structure. Similarly, $(B)_\Lambda^{n\times 1}$ is the Hilbert space of $n\times 1$ columns with components in $(B)_\Lambda$.

For $F \in (AP)_\Sigma^{m\times n}$, where Σ is an additive subgroup of $\mathbb{R}$, the *Toeplitz operator* $T(F)_\Sigma$ is defined by

$$T(F)_\Sigma \colon (B)^{n\times 1}_{\Sigma\cap\mathbb{R}_+} \longrightarrow (B)^{m\times 1}_{\Sigma\cap\mathbb{R}_+}, \quad T(F)_\Sigma\phi = \Pi_{\mathbb{R}_+}(F\phi), \quad \phi \in (B)^{n\times 1}_{\Sigma\cap\mathbb{R}_+}.$$

Theorem 3.1. *Let Σ be an additive subgroup of $\mathbb{R}$ and let $G \in (APW)_\Sigma^{n\times n}$. Then G admits a canonical $(AP)_\mathbb{R}$ factorization if and only if the operator $T(G')_\mathbb{R}$ is invertible, where the superscript $'$ indicates the transposed matrix. If this is the case, then the operator $T(G')_\Sigma$ is invertible as well, and any $(AP)_\mathbb{R}$ factorization of G automatically is its canonical $(APW)_\Sigma$ factorization.*

Theorem 3.1 is a subset of a lengthy set of equivalent statements constituting Theorem 2.3 of [4]; see [4] also for the history of the matter.

4. Main results

Theorem 4.1. *Let $\mathcal{R}$ be a (not necessarily closed) Hermite subalgebra of the algebra $(APW)_{\Lambda'\cap\mathbb{R}_+}$, where Λ' is a subgroup of $\mathbb{R}$. Suppose that we are given $A \in \mathcal{R}^{p\times m}$, $B \in \mathcal{R}^{p\times p}$ with B invertible in $(APW)^{p\times p}$, and a positive number γ. Assume in addition that A and B are left coprime over $\mathcal{R}$. Then there is an $F \in (APW)^{m\times p}_{\mathbb{R}_+\cap\Lambda'}$ such that*

$$\|F\|_\infty \le \gamma, \quad AF = B \tag{4.1}$$

only if

$$T(A)_{\Lambda'}(T(A)_{\Lambda'})^* \ge \frac{1}{\gamma^2} T(B)_{\Lambda'}(T(B)_{\Lambda'})^*. \tag{4.2}$$

Conversely, if the operator

$$T(A)_{\Lambda'}(T(A)_{\Lambda'})^* - \frac{1}{\gamma^2} T(B)_{\Lambda'}(T(B)_{\Lambda'})^* \tag{4.3}$$

is strictly positive definite, then there exists $F \in (APW)^{m\times p}_{\mathbb{R}_+\cap\Lambda'}$ such that (4.1) holds.

In connection with Theorem 4.1 notice that $B \in (APW)^{p\times p}$ is invertible in $(APW)^{p\times p}$ if and only if B is invertible in $(APW)^{p\times p}_{\Lambda}$, where Λ is the additive subgroup of $\mathbb{R}$ generated by $\sigma(B)$, if and only if $|\det(B(x))| \geq \epsilon > 0$ for all $x \in \mathbb{R}$, where ϵ is independent of x. See, e.g., Proposition 2.2 and Corollary 2.7 of [15] for a proof of this statement.

The special case of Theorem 4.1 where $B = I$ appears as part of Theorem 5.2 in [4], under the more relaxed hypothesis that $A \in (APW)_{\Lambda'\cap\mathbb{R}_+}$. The proof of the general case follows in a similar way, once we verify the following lemma.

Lemma 4.2. *Let A,B be as in Theorem 4.1. Then:*

(i) *the function* $W = B^{-1}A$ *has a right coprime factorization* $W = CD^{-1}$ *over* $\mathcal{R}$ *(with* $C \in \mathcal{R}^{p\times m}$, $D \in \mathcal{R}^{m\times m}$ *and* D *invertible in* $(APW)^{m\times m}_{\Lambda'}$*), and, moreover*

(ii) *the following kernel-range Toeplitz operator identity holds:*

$$\operatorname{Ker}\begin{bmatrix} T(A)_{\Lambda'} & -T(B)_{\Lambda'} \end{bmatrix} = \operatorname{Range}\begin{bmatrix} T(D)_{\Lambda'} \\ T(C)_{\Lambda'} \end{bmatrix}. \tag{4.4}$$

Proof. By assumption, the pair (A, B) is left coprime, and hence $W = B^{-1}A$ is a left coprime factorization. By Proposition 2.1 W admits a right coprime factorization. Hence we may write $B^{-1}A = W = CD^{-1}$ where $C \in \mathcal{R}^{p\times m}$ and $D \in \mathcal{R}^{m\times m}$ with the determinant of D not identically zero. The coprime property of the pair (C, D) gives the existence of an $X \in \mathcal{R}^{m\times m}$ and $Y \in \mathcal{R}^{m\times p}$ such that

$$XD + YC = I. \tag{4.5}$$

Therefore, $D^{-1} = X + YW$, and since $W \in (APW)^{p\times m}_{\Lambda'}$ we conclude that in fact D^{-1} exists in $(APW)^{m\times m}_{\Lambda'}$, and (i) follows.

To prove (ii), let $f \in (B)^{m\times 1}_{\Lambda'\cap\mathbb{R}_+}$ and $g \in (B)^{p\times 1}_{\Lambda'\cap\mathbb{R}_+}$ be such that $0 = Af - Bg$. Then

$$0 = Af - Bg = B(B^{-1}Af - g) = B(CD^{-1}f - g) \tag{4.6}$$

from which we see that

$$\begin{bmatrix} f \\ g \end{bmatrix} = \begin{bmatrix} D \\ C \end{bmatrix} h \quad \text{with} \quad h = D^{-1}f. \tag{4.7}$$

But then, using (4.5) it follows that $h = (XD+YC)h = Xf + Yg$ is in $(B)^{m\times 1}_{\Lambda'\cap\mathbb{R}_+}$, since all of X, f, Y, g have spectrum in $\Lambda' \cap \mathbb{R}_+$. This proves the containment $\subseteq$ in (4.4). The reverse containment follows easily from $B^{-1}A = CD^{-1}$, and hence (4.4) follows. □

Proof of Theorem 4.1. Necessity of the condition (4.2) is straightforward. Indeed, if $F \in (APW)^{m\times p}_{\Lambda'\cap\mathbb{R}_+}$ satisfies $\|F\|_\infty \leq \gamma$ and $AF = B$, then

$$T(B)_{\Lambda'}(T(B)_{\Lambda'})^* = T(A)_{\Lambda'}T(F)_{\Lambda'}(T(F)_{\Lambda'})^*(T(A)_{\Lambda'})^* \leq \gamma^2 T(A)_{\Lambda'}(T(A)_{\Lambda'})^*.$$

We next consider the proof of the converse statement. First observe that, by replacing B with $\gamma^{-1}B$ and F with $\gamma^{-1}F$, we may assume without loss of generality that $\gamma = 1$. The assumption that $T(A)_{\Lambda'}(T(A)_{\Lambda'})^* - T(B)_{\Lambda'}(T(B)_{\Lambda'})^*$ is strictly positive definite has the geometric interpretation that the subspace

$$\mathcal{P} = \text{Range} \begin{bmatrix} (T(A)_{\Lambda'})^* \\ (T(B)_{\Lambda'})^* \end{bmatrix}$$

is a strictly positive subspace in the J_1-inner product on $(B)^{(m+p)\times 1}_{\Lambda'\cap\mathbb{R}_+}$, where

$$J_1 = \begin{bmatrix} I_m & 0 \\ 0 & -I_p \end{bmatrix}.$$

Consequently the J_1-orthogonal complement of this subspace, namely

$$\mathcal{P}^{\perp J_1} = \text{Ker} \begin{bmatrix} T(A)_{\Lambda'} & -T(B)_{\Lambda'} \end{bmatrix},$$

is a regular subspace of $(B)^{(m+p)\times 1}_{\Lambda'\cap\mathbb{R}_+}$ in the J_1 Krein space inner product. By Lemma 4.2 we have the alternative representation

$$\mathcal{P}^{\perp J_1} = \text{Range} \begin{bmatrix} T(D)_{\Lambda'} \\ T(C)_{\Lambda'} \end{bmatrix}.$$

In terms of this representation, the fact that $\mathcal{P}^{\perp J_1}$ is a regular subspace means that

$$(T(D)_{\Lambda'})^* T(D)_{\Lambda'} - (T(C)_{\Lambda'})^* T(C)_{\Lambda'} \text{ is invertible.}$$

By Theorem 3.1, the matrix function $(D^*D - C^*C)'$ admits a canonical factorization. Since $(D^*D - C^*C)'$ is Hermitian-valued, its factorization can be chosen in a special form described in [4, Theorem 2.4]. Passing to transposed matrices, this special form becomes

$$D^*D - C^*C = R^* J_0 R$$

for an appropriate signature matrix J_0 (i..e, J_0 is simultaneously Hermitian and unitary), where $R^{\pm 1} \in (APW)^{m\times m}_{\Lambda'\cap\mathbb{R}_+}$. We then let

$$\Theta = \begin{bmatrix} D \\ C \end{bmatrix} R^{-1}.$$

One computes

$$\begin{aligned} \Theta^* J_1 \Theta &= \left(\begin{bmatrix} D \\ C \end{bmatrix} R^{-1} \right)^* \begin{bmatrix} I_m & 0 \\ 0 & -I_p \end{bmatrix} \left(\begin{bmatrix} D \\ C \end{bmatrix} R^{-1} \right) \\ &= (R^{-1})^*(D^*D - C^*C)R^{-1} = (R^{-1})^*(R^* J_0 R)R^{-1} = J_0. \end{aligned}$$

Thus Θ is (J_0, J_1)-isometric. Moreover since $R^{\pm 1} \in (APW)^{m\times m}_{\Lambda'\cap\mathbb{R}_+}$, we have that

$$\Theta(B)^{m\times 1}_{\Lambda'\cap\mathbb{R}_+} = \begin{bmatrix} D \\ C \end{bmatrix} (B)^{m\times 1}_{\Lambda'\cap\mathbb{R}_+} = \text{Ker} \begin{bmatrix} T(A)_{\Lambda'} & -T(B)_{\Lambda'} \end{bmatrix}.$$

By an analogous argument as in the proof of Theorem 5.2 in [4], the matrix J_0 has exactly p negative eigenvalues, i.e., we can take $J_0 = \begin{bmatrix} I_{m-p} & 0 \\ 0 & -I_p \end{bmatrix}$. Indeed, the number of negative eigenvalues in J_0 is equal to the minimal number

p' of functions $\psi_1, \dots, \psi_{p'}$ one needs to generate (via linear combinations with coefficients in $(AP)_{\Lambda' \cap \mathbb{R}_+}$) a maximal negative subspace $\mathcal{N}$ of $(B)^{m \times 1}_{\Lambda' \cap \mathbb{R}_+}$ (with the Krein space structure defined by J_0) in the Hilbert space topology:

$$\mathcal{N} = \text{closure of span } \{ e^{ir_1 \cdot}\psi_1, \dots, e^{ir_{p'} \cdot}\psi_{p'} : r_1, \dots, r_{p'} \in \Lambda' \cap \mathbb{R}_+ \}.$$

Since $T(\Theta)_{\Lambda'}$ (where $\Theta = \begin{bmatrix} D \\ C \end{bmatrix} R^{-1}$) is a Krein space isomorphism from $(B)^{m \times 1}_{\Lambda' \cap \mathbb{R}_+}$ onto

$$\left(\text{Range} \left(T \left(\begin{bmatrix} D \\ C \end{bmatrix} \right)_{\Lambda'} \right), J_1 \right)$$

which commutes with multiplication by scalar functions from $(AP)_{\Lambda' \cap \mathbb{R}_+}$, the same number p' is the minimal number of generators required to generate a Range $\left(T \left(\begin{bmatrix} D \\ C \end{bmatrix} \right)_{\Lambda'} \right)$-maximal negative subspace in the J_1-inner product. However, since

$$\left\{ \text{Range} \left(T \left(\begin{bmatrix} D \\ C \end{bmatrix} \right)_{\Lambda'} \right) \right\}^{\perp J_1}$$

is a Hilbert space, a subspace of Range $\left(T \left(\begin{bmatrix} D \\ C \end{bmatrix} \right)_{\Lambda'} \right)$ is maximal negative as a subspace of the Krein space

$$\left(\text{Range} \left(T \left(\begin{bmatrix} D \\ C \end{bmatrix} \right)_{\Lambda'} \right), J_1 \right)$$

if and only if it is maximal negative as a subspace of $(B)^{(m+p) \times 1}_{\Lambda' \cap \mathbb{R}_+}$, again with respect to J_1. For further details, the reader may consult [4].

By standard arguments, one now sees that the function Θ can be used to parametrize all solutions F in $(APW)^{m \times p}_{\Lambda' \cap \mathbb{R}_+}$ of the interpolation condition $AF = B$ with infinity norm at most 1. Indeed, such functions F stand in one-to-one correspondence with subspaces $\mathcal{G}$ of $(B)^{(m+p) \times 1}_{\Lambda' \cap \mathbb{R}_+}$ satisfying (i) *$\mathcal{G}$ is maximal negative in $(B)^{(m+p) \times 1}_{\Lambda' \cap \mathbb{R}_+}$*, (ii) *$\mathcal{G}$ is contained in $\mathcal{P}^{\perp J_1}$*, and (iii) *$\mathcal{G}$ is invariant under multiplication by scalar functions in $(APW)_{\Lambda' \cap \mathbb{R}_+}$*, via the correspondence

$$F \to \mathcal{G} = \text{Range} \begin{bmatrix} T(F)_{\Lambda'} \\ I \end{bmatrix}. \tag{4.8}$$

Since $\mathcal{P}$ is a positive (even strictly positive) subspace in the J_1-inner product, any subspace of $\mathcal{P}^{\perp J_1}$ which is maximal negative as a subspace of $\mathcal{P}^{\perp J_1}$ is automatically also maximal negative as a subspace of $(B)^{(m+p) \times 1}_{\Lambda' \cap \mathbb{R}_+}$. Moreover, the multiplication operator $T(\Theta)_{\Lambda'}$ is a Krein space isomorphism between $(B)^{m \times 1}_{\Lambda' \cap \mathbb{R}_+}$ in the J_0-inner product and $\mathcal{P}^{\perp J_1}$ in the J_1-inner product, which also intertwines the operator of multiplication by scalar functions in $(APW)_{\Lambda' \cap \mathbb{R}_+}$ on the two spaces. Hence subspaces $\mathcal{G}$ satisfying (i), (ii), (iii) given above are exactly those subspaces of the form

$$\mathcal{G} = \Theta\mathcal{G}_1 \tag{4.9}$$

where $\mathcal{G}_1$ is a subspace of $(B)^{m\times 1}_{\Lambda'\cap\mathbb{R}_+}$ such that (i)′ *$\mathcal{G}_1$ is maximal negative as a subspace of $(B)^{m\times 1}_{\Lambda'\cap\mathbb{R}_+}$ in the J_0-inner product,* and (ii)′ *$\mathcal{G}_1$ is invariant under multiplication by scalar functions in $(APW)_{\Lambda'\cap\mathbb{R}_+}$*. But such subspaces $\mathcal{G}_1$ exist in abundance; namely,

$$\mathcal{G}_1 = \text{Range}\begin{bmatrix} T(G)_{\Lambda'} \\ I \end{bmatrix} \tag{4.10}$$

for any $G \in (APW)^{(m-p)\times p}_{\Lambda'\cap\mathbb{R}_+}$ with $\|G\|_\infty \le 1$. By combining (4.8), (4.9) and (4.10), we see that solutions F of our interpolation problem are in bijective correspondence with free parameter functions $G \in (APW)^{(m-p)\times p}_{\Lambda'\cap\mathbb{R}_+}$ with $\|G\|_\infty \le 1$ according to the formula

$$\begin{bmatrix} F \\ I \end{bmatrix} \cdot (B)^{p\times 1}_{\Lambda'\cap\mathbb{R}_+} = \Theta \begin{bmatrix} G \\ I \end{bmatrix} \cdot (B)^{m\times 1}_{\Lambda'\cap\mathbb{R}_+}.$$

By a standard argument (see e.g. the proof of Theorem 5.2 in [4]), if we decompose Θ as

$$\Theta = \begin{bmatrix} \Theta_{11} & \Theta_{12} \\ \Theta_{21} & \Theta_{22} \end{bmatrix}$$

with the size of Θ_{11} equal to $m \times (m-p)$ and that of Θ_{22} equal to $p \times p$, etc., then we get the linear-fractional parametrization formula

$$F = (\Theta_{11}G + \Theta_{12})(\Theta_{21}G + \Theta_{22})^{-1} \tag{4.11}$$

for the set of all solutions F in $(APW)^{m\times p}_{\Lambda'\cap\mathbb{R}_+}$ of $AF = B$ with $\|F\|_\infty \le 1$ in terms of a free parameter function $G \in (APW)^{(m-p)\times p}_{\Lambda'\cap\mathbb{R}_+}$ with $\|G\|_\infty \le 1$. Theorem 4.1 is proved. □

Note that in the course of the proof we have also obtained a description of all $F \in (APW)^{m\times p}_{\mathbb{R}_+\cap\Lambda'}$ such that (4.1) holds, provided the operator (4.3) is strictly positive definite. The description is given by (4.11). If the operator (4.3) is merely positive semidefinite, then there exists F in the infinity norm Besicovitch space $(B_\infty)^{m\times p}_{\Lambda'\cap\mathbb{R}_+}$ that satisfies (4.1). This statement is completely analogous to the corresponding part of Theorem 5.2 of [4], and can be obtained in the same way. We refer the reader to [4] for more details.

More generally, Theorem 4.1 remains true whenever we can identify the set $\operatorname{Ker} T([A \quad -B])_{\Lambda'}$ as the range of an operator $T\left(\begin{bmatrix} D \\ C \end{bmatrix}\right)_{\Lambda'}$. In the periodic case, i.e., when $\Lambda' = \mathbb{Z}$, this is known to be true via the Beurling-Lax theorem characterization of invariant subspaces of the shift operator. For the general case, it is known that there are invariant subspaces for which the Beurling-Lax representation can fail (see Remark 8.5.4 in [17]).

Corollary 4.3. *Theorem* 4.1 *is valid with the algebra* $(APP)_{\Lambda' \cap \mathbb{R}_+}$ *in place of* $\mathcal{R}$.

For the proof use Corollary 2.6.

Acknowledgement. We are indebted to J. Gubeladze and R. G. Swan for pointing out to us Gubeladze's theorem and its references.

References

[1] R. Arens and I. Singer, *Generalized analytic functions*, Trans. Amer. Math. Soc., **81** (1956), 379–393.

[2] M. Artin, *Algebra*, Prentice Hall, Englewood Cliffs, NJ, 1991.

[3] M. Bakonyi and D. Timotin, *The intertwining lifting theorem for ordered groups*, Technical report Nov-12-01-2, Georgia State University.

[4] J.A. Ball, Yu. Karlovich, L. Rodman and I.M. Spitkovsky, *Sarason interpolation and Toeplitz corona theorem for almost periodic matrix functions*, Integral Equations and Operator Theory, **32** (1998), 243–281.

[5] A. Böttcher, *On the corona theorem for almost periodic functions*, Integral Equations and Operator Theory, **33** (1999), 253–272.

[6] A. Böttcher, Yu.I. Karlovich and I.M. Spitkovsky, *Convolution Operators and Factorization of Almost Periodic Matrix Functions*, Birkhäuser, Basel and Boston, 2002.

[7] C. Corduneanu, *Almost Periodic Functions*, J. Wiley & Sons, New York-London-Sydney, 1968.

[8] J. Gubeladze, *Geometric and algebraic representations of commutative cancellative monoids*, Proc. A. Razmadze Math. Inst., **113** (1995), 31–81.

[9] J.W. Helton, *Optimization over spaces of analytic functions and the corona problem*, Journal of Operator Theory, **15** (1986), 359–375.

[10] Yu.I. Karlovich and I.M. Spitkovsky, *Factorization of almost periodic matrix-valued functions and the Noether theory for certain classes of equations of convolution type*, Mathematics of the USSR, Izvestiya, **34** (1990), 281–316.

[11] B.M. Levitan and V.V. Zhikov, *Almost Periodic Functions and Differential Equations*, Cambridge University Press, Cambridge-New York, 1982.

[12] S. MacLane and G. Birkhoff, *Algebra*, Chelsea, New York, 1988.

[13] B. McDonald, *Linear Algebra over Commutative Rings*, Marcel Dekker, New York, 1984.

[14] L. Rodman and I.M. Spitkovsky, *Almost periodic factorization and corona theorem*, Indiana Univ. Math. J. **47** (1998), 1243–1256.

[15] L. Rodman, I.M. Spitkovsky, and H.J. Woerdeman, *Carathéodory-Toeplitz and Nehari problems for matrix-valued almost periodic functions*, Trans. Amer. Math. Soc., **350** (1998), 2185–2227.

[16] M. Rosenblum, *A corona theorem for countably many functions*, Integral Equations and Operator Theory, **3** (1980), 125–137.

[17] W. Rudin, *Fourier Analysis on Groups*, Interscience (Wiley), New York, 1967.

[18] R.G. Swan, *Gubeladze's proof of Anderson's conjecture*, Contemporary Mathematics, Azumaya algebras, actions, and modules, **124** (1992), 215–250.

[19] V.A. Tolokonnikov, *Estimates in Carleson's corona theorem and finitely generated ideals of the algebra H^∞*, Functional Analysis and Applications, **14** (1981), 320–322.

[20] M. Vidyasagar, *Control Systems Synthesis. A Factorization Approach*, MIT Press, Cambridge, MA, 1985.

[21] J. Xia, *Conditional expectations and the corona problem of ergodic Hardy spaces*, J. Functional Analysis, **64** (1985), 251–274.

J.A. Ball
Department of Mathematics
Virginia Tech
Blacksburg, VA 24061-0123
e-mail: `ball@math.vt.edu`

L. Rodman and I.M. Spitkovsky
Department of Mathematics
College of William & Mary
Williamsburg, VA 23187-8795
e-mail: `lxrodm@math.wm.edu` and `ilya@math.wm.edu`

Received: 27 December 2001

Operator Theory:
Advances and Applications, Vol. 135, 39–60

Sums of Idempotents in the Banach Algebra Generated by the Compact Operators and the Identity

H. Bart, T. Ehrhardt and B. Silbermann

Abstract. The present paper is concerned with sums of idempotents in the Banach algebra generated by the compact operators and the identity in the case when the underlying Banach space is infinite dimensional. These sums are characterized in terms of ranks, traces and dimensions of null spaces. Another, quite different, characterization is given in terms of logarithmic residues, i.e., contour integrals of logarithmic derivatives, of certain analytic operator functions. The functions in question have values in the (non-closed) subalgebra generated by the identity and the finite rank operators. Topological properties of the set of sums of idempotents are considered too.

1. Introduction

This paper is part of a systematic investigation of logarithmic residues in Banach algebras (see [BES2]–[BES8]). Earlier material on this subject can be found in [M], [MS], [GS] and [B2]; cf. also [BKL2] and [GGK], Section XI.9.

A logarithmic residue in a complex Banach algebra $\mathcal{B}$ is a contour integral of a logarithmic derivative of an analytic $\mathcal{B}$-valued function F. There is a left version and there is a right version of this notion. The left version corresponds to the left logarithmic derivative $F'(\lambda)F(\lambda)^{-1}$, the right version to the right logarithmic derivative $F(\lambda)^{-1}F'(\lambda)$.

As is known from [BES2], there is a close relationship between logarithmic residues and sums of idempotents. Here, of course, we have in mind finite sums. Indeed, the norm of a non-zero idempotent is always at least one; hence, a convergent series of idempotents reduces to a finite sum.

A sum of idempotents in a Banach algebra $\mathcal{B}$ is always a logarithmic residue, even of an entire $\mathcal{B}$-valued function (cf. [E]). For several important Banach algebras, the set of logarithmic residues actually coincides with the set of sums of idempotents, but there are also Banach algebras – equally important – for which this is not the case. Another aspect of the relationship is the following. If in $\mathcal{B}$ logarithmic residues only vanish in the (trivial) situation where the function F takes invertible values inside the integration contour, then $\mathcal{B}$ admits only trivial zero sums of idempotents. In other words, a sum of idempotents in $\mathcal{B}$ can then only vanish when so do all its terms (see [BES1] and [BES2]).

An important Banach algebra that we already touched upon in [BES2], Example 4.4, is the Banach algebra $\mathcal{L}_{\mathcal{C}}(X)$ generated by the compact operators and the identity operator on X, where X is a complex Banach space. Here, and in [BES7], we shall study this Banach algebra in a more systematic way. In [BES7], the focus is on logarithmic residues; in the present paper the emphasis is on sums of idempotents.

There is an essential difference between the case when the dimension of X is finite and that where it is infinite. The first case (i.e., the matrix case $\mathcal{B} = \mathbb{C}^{n\times n}$), has been studied in [BES4]. *In the present paper we concentrate on the infinite dimensional situation.*

We shall now give an outline of the paper.

Section 2 is partly of a preliminary nature in the sense that it contains definitions and notations. In another part, Section 2 gives some first observations on (sums of) idempotents in $\mathcal{L}_{\mathcal{C}}(X)$, the Banach algebra under consideration.

The idempotents in $\mathcal{L}_{\mathcal{C}}(X)$ are those projections P on X such that either P itself or the complementary projection $I - P$ has finite rank. Hence, modulo a multiple of the identity operator, a sum of idempotents in $\mathcal{L}_{\mathcal{C}}(X)$ is a difference of sums of finite rank projections on X. Thus, to a large extent, the study of sums of idempotents in $\mathcal{L}_{\mathcal{C}}(X)$ amounts to the investigation of differences of sums of finite rank projections on X. This investigation is carried out in Section 3. The assumption that X is infinite dimensional is crucial here.

Section 4 deals with sums of idempotents in $\mathcal{L}_{\mathcal{C}}(X)$. On the basis of the results obtained in Section 3, these sums are characterized in terms of conditions involving ranks, traces and dimensions of null spaces. We also describe the closure of the set of sums of idempotents. As a result the connected components of the set of sums of idempotents in $\mathcal{L}_{\mathcal{C}}(X)$ and of its closure are identified.

In Section 5, the sums of idempotents in $\mathcal{L}_{\mathcal{C}}(X)$ are characterized in a quite different way, namely as the logarithmic residues of analytic operator functions F with values in the (non-closed) subalgebra of $\mathcal{L}_{\mathcal{C}}(X)$ generated by the identity and the finite rank operators on X. To put this result into perspective, we mention that in general a logarithmic residue in $\mathcal{L}_{\mathcal{C}}(X)$ need not be a sum of idempotents in $\mathcal{L}_{\mathcal{C}}(X)$. For counterexamples and for an analysis of the relationship between the set of logarithmic residues in $\mathcal{L}_{\mathcal{C}}(X)$ and (the closure of) the set of sums of idempotents in $\mathcal{L}_{\mathcal{C}}(X)$, see [BES7] or [BES8].

The results of Sections 3 and 4, play a major role in [BES7] where logarithmic residues in $\mathcal{L}_{\mathcal{C}}(X)$ are (further) investigated. In fact, the article [BES7] can be seen as a continuation of the present paper.

2. Preliminaries and first results

This paper is concerned with the special Banach algebra

$$\mathcal{L}_{\mathcal{C}}(X) = \{\alpha I + C \mid \alpha \in \mathbb{C},\ C \in \mathcal{C}(X)\}.$$

Here X is a complex Banach space, $\mathcal{C}(X)$ denotes the set of all compact bounded linear operators on X and $I = I_X$ is the identity operator on X. Recall that $\mathcal{C}(X)$ is a closed ideal in $\mathcal{L}(X)$, the Banach algebra of all bounded linear operators on X. Hence $\mathcal{L}_\mathcal{C}(X)$ is a Banach subalgebra of $\mathcal{L}(X)$ which contains $\mathcal{C}(X)$ as a closed ideal. Note that $\mathcal{L}_\mathcal{C}(X)$ is inverse closed with respect to $\mathcal{L}(X)$. This can most easily be seen from the formula $(\alpha I + C)^{-1} = \alpha^{-1} I - \alpha^{-1} C(\alpha I + C)^{-1}$. In [BES2], Example 4.4, it was observed that, for X a separable Hilbert space, $\mathcal{L}_\mathcal{C}(X)$ belongs to a specific class of Banach algebras introduced by S. Roch and B. Silbermann [RS]. However, this fact will not play a role in the present paper.

If X has finite dimension n, then $\mathcal{L}(X)$, $\mathcal{C}(X)$ and $\mathcal{L}_\mathcal{C}(X)$ coincide and can be identified with $\mathbb{C}^{n\times n}$. The papers [BES4], [HP] and [Wu] deal with this situation. *Here we shall investigate the Banach algebra $\mathcal{B} = \mathcal{L}_\mathcal{C}(X)$ under the standing assumption that X is infinite dimensional.*

Because of the infinite dimensionality of X, the unit element I of $\mathcal{L}_\mathcal{C}(X)$ is not in $\mathcal{C}(X)$. Hence $\mathcal{C}(X)$ is a complemented closed subspace of $\mathcal{L}_\mathcal{C}(X)$ of codimension 1. In fact, for $T \in \mathcal{L}_\mathcal{C}(X)$ the representation $T = \alpha I + C$ with $\alpha \in \mathbb{C}$ and $C \in \mathcal{C}(X)$ is unique. Moreover, the mapping $\alpha I + C \in \mathcal{L}_\mathcal{C}(X) \mapsto \alpha \in \mathbb{C}$ is Banach algebra homomorphism with kernel $\mathcal{C}(X)$.

We now turn to the study of sums of idempotents in the Banach algebra $\mathcal{L}_\mathcal{C}(X)$, where X is an infinite dimensional (complex) Banach space. An idempotent in $\mathcal{L}_\mathcal{C}(X)$ is, a fortiori, an idempotent in $\mathcal{L}(X)$. In other words, the idempotents in $\mathcal{L}_\mathcal{C}(X)$ are projections on X. Recall that a projection on X is compact if and only if it is of finite rank.

Proposition 2.1. *The idempotents in $\mathcal{L}_\mathcal{C}(X)$ are the projections on X for which either the range or the null space has finite dimension.*

In other words, $P \in \mathcal{L}_\mathcal{C}(X)$ is an idempotent if and only if either P itself or the complementary projection $I - P$ is a finite rank projection on X.

Proof. Suppose P is an idempotent in $\mathcal{L}_\mathcal{C}(X)$ and write $P = \alpha I + C$ with $\alpha \in \mathbb{C}$ and $C \in \mathcal{C}(X)$. Now

$$\alpha I + C = (\alpha I + C)^2 = \alpha^2 I + 2\alpha C + C^2, \tag{1}$$

so $\alpha(\alpha - 1)I$ is compact. Since X is infinite dimensional, it follows that $\alpha = 0$ or $\alpha = 1$. In case $\alpha = 0$, the identity (1) reduces to $C^2 = C$. But then $P = C$ is a compact projection on X, hence of finite rank. In the situation where $\alpha = 1$, we have $I - P = -C$ and the identity (1) becomes $C^2 = -C$; thus $I - P = -C$ is a compact and therefore a finite rank projection on X. □

Corollary 2.2. *The Banach algebra $\mathcal{L}_\mathcal{C}(X)$ has only trivial zero sums of idempotents.*

Thus, if $P_1, \dots, P_n$ are idempotents in $\mathcal{L}_\mathcal{C}(X)$ and $P_1 + \cdots + P_n = 0$, then $P_j = 0$ for all $j = 1, \dots, n$.

Proof. Suppose we have a zero sum of idempotents in $\mathcal{L}_C(X)$. Then there exist non-negative integers n and m and finite rank projections $P_1, \ldots, P_{n+m}$ on X such that

$$\sum_{j=1}^{n}(I - P_j) + \sum_{j=n+1}^{n+m} P_j = 0. \tag{2}$$

Clearly, nI is of finite rank and, as X is infinite dimensional, it follows that $n = 0$. But then (2) comes down to

$$\sum_{j=1}^{m} P_j = 0.$$

Taking traces and using that for finite rank projections trace and rank coincide, we see that $P_j = 0$ for all $j = 1, \ldots, m$. □

There is a standard way of introducing a partial ordering on the set of projections on X (the difference of two projections should be a projection again). Corollary 2.2 makes it possible to introduce a partial ordering on the set of sums of idempotents in $\mathcal{L}_C(X)$. For S_1 and S_2 sums of idempotents in $\mathcal{L}_C(X)$, we write $S_1 \preceq S_2$ if $S_2 - S_1$ is again a sum of idempotents in $\mathcal{L}_C(X)$. A straightforward argument shows that $\preceq$ is a partial ordering indeed. It can also be seen that $\preceq$ extends the standard partial ordering on the set of idempotents in $\mathcal{L}_C(X)$ induced by that on the set of all projections on X, referred to above. In [BES7], the partial ordering $\preceq$ will be used to clarify the situation with respect to logarithmic residues of operator polynomials with compact non-leading coefficients.

From Proposition 2.1 we see that a bounded linear operator S on X is a sum of idempotents in $\mathcal{L}_C(X)$ if and only if it can be written in the form

$$S = \sum_{j=1}^{n}(I - P_j) + \sum_{j=n+1}^{n+m} P_j = nI - \left(\sum_{j=1}^{n} P_j - \sum_{j=n+1}^{n+m} P_j \right) \tag{3}$$

with n and m non-negative integers and $P_1, \ldots, P_{n+m}$ finite rank projections on X. Motivated by the last part of (3), we consider the set $\mathcal{P}(X)$ of all bounded linear operator on X of the form $T = V - W$ where V and W are sums of finite rank projections on X. The operators in $\mathcal{P}(X)$ are of finite rank and therefore belong to $\mathcal{L}_C(X)$.

For given $n = 0, 1, 2, \ldots$, let $\mathcal{P}_n(X)$ be the set of all operators T on X that can be written as

$$T = -\left(\sum_{j=1}^{n} P_j - \sum_{j=n+1}^{n+m} P_j \right) \tag{4}$$

with m a (non-fixed) non-negative integer and $P_1, \ldots, P_{m+n}$ finite rank projections on X. Clearly $\mathcal{P}_0(X) \subset \mathcal{P}_1(X) \subset \mathcal{P}_2(X) \subset \ldots$ and $\mathcal{P}(X)$ is the union of the sets $\mathcal{P}_n(X)$.

Write $\mathcal{S}(X)$ for the set of sums of projections on X with finite dimensional null space or range. In other words, $\mathcal{S}(X)$ is the set of sums of idempotents in $\mathcal{L}_\mathcal{C}(X)$. Clearly, a bounded linear operator S on X belongs to $\mathcal{S}(X)$ if and only if it can be written in the form $S = nI + T$ with n a non-negative integer and $T \in \mathcal{P}_n(X)$. Since X is infinite dimensional, the non-negative integer n in this expression is uniquely determined by S. So,

$$\mathcal{S}(X) = \bigcup_{n=0}^{\infty} \left\{ nI + T \mid T \in \mathcal{P}_n(X) \right\} \tag{5}$$

and this union is disjoint.

This discussion suggests that we consider differences of sums of idempotents first.

3. Differences of sums of finite rank projections on X

We begin by considering $\mathcal{P}_0(X)$. By definition, this is the set of sums of finite rank projections or – what amounts to the same – the set of sums of rank one projections on X. For $\tau = 0, 1, 2, \ldots$, let $\mathcal{P}_{0,\tau}(X)$ denote the set of all $T \in \mathcal{P}_0(X)$ for which $\operatorname{trace} T = \tau$. Obviously, $\mathcal{P}_0(X)$ is the disjoint union of the sets $\mathcal{P}_{0,\tau}(X)$.

The following result is part of Theorem 4.4 from [BES6], slightly reformulated (cf. also the earlier papers [HP] and [Wu] for the matrix case). As before, X will be an infinite dimensional complex Banach space.

Proposition 3.1. *The following statements are true:*

(i) *$\mathcal{P}_0(X)$ consists of all finite rank operators T on X for which*

$$\operatorname{rank} T \leq \operatorname{trace} T \in \mathbb{Z};$$

(ii) *The sets $\mathcal{P}_0(X)$ and $\mathcal{P}_{0,\tau}(X)$ have empty interior;*

(iii) *The zero operator on X is the unique isolated point of $\mathcal{P}_0(X)$; in fact, for non-zero τ, the set $\mathcal{P}_{0,\tau}(X)$ has no isolated points;*

(iv) *For τ and σ non-negative integers, not both zero,*

$$\operatorname{dist}\left(\mathcal{P}_{0,\tau}(X), \mathcal{P}_{0,\sigma}(X)\right) \geq \frac{|\tau - \sigma|}{\tau + \sigma}$$

where the left-hand side in this inequality stands for the distance between $\mathcal{P}_{0,\tau}(X)$ and $\mathcal{P}_{0,\sigma}(X)$;

(v) *The (arcwise) connected components of $\mathcal{P}_0(X)$ are the (different) sets $\mathcal{P}_{0,\tau}(X), \quad \tau = 0, 1, 2, \ldots$.*

Next we turn to $\mathcal{P}_n(X)$ for $n \geq 1$. It is convenient to distinguish between the cases $n = 1$ and $n \geq 2$.

Proposition 3.2. *The following statements are true:*

(i) *$\mathcal{P}_1(X)$ consists of all finite rank operators T on X for which*

$$-\dim \operatorname{Ker}(I + T) \leq \operatorname{trace} T \in \mathbb{Z};$$

(ii) $\mathcal{P}_1(X)$ *has empty interior and no isolated points;*
(iii) $\mathcal{P}_1(X)$ *is arcwise connected.*

With regard to (i) we note that the dimension of $\operatorname{Ker}(I+T)$ is finite and equal to the codimension of $\operatorname{Im}(I+T)$ in X. Indeed, as T is of finite rank, $I+T$ is a Fredholm operator of index zero. Recall in this context that a bounded linear operator $T : X \to X$ is said to be a *Fredholm operator* if its null space $\operatorname{Ker} T$ has finite dimension and its range space $\operatorname{Im} T$ has finite codimension in X (and is therefore closed). The difference of the last and the first number is called the *index* of T. It is well known that if $A \in \mathcal{L}(X)$ is invertible and $C \in \mathcal{C}(X)$, then $A + C$ is a Fredholm operator of index zero.

Proposition 3.3. *The following statements are true:*
(i) $\mathcal{P}_n(X) = \mathcal{P}_2(X) = \mathcal{P}_0(X) - \mathcal{P}_0(X) = \mathcal{P}(X)$ *for all* $n = 2, 3, 4, \ldots;$
(ii) $\mathcal{P}(X)$ *consists of all finite rank operators* T *on* X *for which* $\operatorname{trace} T \in \mathbb{Z}$*;*
(iii) $\mathcal{P}(X)$ *has empty interior and no isolated points;*
(iv) $\mathcal{P}(X)$ *is arcwise connected.*

We prepare for the proofs of Propositions 3.2 and 3.3 with a lemma. In this lemma the (standing) assumption that X is infinite dimensional is essential. The lemma will also be used in Section 4. We denote closure of the set of finite rank operators in $\mathcal{L}(X)$ by $\mathcal{C}_{\mathcal{F}}(X)$.

Lemma 3.4. *Let* V *be a subset of* $\mathcal{C}_{\mathcal{F}}(X)$ *containing all finite rank operators on* X *with zero trace. Then* V *is arcwise connected.*

An immediate consequence is that V is dense in $\mathcal{C}_{\mathcal{F}}(X)$.

The lemma can be reformulated as follows: *Given* $A \in \mathcal{C}_{\mathcal{F}}(X)$*, there exists a continuous function* $\Phi : [0,1] \to \mathcal{L}(X)$ *such that*
(i) *For all* t *in the half open interval* $[0,1)$*, the operator* $\Phi(t)$ *has finite rank and zero trace;*
(ii) $\Phi(0) = 0$ *and* $\Phi(1) = A$.

Proof. Take A in $\mathcal{C}_{\mathcal{F}}(X)$ and let $A_1, A_2, A_3, \ldots$ be a sequence of finite rank operators on X converging to A. For $n = 1, 2, 3, \ldots$, put $\tau_n = \operatorname{trace} A_n$ and let m_n be a positive integer larger than $n^2|\tau_n|^2$. From [KS] – see also [Wo], Ch.3.B – we know that there exists a projection P_n on the (infinite dimensional) Banach space X such that $\operatorname{trace} P_n = \operatorname{rank} P_n = m_n$ and $\|P_n\| \le \sqrt{m_n}$. Introduce

$$B_n = A_n - \frac{\tau_n}{m_n} P_n.$$

Then $\operatorname{trace} B_n = 0$ and $B_n \to A$ for $n \to \infty$.

We now define $\Phi : [0,1] \to \mathcal{L}(X)$ as follows. First we put $\Phi(1) = A$, so that the second part of (ii) in the reformulation of the lemma is met. Next we define Φ on the half open intervals

$$\left[1 - \frac{1}{2^{k-1}}, 1 - \frac{1}{2^k}\right), \quad k = 1, 2, 3, \ldots . \tag{6}$$

The definition is

$$\Phi\left(1-\frac{1+x}{2^k}\right) = xB_{k-1} + (1-x)B_k, \quad x \in (0,1];\ k = 1,2,3,\dots,$$

where $B_0 = 0$. Then $\Phi(1-\frac{1}{2^{k-1}}) = B_{k-1}$ for $k = 1,2,3,\dots$; in particular $\Phi(0) = B_0 = 0$. Thus (ii) is satisfied. Clearly, (i) holds too. It remains to prove that Φ is continuous.

Taking limits (from the left) in the right end points of the intervals (6), one sees that Φ is continuous on the half open interval $[0,1)$. To deal with the right end point of the interval $[0,1]$, we note that, for $k = 1,2,3\dots$ and $x \in (0,1]$,

$$\Phi\left(1-\frac{1+x}{2^k}\right) - A = x(B_{k-1} - A) + (1-x)(B_k - A).$$

Hence, for $k = 1,2,3,\dots$,

$$\|\Phi(t) - A\| \le \|B_{k-1} - A\| + \|B_k - A\|, \quad 1 - \frac{1}{2^{k-1}} \le t < 1 - \frac{1}{2^k}.$$

But then $\Phi(t) \to A$ for $t \to 1$ (from the left), and the proof is complete. □

Proof of Proposition 3.2. Let $T \in \mathcal{P}_1(X)$ and write T as

$$T = S - P_0, \quad S = \sum_{j=1}^{m} P_j$$

with $P_0,\dots,P_m$ projections of finite rank (see (4)). Taking traces and using that for finite rank projections trace and rank coincide, we see that the trace of T is an integer.

To prove that $\operatorname{trace} T$ is larger than or equal to $-\dim\operatorname{Ker}(I+T)$, we argue as follows. With respect to an appropriately chosen decomposition $X = \widetilde{X} \oplus \widehat{X}$, involving a finite dimensional subspace $\widetilde{X}$ of X and a closed subspace $\widehat{X}$ of X, the finite rank projections $P_0,\dots,P_m$ have the form

$$P_j = \begin{pmatrix} \widetilde{P}_j & 0 \\ 0 & 0 \end{pmatrix}.$$

Here the restrictions $\widetilde{P}_0,\dots,\widetilde{P}_m$ to $\widetilde{X}$ of $P_0,\dots,P_m$, respectively, are projections on $\widetilde{X}$. Now

$$T = \begin{pmatrix} \widetilde{T} & 0 \\ 0 & 0 \end{pmatrix},$$

where

$$\widetilde{T} = \widetilde{S} - \widetilde{P}_0, \quad \widetilde{S} = \sum_{j=1}^{m} \widetilde{P}_j.$$

Clearly, $\operatorname{trace} T = \operatorname{trace}\widetilde{T}$ and $\dim\operatorname{Ker}(I+T) = \dim\operatorname{Ker}(\widetilde{I}+\widetilde{T})$, where $\widetilde{I}$ is the identity operator on $\widetilde{X}$. So it is sufficient to consider $\widetilde{T}$ and $\widetilde{S}$ in place of T and

S. This has the advantage that the underlying space $\widetilde{X}$ has finite dimension. Put $d = \dim \widetilde{X}$. Then we get from $\widetilde{I} + \widetilde{T} = \widetilde{S} + \widetilde{I} - \widetilde{P}_0$ that

$$\begin{aligned} d - \dim \operatorname{Ker}(\widetilde{I} + \widetilde{T}) &= \operatorname{rank}(\widetilde{I} + \widetilde{T}) \\ &\le \operatorname{rank} \widetilde{S} + \operatorname{rank}(\widetilde{I} - \widetilde{P}_0) \\ &= \operatorname{rank} \widetilde{S} + d - \operatorname{rank} \widetilde{P}_0 \end{aligned}$$

and so $-\dim \operatorname{Ker}(\widetilde{I} + \widetilde{T}) \le \operatorname{rank} \widetilde{S} - \operatorname{rank} \widetilde{P}_0$. Now $\widetilde{S}$ is a sum of (finite rank) projections on $\widetilde{X}$, hence $\operatorname{rank} \widetilde{S} \le \operatorname{trace} \widetilde{S}$. It follows that

$$-\dim \operatorname{Ker}(\widetilde{I} + \widetilde{T}) \le \operatorname{trace} \widetilde{S} - \operatorname{rank} \widetilde{P}_0 = \operatorname{trace}(\widetilde{S} - \widetilde{P}_0) = \operatorname{trace} \widetilde{T},$$

as desired.

Conversely, assume T has finite rank, integer trace and

$$-\dim \operatorname{Ker}(I + T) \le \operatorname{trace} T.$$

Write $\widetilde{X} = \operatorname{Ker}(I + T)$. Since $I + T$ is a Fredholm operator (of index zero), $\widetilde{X}$ is a finite dimensional space. Let $\widehat{X}$ be a closed complement of $\widetilde{X}$ in X. With respect to the decomposition $X = \widetilde{X} \oplus \widehat{X}$, the operator T has the form

$$T = \begin{pmatrix} -\widetilde{I} & A \\ 0 & \widehat{T} \end{pmatrix}$$

with $\widetilde{I}$ the identity operator on $\widetilde{X}$, $\widehat{T} \in \mathcal{L}(\widehat{X})$ and $A : \widehat{X} \to \widetilde{X}$ a bounded linear operator. Obviously, along with T, the operator $\widehat{T}$ has finite rank. Moreover $\operatorname{trace} T = -d + \operatorname{trace} \widehat{T}$, where $d = \dim \widetilde{X} = \dim \operatorname{Ker}(I + T)$. It follows that $\operatorname{trace} \widehat{T}$ is a non-negative integer. Besides $\widehat{T}$, the operator A, having its range in $\widetilde{X}$, is of finite rank as well. Hence $\operatorname{Ker} A \cap \operatorname{Ker} \widehat{T}$ is a closed subspace of $\widehat{X}$ with finite codimension in $\widehat{X}$. Let $\widehat{P}$ be a projection of $\widehat{X}$ along $\operatorname{Ker} A \cap \operatorname{Ker} \widehat{T}$. Then $\widehat{P}$ is of finite rank, $A = A\widehat{P}$ and $\widehat{T} = \widehat{T}\widehat{P}$. Define the finite rank projection P on $X = \widetilde{X} \oplus \widehat{X}$ by

$$P = \begin{pmatrix} \widetilde{I} & 0 \\ 0 & \widehat{P} \end{pmatrix}.$$

We claim that $T + P$ is a sum of finite rank projections on X, in other words $T + P \in \mathcal{P}_0(X)$. This is the argument.

From

$$T + P = \begin{pmatrix} 0 & A \\ 0 & \widehat{T} + \widehat{P} \end{pmatrix} = \begin{pmatrix} 0 & A\widehat{P} \\ 0 & \widehat{T}\widehat{P} + \widehat{P} \end{pmatrix} = \begin{pmatrix} 0 & A \\ 0 & \widehat{T} + \widehat{I} \end{pmatrix} \begin{pmatrix} 0 & 0 \\ 0 & \widehat{P} \end{pmatrix},$$

where $\widehat{I}$ is the identity operator on $\widehat{X}$, it is clear that $T + P$ has finite rank not exceeding that of $\widehat{P}$. Further

$$\operatorname{trace}(T + P) = \operatorname{trace}(\widehat{T} + \widehat{P}) = \operatorname{trace} \widehat{T} + \operatorname{trace} \widehat{P} = \operatorname{trace} \widehat{T} + \operatorname{rank} \widehat{P}.$$

Now trace $\widehat{T}$ is a non-negative integer, so we may conclude that $T+P$ has integer trace and

$$\operatorname{trace}(T+P) \geq \operatorname{rank} \widehat{P} \geq \operatorname{rank}(T+P).$$

Thus $T+P \in \mathcal{P}_0(X)$ as desired. This finishes the proof of (i).

The trace takes only integer values on $\mathcal{P}_1(X)$. Hence $\mathcal{P}_1(X)$ has empty interior (in the topological space $\mathcal{L}_C(X)$). This covers the first part of (ii). From (i) it is clear that $\mathcal{P}_1(X)$ contains all finite rank operators on X with zero trace. The second part of (ii) and statement (iii) now follow from Lemma 3.4. □

Proof of Proposition 3.3. Mutatis mutandis, the argument for (iii) and (iv) is the same as that for (ii) and (iii) of Proposition 3.2. Note that again Lemma 3.4 – valid only in an infinite dimensional context – is used. It remains to establish (i) and (ii). For this, we argue as follows.

Take $n \geq 2$. Then, as we observed already, $\mathcal{P}_2(X) \subset \mathcal{P}_n(X)$. From the definitions it is clear that $\mathcal{P}_n(X) \subset \mathcal{P}_0(X) - \mathcal{P}_0(X) = \mathcal{P}(X)$. If T belongs to the latter set, then T is of finite rank and it follows from Proposition 3.1(i) and the linearity of the trace that trace T is an integer. Now suppose that T fulfills these conditions on the rank and the trace by assumption. We shall prove that $T \in \mathcal{P}_2(X)$. With this (i) and (ii) are established.

Let r be the largest of the integers 0 and $\operatorname{rank} T - \operatorname{trace} T$. Then r is a non-negative integer and $\operatorname{rank} T \leq r + \operatorname{trace} T$. Choose a projection of X having rank (and hence also trace) equal to r. Note that the possibility to do this – regardless of the value of r – stems from the infinite dimensionality of X. Put $H = T + 2P$. Then H has integer trace. Also

$$\operatorname{rank} H \leq r + \operatorname{rank} T \leq 2r + \operatorname{trace} T = \operatorname{trace} H.$$

So $H \in \mathcal{P}_0(X)$ on account of Proposition 3.1(i). Hence $T = H - 2P \in \mathcal{P}_2(X)$, as desired. □

Elaborating on Propositions 3.1–3.3, we note that

$$\mathcal{P}_0(X) \subset \mathcal{P}_1(X) \subset \mathcal{P}_2(X) = \mathcal{P}_3(X) = \cdots = \mathcal{P}(X) = \mathcal{P}_0(X) - \mathcal{P}_0(X) \qquad (7)$$

and that the two inclusions in (7) are strict. In fact, $\mathcal{P}_1(X) \backslash \mathcal{P}_0(X)$ consists of all finite rank operators T on X for which trace T is an integer satisfying

$$-\dim \operatorname{Ker}(I+T) \leq \operatorname{trace} T < \operatorname{rank} T$$

and $\mathcal{P}_2(X) \backslash \mathcal{P}_1(X)$ is the set of all finite rank operators T on X such that trace T is an integer and

$$\operatorname{trace} T < -\dim \operatorname{Ker}(I+T).$$

So, for instance, when Q is any non-zero finite rank projection on X, then $-Q \in \mathcal{P}_1(X) \backslash \mathcal{P}_0(X)$ and $-2Q \in \mathcal{P}_2(X) \backslash \mathcal{P}_1(X)$.

The decomposition of $\mathcal{P}_0(X)$ into a disjoint union of subsets $\mathcal{P}_{0,\tau}(X)$ on which the trace is constant (cf. the first paragraph of this section), suggests to do likewise for the sets $\mathcal{P}_n(X)$. Thus we write

$$\mathcal{P}_n(X) = \bigcup_{\tau=-\infty}^{\infty} \mathcal{P}_{n,\tau}(X), \tag{8}$$

where, for $n = 0, 1, 2, \ldots$ and $\tau \in \mathbb{Z}$, $\mathcal{P}_{n,\tau}(X) = \{T \in \mathcal{P}_n(X) \mid \operatorname{trace} T = \tau\}$. Note that for non-negative τ, the expression $\mathcal{P}_{0,\tau}(X)$ has the same meaning as before, while $\mathcal{P}_{0,\tau}(X)$ is empty whenever τ is negative. The union (8) is obviously disjoint and in [BES7] it will be proved that its "constituents", i.e., the sets $\mathcal{P}_{n,\tau}(X)$, are arcwise connected (cf. also [BES8]). For a continuation of this discussion and an indication of the context in which it is relevant, see the end of the next section.

4. The set of sums of idempotents in $\mathcal{L}_C(X)$: characterization and topological properties

We now return to $\mathcal{S}(X)$, the set of sums of idempotents in $\mathcal{L}_C(X)$. To facilitate the discussion, we rewrite (5) as

$$\mathcal{S}(X) = \bigcup_{n=0}^{\infty} \mathcal{S}_n(X), \tag{9}$$

where $\mathcal{S}_n(X) = \{nI + T \mid T \in \mathcal{P}_n(X)\}$. Recall that the union in (9) is disjoint. In fact, for n and m non-negative integers,

$$\operatorname{dist}\left(\mathcal{S}_n(X), \mathcal{S}_m(X)\right) = |n - m|, \tag{10}$$

the left-hand side in this identity standing for the distance of $\mathcal{S}_n(X)$ and $\mathcal{S}_m(X)$. To prove (10), we argue as follows. As X is infinite dimensional, there are no finite rank operators T on X such that $\|T - I\| < 1$. This implies that the right-hand side of (10) does not exceed the left-hand side. On the other hand it is obvious that the left-hand side of (10) does not exceed the right-hand side, for $nI \in \mathcal{S}_n(X)$ and $mI \in \mathcal{S}_m(X)$.

Since $\mathcal{S}_0(X) = \mathcal{P}_0(X)$, the identity (9) can be rewritten as

$$\mathcal{S}(X) = \left(\bigcup_{\tau=0}^{\infty} \mathcal{P}_{0,\tau}(X) \right) \cup \left(\bigcup_{n=1}^{\infty} \{nI + T \mid T \in \mathcal{P}_n(X)\} \right)$$

which, with the help of Propositions 3.2 and 3.3, can be transformed into

$$\mathcal{S}(X) = \left(\bigcup_{\tau=0}^{\infty} \mathcal{P}_{0,\tau}(X) \right) \cup \{I + T \mid T \in \mathcal{P}_1(X)\} \cup \left(\bigcup_{n=2}^{\infty} \{nI + T \mid T \in \mathcal{P}(X)\} \right).$$

To make the picture complete, we mention that, for $n = 1, 2, 3, \ldots$, $\tau = 0, 1, 2, \ldots$,

$$\operatorname{dist}\left(\mathcal{S}_n(X), \mathcal{P}_{0,\tau}(X)\right) = n.$$

This one verifies without difficulty, using that for each T in $\mathcal{P}_{0,\tau}(X)$, the operator $nI + T$ belongs to $\mathcal{S}_n(X) = \{nI + T \mid T \in \mathcal{P}_n(X)\}$.

The next result is now an immediate consequence of Propositions 3.1–3.3.

Theorem 4.1. *A bounded linear operator S on X is a sum of idempotents in $\mathcal{L}_\mathcal{C}(X)$ if and only if one of the following three mutually exclusive conditions is satisfied:*

(i) *S is a sum of finite rank projections or – what amounts to the same – rank one projections on X; equivalently, S has finite rank and*

$$\operatorname{rank} S \leq \operatorname{trace} S \in \mathbb{Z};$$

(ii) *$S - I$ has finite rank and*

$$-\dim \operatorname{Ker} S \leq \operatorname{trace}(S - I) \in \mathbb{Z};$$

(iii) *There exists an integer n, $n \geq 2$, such that $S - nI$ has finite rank and integer trace; equivalently, there exists an integer n, $n \geq 2$, such that $S - nI$ is the difference of two operators on X that both can be written as sums of finite rank projections on X.*

Moreover, the zero operator on X is the unique isolated point of $\mathcal{S}(X)$, and $\mathcal{S}(X)$ has empty interior. Finally, the (arcwise) connected components of $\mathcal{S}(X)$ are the (different) sets $\mathcal{P}_{0,\tau}(X)$ and $\mathcal{S}_n(X)$, where $\tau = 0, 1, 2, \ldots$ and $n = 1, 2, 3, \ldots$.

A few comments are in order. The conditions (i)–(iii) are mutually exclusive, indeed. This corresponds to the fact that the union in (5), written also as (9), is disjoint. With regard to (ii) we note that, since in this case S is a Fredholm operator of index zero, the dimension of $\operatorname{Ker} S$ is finite and equal to the codimension of $\operatorname{Im} S$ in X. For operators on finite dimensional spaces, the conditions (i) and (ii) would amount to the same: in that situation $\operatorname{trace}(S - I) = \operatorname{trace} S - d$ and $-\dim \operatorname{Ker} S = \operatorname{rank} S - d$. For underlying finite dimensional spaces, condition (iii) would mean nothing else than that $\operatorname{trace} S$ is an integer. So the validity of (iii) depends crucially on the assumption that X is infinite dimensional.

Partly because it is interesting in its own right and partly because it is needed for [BES7], we now look at the closure $\overline{\mathcal{S}(X)}$ of $\mathcal{S}(X)$ in $\mathcal{L}_\mathcal{C}(X)$. From (9) and (10), one immediately derives that

$$\overline{\mathcal{S}(X)} = \bigcup_{n=0}^{\infty} \overline{\mathcal{S}_n(X)}$$

and

$$\operatorname{dist}\left(\overline{\mathcal{S}_n(X)}, \overline{\mathcal{S}_m(X)}\right) = |n - m|.$$

Now $\overline{\mathcal{S}_n(X)} = \{nI + T \mid T \in \overline{\mathcal{P}_n(X)}\}$. So, in order to proceed, we need to determine $\overline{\mathcal{P}_n(X)}$.

Recall that $\mathcal{C}_\mathcal{F}(X)$ stands for the closure of the set of finite rank operators
that $\mathcal{C}_\mathcal{F}(X)$ is a (closed) ideal in $\mathcal{L}(X)$ and $\mathcal{L}_\mathcal{C}(X)$, which is con-
For many important Banach spaces X, the ideals $\mathcal{C}_\mathcal{F}(X)$ and $\mathcal{C}(X)$

Proposition 4.2. *The following statements are true:*

(i) *The set $\mathcal{P}_0(X)$ is closed in $\mathcal{L}_{\mathcal{C}}(X)$;*

(ii) *For $n = 1, 2, 3, \ldots$, the set $\mathcal{P}_n(X)$ is not closed in $\mathcal{L}_{\mathcal{C}}(X)$; its closure coincides with $\mathcal{C}_{\mathcal{F}}(X)$ and is an arcwise connected (even convex) subset of $\mathcal{L}_{\mathcal{C}}(X)$.*

It is worthwhile to note that the trace is a continuous function on $\mathcal{P}_0(X)$. For a proof of this (and an even more general result), see [BES6].

Proof. Statement (i) is part of Theorem 4.4 from [BES6]. For $n = 1, 2, 3, \ldots$, the set $\mathcal{P}_n(X)$ contains all finite rank operators with zero trace. This we see from Propositions 3.2 and 3.3. Hence $\mathcal{P}_n(X)$ is dense in $\mathcal{C}_{\mathcal{F}}(X)$ by Lemma 3.4. From this it is clear that $\mathcal{P}_n(X)$ is not closed. It is equally obvious that $\mathcal{C}_{\mathcal{F}}(X)$ is a convex (hence arcwise connected) set. □

On account of Proposition 4.2 we can now conclude that

$$\overline{\mathcal{S}_n(X)} = \{nI + T \mid T \in \mathcal{C}_{\mathcal{F}}(X)\}, \qquad n = 1, 2, 3, \ldots .$$

Also, $\mathcal{P}_0(X)$ is closed, so $\overline{\mathcal{S}_0(X)} = \mathcal{S}_0(X) = \mathcal{P}_0(X)$. Thus

$$\begin{aligned} \overline{\mathcal{S}(X)} &= \mathcal{P}_0(X) \cup \bigcup_{n=1}^{\infty} \{nI + T \mid T \in \mathcal{C}_{\mathcal{F}}(X)\} \\ &= \left(\bigcup_{\tau=0}^{\infty} \mathcal{P}_{0,\tau}(X) \right) \cup \left(\bigcup_{n=1}^{\infty} \{nI + T \mid T \in \mathcal{C}_{\mathcal{F}}(X)\} \right). \end{aligned}$$

For completeness we note that, for $n = 1, 2, 3, \ldots$ and $\tau = 0, 1, 2, \ldots$,

$$\operatorname{dist}\left(\overline{\mathcal{S}_n(X)}, \mathcal{P}_{0,\tau}(X)\right) = n.$$

The following theorem is now immediate.

Theorem 4.3. *A bounded linear operator S on X belongs to the closure $\overline{\mathcal{S}(X)}$ of the set $\mathcal{S}(X)$ of sums of idempotents in $\mathcal{L}_{\mathcal{C}}(X)$ if and only if one of the following two mutually exclusive conditions is satisfied:*

(i) *S is a sum of finite rank projections or – what amounts to the same – rank one projections on X; equivalently, S has finite rank and*

$$\operatorname{rank} S \leq \operatorname{trace} S \in \mathbb{Z};$$

(ii) *there exists an integer $n, n \geq 1$, such that $S - nI \in \mathcal{C}_{\mathcal{F}}(X)$, i.e., $S - nI$ is a limit of finite rank operators on X.*

Moreover, the zero operator on X is the unique isolated point of $\overline{\mathcal{S}(X)}$, and $\overline{\mathcal{S}(X)}$ has empty interior. Finally, the (arcwise) connected components of $\overline{\mathcal{S}(X)}$ are the (different) sets $\mathcal{P}_{0,\tau}(X)$ and $\{nI + T \mid T \in \mathcal{C}_{\mathcal{F}}(X)\}$, where $\tau = 0, 1, 2, \ldots$ and $n =$ $1, 2, 3, \ldots$.

Note that the sets $\{nI + T \mid T \in \mathcal{C}_{\mathcal{F}}(X)\}$ are even convex.

Suppose X has the *approximation property*. By this we mean here that each compact operator on X is the limit of a sequence of finite rank operators on X, in other words $\mathcal{C}(X) = \mathcal{C}_{\mathcal{F}}(X)$. Then the expressions for $\overline{\mathcal{S}(X)}$ given just before Theorem 4.3 can be rewritten as

$$\overline{\mathcal{S}(X)} = \left(\bigcup_{\tau=0}^{\infty} \mathcal{P}_{0,\tau}(X)\right) \cup \left(\bigcup_{n=1}^{\infty} \{nI + T \mid T \in \mathcal{C}(X)\}\right).$$

Returning to the general situation (where X does not necessarily have the approximation property), we now follow up on the remarks made in the last paragraph of Section 3. Recall that $\mathcal{P}_{n,\tau}(X) = \{T \in \mathcal{P}_n(X) \mid \operatorname{trace} T = \tau\}$ and that the (disjoint) union of these sets, taken over $\tau \in \mathbb{Z}$, is $\mathcal{P}_n(X)$. So, in view of (5), we can write $\mathcal{S}(X)$ as

$$\mathcal{S}(X) = \bigcup_{n=0}^{\infty} \bigcup_{\tau=-\infty}^{\infty} \{nI + T \mid T \in \mathcal{P}_{n,\tau}(X)\} \tag{11}$$

(where it should be recalled that $\mathcal{P}_{n,\tau}(X) = \emptyset$ whenever $n = 0$ and $\tau < 0$). With the help of the results of Section 3, the sets $\{nI + T \mid T \in \mathcal{P}_{n,\tau}(X)\}$ in the right-hand side of (11) can be described in terms of ranks, traces and dimensions of null spaces. For details, see [BES7] (or [BES8]), where also a relationship is revealed with the issue of left versus right logarithmic residues in $\mathcal{L}_{\mathcal{C}}(X)$. Note that (11) is a disjoint union of arcwise connected sets (cf. the end of Section 3).

5. Sums of idempotents in $\mathcal{L}_{\mathcal{C}}(X)$: a characterization as logarithmic residues

We continue the study of sums of idempotents in the Banach algebra $\mathcal{L}_{\mathcal{C}}(X)$, making a connection with logarithmic residues. This connection will be further elaborated on in [BES7] (cf. [BES8]). Here the emphasis is on the characterization of the set $\mathcal{S}(X)$.

First we recall some definitions and notations and we present two results that will be employed in the proof of the main theorem in this section. It is convenient to do this in a general Banach algebra setting.

Let $\mathcal{B}$ be a (complex) Banach algebra with unit element. If F is a $\mathcal{B}$-valued function with domain Δ, then F^{-1} stands for the function given by $F^{-1}(\lambda) = F(\lambda)^{-1}$ with domain the set of all $\lambda \in \Delta$ such that $F(\lambda)$ is invertible. If Δ is an open subset of the complex plane $\mathbb{C}$ and $F : \Delta \to \mathcal{B}$ is analytic, then so is F^{-1} on its domain. The derivative of F will be denoted by F'. The *left*, respectively *right*, *logarithmic derivative* of F is the function given by $F'(\lambda)F^{-1}(\lambda)$, respectively $F^{-1}(\lambda)F'(\lambda)$, with the same domain as F^{-1}.

Logarithmic residues are contour integrals of logarithmic derivatives. To make this notion more precise, we shall employ bounded Cauchy domains in $\mathbb{C}$ and their positively oriented boundaries. For a discussion of these notions, see, for instance [TL].

Let D be a bounded Cauchy domain in $\mathbb{C}$. The (positively oriented) boundary of D will be denoted by ∂D. We write $\mathcal{A}_\partial(D;\mathcal{B})$ for the set of all $\mathcal{B}$-valued functions F with the following properties: F is defined and analytic on a neighborhood of the closure $\overline{D} = D \cup \partial D$ of D and F takes invertible values on all of ∂D (hence F^{-1} is analytic on a neighborhood of ∂D). For $F \in \mathcal{A}_\partial(D;\mathcal{B})$, one can define the contour integrals

$$LR_{left}(F;D) = \frac{1}{2\pi i}\int_{\partial D} F'(\lambda)F^{-1}(\lambda)d\lambda, \tag{12}$$

$$LR_{right}(F;D) = \frac{1}{2\pi i}\int_{\partial D} F^{-1}(\lambda)F'(\lambda)d\lambda. \tag{13}$$

The elements of the form (12) or (13) are called logarithmic residues in $\mathcal{B}$. More specifically, we call $LR_{left}(F;D)$ the *left* and $LR_{right}(F;D)$ the *right logarithmic residue of F with respect to D.*

It is convenient to introduce a local version of these concepts too. Given a complex number λ_0, we let $\mathcal{A}(\lambda_0;\mathcal{B})$ be the set of all $\mathcal{B}$-valued functions F with the following properties: F is defined and analytic on an open neighborhood of λ_0 and F takes invertible values on a deleted neighborhood of λ_0. For $F \in \mathcal{A}(\lambda_0;\mathcal{B})$, one can introduce

$$LR_{left}(F;\lambda_0) = \frac{1}{2\pi i}\int_{|\lambda-\lambda_0|=\varrho} F'(\lambda)F^{-1}(\lambda)d\lambda, \tag{14}$$

$$LR_{right}(F;\lambda_0) = \frac{1}{2\pi i}\int_{|\lambda-\lambda_0|=\varrho} F^{-1}(\lambda)F'(\lambda)d\lambda, \tag{15}$$

where ϱ is a positive number such that both F and F^{-1} are analytic on an open neighborhood of the punctured closed disc with center λ_0 and radius ϱ. The orientation of the integration contour $|\lambda - \lambda_0| = \varrho$ is, of course, taken positively, that is counterclockwise. Note that the right-hand sides of (14) and (15) do not depend on the choice of ϱ. In fact, (14) and (15) are equal to the coefficient of $(\lambda - \lambda_0)^{-1}$ in the Laurent expansion at λ_0 of the left and right logarithmic derivative of F at λ_0, respectively. Obviously, $LR_{left}(F;\lambda_0)$, respectively $LR_{right}(F;\lambda_0)$, is a left, respectively right, logarithmic residue of F in the sense of the definitions given in the preceding paragraph (take for D the open disc with radius ϱ centered at λ_0). We call $LR_{left}(F;\lambda_0)$ the *left* and $LR_{right}(F;\lambda_0)$ the *right logarithmic residue of F at λ_0.*

In certain cases, the study of logarithmic residues with respect to bounded Cauchy domains can be reduced to the study of logarithmic residues with respect to single points. The typical situation is as follows. Let D be a bounded Cauchy domain, let $F \in \mathcal{A}_\partial(D;\mathcal{B})$ and suppose F takes invertible values on D except in a finite number of distinct points $\lambda_1, \dots, \lambda_n \in D$. Then

$$LR_{left}(F; D) = \sum_{j=1}^{n} LR_{left}(F; \lambda_j), \tag{16}$$

$$LR_{right}(F; D) = \sum_{j=1}^{n} LR_{right}(F; \lambda_j). \tag{17}$$

This occurs, in particular, when F^{-1} is meromorphic on D with a finite number of poles in D, a state of affairs that we will encounter in what follows.

Sums of idempotents in a Banach algebra with unit element are always logarithmic residues (cf. [BES2]). This is easy to see when one allows Cauchy domains with an arbitrary number of connected components. Things are considerably more complicated when the Cauchy domains are required to be connected. The following theorem, which is a conclusion of a result due to the second author [E], covers this case. It is formulated here in a way appropriate for our needs later in this section.

A Banach algebra valued function is called *entire* when it is defined and analytic on all of $\mathbb{C}$. A pole is said to be *simple* when it has order one.

Theorem 5.1. *Let $\mathcal{B}$ be a complex Banach algebra with unit element e and let $\mathcal{B}_0$ be a subalgebra of $\mathcal{B}$ (possibly non-closed and not necessarily containing e). Let $p_1, \dots, p_n$ be non-zero idempotents in $\mathcal{B}$ and let $\lambda_1, \dots, \lambda_n$ be distinct (but otherwise arbitrary) points in $\mathbb{C}$. Assume that for each $j = 1, \dots, n$, either p_j or $e - p_j$ belongs to $\mathcal{B}_0$. Then there exists an entire function $F : \mathbb{C} \to \mathcal{B}$ such that the following is satisfied:*

(i) *F takes invertible values on $\mathbb{C}$, except in the points $\lambda_1, \dots, \lambda_n$, where F^{-1} has simple poles;*
(ii) *$LR_{left}(F; \lambda_j) = LR_{right}(F; \lambda_j) = p_j$ for all $j = 1, \dots, n$;*
(iii) *F admits a representation $F(\lambda) = f(\lambda)e + F_0(\lambda)$, where $f : \mathbb{C} \to \mathbb{C}$ and $F_0 : \mathbb{C} \to \mathcal{B}$ are entire while, moreover, F_0 takes its values in $\mathcal{B}_0$.*

In case all idempotents $p_1, \dots, p_n$ belong to $\mathcal{B}_0$, the scalar function f can be chosen to be constant with value 1.

The theorem is stated in terms of logarithmic residues at points. In combination with (16) and (17) it can be used to obtain results about logarithmic residues with respect to bounded Cauchy domains. We shall apply Theorem 5.1 in a situation where $e \notin \mathcal{B}_0$. A decomposition of F into f and F_0 as indicated in (iii) is then unique.

Proof. Let $p_1, \dots, p_n$ be non-zero idempotents in $\mathcal{B}$ and let $\lambda_1, \dots, \lambda_n$ be distinct (but otherwise arbitrary) points in $\mathbb{C}$. By [E], there exists an entire $\mathcal{B}$-valued function F such that (i) and (ii) are satisfied. The function F as constructed in [E] is a (possibly non-commutative) product of $3n$ functions of the type $e - p + \phi(\lambda)p$, where $p \in \{p_1, \dots, p_n\}$ and ϕ is an entire scalar function. Now

$$e - p + \phi(\lambda)p = e + (\phi(\lambda) - 1)p = \phi(\lambda)e + (1 - \phi(\lambda))(e - p)$$

and either p or $e-p$ is in $\mathcal{B}_0$. So each of the functions in the product representing F has the form $\alpha(\lambda)e+\beta(\lambda)q$, where α and β are entire scalar functions and $q \in \mathcal{B}_0$ is an idempotent. But then F can be written as a (non-commutative) product

$$F(\lambda) = \prod_{k=1}^{3n} \big(\alpha_k(\lambda)e + \beta_k(\lambda)q_k\big)$$

involving entire scalar functions α_k, β_k and idempotents q_k from $\mathcal{B}_0$. For $\lambda \in \mathbb{C}$, write

$$f(\lambda) = \prod_{k=1}^{3n} \alpha_k(\lambda), \qquad F_0(\lambda) = F(\lambda) - f(\lambda)e.$$

Then f is an entire scalar function and F_0 is an entire $\mathcal{B}$-valued function. Since $\mathcal{B}_0$ is a subalgebra of $\mathcal{B}$ and $q_1, \ldots, q_n$ belong to $\mathcal{B}_0$, the function F_0 takes its values in $\mathcal{B}_0$. Thus (iii) is satisfied.

The last statement of the theorem follows by observing that, in case all idempotents $p_1, \ldots, p_n$ belong to $\mathcal{B}_0$, one can take $\alpha_1(\lambda) = \cdots = \alpha_{3n}(\lambda) = 1$. □

The second auxiliary result is a simple lemma in which $\mathcal{B}$ denotes again a complex Banach algebra with unit element and λ_0 is a complex number.

Lemma 5.2. *Let $F \in \mathcal{A}(\lambda_0;\mathcal{B})$ and assume F^{-1} has a pole at λ_0 of (positive) order p. Let G be a $\mathcal{B}$-valued function which is defined and analytic on an open neighborhood of λ_0, and suppose that $F-G$ has a zero at λ_0 of order at least $2p$. Then $G \in \mathcal{A}(\lambda_0;\mathcal{B})$, the function G^{-1} has a pole at λ_0 of order p and*

$$LR_{left}(F;\lambda_0) = LR_{left}(G;\lambda_0), \quad LR_{right}(F;\lambda_0) = LR_{right}(G;\lambda_0).$$

The logarithmic residues in the above identities are the coefficients of the term $(\lambda-\lambda_0)^{-1}$ in the Laurent expansion at λ_0 of the appropriate left or right logarithmic derivative of F or G. In fact, as we shall see, under the assumptions of the lemma, the principal parts of the Laurent expansion at λ_0 of the left, respectively right, logarithmic derivatives of F and G coincide.

Proof. We denote the unit element in $\mathcal{B}$ by e. For λ in a deleted neighborhood of λ_0, put $H(\lambda) = e - (F(\lambda) - G(\lambda))F^{-1}(\lambda)$ and write $H(\lambda_0) = e$. Then H is analytic on a neighborhood of λ_0 and the function $H(\lambda) - e$ has a zero at λ_0 of order at least p. Hence, for λ in a neighborhood of λ_0, $H(\lambda)$ is invertible and the function $H(\lambda)^{-1} - e$ also has a zero at λ_0 of order at least p. From the identity $G(\lambda) = H(\lambda)F(\lambda)$ it is now clear that $G \in \mathcal{A}(\lambda_0;\mathcal{B})$ and that the principal part of the Laurent expansion of G^{-1} at λ_0 coincides with that of F^{-1}. So, in particular, G^{-1} has a pole at λ_0 of order p. Observe that $F' - G'$ has a zero at λ_0 of order at least $2p-1$. It follows that the principal parts of the Laurent expansion at λ_0 of the left logarithmic derivatives of F and G coincide and the same conclusion holds for the right logarithmic derivatives. □

We now return to the study of the special Banach algebra $\mathcal{L}_\mathcal{C}(X)$. Notations are as before and – as all the time in this paper – the Banach space X is assumed to be infinite dimensional.

Let F be a function with domain Δ and with values in $\mathcal{L}_\mathcal{C}(X)$. Then there exist unique functions $f : \Delta \to \mathbb{C}$ and $C : \Delta \to \mathcal{C}(X)$ such that

$$F(\lambda) = f(\lambda)I + C(\lambda), \quad \lambda \in \Delta.$$

We call f the *scalar* and C the *compact part* of F. If Δ is an open subset of $\mathbb{C}$ and F is analytic on Δ, then so are f and C. Indeed, for each λ in the domain Δ of F, we obtain that $f(\lambda)$ is the canonical image of $F(\lambda)$ in $\mathcal{L}_\mathcal{C}(X)/\mathcal{C}(X)$ where this quotient algebra is identified with $\mathbb{C}$.

The next theorem characterizes the sums of idempotents in $\mathcal{L}_\mathcal{C}(X)$ as the logarithmic residues of $\mathcal{L}_\mathcal{C}(X)$-valued functions F such that the values of the compact part of F are finite rank operators on X, i.e., as the logarithmic residues of functions taking their values in the (non-closed) subalgebra of $\mathcal{L}_\mathcal{C}(X)$ generated by the identity operator and the finite rank operators on X. In general, a logarithmic residue in $\mathcal{L}_\mathcal{C}(X)$ need not be a sum of idempotents. For a counterexample involving a Hilbert space X, see [BES7] or [BES8]. When X fails to have the approximation property, it may even occur that there are logarithmic residues in $\mathcal{L}_\mathcal{C}(X)$ lying outside the closure of $\mathcal{S}(X)$ (see again [BES7] or [BES8]).

Theorem 5.3. *Let D be a bounded Cauchy domain in $\mathbb{C}$ and let S be a bounded linear operator on the infinite dimensional Banach space X. The following statements are equivalent:*

(i) *S is a sum of idempotents in $\mathcal{L}_\mathcal{C}(X)$, in other words $S \in \mathcal{S}(X)$;*
(ii) *S is the left logarithmic residue with respect to D of a function F in $\mathcal{A}_\partial(D; \mathcal{L}_\mathcal{C}(X))$ whose values on D belong to the subalgebra of $\mathcal{L}_\mathcal{C}(X)$ generated by the identity operator and the finite rank operators on X;*
(iii) *S is the right logarithmic residue with respect to D of a function F in $\mathcal{A}_\partial(D; \mathcal{L}_\mathcal{C}(X))$ whose values on D belong to the subalgebra of $\mathcal{L}_\mathcal{C}(X)$ generated by the identity operator and the finite rank operators on X.*

In this result the Cauchy domain D is given. It may or may not be connected. In connection with this, we note that the proof of the implications (i)⇒(ii) and (i)⇒(iii) will provide additional information about the freedom one has in choosing the function F. Among other things it will become clear that F can always be chosen to be an entire function such that F^{-1} has only a finite number of poles which are all simple.

Proof. We begin by proving the implications (i)⇒(ii) and (i)⇒(iii). The complexity of the argument depends on the "shape" of D.

Let $P_1, \ldots, P_n$ be idempotents in $\mathcal{L}_\mathcal{C}(X)$ and let $D_1, \ldots, D_k$ be the connected components of D. When $k \geq n$, the situation is rather simple and the argument is just a slight modification of the proof of [BES3], Proposition 2.1. Indeed, choose distinct points $\lambda_1, \ldots, \lambda_n$ in $D_1, \ldots, D_n$ respectively, and let $F \in \mathcal{A}_\partial(D; B)$ be such that

$$F(\lambda) = \begin{cases} I - P_j + (\lambda - \lambda_j)P_j, & \lambda \in \overline{D}_j; j = 1, \ldots, n, \\ I, & \lambda \in \overline{D}_j; j = n+1, \ldots, k. \end{cases}$$

Then one verifies without difficulty that

$$LR_{left}(F; \lambda_j) = LR_{right}(F; \lambda_j) = P_j, \qquad j = 1, \ldots, n$$

and hence, see (16) and (17),

$$LR_{left}(F; D) = LR_{right}(F; D) = \sum_{j=1}^{n} P_j. \tag{18}$$

For each j, either the projection P_j or the complementary projection $I - P_j$ is of finite rank. Consequently, the function F has its values in the subalgebra of $\mathcal{L}_C(X)$ generated by the identity operator and the finite rank operators on X.

Things are considerably more complicated when $k < n$. The key to the solution is then Ehrhardt's theorem as formulated in Theorem 5.1. Applying this theorem to the situation where $\mathcal{B} = \mathcal{L}_C(X)$ and $\mathcal{B}_0$ is the subalgebra of $\mathcal{L}_C(X)$ consisting of all finite rank operators on X, one immediately gets the following result. *Let $P_1, \ldots, P_n$ be non-zero idempotents in $\mathcal{L}_C(X)$ and let $\lambda_1, \ldots, \lambda_n$ be distinct (but otherwise arbitrary) points in $\mathbb{C}$. Then there exists an entire function $F : \mathbb{C} \to \mathcal{L}_C(X)$ with the following properties:*

(a) *F takes invertible values on $\mathbb{C}$, except in the points $\lambda_1, \ldots, \lambda_n$, where F^{-1} has simple poles;*
(b) *$LR_{left}(F; \lambda_j) = LR_{right}(F; \lambda_j) = P_j$, for all $j = 1, \ldots, n$;*
(c) *The values of F on $\mathbb{C}$ belong to the subalgebra of $\mathcal{L}_C(X)$ generated by the identity operator and the finite rank operators on X.*

Taking into account (16) and (17) and choosing the points $\lambda_1, \ldots, \lambda_n$ in the given Cauchy domain D, one gets (18). This settles the implications (i)⇒(ii) and (i)⇒(iii).

Next we prove that (ii) implies (i). Let $S = LR_{left}(F; D)$ be the left logarithmic residue with respect to D of a function $F \in \mathcal{A}_\partial(D; \mathcal{L}_C(X))$ whose values on D belong to the subalgebra of $\mathcal{L}_C(X)$ generated by the identity operator and the finite rank operators on X. Write f and C for the scalar and the compact part of F, respectively. Then $f \in \mathcal{A}_\partial(D; \mathbb{C})$ and $C(\lambda)$ is of finite rank for each $\lambda \in D$. The function f does not vanish on ∂D and so f has only a finite number of zeros in D. We denote these zeros by $\mu_1, \ldots, \mu_k$. Since X is infinite dimensional, the operators $F(\mu_1), \ldots, F(\mu_k)$ are not invertible. For λ satisfying $f(\lambda) \neq 0$, define $H(\lambda)$ by

$$H(\lambda) = \frac{1}{f(\lambda)} F(\lambda) = I + \frac{1}{f(\lambda)} C(\lambda)$$

and put $D_0 = D \setminus \{\mu_1, \ldots, \mu_k\}$. Then H is analytic on D_0 and has poles or removable singularities at the points $\mu_1, \ldots, \mu_k$. We shall prove that there exist $\mu_{k+1}, \ldots, \mu_l \in D_0$ such that $H(\lambda)$ is invertible for λ in $D_0 \setminus \{\mu_{k+1}, \ldots, \mu_l\} =$

$D\setminus\{\mu_1,\ldots,\mu_l\}$ and that the function H^{-1} has poles or removable singularities at the points $\mu_1,\ldots,\mu_l$. The argument – which draws heavily on [H] and [B1] (cf. also [BKL1], Section 7) – is as follows.

By assumption, $C(\lambda)$ is a finite rank operator for each $\lambda \in D$. Maybe somewhat surprising at first sight, this implies that there exists a finite upper bound for the rank of $C(\lambda)$ when λ ranges through D. To be precise, this holds on each connected component of D. The extension to all of D follows by noting that D, being a Cauchy domain, has only a finite number of connected components. As a consequence of the boundedness of the rank (and using the lower semi-continuity of the rank), we have that for each $\lambda \in D$, the values $C'(\lambda)$ of the derivative of C are of finite rank again and the same conclusion holds for the higher order derivatives of C.

Thus the coefficients in the Taylor expansions of C at points of D are always of finite rank. It follows that the Laurent expansion of H at a point in D has a constant term which is a Fredholm operator of index zero while all other coefficients are of finite rank. In particular, H is what is called finitely meromorphic on D (cf. [GGK]). Along with F, the function H takes invertible values on the boundary of D, and hence also on a neighborhood of ∂D. Such a neighborhood has a non-empty intersection with each component of D. Thus we may conclude that H^{-1} is also finitely meromorphic on D and that H^{-1} has a finite number of poles in D (see [GGK], Section XI.8). In particular there exist $\mu_{k+1},\ldots,\mu_l$ in D_0 with the properties indicated above.

Let us return to the function F. Clearly

$$F(\lambda) = f(\lambda)H(\lambda), \qquad \lambda \in D\setminus\{\mu_1,\ldots,\mu_k\}$$

and the scalar function f does not vanish on $D\setminus\{\mu_1,\ldots,\mu_k\}$. Further H takes invertible values on $D\setminus\{\mu_1,\ldots,\mu_l\}$ which is a subset of $D\setminus\{\mu_1,\ldots,\mu_k\}$. It follows that F takes invertible values on $D\setminus\{\mu_1,\ldots,\mu_l\}$ and

$$F^{-1}(\lambda) = \frac{1}{f(\lambda)}H^{-1}(\lambda), \qquad \lambda \in D\setminus\{\mu_1,\ldots,\mu_l\}.$$

As H^{-1} has poles or removable singularities at the points $\mu_1,\ldots,\mu_l$, so does F^{-1}.

The upshot of all of this is that F takes invertible values on D except in a finite number of distinct points $\lambda_1,\ldots,\lambda_n$ where F^{-1} has poles. In view of the identities (16) and (17), things can now be reduced to the case $n = 1$, where D contains only one point λ_0 at which F is not invertible and $S = LR_{left}(F;\lambda_0)$. This also means that f has at most one zero in D which is then located at λ_0.

If f does not vanish at λ_0, then F is Fredholm operator valued on D and we know from [BES4], Theorem 5.1 or [BES6], Theorem 3.4 that S is a sum of finite rank projections; in particular, S is sum of idempotents in $\mathcal{L}_C(X)$. It remains to tackle the (more interesting) situation where $f(\lambda_0) = 0$ and $F(\lambda_0)$ is of finite rank. We shall first show that it suffices to consider the case when F is a function of polynomial type.

Let p be the order of λ_0 as a pole of F^{-1} and let q be the order of λ_0 as a zero of f. Since X is infinite dimensional, a compact operator cannot cancel the identity. Hence $q \leq p$. Introduce

$$G(\lambda) = \sum_{k=0}^{2p-1} (\lambda - \lambda_0)^k F_k,$$

where F_k stands for the coefficient of $(\lambda - \lambda_0)^k$ in the Taylor expansion of F at λ_0. So G is the $(2p-1)$-th order approximation of F at λ_0. The scalar part of G is then the $(2p-1)$-th order approximation of f at λ_0 and has therefore a zero at λ_0 of order q where $q \leq 2p-1$. Obviously, the function $F - G$ has a zero at λ_0 of order at least $2p$. Thus, by Lemma 5.2, G takes invertible values in a deleted neighborhood of λ_0, G^{-1} has a pole at λ_0 of order p and

$$LR_{left}(G;\lambda_0) = LR_{left}(F;\lambda_0) = S.$$

So, as claimed above, we may assume F to be a function of polynomial type.

Now, if F is a function of polynomial type, then so are its scalar and compact part. Write the compact part C as

$$C(\lambda) = \sum_{k=0}^{m} (\lambda - \lambda_0)^k C_k,$$

where $C_0, \ldots, C_m$ are of finite rank. Let $X = \widehat{X} \oplus \widetilde{X}$ be a direct sum decomposition of X, with $\widehat{X}$ finite dimensional and $\widetilde{X}$ closed, such that the operators C_j have a representation of the form

$$C_j = \begin{pmatrix} \widehat{C}_j & 0 \\ 0 & 0 \end{pmatrix} : \widehat{X} \oplus \widetilde{X} \to \widehat{X} \oplus \widetilde{X}.$$

Then F can be written as

$$F(\lambda) = \begin{pmatrix} \widehat{F}(\lambda) & 0 \\ 0 & f(\lambda)\widetilde{I} \end{pmatrix} : \widehat{X} \oplus \widetilde{X} \to \widehat{X} \oplus \widetilde{X},$$

where $\widetilde{I}$ is the identity operator on $\widetilde{X}$ and $\widehat{F} \in \mathcal{A}(\lambda_0; \mathcal{L}(\widehat{X}))$. It follows that

$$LR_{left}(F;\lambda_0) = \begin{pmatrix} LR_{left}(\widehat{F};\lambda_0) & 0 \\ 0 & q\widetilde{I} \end{pmatrix} : \widehat{X} \oplus \widetilde{X} \to \widehat{X} \oplus \widetilde{X},$$

where q is the order of λ_0 as a zero of f. Since $\widehat{X}$ is finite dimensional (hence $\widehat{F}$ may be identified with a matrix function), we know from [BES4] that $LR_{left}(\widehat{F};\lambda_0)$ is a sum of projections on $\widehat{X}$, say

$$LR_{left}(\widehat{F};\lambda_0) = \sum_{j=1}^{k} \widehat{P}_j.$$

But then, with respect to the decomposition $X = \widehat{X} \oplus \widetilde{X}$, the left logarithmic residue of F at λ_0 has the matrix representation

$$LR_{left}(F;\lambda_0) = q\begin{pmatrix} 0 & 0 \\ 0 & \widetilde{I} \end{pmatrix} + \sum_{j=1}^{n} \begin{pmatrix} \widehat{P}_j & 0 \\ 0 & 0 \end{pmatrix}.$$

On account of Proposition 2.1, we may now conclude that $S = LR_{left}(F;\lambda_0)$ is a sum of idempotents in $\mathcal{L}_{\mathcal{C}}(X)$.

With this we have established the implication (ii)⇒(i). Mutatis mutandis, the same argument can be used for (iii)⇒(i). □

References

[B1] H. Bart, *Meromorphic Operator Valued Functions*, Thesis Vrije Universiteit Amsterdam 1973, also in: *Mathematical Centre Tracts* **44**, Mathematical Centre, Amsterdam 1973.

[B2] H. Bart, Spectral properties of locally holomorphic vector-valued functions, *Pacific J. Math.* **52** (1974), 321–329.

[BES1] H. Bart, T. Ehrhardt and B. Silbermann, Zero sums of idempotents in Banach algebras, *Integral Equations and Operator Theory* **19** (1994), 125–134.

[BES2] H. Bart, T. Ehrhardt and B. Silbermann, Logarithmic residues in Banach algebras, *Integral Equations and Operator Theory* **19** (1994), 135–152.

[BES3] H. Bart, T. Ehrhardt and B. Silbermann, Logarithmic residues, generalized idempotents and sums of idempotents in Banach algebras, *Integral Equations and Operator Theory* **29** (1997), 155–186.

[BES4] H. Bart, T. Ehrhardt and B. Silbermann, Sums of idempotents and logarithmic residues in matrix algebras, In: *Operator Theory: Advances and Applications*, Vol. 122, Birkhäuser, Basel 2001, 139–168.

[BES5] H. Bart, T. Ehrhardt and B. Silbermann, Logarithmic residues of analytic Banach algebra valued functions possessing a simply meromorphic inverse, *Linear Algebra Appl.* **341** (2002), 327–344.

[BES6] H. Bart, T. Ehrhardt and B. Silbermann, Logarithmic residues of Fredholm operator valued functions and sums of finite rank projections, In: *Operator Theory: Advances and Applications*, Vol. 130, Birkhäuser, Basel 2001, 83–106.

[BES7] H. Bart, T. Ehrhardt and B. Silbermann, Logarithmic residues in the Banach algebra generated by the compact operators and the identity, forthcoming.

[BES8] H. Bart, T. Ehrhardt and B. Silbermann, Logarithmic residues and sums of idempotents in the Banach algebra generated by the compact operators and the identity, Report EI 2001-43, Econometric Institute, Erasmus University Rotterdam, 2001.

[BKL1] H. Bart, M.A. Kaashoek and D.C. Lay, Stability properties of finite meromorphic operator functions. I, II, III, *Nederl. Akad. Wetensch. Proc. Ser. A* **77** (1974), 217–259.

[BKL2] H. Bart, M.A. Kaashoek and D.C. Lay, The integral formula for the reduced algebraic multiplicity of meromorphic operator functions, *Proceedings Edinburgh Mathematical Society*, **21** (1978), 65–72.

[E] T. Ehrhardt, Finite sums of idempotents and logarithmic residues on connected domains, *Integral Equations and Operator Theory* **21** (1995), 238–242.

[GGK] I. Gohberg, S. Goldberg and M.A. Kaashoek, *Classes of Linear Operators, Vol. I*, Operator Theory: Advances and Applications, Vol. 49, Birkhäuser, Basel 1990.

[GS] I.C. Gohberg and E.I. Sigal, An operator generalization of the logarithmic residue theorem and the theorem of Rouché, *Mat. Sbornik* **84 (126)** (1971), 607–629 (in Russian), English Transl. in: *Math. USSR Sbornik* **13** (1971), 603–625.

[H] J.S. Howland, Analyticity of determinants of operators on a Banach space, *Proc. Amer. Math. Soc.* **28** (1971), 177–180.

[HP] R.E. Hartwig and M.S. Putcha, When is a matrix a sum of idempotents?, *Linear and Multilinear Algebra* **26** (1990), 279–286.

[KS] M.I. Kadets and M.G. Snobar, Certain functionals on the Minkowski compactum, *Math. Notes* **10** (1971), 694–696.

[M] L. Mittenthal, Operator valued analytic functions and generalizations of spectral theory, *Pacific J. Math.* **24** (1968), 119–132.

[MS] A.S. Markus and E.I. Sigal, The multiplicity of the characteristic number of an analytic operator function, *Mat. Issled.* **5** (1970), no.3(17), 129–147 (in Russian).

[RS] S. Roch and B. Silbermann, The Calkin image of algebras of singular integral operators, *Integral Equations and Operator Theory* **12** (1989), 855–897.

[TL] A.E. Taylor and D. C. Lay, *Introduction to Functional Analysis*, Second Edition, John Wiley and Sons, New York 1980.

[Wo] P. Wojtaszczyk, *Banach Spaces for Analysts*, Cambridge Studies in Advanced Mathematics, Vol. 25, Cambridge University Press, Cambridge 1991.

[Wu] P.Y. Wu, Sums of idempotent matrices, *Linear Algebra Appl.* **142** (1990), 43–54.

H. Bart
Econometrisch Instituut
Erasmus Universiteit Rotterdam
Postbus 1738
3000 DR Rotterdam, The Netherlands
e-mail: `bart@few.eur.nl`

T. Ehrhardt and B. Silbermann
Fakultät für Mathematik
Technische Universität Chemnitz
09107 Chemnitz, Germany
e-mail: `tehrhard@mathematik.tu-chemnitz.de`
e-mail: `silbermn@mathematik.tu-chemnitz.de`

Received: 30 November 2001

Operator Theory:
Advances and Applications, Vol. 135, 61–90

Asymptotic Formulas for the Determinants of Symmetric Toeplitz plus Hankel Matrices

Estelle L. Basor and Torsten Ehrhardt

Dedicated to Bernd Silbermann on the occasion of his 60th birthday

Abstract. We establish asymptotic formulas for the determinants of $N \times N$ Toeplitz plus Hankel matrices $T_N(\phi)+H_N(\phi)$ as N goes to infinity for singular generating functions ϕ defined on the unit circle in the special case where ϕ is even, i.e., where the Toeplitz plus Hankel matrices are symmetric.

1. Introduction

In the theory of random matrices, for certain ensembles, one is led to consider the asymptotics of Fredholm operators of the form $I + W + H$ where W is a finite and symmetric Wiener-Hopf operator and H is a finite Hankel operator [13]. This problem arises when investigating the probability distribution function of a random variable thought of as a function of the eigenvalues of a positive Hermitian random matrix. For general information about random matrix theory we refer the reader to [13] and also to [1, 3, 7] for more specific tie-ins to the random variable problem.

The focus of this paper is to study the discrete analogue of this problem. This is not precisely the desired situation for those interested in random matrix theory. However, it is a natural starting place for cases where the random variable is discontinuous, since then the discrete nature of the computations make things a bit more accessible and the mathematical questions that arise are quite interesting in themselves.

The discrete analogue of this problem is to find an asymptotic expansion of the determinants of Toeplitz plus Hankel matrices

$$M_N(\phi) = T_N(\phi) + H_N(\phi) \tag{1.1}$$

in the case where these matrices are symmetric. Here the $N \times N$ Toeplitz and Hankel matrices are defined as usual by

$$T_N(\phi) = \left(\phi_{j-k}\right)_{j,k=0}^{N-1}, \qquad H_N(\phi) = \left(\phi_{j+k+1}\right)_{j,k=0}^{N-1}. \tag{1.2}$$

The first author was supported in part by National Science Foundation grant DMS-9970879.

The entries ϕ_n are the Fourier coefficients

$$\phi_n = \frac{1}{2\pi} \int_0^{2\pi} \phi(e^{i\theta}) e^{-in\theta} \, d\theta \tag{1.3}$$

of a function $\phi \in L^1(\mathbb{T})$ defined on the unit circle $\mathbb{T}$.

The matrices $M_N(\phi)$ are symmetric if and only if the function ϕ is even, i.e., if $\phi(e^{i\theta}) = \phi(e^{-i\theta})$. From the point of view of random matrix theory, one is particularly interested in the asymptotics of $\det M_N(\phi)$ as $N \to \infty$ in the case of even, piecewise continuous functions ϕ. This is related to the problem of finding the distribution function of a random variable that counts the number of eigenvalues of a random matrix that lie in an interval and to finding the distribution function for other random variables. See [3] to see the connections between these problems.

The problem of determining the asymptotics of the determinants of (not necessarily symmetric) matrices $M_N(\phi)$ has been studied intensively in a previous paper [4]. For example, it was shown there that if ϕ is continuous and sufficiently smooth, then the asymptotics are very similar to the ones given in the Strong Szegő-Widom Limit Theorem. Indeed, it is only in the constant, or third order term that the answers differ. This is no surprise since if ϕ is continuous, then the Toeplitz operator is perturbed by a compact Hankel operator only.

If, however, the symbol ϕ is singular, then the problem is much harder to solve. In the case of Toeplitz determinants the answer is provided by the Fisher-Hartwig conjecture, which has been proved under certain smoothness assumptions in all the cases where it is expected to hold [9]. In [4] an asymptotic formula for the determinants $\det M_N(\phi)$ was obtained for piecewise continuous functions ϕ, but under the additional assumption that the function ϕ does not possess a discontinuity at both a point on the unit circle and its complex conjugate. In this case, the asymptotic formula shows that the asymptotics differ from the asymptotics of Toeplitz determinants only in the third order term, i.e., in the constant term, while the second order term is the same.

Unfortunately, the additional assumption on the location of the discontinuities imposed in [4] excludes all even, piecewise continuous functions. Hence the paper [4] does not answer the discrete analogue of the problem motivated by random matrix theory.

It is the purpose of this paper to solve this problem by establishing an asymptotic formula for determinants of matrices (1.1) for even piecewise continuous functions ϕ. For such functions, it turns out that the asymptotics differs also in the second order term in comparison with the asymptotics of Toeplitz determinants.

The paper is organized as follows. In Section 2 we will recall some of the results established in [4] that are of relevance for this paper. In Section 3 we establish an identity which is the key for computing the asymptotics of $\det M_N(\phi)$ for ϕ even. This identity can be formulated as follows:

$$(\det M_N(\boldsymbol{a}))^2 = \det T_{2N}(\boldsymbol{a\sigma}). \tag{1.4}$$

In this identity $\boldsymbol{a}$ can no longer be considered as a function, but has to be understood as a distribution, which satisfies certain properties. Moreover, $\boldsymbol{\sigma}$ is here a certain concrete distribution. This identity was established first in [12] in a formulation that is not based on distributions. The goal of Section 3 is to provide the necessary tools needed for dealing with distributions, in particular, to define a product between $\boldsymbol{a}$ and $\boldsymbol{\sigma}$ in an appropriate way. Having done this, we are able to derive the distributional formulation of this identity from the original one.

The identity (1.4) reduces the asymptotics of $\det M_N(\boldsymbol{a})$ to the asymptotics of the (skewsymmetric) Toeplitz determinant $\det T_{2N}(\boldsymbol{a\sigma})$. In order to analyze this Toeplitz determinant we cannot rely on the (original) Fisher-Hartwig conjecture because it breaks down in this case. However, in Section 4 we will prove a limit theorem saying that the quotient

$$\frac{\det T_{2N}(\boldsymbol{a\sigma})}{\det T_{2N}(\boldsymbol{a})} \tag{1.5}$$

converges – under certain conditions on $\boldsymbol{a}$ – to a nonzero constant. In order to prove this limit theorem we make heavy use of the machinery that has been developed in [9] in order to prove the Fisher-Hartwig conjecture.

Thus, up to this point, we have reduced the asymptotics of $\det M_N(\boldsymbol{a})$ to the asymptotics of $\det T_{2N}(\boldsymbol{a})$. The Toeplitz determinant $\det T_{2N}(\boldsymbol{a})$ is (generically) of a kind for which the Fisher-Hartwig conjecture holds. In Section 5 we will therefore recall the Fisher-Hartwig conjecture in the form as it has been proved in [9]. Moreover, we specialize it to the distributions (namely, even distributions of Fisher-Hartwig type) that we are interested in. In Section 6 we combine all the previous results and obtain the asymptotics of $\det M_N(\boldsymbol{a})$ for even distributions $\boldsymbol{a}$ of Fisher-Hartwig type, which satisfy appropriate conditions on the parameters.

In Section 7 we finally specialize the quite general result of Section 6 to even piecewise continuous functions. We thus obtain the asymptotics of $\det M_N(\phi)$ for a certain class of even piecewise continuous functions. This result together with the results that are known from [4] suggest a conjecture about the asymptotics of $\det M_N(\phi)$ for quite general piecewise continuous, not just those necessarily even. We end the paper with the conjecture.

2. Known results for determinants of Toeplitz plus Hankel matrices

Let us begin by recalling some of the results concerning the asymptotics of the determinants of the matrices $M_N(\phi)$ that have already been established in [4].

We first consider the case of continuous and sufficiently smooth generating functions ϕ, where an analogue to the Strong Szegő-Widom Limit Theorem holds. In order to be more specific about the smoothness condition, let us consider the Besov class B_1^1, which is by definition the set of all functions $b \in L^1(\mathbb{T})$ such that

$$\|b\|_{B_1^1} := \int_{-\pi}^{\pi} \frac{1}{y^2} \int_{-\pi}^{\pi} \left| b(e^{ix+iy}) + b(e^{ix-iy}) - 2b(e^{ix}) \right| dx dy < \infty. \tag{2.1}$$

It is known that B_1^1 forms a Banach algebra with the above norm and is continuously embedded into the Banach algebra of all continuous functions on $\mathbb{T}$. By $G_1B_1^1$ we denote the set of all nonvanishing functions in B_1^1 with winding number zero. The set $G_1B_1^1$ can also be characterized as the set of all functions b which possess a logarithm $\log b$ in B_1^1.

For $b \in G_1B_1^1$, the constants

$$G[b] = \exp\left([\log b]_0\right), \tag{2.2}$$

$$E[b] = \exp\left(\sum_{k=1}^{\infty} k[\log b]_k[\log b]_{-k}\right), \tag{2.3}$$

$$F[b] = \exp\left(\sum_{k=1}^{\infty}[\log b]_{2k-1} - \frac{1}{2}\sum_{k=1}^{\infty} k[\log b]_k^2\right), \tag{2.4}$$

are well defined, where $[\log b]_n$ stand for the Fourier coefficients of $\log b \in B_1^1$. Moreover, for $b \in G_1B_1^1$, the functions $b_+, b_- \in G_1B_1^1$ are well defined by

$$b_\pm(t) = \exp\left(\sum_{n=1}^{\infty} t^{\pm n}[\log b]_{\pm n}\right), \qquad t \in \mathbb{T}. \tag{2.5}$$

Note that $b(t) = b_-(t)G[b]b_+(t)$, $t \in \mathbb{T}$, is the normalized canonical Wiener-Hopf factorization of the function b.

The analogue to the Strong Szegő-Widom Limit Theorem for the determinants $\det M_N(\phi)$, which has been established in [4, Corollary 2.6], now says that if $b \in G_1B_1^1$, then

$$\det M_N(b) \sim G[b]^N E[b]F[b] \quad \text{as } N \to \infty. \tag{2.6}$$

In the case of even functions $b \in G_1B_1^1$ this simplifies to

$$\det M_N(b) \sim G[b]^N \widehat{E}[b] \quad \text{as } N \to \infty, \tag{2.7}$$

where $\widehat{E}[b]$ is the constant

$$\widehat{E}[b] = \exp\left(\frac{1}{2}\sum_{k=1}^{\infty} k[\log b]_k^2 + \sum_{k=1}^{\infty}[\log b]_{2k-1}\right). \tag{2.8}$$

In order to discuss the asymptotics for the case of piecewise continuous generating functions ϕ, let us introduce the functions

$$t_{\beta,\theta_0}(e^{i\theta}) = e^{i\beta(\theta-\theta_0-\pi)}, \qquad 0 < \theta - \theta_0 < 2\pi, \tag{2.9}$$

where $\beta \in \mathbb{C}$ and $\theta_0 \in (-\pi, \pi]$. The piecewise continuous functions that we consider are of the form

$$\phi(e^{i\theta}) = b(e^{i\theta}) \prod_{r=1}^{R} t_{\beta_r,\theta_r}(e^{i\theta}), \tag{2.10}$$

where $\theta_1, \dots, \theta_R \in (-\pi, \pi]$ are distinct numbers determining the location of the jump discontinuities and $\beta_1, \dots, \beta_R$ are complex parameters determining the "size" of the jumps. The function b is usually assumed to belong to $G_1 B_1^1$.

As is known from the theory of Toeplitz determinants, a key ingredient for the determination of the asymptotics of the determinants are localization theorems (see, e.g., [8]). A localization theorem for the determinants $\det M_N(\phi)$ with ϕ being piecewise continuous has been established in [4, Theorem 5.11]. This localization theorem reduces the asymptotics of $\det M_N(\phi)$ for "general" piecewise continuous functions (2.10) to the asymptotics for particular piecewise continuous functions.

Theorem 2.1 (Localization Theorem). *Let ϕ be a function of the form*

$$\phi(e^{i\theta}) = b(e^{i\theta})\phi_+(e^{i\theta})\phi_-(e^{i\theta}) \prod_{r=1}^{R} \phi_r(e^{i\theta}), \tag{2.11}$$

where $b \in G_1 B_1^1$, $\phi_+ = t_{\beta_+,0}$, $\phi_- = t_{\beta_-,\pi}$ and $\phi_r = t_{\beta_r^+,\theta_r} t_{\beta_r^-,-\theta_r}$ for $1 \le r \le R$. Suppose that $\theta_1, \dots, \theta_R \in (0, \pi)$ are distinct numbers and that $\beta_\pm, \beta_1^\pm, \dots, \beta_R^\pm \in \mathbb{C}$ are such that

(a) $-1/2 < \operatorname{Re} \beta_+ < 1/4$ *and* $-1/4 < \operatorname{Re} \beta_- < 1/2$,
(b) $|\operatorname{Re} \beta_r^+| < 1/2$ *and* $|\operatorname{Re} \beta_r^-| < 1/2$ *and* $|\operatorname{Re}(\beta_r^+ + \beta_r^-)| < 1/2$ *for each* $1 \le r \le R$.

Then

$$\lim_{N\to\infty} \frac{\det M_N(\phi)}{\det M_N(b) \det M_N(\phi_+) \det M_N(\phi_-) \prod_{r=1}^{R} \det M_N(\phi_r)} = H,$$

where

$$\begin{aligned}
H &= b_+(1)^{2\beta_+} b_-(1)^{-\beta_+} b_+(-1)^{2\beta_-} b_-(-1)^{-\beta_-} 2^{3\beta_+ + \beta_-} \\
&\times \prod_{r=1}^{R} b_+(t_r)^{\beta_r^+ + \beta_r^-} b_-(t_r)^{-\beta_r^+} b_+(t_r^{-1})^{\beta_r^+ + \beta_r^-} b_-(t_r^{-1})^{-\beta_r^-} \\
&\times \prod_{r=1}^{R} (1 - t_r)^{\beta_+ (\beta_r^+ + 2\beta_r^-)} (1 - t_r^{-1})^{\beta_+ (2\beta_r^+ + \beta_r^-)} \\
&\times \prod_{r=1}^{R} (1 + t_r)^{\beta_- (\beta_r^+ + 2\beta_r^-)} (1 + t_r^{-1})^{\beta_- (2\beta_r^+ + \beta_r^-)} \\
&\times \prod_{1 \le r < s \le R} (1 - t_r t_s)^{\beta_r^- \beta_s^- + \beta_r^+ \beta_s^- + \beta_r^- \beta_s^+} (1 - t_r^{-1} t_s^{-1})^{\beta_r^+ \beta_s^+ + \beta_r^+ \beta_s^- + \beta_r^- \beta_s^+} \\
&\times \prod_{1 \le r < s \le R} (1 - t_r t_s^{-1})^{\beta_r^+ \beta_s^+ + \beta_r^- \beta_s^- + \beta_r^- \beta_s^+} (1 - t_r^{-1} t_s)^{\beta_r^+ \beta_s^+ + \beta_r^- \beta_s^- + \beta_r^+ \beta_s^-}.
\end{aligned}$$

Here $t_r = e^{i\theta_r}$, $1 \le r \le R$, and $b_\pm$ are the functions (2.5).

The asymptotic behavior of $\det M_N(\phi)$ with the generating function $\phi_+ = t_{\beta_+,0}$ and $\phi_- = t_{\beta_-,\pi}$, respectively, has also been determined (see Theorem 6.2 and Theorem 6.3 in [4]). These functions have one single jump discontinuity at the points 1 and -1, respectively.

Theorem 2.2. *Let $\beta \in \mathbb{C} \setminus \mathbb{Z}$. Then*

$$\text{(a)}\ \lim_{N\to\infty} \frac{\det M_N(t_{\beta,0})}{N^{-\frac{3\beta^2}{2}-\frac{\beta}{2}}} = (2\pi)^{\frac{\beta}{2}} 2^{\frac{3\beta^2}{2}} \frac{G(\frac{1}{2}-\beta)G(1-\beta)G(1+\beta)}{G(\frac{1}{2})},$$

$$\text{(b)}\ \lim_{N\to\infty} \frac{\det M_N(t_{\beta,\pi})}{N^{-\frac{3\beta^2}{2}+\frac{\beta}{2}}} = (2\pi)^{\frac{\beta}{2}} 2^{\frac{3\beta^2}{2}} \frac{G(\frac{3}{2}-\beta)G(1-\beta)G(1+\beta)}{G(\frac{3}{2})}.$$

In these asymptotic formulas the Barnes G-function $G(z)$ appears [2, 15], which is an entire function defined by

$$G(1+z) = (2\pi)^{\frac{z}{2}} e^{-\frac{(z+1)z}{2} - C_E\frac{z^2}{2}} \prod_{k=1}^{\infty} \left(\left(1+\frac{z}{k}\right)^k e^{-z+\frac{z^2}{2k}} \right) \tag{2.12}$$

with C_E being Euler's constant.

In the case of the generating functions $\phi_r = t_{\beta_r^+,\theta_r} t_{\beta_r^-,-\theta_r}$, which have two jump discontinuities at a point of the unit circle and its complex conjugate, the asymptotic behavior is only known in particular cases. One case is that where either $\beta_r^+ = 0$ or $\beta_r^- = 0$, i.e., where the function has exactly one jump discontinuity at a point in $\mathbb{T} \setminus \{1,-1\}$. Here the result is taken from [4, Theorem 4.5].

Theorem 2.3. *Let $\theta_0 \in (-\pi,0) \cup (0,\pi)$ and $\beta \in \mathbb{C}$ be such that $|\operatorname{Re}\beta| < 1/2$. Put $t_0 = e^{i\theta_0}$. Then*

$$\lim_{N\to\infty} \frac{\det M_N(t_{\beta,\theta_0})}{N^{-\beta^2}} = G(1-\beta)G(1+\beta)\left(1-t_0^{-1}\right)^{\frac{\beta^2}{2}+\frac{\beta}{2}} \left(1+t_0^{-1}\right)^{\frac{\beta^2}{2}-\frac{\beta}{2}}.$$

Finally, another, even more particular case of a function with two jump discontinuities at i and $-i$ and the same size of the jumps has been established [4] if one combines Theorem 7.4 and Theorem 7.5.

Theorem 2.4. *Let $\beta \in \mathbb{C} \setminus \mathbb{Z}$. Then*

$$\lim_{N\to\infty} \frac{\det M_N(t_{\beta,\frac{\pi}{2}} t_{\beta,-\frac{\pi}{2}})}{N^{-3\beta^2}} = 2^{4\beta^2} G(1-2\beta)G(1+\beta)^2.$$

For later use, let us specialize the localization theorem (Theorem 2.1) to the case of even functions ϕ which are of the form (2.11).

Corollary 2.5. *Let ϕ be a function of the form*

$$\phi(e^{i\theta}) = b(e^{i\theta}) \prod_{r=1}^{R} \phi_r(e^{i\theta}), \tag{2.13}$$

where $b \in G_1 B_1^1$ is an even function and $\phi_r = t_{\beta_r,\theta_r} t_{-\beta_r,-\theta_r}$ for $1 \le r \le R$. Suppose that $\theta_1, \dots, \theta_R \in (0,\pi)$ are distinct numbers and that $\beta_1, \dots, \beta_R \in \mathbb{C}$ are *such that* $|\operatorname{Re}\beta_r| < 1/2$ *for each* $1 \le r \le R$. *Then*

$$\lim_{N\to\infty} \frac{\det M_N(\phi)}{\det M_N(b) \prod_{r=1}^{R} \det M_N(\phi_r)} = H,$$

where

$$H = \prod_{r=1}^{R} b_+(t_r)^{\beta_r} b_-(t_r)^{-\beta_r}$$
$$\times \prod_{1\le r<s\le R} (1 - t_r t_s)^{-\beta_r\beta_s} (1 - t_r^{-1} t_s^{-1})^{-\beta_r\beta_s}$$
$$\times \prod_{1\le r<s\le R} (1 - t_r t_s^{-1})^{\beta_r\beta_s} (1 - t_r^{-1} t_s)^{\beta_r\beta_s}.$$

Here $t_r = e^{i\theta_r}$, $1 \le r \le R$, *and* $b_\pm$ *are the functions* (2.5).

Proof. We apply Theorem 2.1 with the parameters $\beta_\pm = 0$ and $\beta_r^\pm = \pm\beta_r$. We remark also that $b_+ = \tilde{b}_-$. □

3. Preliminary results for determinants of symmetric Toeplitz plus Hankel matrices

The first step in order to determine the asymptotics of the determinants of symmetric Toeplitz plus Hankel matrices $M_N(\phi)$ is to express these determinants by means of determinants of skewsymmetric Toeplitz matrices. The identity as it appears in the following theorem has been stated explicitly in [12, Lemma 18], but is already implicitly contained in [11, Lemma 1] and [14, Proof of Theorem 7.1(a)], where it has been proved. A different, self-contained proof has been given by the authors in [5].

Theorem 3.1. *Let* $\{a_n\}_{n=-\infty}^{\infty}$ *be a sequence of complex numbers such that* $a_{-n} = a_n$. *Let*

$$c_n = \sum_{k=-n+1}^{n} a_k \qquad \text{for } n > 0, \tag{3.1}$$

and put $c_0 = 0$ *and* $c_{-n} = -c_n$. *Then*

$$\left(\det \left(a_{j-k} + a_{j+k+1}\right)_{j,k=0}^{N-1}\right)^2 = \det \left(c_{j-k}\right)_{j,k=0}^{2N-1}. \tag{3.2}$$

The matrices appearing in (3.2) are a symmetric Toeplitz plus Hankel matrix of the same kind as (1.1) and a skewsymmetric Toeplitz matrix. If we are trying to rewrite this identity by using the standard notation (1.2) for Toeplitz and Hankel matrices where the sequences $\{a_n\}$ and $\{c_n\}$ are the Fourier coefficients of functions $a, c \in L^1(\mathbb{T})$, we face the difficulty that this is in general not possible. Consider for instance the simplest case where $a(t) = 1$, i.e., $a_0 = 1$ and $a_n = 0$ if $n \neq 0$. Then we obtain $c_n = \operatorname{sign}(n)$, and obviously, there does not exist a function $c \in L^1(\mathbb{T})$ with such Fourier coefficients.

A way out of this situation is to consider distributions on the unit circle in place of functions in $L^1(\mathbb{T})$ and to take their Fourier coefficients as the entries of the Toeplitz and Hankel matrices. For this purpose we need several preliminary results. Apart from basic issues, the following has all been stated in [10] and proved in [9].

Let $\mathcal{D} = C^\infty(\mathbb{T})$ be the linear topological space of all infinitely differentiable functions defined on the unit circle. By $\mathcal{D}'$ we denote the set of all distributions on the unit circle, i.e., linear and continuous functionals on $\mathcal{D}$. The Fourier coefficients of a distribution $\boldsymbol{a} \in \mathcal{D}$ are defined as

$$\boldsymbol{a}_n = \boldsymbol{a}(\chi_{-n}), \tag{3.3}$$

where $\chi_n \in \mathcal{D}$ is the function $\chi_n(t) = t^n$. There is a natural identification of functions $a \in L^1(\mathbb{T})$ with a subset of distributions. It is established by the mapping $a \in L^1(\mathbb{T}) \mapsto \boldsymbol{a} \in \mathcal{D}'$ where

$$\boldsymbol{a}(f) = \frac{1}{2\pi}\int_0^{2\pi} a(e^{i\theta})f(e^{i\theta})\,d\theta, \qquad f \in \mathcal{D}. \tag{3.4}$$

This definition ensures that the Fourier coefficients of a and $\boldsymbol{a}$ are the same. We also remark that there is a one-to-one correspondence between $\mathcal{D}'$ and the set of all at most polynomially increasing sequences $\{a_n\}_{n=-\infty}^{\infty}$, which is given by associating to $\boldsymbol{a} \in \mathcal{D}'$ the series $\{\boldsymbol{a}_n\}_{n=-\infty}^{\infty}$ of its Fourier coefficients.

Let $a \in \mathcal{D}$ and $\boldsymbol{b} \in \mathcal{D}'$. Then the product of a and $\boldsymbol{b}$ is the distribution $a\boldsymbol{b} \in \mathcal{D}'$ which is defined by

$$(a\boldsymbol{b})(f) = \boldsymbol{b}(af), \qquad f \in \mathcal{D}. \tag{3.5}$$

Let K be a compact subset of $\mathbb{T}$. We denote by $C^\infty(\mathbb{T} \setminus K)$ the set of all infinitely differentiable functions on $\mathbb{T} \setminus K$. By $C_K^\infty(\mathbb{T})$ we refer to the set of all functions $f \in \mathcal{D}$ which vanish on an open neighborhood of K. The product of a function $f \in C^\infty(\mathbb{T}\setminus K)$ with a function $g \in C_K^\infty(\mathbb{T})$ is a function $fg \in C_K^\infty(\mathbb{T}) \subseteq \mathcal{D}$ by putting $(fg)(t) = 0$ for $t \in K$.

We will proceed with some definitions that are not so quite common, but necessary for our considerations. They are taken from [9, 10]. Let $\mathcal{D}'(K)$ stand for the set of all distributions $\boldsymbol{a}$ for which there exists a function $a \in C^\infty(\mathbb{T} \setminus K)$ such that

$$\boldsymbol{a}(f) = \frac{1}{2\pi}\int_0^{2\pi} a(e^{i\theta})f(e^{i\theta})\,d\theta \tag{3.6}$$

for all $f \in C_K^\infty(\mathbb{T})$. The function $a \in C^\infty(\mathbb{T} \setminus K)$ is uniquely determined by $\boldsymbol{a}$ and called the *smooth part of the distribution* $\boldsymbol{a}$. Definition (3.6) can be rephrased by saying that $f\boldsymbol{a} = fa$ for all $f \in C_K^\infty(\mathbb{T})$, where the left-hand side of this equation is a distribution $\mathcal{D}'$ and the right-hand side is a function in $\mathcal{D}$, which are identified in the sense of (3.4).

Next we are going to show that one can define – under certain assumptions – the product of two distributions. Let M and N be compact and disjoint subsets of the unit circle. Given $\boldsymbol{a} \in \mathcal{D}'(M)$ and $\boldsymbol{b} \in \mathcal{D}'(N)$ with smooth parts $a \in C^\infty(\mathbb{T}\setminus M)$ and $b \in C^\infty(\mathbb{T} \setminus N)$, let $\boldsymbol{ab} \in \mathcal{D}'(M \cup N)$ be defined as

$$\boldsymbol{ab} = (bf_b)\boldsymbol{a} + (af_a)\boldsymbol{b}, \tag{3.7}$$

where $f_a \in C^\infty_M(\mathbb{T})$ and $f_b \in C^\infty_N(\mathbb{T})$ are such that $f_a + f_b = 1$. This definition is independent of the particular choice of f_a and f_b. Moreover, $\boldsymbol{ab}$ has the smooth part $ab \in C^\infty(\mathbb{T} \setminus (M \cup N))$.

Given a function a defined on (a subset of) the unit circle, we define the function $\tilde{a}$ by $\tilde{a}(t) = a(t^{-1})$, $t \in \mathbb{T}$. In accordance with this definition, given a distribution $\boldsymbol{a} \in \mathcal{D}'$, let $\tilde{\boldsymbol{a}} \in \mathcal{D}'$ stand for the distribution with Fourier coefficients $\tilde{\boldsymbol{a}}_{-n} = \boldsymbol{a}_n$. A distribution $\boldsymbol{a}$ will be called *even* (*odd*) if $\boldsymbol{a} = \pm\tilde{\boldsymbol{a}}$. A function a will be called *even* (*odd*) if $a = \pm\tilde{a}$. Finally, if K is a subset of $\mathbb{T}$, put $\widetilde{K} = \{t \in \mathbb{T} : t^{-1} \in K\}$. If $\widetilde{K} = K$, we call K a *symmetric subset of the unit circle.*

In the reformulation of Theorem 3.1, the following distribution will play a role. Let $\boldsymbol{\sigma} \in \mathcal{D}'$ be the distribution which has the Fourier coefficients

$$\boldsymbol{\sigma}_n = \operatorname{sign}(n). \tag{3.8}$$

Moreover, let $\sigma \in C^\infty(\mathbb{T} \setminus \{1\})$ be the function

$$\sigma(t) = \frac{1+t}{1-t}. \tag{3.9}$$

Remark that both the distribution $\boldsymbol{\sigma}$ and the function σ are odd.

Proposition 3.2. *The distribution $\boldsymbol{\sigma}$ is in $\mathcal{D}'(\{1\})$ and has the smooth part σ.*

Proof. For $f \in C^\infty_{\{1\}}(\mathbb{T})$ we can write $f(t) = (1-t)h(t)$ where $h \in \mathcal{D}$. Then

$$\begin{aligned}
[f\boldsymbol{\sigma}]_n &= \sum_{k>0} f_{n-k} - \sum_{k<0} f_{n-k} \\
&= \sum_{k>0} (h_{n-k} - h_{n-k-1}) - \sum_{k<0} (h_{n-k} - h_{n-k-1}) \\
&= h_{n-1} + h_n = [h(t)(1+t)]_n = [f\sigma]_n.
\end{aligned}$$

Note that h_k converges to zero sufficiently fast. This completes the proof. □

Given a distribution $\boldsymbol{a} \in \mathcal{D}'$ with Fourier coefficients $\{\boldsymbol{a}_n\}_{n=-\infty}^{\infty}$, define the $N \times N$ Toeplitz and Hankel matrices by

$$T_N(\boldsymbol{a}) = \left(\boldsymbol{a}_{j-k}\right)_{j,k=0}^{N-1}, \qquad H_N(\boldsymbol{a}) = \left(\boldsymbol{a}_{j+k+1}\right)_{j,k=0}^{N-1}. \tag{3.10}$$

This definition is in accordance with (1.2). Moreover define

$$M_N(\boldsymbol{a}) = T_N(\boldsymbol{a}) + H_N(\boldsymbol{a}). \tag{3.11}$$

Now we are ready to give the desired reformulation of Theorem 3.1.

Theorem 3.3. *Let K be a compact and symmetric subset of $\mathbb{T} \setminus \{1\}$, and assume that $\boldsymbol{a} \in \mathcal{D}'(K)$ is an even distribution. Then*

$$\det T_{2N}(\boldsymbol{a\sigma}) = (\det M_N(\boldsymbol{a}))^2. \tag{3.12}$$

In the above $\mathbf{c} = \boldsymbol{a\sigma} \in \mathcal{D}'(K \cup \{1\})$ is an odd distribution.

Proof. Since K and $\{1\}$ are disjoint compact sets, the distribution $\boldsymbol{c} = \boldsymbol{a\sigma}$ is well defined. In the definition $\boldsymbol{c} = (\sigma f_\sigma)\boldsymbol{a} + (af_a)\boldsymbol{\sigma}$ we may assume without loss of generality that f_σ and f_a are even functions. From this it follows easily that σf_σ is odd and af_a is even; thus both $(\sigma f_\sigma)\boldsymbol{a}$ and $(af_a)\boldsymbol{\sigma}$ are odd. Hence $\boldsymbol{c}$ is odd.

Next write $f_\sigma(t) = g(t)(1-t)(1-t^{-1})$. Then $(\sigma f_\sigma)(t) = (1+t)(1-t^{-1})g(t) = (t-t^{-1})g(t)$. We obtain that

$$[(\sigma f_\sigma)\boldsymbol{a}]_n = [((t-t^{-1})g)\boldsymbol{a}]_n = [g\boldsymbol{a}]_{n-1} - [g\boldsymbol{a}]_{n+1}.$$

Moreover, using the fact that $f_\sigma(t) = g(t)(1-t)(1-t^{-1})$ and keeping track of the cancellation, it follows

$$\begin{aligned}\sum_{k=-n+1}^{n} [f_\sigma \boldsymbol{a}]_k &= \sum_{k=-n+1}^{n} \Big(-[g\boldsymbol{a}]_{k-1} + 2[g\boldsymbol{a}]_k - [g\boldsymbol{a}]_{k+1} \Big) \\ &= -[g\boldsymbol{a}]_n + [g\boldsymbol{a}]_{n-1} + [g\boldsymbol{a}]_{-n} - [g\boldsymbol{a}]_{-n-1} \\ &= [g\boldsymbol{a}]_{n-1} - [g\boldsymbol{a}]_{n+1}.\end{aligned}$$

Here we have also used that $g\boldsymbol{a}$ is even. From these two identities we obtain

$$[(\sigma f_\sigma)\boldsymbol{a}]_n = \sum_{k=-n+1}^{n} [f_\sigma \boldsymbol{a}]_k. \tag{3.13}$$

On the other hand, since $af_a = \boldsymbol{a}f_a$ is even,

$$[(af_a)\boldsymbol{\sigma})]_n = \sum_{k>0} [af_a]_{n-k} - \sum_{k<0} [af_a]_{n-k} = \sum_{k=-n+1}^{n} [af_a]_k. \tag{3.14}$$

Combining (3.13) and (3.14) yields

$$\boldsymbol{c}_n = \sum_{k=-n+1}^{n} \boldsymbol{a}_k.$$

Together with Theorem 3.1 this completes the proof. □

Finally we will need the following result.

Proposition 3.4. *For each $N \geq 1$ we have $\det T_{2N}(\boldsymbol{\sigma}) = 1$ and*

$$T_{2N}^{-1}(\boldsymbol{\sigma}) = T_{2N}(\boldsymbol{\nu}) \tag{3.15}$$

where $\boldsymbol{\nu}$ is the distribution with Fourier coefficients $\boldsymbol{\nu}_n = \operatorname{sign}(n)(-1)^n$.

Proof. By Theorem 3.3 we have $\det T_{2N}(\boldsymbol{\sigma}) = (\det M_N(1))^2$, where obviously $M_N(1) = I_N$. The formula for the inverse of $T_{2N}(\boldsymbol{\sigma})$ can be easily checked. □

4. A limit theorem for determinants of skewsymmetric Toeplitz matrices

Theorem 3.3 reduces the computation of the asymptotics of $\det M_N(\boldsymbol{a})$, which is what we are interested in, for certain distributions $\boldsymbol{a}$, to the computation of the asymptotics of $\det T_{2N}(\boldsymbol{\sigma a})$.

At first glance one might think that the asymptotics of $\det T_{2N}(\boldsymbol{\sigma a})$ could be obtained from the predictions of the Fisher-Hartwig conjecture, which was proved in [9]. Unfortunately, since $T_{2N}(\boldsymbol{\sigma a})$ is a skewsymmetric Toeplitz matrix, the Toeplitz determinant belongs to those classes of functions where the Fisher-Hartwig conjecture breaks down. It might be the case that the asymptotic behavior fits with the still unproved generalized conjecture [7, 10]. However, distributions of the kind $\boldsymbol{\sigma a}$ appear here probably for the first time in connection with Toeplitz determinants and since previous results do not include this setting, previous techniques must be modified.

To that end, the purpose of this section is to prove that under certain assumptions on the distribution $\boldsymbol{a}$ the expression

$$\frac{\det T_{2N}(\boldsymbol{a\sigma})}{\det T_{2N}(\boldsymbol{a})} \tag{4.1}$$

converges to a certain (explicitly given) nonzero constant as $N \to \infty$. Although we cannot rely on the main results of [9] (see also [10]), i.e., the Fisher-Hartwig conjecture, we will very heavily rely on the machinery and several auxiliary results established in [9].

Let us proceed with recalling the necessary definitions. For $\mu \in \mathbb{R}$ let ℓ^2_μ stand for the Hilbert space of all sequences $\{x_n\}_{n=0}^\infty$ of complex numbers for which

$$\left\|\{x_n\}_{n=0}^\infty\right\|_\mu := \left(\sum_{n=0}^\infty (1+n)^{2\mu}|x_n|^2\right)^{1/2} < \infty. \tag{4.2}$$

For $\mu_1 > \mu_2$ the space $\ell^2_{\mu_1}$ is continuously and densely embedded in $\ell^2_{\mu_2}$.

The Toeplitz and Hankel operator generated by $\boldsymbol{a} \in \mathcal{D}'$ are the one-sided infinite matrices

$$T(\boldsymbol{a}) = (\boldsymbol{a}_{j-k})_{j,k=0}^\infty, \qquad H(\boldsymbol{a}) = (\boldsymbol{a}_{j+k+1})_{j,k=0}^\infty, \tag{4.3}$$

where $\boldsymbol{a}_n$ are the Fourier coefficients of the distribution $\boldsymbol{a}$. For each $\boldsymbol{a} \in \mathcal{D}'$ there exist a (sufficiently large) μ_1 and a (sufficiently small) μ_2 such that the matrices $T(\boldsymbol{a})$ and $H(\boldsymbol{a})$ represent linear bounded operators acting from $\ell^2_{\mu_1}$ into $\ell^2_{\mu_2}$.

The situation of the boundedness of Toeplitz and Hankel operators generated by functions in $\mathcal{D}$ was established in the following lemma taken from [9, Sect. 6.2].

Lemma 4.1. *For each $\mu, \mu_1, \mu_2 \in \mathbb{R}$ and $a \in \mathcal{D}$, the operator $T(a)$ is bounded on $\mathcal{L}(\ell^2_\mu, \ell^2_\mu)$ and the operator $H(a)$ is bounded on $\mathcal{L}(\ell^2_{\mu_1}, \ell^2_{\mu_2})$.*

We define the following finite rank operators acting on ℓ^2_μ:

$$P_N : (x_0, x_1, x_2, \dots) \mapsto (x_0, x_1, \dots, x_{N-2}, x_{N-1}, 0, 0, \dots), \tag{4.4}$$

$$W_N : (x_0, x_1, x_2, \dots) \mapsto (x_{N-1}, x_{N-2}, \dots, x_1, x_0, 0, 0, \dots). \tag{4.5}$$

Obviously, $P_N^2 = W_N^2 = P_N$ and $W_N P_N = P_N W_N = W_N$. If we consider the matrix $T_N(\boldsymbol{a})$ as acting on the image of the projection P_N in the space ℓ^2_μ, then $T_N(\boldsymbol{a}) = P_N T(\boldsymbol{a}) P_N$. Moreover,

$$W_N T_N(\boldsymbol{a}) W_N = T_N(\tilde{\boldsymbol{a}}). \tag{4.6}$$

Recall that $\tilde{\boldsymbol{a}}$ is the distribution with the Fourier coefficients $\tilde{\boldsymbol{a}}_n = \boldsymbol{a}_{-n}$.

For our purposes we need to single out two additional classes of distributions. Let $\mathcal{D}'_+$ ($\mathcal{D}'_-$, resp.) stand for the set of all distributions $\boldsymbol{a} \in \mathcal{D}'$ for which $\boldsymbol{a}_n = 0$ for all $n < 0$ ($n > 0$, resp.). These two sets form commutative algebras with a unit element $e(t) \equiv 1$. For $\boldsymbol{a}, \boldsymbol{b} \in \mathcal{D}'_+$, the product $\boldsymbol{c} = \boldsymbol{ab}$ is defined by stipulating $\boldsymbol{c}_n = 0$ for $n < 0$ and

$$\boldsymbol{c}_n = \sum_{k=0}^{n} \boldsymbol{a}_{n-k} \boldsymbol{b}_k \qquad \text{for } n \geq 0. \tag{4.7}$$

For $\boldsymbol{a}, \boldsymbol{b} \in \mathcal{D}'_-$, the product $\boldsymbol{c} = \boldsymbol{ab}$ is defined by stipulating $\boldsymbol{c}_n = 0$ for $n > 0$ and

$$\boldsymbol{c}_n = \sum_{k=n}^{0} \boldsymbol{a}_{n-k} \boldsymbol{b}_k \qquad \text{for } n \leq 0. \tag{4.8}$$

This definition of a multiplication is compatible with that of (3.7) whenever both are defined. Let $\mathcal{GD}'_\pm$ stand for the group of all invertible distributions in $\mathcal{D}'_\pm$. Moreover, we put $\mathcal{D}'_\pm(K) = \mathcal{D}'_\pm \cap \mathcal{D}'(K)$ and let $\mathcal{GD}'_\pm(K)$ stand for the group of all invertible elements in $\mathcal{D}'_\pm(K)$. One can show that if $\boldsymbol{a} \in \mathcal{GD}'_\pm(K)$ has the smooth part a, then a is an invertible element of $C^\infty(\mathbb{T} \setminus K)$.

There are some obvious relations between the distributions $\boldsymbol{a}$ and $\tilde{\boldsymbol{a}}$. For instance, if $\boldsymbol{a} \in \mathcal{GD}'_+(K)$, then $\tilde{\boldsymbol{a}} \in \mathcal{GD}'_-(\widetilde{K})$. Moreover, if $\boldsymbol{a}$ has the smooth part a, then $\tilde{\boldsymbol{a}}$ has the smooth part $\tilde{a}$.

Let H_1 and H_2 be Hilbert spaces. We consider sequences $\{C_N\}_{N=1}^\infty$ the elements of which are well defined linear bounded operators (or matrices) $C_N : H_1 \to H_2$ for all sufficiently large N. Let $\mathcal{O}(\varrho)$ with $\varrho \in \mathbb{R}$ stand for the set of all such sequences for which

$$\|C_N\|_{\mathcal{L}(H_1,H_2)} = O(N^\varrho) \qquad \text{as } N \to \infty. \tag{4.9}$$

The dependence of $\mathcal{O}(\varrho)$ on H_1 and H_2 will not be displayed in the notation. We also use the notation $\mathcal{O}(\varrho)$ in order to denote any sequence of this type. In this sense, $C_N = C + \mathcal{O}(\varrho)$ means that $\{C_N - C\}_{N=1}^\infty \in \mathcal{O}(\varrho)$.

Now let H_1, H_2, $\widetilde{H}_1$ and $\widetilde{H}_2$ be Hilbert spaces and $\varrho_0, \varrho_1, \varrho_2 \in \mathbb{R}$. We denote by $\mathcal{O}(\varrho_0, \varrho_1, \varrho_2)$ the set of all sequences $\{C_N\}_{N=1}^\infty$ of 2×2 block operators for which

$$C_N = \begin{pmatrix} \mathcal{O}(\varrho_0) & \mathcal{O}(\varrho_1) \\ \mathcal{O}(\varrho_2) & \mathcal{O}(\varrho_0) \end{pmatrix} : H_1 \oplus \widetilde{H}_1 \to H_2 \oplus \widetilde{H}_2. \tag{4.10}$$

We will also use the notations $\mathcal{O}_2(\varrho)$ and $\mathcal{O}_1(\varrho)$. The only difference in comparison with $\mathcal{O}(\varrho)$ is that we consider the convergence in (4.9) in the Hilbert-Schmidt and in the trace class norm, respectively. Likewise, we will use the notations $\mathcal{O}_2(\varrho_0, \varrho_1, \varrho_2)$ and $\mathcal{O}_1(\varrho_0, \varrho_1, \varrho_2)$.

Given $\boldsymbol{a} \in \mathcal{D}'$, assume that $T_N(\boldsymbol{a})$ is invertible for all sufficiently large N, and introduce the following sequences of operators of 2×2 block form:

$$R_N(\boldsymbol{a}) = \begin{pmatrix} T_N^{-1}(\boldsymbol{a}) & T_N^{-1}(\boldsymbol{a})W_N \\ T_N^{-1}(\tilde{\boldsymbol{a}})W_N & T_N^{-1}(\tilde{\boldsymbol{a}}) \end{pmatrix}, \tag{4.11}$$

$$RH_N(\boldsymbol{a}) = \begin{pmatrix} T_N^{-1}(\boldsymbol{a})P_N H(\boldsymbol{a}) & T_N^{-1}(\boldsymbol{a})W_N H(\tilde{\boldsymbol{a}}) \\ T_N^{-1}(\tilde{\boldsymbol{a}})W_N H(\boldsymbol{a}) & T_N^{-1}(\tilde{\boldsymbol{a}})P_N H(\tilde{\boldsymbol{a}}) \end{pmatrix}, \tag{4.12}$$

$$HR_N(\boldsymbol{a}) = \begin{pmatrix} H(\tilde{\boldsymbol{a}})P_N T_N^{-1}(\boldsymbol{a}) & H(\tilde{\boldsymbol{a}})W_N T_N^{-1}(\tilde{\boldsymbol{a}}) \\ H(\boldsymbol{a})W_N T_N^{-1}(\boldsymbol{a}) & H(\boldsymbol{a})P_N T_N^{-1}(\tilde{\boldsymbol{a}}) \end{pmatrix}, \tag{4.13}$$

$$HRH_N(\boldsymbol{a}) = \begin{pmatrix} H(\tilde{\boldsymbol{a}})P_N T_N^{-1}(\boldsymbol{a})P_N H(\boldsymbol{a}) & H(\tilde{\boldsymbol{a}})P_N T_N^{-1}(\boldsymbol{a})W_N H(\tilde{\boldsymbol{a}}) \\ H(\boldsymbol{a})P_N T_N^{-1}(\tilde{\boldsymbol{a}})W_N H(\boldsymbol{a}) & H(\boldsymbol{a})P_N T_N^{-1}(\tilde{\boldsymbol{a}})P_N H(\tilde{\boldsymbol{a}}) \end{pmatrix} - \begin{pmatrix} T(\tilde{\boldsymbol{a}}) & H(\tilde{\boldsymbol{a}}\chi_{-N}) \\ H(\boldsymbol{a}\chi_{-N}) & T(\boldsymbol{a}) \end{pmatrix}. \tag{4.14}$$

Here, as before, $\chi_{-N}(t) = t^{-N}$, $t \in \mathbb{T}$. These sequences of operators are considered from $\ell^2_{\mu_1} \oplus \ell^2_{\mu_1}$ into $\ell^2_{\mu_2} \oplus \ell^2_{\mu_2}$ with μ_1 sufficiently large and μ_2 sufficiently small, which ensures the boundedness of the operators.

Now we are prepared to define the notion of $\mathcal{R}$-convergence, which has been introduced in [9, 10]. Let $\boldsymbol{a} \in \mathcal{D}'$, $\boldsymbol{a}_+ \in \mathcal{GD}'_+$ and $\boldsymbol{a}_- \in \mathcal{GD}'_-$ be distributions, and let $\varrho_0, \varrho_1, \varrho_2 \in \mathbb{R}$. We say that the distribution $\boldsymbol{a}$ *effects $\mathcal{R}$-convergence with respect to* $[\boldsymbol{a}_+, \boldsymbol{a}_-]$ *and* $(\varrho_0, \varrho_1, \varrho_2)$ if there exist $\mu_1 \geq 0$ and $\mu_2 \leq 0$ such that

$$R_N(\boldsymbol{a}) = \operatorname{diag}\left(T(\boldsymbol{a}_+^{-1})T(\boldsymbol{a}_-^{-1}),\, T(\tilde{\boldsymbol{a}}_-^{-1})T(\tilde{\boldsymbol{a}}_+^{-1})\right) + \mathcal{O}(\varrho_0, \varrho_1, \varrho_2), \tag{4.15}$$

$$RH_N(\boldsymbol{a}) = \operatorname{diag}\left(T(\boldsymbol{a}_+^{-1})H(\boldsymbol{a}_+),\, T(\tilde{\boldsymbol{a}}_-^{-1})H(\tilde{\boldsymbol{a}}_-)\right) + \mathcal{O}(\varrho_0, \varrho_1, \varrho_2), \tag{4.16}$$

$$HR_N(\boldsymbol{a}) = \operatorname{diag}\left(H(\tilde{\boldsymbol{a}}_-)T(\boldsymbol{a}_-^{-1}),\, H(\boldsymbol{a}_+)T(\tilde{\boldsymbol{a}}_+^{-1})\right) + \mathcal{O}(\varrho_0, \varrho_1, \varrho_2), \tag{4.17}$$

$$HRH_N(\boldsymbol{a}) = -\operatorname{diag}\left(T(\tilde{\boldsymbol{a}}_-)T(\tilde{\boldsymbol{a}}_+),\, T(\boldsymbol{a}_+)T(\boldsymbol{a}_-)\right) + \mathcal{O}(\varrho_0, \varrho_1, \varrho_2), \tag{4.18}$$

where these sequences are considered from $\ell^2_{\mu_1} \oplus \ell^2_{\mu_1}$ into $\ell^2_{\mu_2} \oplus \ell^2_{\mu_2}$.

We will need the concept of $\mathcal{R}$-convergence for a distribution $\boldsymbol{a}$ which is even. This particular case gives rise to some simplifications. For any distribution $\boldsymbol{a} \in \mathcal{D}'$, the following statements are equivalent:

(1) $\boldsymbol{a}$ effects $\mathcal{R}$-convergence with respect to $[\boldsymbol{a}_+, \boldsymbol{a}_-]$ and $(\varrho_0, \varrho_1, \varrho_2)$;
(2) $\tilde{\boldsymbol{a}}$ effects $\mathcal{R}$-convergence with respect to $[\tilde{\boldsymbol{a}}_-, \tilde{\boldsymbol{a}}_+]$ and $(\varrho_0, \varrho_2, \varrho_1)$.

In fact, in order to prove this equivalence, one has only to pass to the transpose in equations (4.15)–(4.18). Hence for an even distribution $\boldsymbol{a}$ we can replace ϱ_1 and

ϱ_2 by $\min\{\varrho_1, \varrho_2\}$, i.e., we may assume that $\varrho_1 = \varrho_2 =: \varrho$. Moreover, since the distributions $\boldsymbol{a}_+$ and $\boldsymbol{a}_-$ are uniquely determined up to a nonzero multiplicative constant, we may assume without loss of generality that $\boldsymbol{a}_- = \tilde{\boldsymbol{a}}_+$.

This last remark gives some motivation for the assumptions in the following theorem. In this theorem we establish the asymptotic formula for (4.1).

Theorem 4.2 (Limit Theorem). *Let K be a symmetric and compact subset of $\mathbb{T} \setminus \{1, -1\}$. Moreover, assume that*

(i) *$\boldsymbol{a} \in \mathcal{D}'(K)$ is an* even *distribution with the smooth part $a \in C^\infty(\mathbb{T} \setminus K)$;*
(ii) *$\boldsymbol{a}_+ \in \mathcal{GD}'_+(K)$ is a distribution with the smooth part $a_+ \in C^\infty(\mathbb{T} \setminus K)$;*
(iii) *$a(t) = a_+(t)\tilde{a}_+(t)$ for all $t \in \mathbb{T} \setminus K$;*
(iv) *$\varrho_0 < 0$ and $\varrho < 0$;*
(v) *$\boldsymbol{a}$ effects $\mathcal{R}$-convergence with respect to $[\boldsymbol{a}_+, \tilde{\boldsymbol{a}}_+]$ and $(\varrho_0, \varrho, \varrho)$.*

Then

$$\frac{\det T_{2N}(\boldsymbol{a\sigma})}{\det T_{2N}(\boldsymbol{a})} = \frac{a_+(1)}{a_+(-1)} + O(N^{\max\{\varrho_0, \varrho\}}) \qquad \text{as } N \to \infty. \tag{4.19}$$

The rest of this section is devoted to the proof of this theorem. Once again we need to quote some auxiliary results.

Let us introduce the notation

$$X^{\mu_1}_{\mu_2} = \ell^2_{\mu_1} \oplus \ell^2_{\mu_2}, \qquad X^{\mu_1} = \ell^2_{\mu_1}, \qquad X_{\mu_2} = \ell^2_{\mu_2}. \tag{4.20}$$

This notation is convenient in the sense that it reflects the condition that μ_1 is a sufficiently large and μ_2 a sufficiently small real number, which we will encounter in what follows.

The following proposition is taken from [9, Proposition 8.1] with a slight change of notation. As before, we will denote a distribution by a bold letter and its smooth part by the same non-bold letter without mentioning it explicitly.

Proposition 4.3. *Let K be a compact subset of $\mathbb{T}$, and let $\boldsymbol{a} \in \mathcal{D}'(K)$ and $f \in C^\infty_K(\mathbb{T})$. Then for all sufficiently large $\mu_1 \geq 0$ and all sufficiently small $\mu_2 \leq 0$ the linear operators*

$$H_1(\boldsymbol{a}, f) = \Big(H(\boldsymbol{a}),\ H(af) - T(\boldsymbol{a})H(f) \Big), \tag{4.21}$$

$$H_2(f, \boldsymbol{a}) = \begin{pmatrix} H(\tilde{f}\tilde{a}) - H(\tilde{f})T(\boldsymbol{a}) \\ H(\tilde{\boldsymbol{a}}) \end{pmatrix} \tag{4.22}$$

are bounded on the spaces

$$H_1(\boldsymbol{a}, f) : X^{\mu_1}_{\mu_2} \to X_{\mu_2}, \qquad H_2(f, \boldsymbol{a}) : X^{\mu_1} \to X^{\mu_1}_{\mu_2}. \tag{4.23}$$

In order to provide some meaning to these operators we remark that with appropriately chosen μ_1 and μ_2, these operators can be embedded into the spaces

$$H_1(\boldsymbol{a}, f) : X^{\mu_1}_{\mu_1} \to X_{\mu_2}, \qquad H_2(f, \boldsymbol{a}) : X^{\mu_1} \to X^{\mu_2}_{\mu_2}. \tag{4.24}$$

In this case, these operators can be written as a product:

$$H_1(\boldsymbol{a}, f) = H(\boldsymbol{a})\Big(I,\ T(\tilde{f})\ \Big), \tag{4.25}$$

$$H_2(f, \boldsymbol{a}) = \begin{pmatrix} T(\tilde{f}) \\ I \end{pmatrix} H(\tilde{\boldsymbol{a}}). \tag{4.26}$$

The next result is taken from [9, Corollary 8.8].

Proposition 4.4. *Let K_1 and K_2 be disjoint and compact subsets of $\mathbb{T}$, and let $\boldsymbol{a}_i \in \mathcal{D}'(K_i)$ and $f_i \in C^\infty_{K_i}(\mathbb{T})$ $(i = 1, 2)$ be such that $f_1 + f_2 = 1$. Then*

$$T_N(\boldsymbol{a}_1\boldsymbol{a}_2) = T_N(\boldsymbol{a}_1)T_N(\boldsymbol{a}_2) + P_N H_1(\boldsymbol{a}_1, f_1)H_2(f_2, \boldsymbol{a}_2)P_N$$
$$+ W_N H_1(\tilde{\boldsymbol{a}}_1, \tilde{f}_1)H_2(\tilde{f}_2, \tilde{\boldsymbol{a}}_2)W_N. \tag{4.27}$$

We remark that the linear operators occurring in (4.27) are bounded on appropriately chosen spaces. Moreover, (4.27) represents a generalization of the well-known identity

$$T_N(a_1a_2) = T_N(a_1)T_N(a_2) + P_N H(a_1)H(\tilde{a}_2)P_N + W_N H(\tilde{a}_1)H(a_2)W_N$$

due to Widom [16], which holds for functions $a_1, a_2 \in L^\infty(\mathbb{T})$.

Next, we consider the functions $\xi_1(t) = 1 - t^{-1}$ and $\xi_{-1}(t) = 1 + t^{-1}$. These functions can be identified with distributions $\boldsymbol{\xi}_1 \in \mathcal{D}'_-$ and $\boldsymbol{\xi}_{-1} \in \mathcal{D}'_-$, respectively, in the sense of (3.4). Obviously, both $\boldsymbol{\xi}_1$ and $\boldsymbol{\xi}_{-1}$ belong to $\mathcal{GD}'_-$. In fact, the inverse distributions $\boldsymbol{\xi}_1^{-1}$ and $\boldsymbol{\xi}_{-1}^{-1}$ are given as follows by their Fourier coefficients:

$$\left[\boldsymbol{\xi}_1^{-1}\right]_n = \begin{cases} 0 & \text{if } n > 0 \\ 1 & \text{if } n \le 0, \end{cases} \tag{4.28}$$

$$\left[\boldsymbol{\xi}_{-1}^{-1}\right]_n = \begin{cases} 0 & \text{if } n > 0 \\ (-1)^n & \text{if } n \le 0. \end{cases} \tag{4.29}$$

Proposition 4.5. *The following statements are true:*

(a) *$\boldsymbol{\xi}_1 \in \mathcal{GD}'_-(\{1\})$, and $\boldsymbol{\xi}_1^{\pm 1}$ has the smooth part $\xi_1^{\pm 1}$;*

(b) *$\boldsymbol{\xi}_{-1} \in \mathcal{GD}'_-(\{-1\})$, and $\boldsymbol{\xi}_{-1}^{\pm 1}$ has the smooth part $\xi_{-1}^{\pm 1}$.*

Proof. Since $\boldsymbol{\xi}_1 = \xi_1$ in the sense of (3.4), we even have $\boldsymbol{\xi}_1 \in \mathcal{D}'(\emptyset)$ and $\boldsymbol{\xi}_1$ has the smooth part ξ_1. It remains to show that $\boldsymbol{\xi}_1^{-1}$ is contained in $\mathcal{D}'(\{1\})$ and has the smooth part ξ_1^{-1}. Indeed, let $f \in C^\infty_{\{1\}}(\mathbb{T})$. Then we can write $f = \xi_1 g$ where $g \in C^\infty(\mathbb{T})$. It can be checked easily that $\boldsymbol{\xi}_1^{-1}\xi_1 = 1$. Hence $\boldsymbol{\xi}_1^{-1} f = g = \xi_1^{-1} f$. This completes the proof of part (a). Part (b) can be proved analogously. □

The following corollary, in which we define another distribution $\boldsymbol{h}$, is a simple consequence of the previous proposition.

Corollary 4.6. *The distribution $\boldsymbol{h} = \boldsymbol{\xi}_1^{-1}\boldsymbol{\xi}_{-1}$ is contained in $\mathcal{GD}'_-(\{-1, 1\})$ and has the smooth part $h = \xi_1^{-1}\xi_{-1}$. The inverse distribution $\boldsymbol{h}^{-1}$ equals $\boldsymbol{\xi}_1\boldsymbol{\xi}_{-1}^{-1}$ and has the smooth part $h^{-1} = \xi_1\xi_{-1}^{-1}$.*

It is easy to compute the Fourier coefficients of $\boldsymbol{h}$ and $\boldsymbol{h}^{-1}$:

$$[\boldsymbol{h}]_n = \begin{cases} 0 & \text{if } n > 0 \\ 1 & \text{if } n = 0 \\ 2 & \text{if } n < 0, \end{cases} \tag{4.30}$$

$$[\boldsymbol{h}^{-1}]_n = \begin{cases} 0 & \text{if } n > 0 \\ 1 & \text{if } n = 0 \\ 2(-1)^n & \text{if } n < 0. \end{cases} \tag{4.31}$$

There is (for our purposes) an important relation between operators containing the distributions $\boldsymbol{\sigma}$ and $\boldsymbol{h}$, which is given in the following proposition.

Proposition 4.7. *Let $f \in C^\infty_{\{-1,1\}}(\mathbb{T})$. Then*

$$\begin{aligned} &H_2(f,\boldsymbol{\sigma})P_{2N}T_{2N}^{-1}(\boldsymbol{\sigma}) = H_2(f,\boldsymbol{\sigma})W_{2N}T_{2N}^{-1}(\boldsymbol{\sigma}) = \\ &= \frac{1}{2}H_2(f,\boldsymbol{h})P_{2N}T_{2N}^{-1}(\boldsymbol{h}) = -\frac{1}{2}H_2(f,\boldsymbol{h})P_{2N}T_{2N}^{-1}(\boldsymbol{h})W_{2N}. \end{aligned} \tag{4.32}$$

Proof. We first prove that

$$\begin{aligned} &H(\tilde{\boldsymbol{\sigma}})P_{2N}T_{2N}^{-1}(\boldsymbol{\sigma}) = H(\tilde{\boldsymbol{\sigma}})W_{2N}T_{2N}^{-1}(\boldsymbol{\sigma}) = \\ &= \frac{1}{2}H(\tilde{\boldsymbol{h}})P_{2N}T_{2N}^{-1}(\boldsymbol{h}) = -\frac{1}{2}H(\tilde{\boldsymbol{h}})P_{2N}T_{2N}^{-1}(\boldsymbol{h})W_{2N}. \end{aligned} \tag{4.33}$$

In this identity we do not need to worry about the boundedness on certain spaces since both the left- and right-hand side is an infinite Hankel matrix times a finite rank matrix. Let $x = (1,1,\dots)^T$ denote an infinite column vector and $x_{2N} = (1,\dots,1)^T$ a finite column vector of size $2N$. Then

$$H(\tilde{\boldsymbol{\sigma}})P_{2N} = H(\tilde{\boldsymbol{\sigma}})W_{2N} = -xx_{2N}^T, \qquad H(\tilde{\boldsymbol{h}})P_{2N} = 2xx_{2N}^T. \tag{4.34}$$

A moments thought shows that (4.33) is proved as soon as

$$-x_{2N}^T T_{2N}^{-1}(\boldsymbol{\sigma}) = x_{2N}^T T_{2N}^{-1}(\boldsymbol{h}) = -x_{2N}^T T_{2N}^{-1}(\boldsymbol{h})W_{2N} \tag{4.35}$$

is established. However, this is just a straightforward calculation. We have to observe that $T_{2N}^{-1}(\boldsymbol{h}) = T_{2N}(\boldsymbol{h}^{-1})$ with the Fourier coefficients given by (4.31) and moreover that $T_{2N}^{-1}(\boldsymbol{\sigma}) = T_{2N}(\boldsymbol{\nu})$ by Proposition 3.4.

Having proved (4.33), we take into account the identity (4.26) and formula (4.32) follows by a density argument of the Hilbert spaces under consideration. □

In [9, Formula (8.64)], the following 2×2 block operators acting on $X^{\mu_1}_{\mu_2} \oplus X^{\mu_1}_{\mu_2}$ with sufficiently large μ_1 and small μ_2 were defined:

$$\begin{aligned} &HSH_N(f_2,\boldsymbol{a}_2,\boldsymbol{a}_1,f_1) \\ &= \begin{pmatrix} H_2(f_2,\boldsymbol{a}_2)P_N \\ H_2(\tilde{f}_2,\tilde{\boldsymbol{a}}_2)W_N \end{pmatrix} T_N^{-1}(\boldsymbol{a}_2)T_N^{-1}(\boldsymbol{a}_1)\Big(P_N H_1(\boldsymbol{a}_1,f_1),\ W_N H_1(\tilde{\boldsymbol{a}}_1,\tilde{f}_1)\Big). \end{aligned} \tag{4.36}$$

Moreover, it has been shown [9, Formula (8.139)] that

$$HSH_N(f_2,\boldsymbol{a}_2,\boldsymbol{a}_1,f_1) = HZ_N(f_2,\boldsymbol{a}_2,f_2)YH_N(f_1,\boldsymbol{a}_1,f_1), \tag{4.37}$$

where $HZ_N(\dots)$ and $YH_N(\dots)$ are linear bounded 2×2 block operators acting on $X_{\mu_2}^{\mu_1}\oplus X_{\mu_2}^{\mu_1}$, which we are not going to define here. The important result concerning these operators is the following asymptotic formula, which is taken from [9, Proposition 10.4 and Proposition 10.5].

Proposition 4.8. *Let K be a compact subset of $\mathbb{T}$, $f\in C_K^\infty(\mathbb{T})$, and assume that $\boldsymbol{a}\in\mathcal{D}'(K)$, $\boldsymbol{a}_\pm\in\mathcal{GD}'_\pm(K)$ such that $a=a_+a_-$ holds for their smooth parts. If $\boldsymbol{a}$ effects $\mathcal{R}$-convergence with respect to $[\boldsymbol{a}_+,\boldsymbol{a}_-]$ and $(\varrho_0,\varrho_1,\varrho_2)\in\mathbb{R}^3$, then*

$$YH_N(f,\boldsymbol{a},f)=\operatorname{diag}\left(TH(f,\boldsymbol{a}_+,f),\ TH(\tilde{f},\tilde{\boldsymbol{a}}_-,\tilde{f})\right)+\mathcal{O}_2(\varrho_0,\varrho_1,\varrho_2),$$
$$HZ_N(f,\boldsymbol{a},f)=\operatorname{diag}\left(HT(f,\boldsymbol{a}_-,f),\ HT(\tilde{f},\tilde{\boldsymbol{a}}_+,\tilde{f})\right)+\mathcal{O}_2(\varrho_0,\varrho_1,\varrho_2).$$

In the previous proposition linear operators $TH(\dots)$ and $HT(\dots)$ appear, which were defined in [9, Formulas (9.11) and (9.12)]. These operators are Hilbert-Schmidt operators on the space $X_{\mu_2}^{\mu_1}$ for all sufficiently large $\mu_1\geq 0$ and all sufficiently small $\mu_2\leq 0$ (see [9, Proposition 9.2]).

Moreover, the asymptotic operator relation stated in the previous proposition has to be understood in the way that the operators act on the space $X_{\mu_2}^{\mu_1}\oplus X_{\mu_2}^{\mu_1}$, where $\mu_1\geq 0$ is fixed and sufficiently large and $\mu_2\leq 0$ is fixed and sufficiently small.

Now we are prepared to give the proof of Theorem 4.2.

Proof of Theorem 4.2. We start from Proposition 4.4 with $\boldsymbol{a}_1=\boldsymbol{a}$, $K_1=K$, $\boldsymbol{a}_2=\boldsymbol{\sigma}$, $K_2=\{-1,1\}$. Since K and $\{1,-1\}$ are symmetric subsets of K, we can assume without loss of generality that f_1 and f_2 are even functions. Then

$$\begin{aligned}T_{2N}(\boldsymbol{a}\boldsymbol{\sigma})&=T_{2N}(\boldsymbol{a})T_{2N}(\boldsymbol{\sigma})+P_{2N}H_1(\boldsymbol{a},f_1)H_2(f_2,\boldsymbol{\sigma})P_{2N}\\&\quad-W_{2N}H_1(\boldsymbol{a},f_1)H_2(f_2,\boldsymbol{\sigma})W_{2N},\end{aligned}\tag{4.38}$$

where we have also used that $\boldsymbol{a}$ is even and $\boldsymbol{\sigma}$ is odd. By Proposition 3.4 the inverses of $T_{2N}(\boldsymbol{\sigma})$ exist for all N. Since the distribution $\boldsymbol{a}$ effects $\mathcal{R}$-convergence, the inverses of $T_{2N}(\boldsymbol{a})$ exist for all sufficiently large N. Hence

$$\begin{aligned}T_{2N}^{-1}(\boldsymbol{a})T_{2N}(\boldsymbol{a}\boldsymbol{\sigma})T_{2N}^{-1}(\boldsymbol{\sigma})&=P_{2N}+T_{2N}^{-1}(\boldsymbol{a})P_{2N}H_1(\boldsymbol{a},f_1)H_2(f_2,\boldsymbol{\sigma})P_{2N}T_{2N}^{-1}(\boldsymbol{\sigma})\\&\quad-T_{2N}^{-1}(\boldsymbol{a})W_{2N}H_1(\boldsymbol{a},f_1)H_2(f_2,\boldsymbol{\sigma})W_{2N}T_{2N}^{-1}(\boldsymbol{\sigma}).\end{aligned}$$

From Proposition 4.7 it now follows that

$$\begin{aligned}T_{2N}^{-1}(\boldsymbol{a})T_{2N}(\boldsymbol{a}\boldsymbol{\sigma})T_{2N}^{-1}(\boldsymbol{\sigma})&=P_{2N}+\frac{1}{2}T_{2N}^{-1}(\boldsymbol{a})P_{2N}H_1(\boldsymbol{a},f_1)H_2(f_2,\boldsymbol{h})P_{2N}T_{2N}^{-1}(\boldsymbol{h})\\&\quad-\frac{1}{2}T_{2N}^{-1}(\boldsymbol{a})W_{2N}H_1(\boldsymbol{a},f_1)H_2(f_2,\boldsymbol{h})P_{2N}T_{2N}^{-1}(\boldsymbol{h}).\end{aligned}$$

Taking determinants, observing that $\det T_{2N}(\boldsymbol{\sigma}) = 1$ by Proposition 3.4 and using the formula $\det(I + AB) = \det(I + BA)$ for determinants, we obtain

$$\begin{aligned}
&\frac{\det T_{2N}(\boldsymbol{a}\boldsymbol{\sigma})}{\det T_{2N}(\boldsymbol{a})} \\
&= \det\left(P_{2N} + \frac{1}{2}T_{2N}^{-1}(\boldsymbol{a})(P_{2N} - W_{2N})H_1(\boldsymbol{a}, f_1)H_2(f_2, \boldsymbol{h})P_{2N}T_{2N}^{-1}(\boldsymbol{h})\right) \\
&= \det\left(I + \frac{1}{2}H_2(f_2, \boldsymbol{h})P_{2N}T_{2N}^{-1}(\boldsymbol{h})T_{2N}^{-1}(\boldsymbol{a})(P_{2N} - W_{2N})H_1(\boldsymbol{a}, f_1)\right) \\
&= \det\left(I + \frac{1}{2}H_2(f_2, \boldsymbol{h})P_{2N}T_{2N}^{-1}(\boldsymbol{h})(P_{2N} - W_{2N})T_{2N}^{-1}(\boldsymbol{a})P_{2N}H_1(\boldsymbol{a}, f_1)\right).
\end{aligned}$$

Here we have also used formula (4.6) and the fact that $\boldsymbol{a}$ is even. Again from Proposition 4.7 it follows that

$$\begin{aligned}
&\frac{\det T_{2N}(\boldsymbol{a}\boldsymbol{\sigma})}{\det T_{2N}(\boldsymbol{a})} \\
&= \det\left(I + H_2(f_2, \boldsymbol{h})P_{2N}T_{2N}^{-1}(\boldsymbol{h})T_{2N}^{-1}(\boldsymbol{a})P_{2N}H_1(\boldsymbol{a}, f_1)\right) \\
&= \det\left(I + \begin{pmatrix} I, & 0 \end{pmatrix} HSH_{2N}(f_2, \boldsymbol{h}, \boldsymbol{a}, f_1)\begin{pmatrix} I \\ 0 \end{pmatrix}\right).
\end{aligned} \tag{4.39}$$

Note that in the last formula the (1,1)-block entry of $HSH_{2N}(f_2, \boldsymbol{h}, \boldsymbol{a}, f_1)$ appears (see formula (4.36)).

From the assumption on $\boldsymbol{a}$ and – as concerns the distribution $\boldsymbol{h}$ – from Proposition 4.5, Corollary 4.6 and [9, Theorem 13.1] (see also the remark made after Theorem 5.2 below) we know that

(1) $\boldsymbol{a}$ effects $\mathcal{R}$-convergence with respect to $[\boldsymbol{a}_+, \tilde{\boldsymbol{a}}_+]$ and $(\varrho_0, \varrho, \varrho)$;
(2) $\boldsymbol{h}$ effects $\mathcal{R}$-convergence with respect to $[1, \boldsymbol{h}]$ and $(\omega, 0, \omega)$ for each $\omega \in \mathbb{R}$.

Hence, by Proposition 4.8 we can conclude that

$$YH_{2N}(f_1, \boldsymbol{a}, f_1) = \begin{pmatrix} TH(f_1, \boldsymbol{a}_+, f_1) + \mathcal{O}_2(\varrho_0) & \mathcal{O}_2(\varrho) \\ \mathcal{O}_2(\varrho) & TH(\tilde{f}_1, \tilde{\boldsymbol{a}}_+, \tilde{f}_1) + \mathcal{O}_2(\varrho_0) \end{pmatrix},$$

$$HZ_{2N}(f_1, \boldsymbol{h}, f_2) = \begin{pmatrix} HT(f_2, \boldsymbol{h}, f_2) + \mathcal{O}_2(\omega) & \mathcal{O}_2(0) \\ \mathcal{O}_2(\omega) & HT(\tilde{f}_2, 1, \tilde{f}_2) + \mathcal{O}_2(\omega) \end{pmatrix}.$$

This in connection with (4.37) yields

$$\begin{aligned}
&\begin{pmatrix} I, & 0 \end{pmatrix} HSH_{2N}(f_2, \boldsymbol{h}, \boldsymbol{a}, f_1)\begin{pmatrix} I \\ 0 \end{pmatrix} \\
&= \Big(HT(f_2, \boldsymbol{h}, f_2) + \mathcal{O}_2(\omega)\Big)\Big(TH(f_1, \boldsymbol{a}_+, f_1) + \mathcal{O}_2(\varrho_0)\Big) + \mathcal{O}_2(0)\mathcal{O}_2(\varrho).
\end{aligned}$$

Noting that the operators $HT(\dots)$ and $TH(\dots)$ are Hilbert-Schmidt and choosing ω sufficiently small, this implies

$$\begin{aligned}&\begin{pmatrix} I, & 0\end{pmatrix} HSH_{2N}(f_2, \boldsymbol{h}, \boldsymbol{a}, f_1) \begin{pmatrix} I \\ 0 \end{pmatrix}\\ &= HT(f_2, \boldsymbol{h}, f_2) TH(f_1, \boldsymbol{a}_+, f_1) + \mathcal{O}_1(\max\{\varrho_0, \varrho\}). \end{aligned} \tag{4.40}$$

Formulas (4.39) and (4.40) give

$$\frac{\det T_{2N}(\boldsymbol{a}\boldsymbol{\sigma})}{\det T_{2N}(\boldsymbol{a})} = \det\left(I + HT(f_2, \boldsymbol{h}, f_2) TH(f_1, \boldsymbol{a}_+, f_1)\right) + O(N^{\max\{\varrho_0, \varrho\}}).$$

It remains to show that the above operator determinant equals the constant $a_+(1)/a_+(-1)$. From [9, Proposition 9.10(b)] we obtain

$$\det\left(I + HT(f_2, \boldsymbol{h}, f_2) TH(f_1, \boldsymbol{a}_+, f_1)\right) = \lim_{r \to 1-0} E(h_r \boldsymbol{a}_+, h_r \boldsymbol{h}) \tag{4.41}$$

where $h_r \boldsymbol{a}_+$ and $h_r \boldsymbol{h}$ $(0 \le r < 1)$ are the harmonic extensions of the distributions $\boldsymbol{a}_+$ and $\boldsymbol{h}$, i.e.,

$$(h_r \boldsymbol{a}_+)(t) = \sum_{n=0}^{\infty} r^n t^n [\boldsymbol{a}_+]_n, \qquad (h_r \boldsymbol{h})(t) = \sum_{n=0}^{\infty} r^n t^{-n} \boldsymbol{h}_{-n}, \tag{4.42}$$

and $E(\dots)$ is the constant defined by

$$E(\phi_+, \phi_-) = \exp\left(\sum_{n=1}^{\infty} n [\log \phi_+]_n [\log \phi_-]_{-n}\right). \tag{4.43}$$

In this connection we remark that the harmonic extensions of distributions contained in $\mathcal{GD}'_+$ or in $\mathcal{GD}'_-$ are always functions in $\mathcal{D} = C^\infty(\mathbb{T})$, which possess a continuous logarithm on $\mathbb{T}$ for each $0 \le r < 1$. Indeed, if $\boldsymbol{b} \in \mathcal{GD}'_\pm$, then the harmonic extensions are multiplicative, and consequently $(h_r \boldsymbol{b})(h_r \boldsymbol{b}^{-1}) \equiv 1$. The harmonic extensions depend uniformly on r and are constants for $r = 0$. Hence the functions $h_r \boldsymbol{b}$ are nonzero on all of $\mathbb{T}$ and have winding number zero.

In order to compute $E(h_r \boldsymbol{a}_+, h_r \boldsymbol{h})$, observe first that

$$h_r \boldsymbol{h} = (h_r \boldsymbol{\xi}_1^{-1})(h_r \boldsymbol{\xi}_{-1}) = \left(1 - \frac{r}{t}\right)^{-1} \left(1 + \frac{r}{t}\right).$$

Hence

$$[\log(h_r \boldsymbol{h})]_{-n} = \frac{r^n}{n} - \frac{(-r)^n}{n}, \qquad n \ge 1. \tag{4.44}$$

We obtain that $E(h_r \boldsymbol{a}_+, h_r \boldsymbol{h})$ is the exponential of

$$\begin{aligned}&\sum_{n=1}^{\infty} \left([\log h_r \boldsymbol{a}_+]_n r^n - [\log h_r \boldsymbol{a}_+]_n (-r)^n\right)\\ &\qquad = (h_r(\log h_r \boldsymbol{a}_+))(1) - (h_r(\log h_r a_+))(-1).\end{aligned}$$

Now notice that for $b \in C^\infty(\mathbb{T}) \cap \mathcal{D}'_+$ we have $h_r e^b = \exp(h_r b)$. With $b = \log h_r \boldsymbol{a}_+$ we obtain $\exp(h_r(\log h_r \boldsymbol{a}_+)) = h_r(h_r \boldsymbol{a}) = h_{r^2}\boldsymbol{a}_+$. Hence

$$E(h_r \boldsymbol{a}_+, h_r \boldsymbol{h}) = \exp\left((\log h_{r^2}\boldsymbol{a}_+)(1) - (\log h_{r^2}\boldsymbol{a}_+)(-1)\right). \tag{4.45}$$

From [9, Proposition 4.4] it follows that this converges to $a_+(1)/a_+(-1)$. Thus the proof is complete. □

5. Asymptotics of the determinants of symmetric Toeplitz matrices with Fisher-Hartwig distributions

In this section we recall the known results about the asymptotic behavior of Toeplitz determinants and specialize them to the case of determinants of symmetric Toeplitz matrices. Such an asymptotic formula is provided by the Fisher-Hartwig conjecture, which – in the so far most general setting – has been proved in [9] (see also [10]). We also recall some known results about the $\mathcal{R}$-convergence of certain classes of distributions, which are later on needed in order to be able to use Theorem 4.2. The underlying classes of distributions are defined and their properties are stated next.

For $\alpha, \beta, \gamma, \delta \in \mathbb{C}$ and $\theta_0 \in (-\pi, \pi]$, we introduce the functions

$$\omega_{\alpha,\beta,\theta_0}(e^{i\theta}) = (2 - 2\cos(\theta - \theta_0))^{\alpha} e^{i\beta(\theta-\theta_0-\pi)}, \qquad 0 < \theta - \theta_0 < 2\pi. \tag{5.1}$$

$$\eta_{\gamma,\theta_0}(e^{i\theta}) = (1 - e^{i(\theta-\theta_0)})^{\gamma}, \tag{5.2}$$

$$\xi_{\delta,\theta_0}(e^{i\theta}) = (1 - e^{i(\theta_0-\theta)})^{\delta}. \tag{5.3}$$

This implies

$$\omega_{\alpha,\beta,\theta_0}(e^{i\theta}) = \eta_{\alpha+\beta,\theta_0}(e^{i\theta})\xi_{\alpha-\beta,\theta_0}(e^{i\theta}). \tag{5.4}$$

For $\alpha, \beta, \gamma, \delta \in \mathbb{C}$ with $2\alpha \notin \mathbb{Z}_- := \{-1, -2, -3, \dots\}$ and $\theta_0 \in (-\pi, \pi]$, we introduce the distributions $\boldsymbol{\omega}_{\alpha,\beta,\theta_0}$, $\boldsymbol{\eta}_{\gamma,\theta_0}$ and $\boldsymbol{\xi}_{\delta,\theta_0}$ in terms of their Fourier coefficients

$$[\boldsymbol{\omega}_{\alpha,\beta,\theta_0}]_n = \frac{e^{in(\pi-\theta_0)}\Gamma(1+2\alpha)}{\Gamma(1+\alpha+\beta-n)\Gamma(1+\alpha-\beta+n)}, \qquad n \in \mathbb{Z}, \tag{5.5}$$

$$[\boldsymbol{\eta}_{\gamma,\theta_0}]_n = \begin{cases} e^{in(\pi-\theta_0)}\binom{\gamma}{n} & \text{if } n \geq 0 \\ 0 & \text{if } n < 0, \end{cases} \tag{5.6}$$

$$[\boldsymbol{\xi}_{\delta,\theta_0}]_n = \begin{cases} 0 & \text{if } n > 0 \\ e^{in(\pi-\theta_0)}\binom{\delta}{-n} & \text{if } n \leq 0. \end{cases} \tag{5.7}$$

It can be checked straightforwardly that if $2\alpha \notin \mathbb{Z}_-$, then

$$\boldsymbol{\omega}_{\alpha,\alpha,\theta_0} = \boldsymbol{\eta}_{2\alpha,\theta_0}, \qquad \boldsymbol{\omega}_{\alpha,-\alpha,\theta_0} = \boldsymbol{\xi}_{2\alpha,\theta_0}. \tag{5.8}$$

In what follows let $G_1C^\infty(\mathbb{T})$ stand for the set of all functions in $C^\infty(\mathbb{T})$ which are nonzero on all of $\mathbb{T}$ and have winding number zero. In other words,

$G_1C^\infty(\mathbb{T})$ is the set of all complex-valued functions defined on $\mathbb{T}$ which possess a logarithm that belongs to $C^\infty(\mathbb{T})$. Moreover, let $C^\infty_\pm(\mathbb{T})$ stand for the set of all $f \in C^\infty(\mathbb{T})$ for which $f_n = 0$ for all $n < 0$ ($n > 0$, resp.). We denote by $GC^\infty_\pm(\mathbb{T})$ the set of all invertible functions in $C^\infty_\pm(\mathbb{T})$.

A *distribution of Fisher-Hartwig type* is a distribution of the form

$$\mathbf{c} = b \prod_{r\in M_0} \boldsymbol{\omega}_{\alpha_r,\beta_r,\theta_r} \prod_{r\in M_+} \boldsymbol{\eta}_{\gamma_r,\theta_r} \prod_{r\in M_-} \boldsymbol{\xi}_{\delta_r,\theta_r}, \tag{5.9}$$

where

(i) $R \geq 0$ and $\{1,\ldots,R\} = M_0 \cup M_+ \cup M_-$ is a decomposition into disjoint subsets;
(ii) $\theta_1,\ldots,\theta_R \in (-\pi,\pi]$ are distinct numbers;
(iii) $b \in G_1C^\infty(\mathbb{T})$;
(iv) $\alpha_r, \beta_r \in \mathbb{C}$ and $2\alpha_r \notin \mathbb{Z}_-$ for all $r \in M_0$;
(v) $\gamma_r \in \mathbb{C}$ for all $r \in M_+$ and $\delta_r \in \mathbb{C}$ for all $r \in M_-$.

The product (5.9) of the distributions has to be understood in the sense of (3.7). To this distribution we associate the function

$$c(e^{i\theta}) = b(e^{i\theta}) \prod_{r\in M_0} \omega_{\alpha_r,\beta_r,\theta_r}(e^{i\theta}) \prod_{r\in M_+} \eta_{\gamma_r,\theta_r}(e^{i\theta}) \prod_{r\in M_-} \xi_{\delta_r,\theta_r}(e^{i\theta}). \tag{5.10}$$

Such a function will be called a *function of Fisher-Hartwig type*.

The following result has been proved in [9, Proposition 5.5].

Proposition 5.1. *Let* $\mathbf{c}$ *be the distribution* (5.9) *and* c *be the function* (5.10). *Put* $K = \{e^{i\theta_r} : 1 \leq r \leq R\}$. *Then*

(a) $\mathbf{c} \in \mathcal{D}'(K)$ *and* $\mathbf{c}$ *has the smooth part* c;
(b) *if* $M_0 = M_- = \emptyset$ *and* $b \in GC^\infty_+(\mathbb{T})$, *then* $\mathbf{c} \in \mathcal{GD}'_+(K)$;
(c) *if* $M_0 = M_+ = \emptyset$ *and* $b \in GC^\infty_-(\mathbb{T})$, *then* $\mathbf{c} \in \mathcal{GD}'_-(K)$.

Moreover, if $\operatorname{Re}\alpha > -1/2$, $\operatorname{Re}\gamma > -1$ *and* $\operatorname{Re}\delta > -1$, *then the distribution* $\mathbf{c}$ *can be identified with* $c \in L^1(\mathbb{T})$ *in the sense of* (3.4).

In what follows, we agree on the following conventions. We say that

(i) $\boldsymbol{a}$ effects $\mathcal{R}$-convergence w.r.t. $[\boldsymbol{a}_+, \boldsymbol{a}_-]$ and $(-\infty,-\infty,-\infty)$ if and only if for each $\varrho \in \mathbb{R}$, $\boldsymbol{a}$ effects $\mathcal{R}$-convergence w.r.t. $[\boldsymbol{a}_+, \boldsymbol{a}_-]$ and $(\varrho,\varrho,\varrho)$;
(ii) $\boldsymbol{a}$ effects $\mathcal{R}$-convergence w.r.t. $[\boldsymbol{a}_+, a_-]$ and $(-\infty,-\infty,\mu)$ if and only if for each $\varrho \in \mathbb{R}$, $\boldsymbol{a}$ effects $\mathcal{R}$-convergence w.r.t. $[\boldsymbol{a}_+, \boldsymbol{a}_-]$ and (ϱ,ϱ,μ);
(iii) $\boldsymbol{a}$ effects $\mathcal{R}$-convergence w.r.t. $[\boldsymbol{a}_+, \boldsymbol{a}_-]$ and $(-\infty,\mu,-\infty)$ if and only if for each $\varrho \in \mathbb{R}$, $\boldsymbol{a}$ effects $\mathcal{R}$-convergence w.r.t. $[\boldsymbol{a}_+, \boldsymbol{a}_-]$ and (ϱ,μ,ϱ).

Finally, let $O(N^{-\infty})$ stand for a sequence of complex numbers which is $O(N^\varrho)$ for each $\varrho \in \mathbb{R}$. A maximum taken over an empty set is considered to be $-\infty$.

Under certain conditions on the parameters, the asymptotic behavior of the determinants $\det T_N(\mathbf{c})$ with $\mathbf{c}$ given by (5.9) is described by the Fisher-Hartwig conjecture. The proof of this conjecture together with the statement that such distributions effect $\mathcal{R}$-convergence was the main result of [9] (see Section 13 therein or [10, Section 6]).

Recall that $\mathbb{Z}_- := \{-1,-2,-3,\dots\}$. Moreover, given $b \in G_1C^\infty(\mathbb{T})$, let $G[b]$ and $E[b]$ stand for the constants (2.2) and (2.3), and let $b_\pm \in GC^\infty_\pm(\mathbb{T})$ stand for the functions (2.5).

Theorem 5.2. *Let $\theta_1,\dots,\theta_R \in (-\pi,\pi]$ be distinct numbers, $R \geq 0$, put $t_r = e^{i\theta_r}$, let $\{1,\dots,R\} = M_0 \cup M_+ \cup M_+^* \cup M_- \cup M_-^*$ be a decomposition into disjoint subsets, let $\mathbf{c}$ be the distribution*

$$\mathbf{c} = b \prod_{r\in M_0} \boldsymbol{\omega}_{\alpha_r,\beta_r,\theta_r} \prod_{r\in M_+\cup M_+^*} \boldsymbol{\eta}_{\gamma_r,\theta_r} \prod_{r\in M_-\cup M_-^*} \boldsymbol{\xi}_{\delta_r,\theta_r}, \tag{5.11}$$

and assume that the following conditions are satisfied:

(a) $b \in G_1C^\infty(\mathbb{T})$;
(b) $2\alpha_r \notin \mathbb{Z}_-$, $\alpha_r + \beta_r \notin \mathbb{Z}_- \cup \{0\}$, $\alpha_r - \beta_r \notin \mathbb{Z}_- \cup \{0\}$ *for each* $r \in M_0$;
(c) $\gamma_r \notin \mathbb{Z}_- \cup \{0\}$ *for each* $r \in M_+$;
(d) $\gamma_r \in \mathbb{Z}_-$ *for each* $r \in M_+^*$;
(e) $\delta_r \notin \mathbb{Z}_- \cup \{0\}$ *for each* $r \in M_-$;
(f) $\delta_r \in \mathbb{Z}_-$ *for each* $r \in M_-^*$.
(g) $\varrho_0 < 0$ *(or, equivalently,* $\varrho_1 + \varrho_2 < 0$*), where*

$$\varrho_1 = \max\{-1 - 2\operatorname{Re}\beta_r \ : \ r \in M_0\} \cup \{-1 + \operatorname{Re}\delta_r \ : \ r \in M_-\}, \tag{5.12}$$

$$\varrho_2 = \max\{-1 + 2\operatorname{Re}\beta_r \ : \ r \in M_0\} \cup \{-1 + \operatorname{Re}\gamma_r \ : \ r \in M_+\}, \tag{5.13}$$

$$\varrho_0^* = \begin{cases} -1 & \text{if } M_0 \neq \emptyset \\ -\infty & \text{if } M_0 = \emptyset, \end{cases} \tag{5.14}$$

$$\varrho_0 = \max\{\varrho_0^*, \varrho_1 + \varrho_2\}. \tag{5.15}$$

Finally, define the following distributions and constants:

$$\mathbf{c}_+ = G[b]^{\frac{1}{2}} b_+ \prod_{r\in M_0\cup M_+\cup M_+^*} \boldsymbol{\eta}_{\gamma_r,\theta_r}, \quad \mathbf{c}_- = G[b]^{\frac{1}{2}} b_- \prod_{r\in M_0\cup M_-\cup M_-^*} \boldsymbol{\xi}_{\delta_r,\theta_r}, \tag{5.16}$$

$$\Omega_T = \sum_{r\in M_0} (\alpha_r^2 - \beta_r^2), \tag{5.17}$$

$$E_T = E[b] \prod_{r\in M_0} \frac{G(1+\alpha_r+\beta_r)G(1+\alpha_r-\beta_r)}{G(1+2\alpha_r)} \prod_{\substack{r\in M_0\cup M_+\cup M_+^* \\ s\in M_0\cup M_-\cup M_-^* \\ r\neq s}} (1 - t_s t_r^{-1})^{-\gamma_r\delta_s}$$

$$\times \prod_{r\in M_0\cup M_-\cup M_-^*} b_+(t_r)^{-\delta_r} \prod_{r\in M_0\cup M_+\cup M_+^*} b_-(t_r)^{-\gamma_r}, \tag{5.18}$$

where $\gamma_r = \alpha_r + \beta_r$ and $\delta_r = \alpha_r - \beta_r$ for $r \in M_0$.

Then

(i) $T_N(\mathbf{c})$ *is invertible for all sufficiently large* N*;*
(ii) $\mathbf{c}$ *effects* $\mathcal{R}$*-convergence with respect to* $[\mathbf{c}_+, \mathbf{c}_-]$ *and* $(\varrho_0, \varrho_1, \varrho_2)$*;*
(iii) $\det T_N(\mathbf{c}) = G[b]^N N^{\Omega_T} E_T(1 + O(N^{\varrho_0}))$.

Since $\boldsymbol{\xi}_1^{-1} = \boldsymbol{\xi}_{-1,0}$ and $\boldsymbol{\xi}_{-1} = \boldsymbol{\xi}_{1,\pi}$, where $\boldsymbol{\xi}_{\pm 1}$ are the distributions defined in the paragraph before Proposition 4.5, the previous theorem implies that the distribution $\boldsymbol{h} = \boldsymbol{\xi}_1^{-1}\boldsymbol{\xi}_{-1}$ effects $\mathcal{R}$-convergence with respect to $[1, \boldsymbol{h}]$ and $(-\infty, 0, -\infty)$. This is what we have used in the proof of Theorem 4.2.

Now we specialize the above theorem to the case of distributions which are even and which have no singularities at -1 and 1. Note that

$$\tilde{\boldsymbol{\omega}}_{\alpha_r,\beta_r,\theta_r} = \boldsymbol{\omega}_{\alpha_r,-\beta_r,-\theta_r} \qquad \tilde{\boldsymbol{\eta}}_{\gamma_r,\theta_r} = \boldsymbol{\xi}_{\gamma_r,-\theta_r}. \tag{5.19}$$

With this observation, it is easy to single out the class of distributions of Fisher-Hartwig type which are even.

Corollary 5.3. *Let* $\theta_1, \ldots, \theta_R \in (-\pi, 0) \cup (0, \pi)$ *be such that* $|\theta_r| \neq |\theta_s|$ *for all* $1 \leq r < s \leq R$, $R \geq 0$, *put* $t_r = e^{i\theta_r}$, *let* $\{1, \ldots, R\} = M_0 \cup M_\pm \cup M_\pm^*$ *be a decomposition into three disjoint subsets, let* $\mathbf{c}$ *be the distribution*

$$\mathbf{c} = b \prod_{r \in M_0} \boldsymbol{\omega}_{\alpha_r,\beta_r,\theta_r}\boldsymbol{\omega}_{\alpha_r,-\beta_r,-\theta_r} \prod_{r \in M_\pm \cup M_\pm^*} \boldsymbol{\eta}_{\gamma_r,\theta_r}\boldsymbol{\xi}_{\gamma_r,-\theta_r}, \tag{5.20}$$

and assume that the following conditions are satisfied:

(a) $b \in G_1C^\infty(\mathbb{T})$ *is even;*
(b) $2\alpha_r \notin \mathbb{Z}_-$, $\alpha_r + \beta_r \notin \mathbb{Z}_- \cup \{0\}$, $\alpha_r - \beta_r \notin \mathbb{Z}_- \cup \{0\}$ *and* $|\operatorname{Re}\beta_r| < 1/2$ *for each* $r \in M_0$*;*
(c) $\gamma_r \notin \mathbb{Z}_- \cup \{0\}$ *and* $\operatorname{Re}\gamma_r < 1$ *for each* $r \in M_\pm$*;*
(d) $\gamma_r \in \mathbb{Z}_-$ *for each* $r \in M_\pm^*$.

Define the numbers

$$\varrho_{12} = \max\{-1 + 2|\operatorname{Re}\beta_r| \ : \ r \in M_0\} \cup \{-1 + \operatorname{Re}\gamma_r \ : \ r \in M_\pm\}, \tag{5.21}$$

$$\varrho_0^* = \begin{cases} -1 & \text{if } M_0 \neq \emptyset \\ -\infty & \text{if } M_0 = \emptyset, \end{cases} \tag{5.22}$$

$$\varrho_0 = \max\{\varrho_0^*, 2\varrho_{12}\}, \tag{5.23}$$

and the following distribution and constants:

$$\mathbf{c}_+ = G[b]^{\frac{1}{2}} b_+ \prod_{r \in M_0} \boldsymbol{\eta}_{\alpha_r+\beta_r,\theta_r}\boldsymbol{\eta}_{\alpha_r-\beta_r,-\theta_r} \prod_{r \in M_\pm \cup M_\pm^*} \boldsymbol{\eta}_{\gamma_r,\theta_r} \tag{5.24}$$

$$\Omega_T^{\mathrm{sym}} = 2 \sum_{r \in M_0} (\alpha_r^2 - \beta_r^2), \tag{5.25}$$

$$E_T^{\rm sym} = E[b] \prod_{r\in M_0} \frac{G^2(1+\alpha_r+\beta_r)G^2(1+\alpha_r-\beta_r)}{G^2(1+2\alpha_r)}$$
$$\times \prod_{r\in M_0} b_+(t_r)^{-2(\alpha_r-\beta_r)} b_-(t_r)^{-2(\alpha_r+\beta_r)} \prod_{r\in M_\pm\cup M_\pm^*} b_-(t_r)^{-2\gamma_r}$$
$$\times \prod_{r,s\in M_0} (1-t_rt_s)^{-(\alpha_r-\beta_r)(\alpha_s-\beta_s)}(1-t_r^{-1}t_s^{-1})^{-(\alpha_r+\beta_r)(\alpha_s+\beta_s)}$$
$$\times \prod_{\substack{r,s\in M_0\\ r\neq s}} (1-t_rt_s^{-1})^{-(\alpha_r-\beta_r)(\alpha_s+\beta_s)}(1-t_r^{-1}t_s)^{-(\alpha_r+\beta_r)(\alpha_s-\beta_s)}$$
$$\times \prod_{\substack{r\in M_0\\ s\in M_\pm\cup M_\pm^*}} (1-t_r^{-1}t_s^{-1})^{-2(\alpha_r+\beta_r)\gamma_s}(1-t_rt_s^{-1})^{-2(\alpha_r-\beta_r)\gamma_s}$$
$$\times \prod_{r,s\in M_\pm\cup M_\pm^*} (1-t_r^{-1}t_s^{-1})^{-\gamma_r\gamma_s}. \tag{5.26}$$

Then

(i) $T_N(\boldsymbol{c})$ *is invertible for all sufficiently large* N;
(ii) $\boldsymbol{c}$ *effects* $\mathcal{R}$*-convergence with respect to* $[\boldsymbol{c}_+,\tilde{\boldsymbol{c}}_+]$ *and* $(\varrho_0,\varrho_{12},\varrho_{12})$;
(iii) $\det T_N(\boldsymbol{c}) = G[b]^N N^{\Omega_T^{\rm sym}} E_T^{\rm sym}(1+O(N^{\varrho_0}))$.

Proof. This corollary is just a special case of Theorem 5.2. The setting in which we have to apply Theorem 5.2 is the following. The number R of Theorem 5.2 is twice the number R of this corollary. More precisely, the set M_0 of Theorem 5.2 has to be identified with two copies of the present set M_0. Let us denote these two copies by $M_0^{(1)}$ and $M_0^{(2)}$. The set M_+ (M_+^*, resp.) of Theorem 5.2 has to be identified with one copy of the present set $M_\pm$ ($M_\pm^*$, resp.). In the same way, the set M_- (M_-^*, resp.) of Theorem 5.2 has also to be identified with one copy of the present set $M_\pm$ ($M_\pm^*$, resp.). The parameters corresponding to the index set

- $M_0^{(1)}$ are $(\alpha_r,\beta_r,\theta_r)$ for $r\in M_0$;
- $M_0^{(2)}$ are $(\alpha_r,-\beta_r,-\theta_r)$ for $r\in M_0$;
- $M_+\cup M_+^*$ are (γ_r,θ_r) for $r\in M_\pm\cup M_\pm^*$;
- $M_-\cup M_-^*$ are $(\gamma_r,-\theta_r)$ for $r\in M_\pm\cup M_\pm^*$.

From this it is readily seen that $\varrho_1=\varrho_2=\varrho_{12}$. In order for ϱ_0, or equivalently $\varrho_1+\varrho_2=2\varrho_{12}$, to be negative, it is necessary and sufficient that $|\operatorname{Re}\beta_r|<1/2$ and $\operatorname{Re}\gamma_r<1$. Hence conditions (a)–(d) of this corollary imply conditions (a)–(g) of Theorem 5.2.

The distributions $\boldsymbol{c}_+$ and $\boldsymbol{c}_-$ defined in (5.16) are given by (5.24) and

$$\boldsymbol{c}_- = G[b]^{\frac12} b_- \prod_{r\in M_0} \boldsymbol{\xi}_{\alpha_r+\beta_r,-\theta_r}\boldsymbol{\xi}_{\alpha_r-\beta_r,\theta_r} \prod_{r\in M_\pm\cup M_\pm^*} \boldsymbol{\xi}_{\gamma_r,-\theta_r}. \tag{5.27}$$

Since b is odd, we have $b_- = \tilde{b}_+$ and thus $\mathbf{c}_- = \tilde{\mathbf{c}}_+$. Because of the above parameters corresponding to $M_0^{(1)}$ and $M_0^{(2)}$, it is easily seen that Ω_T as given in (5.17) becomes Ω_T^{sym} as given above. It is somewhat troublesome, but straightforward to verify that E_T as given in (5.18) becomes E_T^{sym}. This completes the proof. □

6. Asymptotics of determinants of symmetric Toeplitz plus Hankel matrices with Fisher-Hartwig distributions

In this section we finally combine the Limit Theorem (Theorem 4.2) and Theorem 3.3 with Corollary 5.3 in order to obtain an asymptotic formula for the determinants of symmetric Toeplitz plus Hankel matrices. This result is based on the asymptotic formula for determinants of (symmetric) Toeplitz matrices.

In what follows, let $G[b]$ and $\widehat{E}[b]$ be the constants (2.2) and (2.8), and let $b_\pm$ be the functions (2.5).

Theorem 6.1. *Let* $\mathbf{c}$ *be a distribution that fulfills the assumptions of Corollary* 5.3. *Define the number* ϱ_{12} *by* (5.21) *and the constants*

$$\Omega_M^{\mathrm{sym}} = \sum_{r\in M_0} \left(\alpha_r^2 - \beta_r^2\right), \tag{6.1}$$

$$\begin{aligned} E_M^{\mathrm{sym}} = \widehat{E}[b] &\prod_{r\in M_0} \frac{G(1+\alpha_r+\beta_r)G(1+\alpha_r-\beta_r)}{G(1+2\alpha_r)} \\ &\times \prod_{r\in M_0} \frac{(1-t_r^{-1})^{(\alpha_r+\beta_r)/2}(1-t_r)^{(\alpha_r-\beta_r)/2}}{(1+t_r^{-1})^{(\alpha_r+\beta_r)/2}(1+t_r)^{(\alpha_r-\beta_r)/2}} 2^{\alpha_r^2-\beta_r^2} \prod_{r\in M_\pm\cup M_\pm^*} \frac{(1-t_r^{-1})^{\gamma_r/2}}{(1+t_r^{-1})^{\gamma_r/2}} \\ &\times \prod_{r\in M_0} b_+(t_r)^{-(\alpha_r-\beta_r)} b_-(t_r)^{-(\alpha_r+\beta_r)} \prod_{r\in M_\pm\cup M_\pm^*} b_-(t_r)^{-\gamma_r} \\ &\times \prod_{r,s\in M_0} (1-t_rt_s)^{-(\alpha_r-\beta_r)(\alpha_s-\beta_s)/2}(1-t_r^{-1}t_s^{-1})^{-(\alpha_r+\beta_r)(\alpha_s+\beta_s)/2} \\ &\times \prod_{\substack{r,s\in M_0\\ r\neq s}} (1-t_rt_s^{-1})^{-(\alpha_r-\beta_r)(\alpha_s+\beta_s)/2}(1-t_r^{-1}t_s)^{-(\alpha_r+\beta_r)(\alpha_s-\beta_s)/2} \\ &\times \prod_{\substack{r\in M_0\\ s\in M_\pm\cup M_\pm^*}} (1-t_r^{-1}t_s^{-1})^{-(\alpha_r+\beta_r)\gamma_s}(1-t_rt_s^{-1})^{-(\alpha_r-\beta_r)\gamma_s} \\ &\times \prod_{r,s\in M_\pm\cup M_\pm^*} (1-t_r^{-1}t_s^{-1})^{-\gamma_r\gamma_s/2}. \end{aligned} \tag{6.2}$$

Then

$$\det M_N(\mathbf{c}) = G[b]^N N^{\Omega_M^{\mathrm{sym}}} E_M^{\mathrm{sym}} (1+O(N^{\varrho_{12}})). \tag{6.3}$$

Proof. The distribution $\boldsymbol{c}$ belongs to $\mathcal{D}'(K \cup \widetilde{K})$, where $K = \{t_r : 1 \le r \le R\}$. Thus the assumptions of Theorem 3.3 are fulfilled and we obtain

$$\det T_{2N}(\boldsymbol{\sigma c}) = (\det M_N(\boldsymbol{c}))^2.$$

Moreover, the distribution $\boldsymbol{c}$ has the smooth part

$$c(e^{i\theta}) = b(e^{i\theta}) \prod_{r \in M_0} \omega_{\alpha_r,\beta_r,\theta_r}(e^{i\theta})\omega_{\alpha_r,-\beta_r,-\theta_r}(e^{i\theta}) \prod_{r \in M_\pm \cup M_\pm^*} \eta_{\gamma_r,\theta_r}(e^{i\theta})\xi_{\gamma_r,-\theta_r}(e^{i\theta}),$$

From Corollary 5.3 (ii) it follows that $\boldsymbol{c}$ effects $\mathcal{R}$-convergence with respect to $[\boldsymbol{c}_+, \tilde{\boldsymbol{c}}_+]$ and $(\varrho_0, \varrho_{12}, \varrho_{12})$, where ϱ_0 is defined by (5.22) and (5.23) and $\boldsymbol{c}_+$ is the distribution given by (5.24). Notice that ϱ_0 and ϱ_{12} are negative real numbers. Obviously, $\boldsymbol{c}_+$ belongs to $\mathcal{GD}'_+(K \cup \widetilde{K})$ and has the smooth part

$$c_+(e^{i\theta}) = G[b]^{\frac{1}{2}} b_+(e^{i\theta}) \prod_{r \in M_0} \eta_{\alpha_r+\beta_r,\theta_r}(e^{i\theta})\eta_{\alpha_r-\beta_r,-\theta_r}(e^{i\theta}) \prod_{r \in M_\pm \cup M_\pm^*} \eta_{\gamma_r,\theta_r}(e^{i\theta}).$$

It is readily seen that $c(t) = c_+(t)\tilde{c}_+(t)$. Hence the assumptions of Theorem 4.2 are fulfilled with $\boldsymbol{a}$ and $\boldsymbol{a}_+$ replaced by $\boldsymbol{c}$ and $\boldsymbol{c}_+$, respectively, and K replaced by $K \cup \widetilde{K}$. Thus

$$\det T_{2N}(\boldsymbol{\sigma c}) = \det T_{2N}(\boldsymbol{c}) \left(\frac{c_+(1)}{c_+(-1)} + O\left(N^{\max\{\varrho_0,\varrho_{12}\}}\right)\right).$$

Note that $\max\{\varrho_0, \varrho_{12}\} = \max\{\varrho_0^*, 2\varrho_{12}, \varrho_{12}\} = \max\{\varrho_0^*, \varrho_{12}\} = \varrho_{12}$. Combining the previous formulas we obtain

$$(\det M_N(\boldsymbol{c}))^2 = \det T_{2N}(\boldsymbol{c}) \frac{c_+(1)}{c_+(-1)} \left(1 + O(N^{\varrho_{12}})\right).$$

The asymptotics of $\det T_{2N}(\boldsymbol{c})$ follows from Corollary 5.3 (iii). Observe the change from N to $2N$ and that $\varrho_0 \le \varrho_{12}$. Thus

$$(\det M_N(\boldsymbol{c}))^2 = G[b]^{2N} (2N)^{\Omega_T^{\mathrm{sym}}} E_T^{\mathrm{sym}} \frac{c_+(1)}{c_+(-1)} \left(1 + O\left(N^{\varrho_{12}}\right)\right),$$

where Ω_T^{sym} and E_T^{sym} are the constants (5.25) and (5.26). Notice also that

$$2^{\Omega_T^{\mathrm{sym}}} \frac{c_+(1)}{c_+(-1)} = \prod_{r \in M_0} 2^{2(\alpha_r^2-\beta_r^2)} \frac{(1-t_r^{-1})^{\alpha_r+\beta_r}(1-t_r)^{\alpha_r-\beta_r}}{(1+t_r^{-1})^{\alpha_r+\beta_r}(1+t_r)^{\alpha_r-\beta_r}}$$

$$\times \frac{b_+(1)}{b_+(-1)} \prod_{r \in M_\pm \cup M_\pm^*} \frac{(1-t_r^{-1})^{\gamma_r}}{(1+t_r^{-1})^{\gamma_r}}.$$

Taking into account that

$$\widehat{E}[b]^2 = E[b] \frac{b_+(1)}{b_+(-1)},$$

it is readily seen that

$$2^{\Omega_T^{\mathrm{sym}}} E_T^{\mathrm{sym}} \frac{c_+(1)}{c_+(-1)} = (E_M^{\mathrm{sym}})^2.$$

Obviously, $\Omega_T^{\rm sym} = 2\Omega_M^{\rm sym}$. Hence the last asymptotic formula becomes

$$(\det M_N(\mathbf{c}))^2 = G[b]^{2N} N^{2\Omega_M^{\rm sym}} (E_M^{\rm sym})^2 (1 + O(N^{\varrho_{12}})). \tag{6.4}$$

From this the desired asymptotic formula follows, up to a sign, which will be determined by the following argument. Let $U \subset \mathbb{C}^d$, where $d = 2|M_0| + |M_\pm \cup M_\pm^*|$, be the set of all d-tuples

$$z = \Big((\alpha_r, \beta_r)_{r\in M_0}, (\gamma_r)_{r\in M_\pm}, (\gamma_r)_{r\in M_\pm^*} \Big) \tag{6.5}$$

such that

(a) $2\alpha_r \notin \mathbb{Z}_-$, $\alpha_r \pm \beta_r \notin \mathbb{Z}_- \cup \{0\}$, $|\operatorname{Re}\beta_r| < 1/2$ for all $r \in M_0$;
(b) $\gamma_r \notin \mathbb{Z}_- \cup \{0\}$, $\operatorname{Re}\gamma_r < 1$ for all $r \in M_\pm \cup M_\pm^*$.

Let the even function $b_1 \in GC^\infty(\mathbb{T})$ be arbitrary but fixed, and introduce

$$f_N(z) = \frac{\det M_N(\mathbf{c})}{G[b]^N N^{\Omega_M^{\rm sym}}}, \qquad f(z) = E_M^{\rm sym},$$

where $\mathbf{c}$ is the distribution (5.20) with the parameters given by (6.5), and the constants $\Omega_M^{\rm sym}$ and $E_M^{\rm sym}$ are defined correspondingly.

From what has been proved so far, it follows that

$$(f_N(z))^2 \to (f(z))^2 \tag{6.6}$$

for $z \in U$. Moreover, the convergence is uniform on compact subsets of U. Since U is connected and since $f(z) \neq 0$ for all $z \in U$, it follows that either $f_N(z) \to f(z)$ on all of U or $f_N(z) \to -f(z)$ on all of U.

Now let U_0 stand for the set of all parameters (6.5) for which

(a) $2\alpha_r \notin \mathbb{Z}_-$, $|\operatorname{Re}\beta_r| < 1/2$ for all $r \in M_0$;
(b) $\operatorname{Re}\gamma_r < 1$ for all $r \in M_\pm \cup M_\pm^*$.

Because $f_N(z)$ and $f(z)$ are functions that depend analytically on z, because of the uniform convergence on compact subsets of U and because of the concrete structure of U and U_0, it follows that either $f_N(z) \to f(z)$ on all of U_0 or $f_N(z) \to -f(z)$ on all of U_0.

For $z = 0$ we know that $0 \in U_0$ and that $f_N(0) \to f(0)$. Thus we can conclude that $f_N(z) \to f(z)$ on all of U_0. Hence the desired asymptotic formula with the correct sign follows. □

7. Asymptotics of determinants of symmetric Toeplitz plus Hankel matrices with piecewise continuous functions

From Theorem 6.1 we obtain immediately the following result concerning the asymptotics of the determinants of symmetric Toeplitz plus Hankel matrices $M_N(\phi)$ with a particular piecewise continuous generating function.

Theorem 7.1. *Let $\theta_0 \in (0,\pi)$ and $\beta \in \mathbb{C}$ be such that $|\operatorname{Re}\beta| < 1/2$. Put $t_0 = e^{i\theta_0}$. Then*

$$\lim_{N\to\infty} \frac{\det M_N(t_{\beta,\theta_0} t_{-\beta,-\theta_0})}{N^{-\beta^2}} = E,$$

where

$$E = 2^{-\beta^2}(1-t_0^2)^{-\frac{\beta^2}{2}}(1-t_0^{-2})^{-\frac{\beta^2}{2}} \frac{(1-t_0^{-1})^{\frac{\beta}{2}}(1-t_0)^{-\frac{\beta}{2}}}{(1+t_0^{-1})^{\frac{\beta}{2}}(1+t_0)^{-\frac{\beta}{2}}} G(1+\beta)G(1-\beta).$$

Proof. In Theorem 6.1 we put $M_\pm = M_\pm^* = \emptyset$, $M_0 = \{1\}$, $\alpha_1 = 0$, $\beta_1 = \beta$, $\theta_1 = \theta_0$ and $b(t) = 1$. Observe that $\omega_{0,\beta,\theta_0}\omega_{0,-\beta,-\theta_0} = t_{\beta,\theta_0}t_{-\beta,-\theta_0}$. □

The main result concerning the asymptotics of the determinants of symmetric Toeplitz plus Hankel matrices $M_N(\phi)$ with "general" piecewise continuous generating functions is the following theorem.

Theorem 7.2. *Let*

$$c(e^{i\theta}) = b(e^{i\theta}) \prod_{r=1}^{R} t_{\beta_r,\theta_r}(e^{i\theta}) t_{-\beta_r,-\theta_r}(e^{i\theta}) \tag{7.1}$$

where $b \in G_1 B_1^1$ is an even function, $\theta_1, \dots, \theta_R \in (0,\pi)$ are distinct numbers, and $\beta_1, \dots, \beta_R \in \mathbb{C}$ are such that $|\operatorname{Re}\beta_r| < 1/2$ for all $1 \le r \le R$. Let $G[b]$ and $\widehat{E}[b]$ be the constants (2.2) and (2.8), $b_\pm$ be the functions (2.5), $t_r = e^{i\theta_r}$, $1 \le r \le R$, and introduce the constants

$$\Omega_M^{\text{sym}} = -\sum_{r=1}^{R} \beta_r^2, \tag{7.2}$$

$$\begin{aligned} E_M^{\text{sym}} &= \widehat{E}[b] \prod_{r=1}^{R} G(1+\beta_r)G(1-\beta_r)(1-t_r^2)^{-\beta_r^2/2}(1-t_r^{-2})^{-\beta_r^2/2} \\ &\times \prod_{1\le r<s\le R} (1-t_r t_s)^{-\beta_r\beta_s}(1-t_r^{-1}t_s^{-1})^{-\beta_r\beta_s}(1-t_r t_s^{-1})^{\beta_r\beta_s}(1-t_r^{-1}t_s)^{\beta_r\beta_s} \\ &\times \prod_{r=1}^{R} 2^{-\beta_r^2} \frac{(1-t_r^{-1})^{\beta_r/2}(1-t_r)^{-\beta_r/2}}{(1+t_r^{-1})^{\beta_r/2}(1+t_r)^{-\beta_r/2}} \prod_{r=1}^{R} b_+(t_r)^{\beta_r} b_-(t_r)^{-\beta_r}. \end{aligned} \tag{7.3}$$

Then

$$\lim_{N\to\infty} \frac{\det M_N(c)}{G[b]^N N^{\Omega_M^{\text{sym}}}} = E_M^{\text{sym}}. \tag{7.4}$$

Proof. The asymptotic formula follows from Corollary 2.5), Theorem 7.1 and the asymptotic formula (2.7). □

We remark that in the case $b \in G_1 C^\infty(\mathbb{T})$, the previous theorem follows also directly from Theorem 6.1.

In view of the asymptotic formulas established in Theorem 2.3, Theorem 2.4 and Theorem 7.1 we now raise the following conjecture.

Conjecture 7.3. *Let $\theta_0 \in (0, \pi)$ and $\beta_1, \beta_2 \in \mathbb{C}$ be such that $|\operatorname{Re}\beta_1| < 1/2$ and $|\operatorname{Re}\beta_2| < 1/2$ and $|\operatorname{Re}\beta_1 + \operatorname{Re}\beta_2| < 1/2$. Put $t_0 = e^{i\theta_0}$. Then*

$$\lim_{N\to\infty} \frac{\det M_N(t_{\beta_1,\theta_0} t_{\beta_2,-\theta_0})}{N^{\Omega}} = E, \tag{7.5}$$

where

$$\Omega = -\beta_1^2 - \beta_1\beta_2 - \beta_2^2, \tag{7.6}$$

$$\begin{aligned} E = {} & G(1+\beta_1)G(1+\beta_2)G(1-\beta_1-\beta_2)2^{\beta_1\beta_2} \\ & \times (1-t_0^{-2})^{\beta_1^2/2+\beta_1\beta_2}(1-t_0^2)^{\beta_2^2/2+\beta_1\beta_2} \\ & \times \frac{(1-t_0^{-1})^{\beta_1/2}(1-t_0)^{\beta_2/2}}{(1+t_0^{-1})^{\beta_1/2}(1+t_0)^{\beta_2/2}}. \end{aligned} \tag{7.7}$$

This conjecture fits with the results established in Theorem 2.3, Theorem 2.4 and Theorem 7.1.

References

[1] T.H. Baker and P.J. Forrester, Finite N fluctuations formulas for random matrices, *J. Stat. Phys.* **88** (1997), 1371–1386.

[2] E.W. Barnes, The theory of the G-function, *Quart. J. Pure and Appl. Math.* **31** (1900), 264–313.

[3] E.L. Basor, Distribution functions for random variables for ensembles of positive Hermitian matrices, *Comm. Math. Phys.* **188**, no. 2 (1997), 327–350.

[4] E.L. Basor and T. Ehrhardt, Asymptotic formulas for determinants of a sum of finite Toeplitz and Hankel matrices, *Math. Nachr.* **122** (2001), 5–45.

[5] E.L. Basor and T. Ehrhardt, Some identities for determinants of structured matrices, *Linear Algebra Appl.* **343–344** (2002), 5–19.

[6] E.L. Basor and C.A. Tracy, The Fisher-Hartwig conjecture and generalizations, *Physica A* **177** (1991), 167–173.

[7] E.L. Basor and C.A. Tracy, Variance calculations and the Bessel kernel, *J. Statist. Phys.* **73**, no. 1–2 (1993), 415–421.

[8] A. Böttcher and B. Silbermann, *Analysis of Toeplitz operators*, Springer, Berlin 1990.

[9] T. Ehrhardt, *Toeplitz determinants with several Fisher-Hartwig singularities*, Dissertation, Technische Universität Chemnitz, Chemnitz 1997.

[10] T. Ehrhardt, A status report on the asymptotic behavior of Toeplitz determinants with Fisher-Hartwig singularities, In: *Oper. Theory Adv. Appl.*, Vol. 122, Birkhäuser, Basel 2001, 217–241.

[11] B. Gordon, Notes on Plane Partitions. V, *J. Combinatorial Theory Ser. B* **11** (1971), 157–168.

[12] C. Krattenthaler, Advanced determinant calculus, *Sém. Lothar. Comb.* **42** (1999), Art. B42q, 67 pp. (electronic).

[13] M.L. Mehta, *Random Matrices*, Academic Press, Rev. and enlarged 2nd ed., San Diego 1991.

[14] J.R. Stembridge, Nonintersection Paths, Pfaffians, and Plane Partitions, *Adv. Math.* **83** (1990), no. 1, 96–131.

[15] E.T. Whittaker and G.N. Watson, *A Course of Modern Analysis*, 4th ed., Cambridge Univ. Press, London/New York 1952.

[16] H. Widom, Asymptotic behavior of block Toeplitz determinants. II, *Adv. Math.* **21** (1976), 1–29.

E.L. Basor
Department of Mathematics
California Polytechnic State University
San Luis Obispo, CA 93407, USA
e-mail: `ebasor@calpoly.edu`

T. Ehrhardt
Fakultät für Mathematik
Technische Universität Chemnitz
09107 Chemnitz, Germany
e-mail: `tehrhard@mathematik.tu-chemnitz.de`

Received: 30 November 2001

Operator Theory:
Advances and Applications, Vol. 135, 91–99

On the Determinant Formulas by Borodin, Okounkov, Baik, Deift and Rains

Albrecht Böttcher

To my Teacher Bernd Silbermann on His Sixtieth Birthday

Abstract. We give alternative proofs to (block case versions of) some formulas for Toeplitz and Fredholm determinants established recently by the authors listed in the title. Our proof of the Borodin-Okounkov formula is very short and direct. The proof of the Baik-Deift-Rains formulas is based on standard manipulations with Wiener-Hopf factorizations.

1. The formulas

Let $\mathbf{T}$ be the complex unit circle and let $L^\infty := L^\infty_{N\times N}$ stand for the algebra of all $N \times N$ matrix functions with entries in $L^\infty(\mathbf{T})$. Given $a \in L^\infty$, we denote by $\{a_k\}_{k\in\mathbf{Z}}$ the sequence of the Fourier coefficients,

$$a_k = \frac{1}{2\pi}\int_0^{2\pi} a(e^{i\theta})e^{-ik\theta}\,d\theta = \frac{1}{2\pi i}\int_{\mathbf{T}} a(z)z^{-k}\frac{dz}{z}.$$

The matrix function a generates several structured (block) matrices:

$$\begin{aligned}
T(a) &= (a_{j-k})_{j,k=0}^{\infty} \quad \text{(infinite block Toeplitz)},\\
T_n(a) &= (a_{j-k})_{j,k=0}^{n-1} \quad \text{(finite block Toeplitz)},\\
H(a) &= (a_{j+k+1})_{j,k=0}^{\infty} \quad \text{(block Hankel)},\\
H(\tilde{a}) &= (a_{-j-k-1})_{j,k=0}^{\infty} \quad \text{(block Hankel)},\\
L(a) &= (a_{j-k})_{j,k=-\infty}^{\infty} \quad \text{(block Laurent)},\\
L(\tilde{a}) &= (a_{k-j})_{j,k=-\infty}^{\infty} \quad \text{(block Laurent)}.
\end{aligned}$$

The matrices $T(a), H(a), H(\tilde{a})$ induce bounded operators on $\ell^2(\mathbf{Z}_+, \mathbf{C}^N)$, and the matrices $L(a), L(\tilde{a})$ define bounded operators on $\ell^2(\mathbf{Z}, \mathbf{C}^N)$.

Let $\|\cdot\|$ be any matrix norm on $\mathbf{C}^{N\times N}$. We need the following classes of matrix functions:

$$W = \{a \in L^\infty : \sum_{n\in\mathbf{Z}} \|a_n\| < \infty\} \quad \text{(Wiener algebra)},$$

$$K_1^1 = \{a \in L^\infty : \sum_{n\in\mathbf{Z}} (|n|+1)\|a_n\| < \infty\} \quad \text{(weighted Wiener algebra)},$$

$$K_2^{1/2} = \{a \in L^\infty : \sum_{n\in\mathbf{Z}} (|n|+1)\|a_n\|^2 < \infty\} \quad \text{(Krein algebra)},$$

$$H_\pm^\infty = \{a \in L^\infty : a_{\mp n} = 0 \text{ for } n > 0\} \quad \text{(Hardy space)}.$$

Clearly, $K_1^1 \subset K_2^{1/2}$. Given a subset E of L^∞, we say that a matrix function $a \in L^\infty$ has a right (resp. left) canonical Wiener-Hopf factorization in E and write $a \in \Phi_r(E)$ (resp. $a \in \Phi_l(E)$) if a can be represented in the form $a = u_- u_+$ (resp. $a = v_+ v_-$) with

$$u_-, v_-, u_-^{-1}, v_-^{-1} \in E \cap H_-^\infty, \quad u_+, v_+, u_+^{-1}, v_+^{-1} \in E \cap H_+^\infty.$$

It is well known (see, e.g., [5], [7]) that if $a \in \Phi_r(L^\infty)$ then $T(a)$ is invertible and $T^{-1}(a) = T(u_+^{-1})T(u_-^{-1})$ and that for $a \in K_1^1$ (resp. $a \in W \cap K_2^{1/2}$) we have

$$a \in \Phi_r(K_1^1) \quad (\text{resp. } a \in \Phi_r(K_2^{1/2})) \quad \Longleftrightarrow \quad T(a) \text{ is invertible.}$$

If $a \in K_1^1$ then $H(a)$ and $H(\tilde{a})$ are trace class operators, and if $a \in K_2^{1/2}$, then $H(a)$ and $H(\tilde{a})$ are Hilbert-Schmidt.

We define the projections P, Q, Q_n ($n \in \mathbf{Z}$) on the space $\ell^2(\mathbf{Z}, \mathbf{C}^N)$ by

$$(Px)_k = \begin{cases} x_k & \text{for } k \ge 0, \\ 0 & \text{for } k < 0, \end{cases} \qquad (Qx)_k = \begin{cases} 0 & \text{for } k \ge 0, \\ x_k & \text{for } k < 0, \end{cases}$$

$$(Q_n x)_k = \begin{cases} x_k & \text{for } k \ge n, \\ 0 & \text{for } k < n. \end{cases}$$

For $n \ge 1$, we let P_n denote the projection on $\ell^2(\mathbf{Z}_+, \mathbf{C}^N)$ given by

$$(P_n x)_k = \begin{cases} x_k & \text{for } 0 \le k \le n-1, \\ 0 & \text{for } k \ge n. \end{cases}$$

If $n \ge 0$, we can also think of Q_n as an operator on $\ell^2(\mathbf{Z}_+, \mathbf{C}^N)$. Note that the notation used here differs from the one of [1], but that our notation is standard in the Toeplitz business.

On defining the flip operator J on $\ell^2(\mathbf{Z}, \mathbf{C}^N)$ by $(Jx)_k = x_{-k-1}$, we can write

$$T(a) = PL(a)P|\operatorname{Im} P, \quad H(a) = PL(a)QJ|\operatorname{Im} P, \quad H(\tilde{a}) = JQL(a)P|\operatorname{Im} P. \quad (1)$$

Moreover, we may identify the operator $L(a)$ on $\ell^2(\mathbf{Z}, \mathbf{C}^N)$ with the operator of multiplication by a on $L^2(\mathbf{T}, \mathbf{C}^N)$. Since P, Q, J are also naturally defined on the space $L^2(\mathbf{T}, \mathbf{C}^N)$, formulas (1) enable us to interpret Toeplitz and Hankel operators as operators on the Hardy space $H^2(\mathbf{T}, \mathbf{C}^N)$.

For $a \in \Phi_l(L^\infty)$, the geometric mean $G(a)$ is the number $(\det v_+)_0 (\det v_-)_0$, where $(\cdot)_k$ stands for the kth Fourier coefficient. Thus, with an appropriately chosen logarithm,

$$G(a) = \exp(\log \det a)_0.$$

Let now $a \in \Phi_r(K_2^{1/2}) \cap \Phi_l(K_2^{1/2})$ and let $a = u_- u_+$ and $a = v_+ v_-$ be canonical Wiener-Hopf factorizations. Put $b = v_- u_+^{-1}$ and $c = u_-^{-1} v_+$. Obviously, $bc = I$. Using (1) it is easily seen that

$$T(b)T(c) + H(b)H(\tilde{c}) = I. \tag{2}$$

Since Hankel operators generated by matrix functions in $K_2^{1/2}$ are Hilbert-Schmidt, the operator $H(b)H(\tilde{c})$ is in the trace class. From (2) we infer that $I - H(b)H(\tilde{c})$ is invertible. We put

$$E(a) = 1/\det(I - H(b)H(\tilde{c})).$$

One can show (again see [5], [7]) that $E(a) = \det T(a)T(a^{-1})$ and that in the scalar case ($N = 1$) we also have

$$E(a) = \exp \sum_{k=1}^{\infty} k (\log a)_k (\log a)_{-k}.$$

Theorem 1.1. (Borodin-Okounkov à la Widom) *If $a \in \Phi_r(K_2^{1/2}) \cap \Phi_l(K_2^{1/2})$ then*

$$\det T_n(a) = G(a)^n E(a) \det(I - Q_n H(b)H(\tilde{c})Q_n) \tag{3}$$

for all $n \geq 1$.

In the scalar case, this beautiful theorem was established by Borodin and Okounkov in [3]. It answered a question raised by Its and Deift. The proof of [3] is rather complicated. Three simpler proofs were subsequently found by Basor and Widom [2] (who also extended the theorem to the block case) and by the author [4]. We here give still another proof, which is very short and direct and is a modification of the second proof of [2].

Now suppose that $a \in \Phi_r(K_1^1) \cap \Phi_l(K_1^1)$ ($\subset \Phi_r(K_2^{1/2}) \cap \Phi_l(K_2^{1/2})$). Define b and c as above. We have

$$\begin{aligned} P - L(c)Q_n L(b) &= (PL(c) - L(c)Q_n)L(b) \\ &= (PL(c)Q - QL(c)P + L(c)(P - Q_n))L(b) \end{aligned}$$

and since $PL(c)Q$ and $QL(c)P$ are trace class operators (notice that $b, c \in K_1^1$) and the operator $P - Q_n$ has finite rank, we see that $P - L(c)Q_n L(b)$ is trace class.

Theorem 1.2. (Baik-Deift-Rains) *If $a \in \Phi_r(K_1^1) \cap \Phi_l(K_1^1)$ then*

$$\det T_n(a) = G(a)^n E(a) 2^{-nN} \det(I + P - L(c)Q_n L(b)) \tag{4}$$

for all $n \geq 1$.

Clearly, to prove Theorem 1.2 it suffices to prove Theorem 1.1 and to verify that

$$\det(I + P - L(c)Q_nL(b)) = 2^{nN} \det(I - Q_nH(b)H(\tilde{c})Q_n) \tag{5}$$

for all $n \geq 1$. By virtue of (1),

$$\det(I - Q_nH(b)H(\tilde{c})Q_n) = \det(I - Q_nL(b)QL(c)Q_n)$$

for all $n \geq 1$. The right-hand side of the last equality makes sense for all $n \in \mathbf{Z}$. In fact, we have the following generalization of (5).

Theorem 1.3. (Baik-Deift-Rains) *If $a \in \Phi_r(K_1^1) \cap \Phi_l(K_1^1)$ then for all $n \in \mathbf{Z}$,*

$$\begin{aligned}
&\det(I + sP - sL(a)Q_nL(a^{-1}))\\
&= (1+s)^{nN} \det(I - s^2Q_nL(a^{-1})QL(a)Q_n) \quad (s \neq -1) &(6)\\
&= (1-s)^{-nN} \det(I - s^2(I - Q_n)L(a^{-1})PL(a)(I - Q_n)) \quad (s \neq 1) &(7)
\end{aligned}$$

Theorems 1.2 and 1.3 are in [1]. The proof given there is as follows: the formulas are easily seen if some operator that is no trace class operator were a trace class operator and to save that insight the authors employ an approximation argument. We here present a proof that is a little more direct and uses Wiener-Hopf factorization.

Theorem 1.1 is proved in Section 2, the proofs of Theorems 1.2 and 1.3 are given in Section 3. In Section 4 we relax the hypothesis of Theorem 1.3 to the requirement that a be in K_1^1 and that $\det a$ have no zeros on the unit circle.

2. Proof of the Borodin-Okounkov formula

If K is an arbitrary trace class operator on $\ell^2(\mathbf{Z}_+, \mathbf{C}^N)$ and $I - K$ is invertible, then

$$\det P_n(I-K)^{-1}P_n = \frac{\det(I - Q_nKQ_n)}{\det(I-K)}. \tag{8}$$

With K replaced by P_mKP_m, this is Jacobi's theorem on the principle $n \times n$ minor of the inverse of a (finite) matrix. In the general case the identity follows from the fact that P_mKP_m converges to K in the trace norm as $m \to \infty$. For $K = H(b)H(\tilde{c})$ we obtain from (2) that

$$\begin{aligned}
&P_n(I-K)^{-1}P_n = P_nT^{-1}(c)T^{-1}(b)P_n\\
&= P_nT(v_+^{-1})T(u_-)T(u_+)T(v_-^{-1})P_n = T_n(v_+^{-1})T_n(a)T_n(v_-^{-1}),
\end{aligned}$$

and since $\det T_n(v_+^{-1})T_n(a)T_n(v_-^{-1}) = G(a)^{-n} \det T_n(a)$, we get (3) from (8). □

3. Proof of the Baik-Deift-Rains formulas

In what follows we abbreviate $L(a)$ to a. Equivalently, we may regard all operators on L^2 instead of ℓ^2 and may therefore think of a as multiplication by a. Notice that if $a \in K_1^1$ is invertible in L^∞, then a^{-1} also belongs to K_1^1.

Lemma 3.1. *If a and a^{-1} are in K_1^1 then*

$$P - aQ_n a^{-1}, \quad Q_n a^{-1} Q a Q_n, \quad (I - Q_n) a^{-1} P a (I - Q_n)$$

are trace class operators for all $n \in \mathbf{Z}$.

Proof. We have

$$\begin{aligned}
&P - aQ_n a^{-1} = (Pa - aQ_n)a^{-1} = (PaQ - QaP + a(P - Q_n))a^{-1}, \\
&Q_n a^{-1} Q a Q_n = -Q_n a^{-1} Q a P + Q_n a^{-1} Q a (P - Q_n), \\
&(I - Q_n) a^{-1} P a (I - Q_n) = (I - Q_n) a^{-1} P a Q + (I - Q_n) a^{-1} P a (P - Q_n),
\end{aligned}$$

and since PaQ and QaP are trace class and $P - Q_n$ has finite rank, we arrive at the assertion. $\square$

We put

$$f_n(s) = \det(I + sP - saQ_n a^{-1}).$$

Proposition 3.2. *If $a \in \Phi_r(K_1^1)$ and $n \geq 0$, then*

$$f_n(s) = (1 + s)^{nN} \det(I - s^2 Q_n a^{-1} Q a Q_n). \tag{9}$$

Proof. Let $a = u_- u_+$ be a right canonical Wiener-Hopf factorization in K_1^1. Then

$$f_n(s) = \det(I + sP - su_- u_+ Q_n u_+^{-1} u_-^{-1}) = \det(I + su_-^{-1} P u_- - su_+ Q_n u_+^{-1}),$$

and as $u_-^{-1} P u_- = P + Q u_-^{-1} P u_- P$ and $u_+ Q_n = P u_+ Q_n$, we get

$$f_n(s) = \det(I + sQu_-^{-1} P u_- P + sP - sPu_+ Q_n u_+^{-1}).$$

The operator $I + sQu_-^{-1}Pu_-P$ has the inverse $I - sQu_-^{-1}Pu_-P$ and its determinant is 1. Hence,

$$\begin{aligned}
f_n(s) &= \det(I + (sP - sPu_+ Q_n u_+^{-1})(I - sQu_-^{-1} P u_- P)) \\
&= \det(I + sP - sPu_+ Q_n u_+^{-1} + s^2 P u_+ Q_n u_+^{-1} Q u_-^{-1} P u_- P).
\end{aligned}$$

Since $\det(I + PA) = \det(I + PAP)$ and

$$Pu_-^{\pm 1} = Pu_-^{\pm 1} P, \quad u_+^{\pm 1} P = P u_+^{\pm 1} P, \quad u_-^{\pm 1} Q = Q u_-^{\pm 1} Q, \quad Q u_+^{\pm 1} = Q u_+^{\pm 1} Q,$$

it follows that

$$\begin{aligned}
f_n(s) &= \det(I + sP - sPu_+ Q_n u_+^{-1} P + s^2 P u_+ Q_n u_+^{-1} Q u_-^{-1} P u_- P) \\
&= \det(I + sP - sPu_+ Q_n u_+^{-1} P - s^2 P u_+ Q_n u_+^{-1} Q u_-^{-1} Q u_- P) \\
&= \det(I + sP - sQ_n - s^2 P u_+^{-1} P u_+ Q_n u_+^{-1} Q u_-^{-1} Q u_- P u_+ P) \\
&= \det(I + sP - sQ_n - s^2 Q_n u_+^{-1} Q u_-^{-1} Q u_- u_+ P) \\
&= \det(I + sP - sQ_n - s^2 Q_n u_+^{-1} u_-^{-1} Q u_- u_+ P) \\
&= \det(I + sP - sQ_n - s^2 Q_n a^{-1} Q a P) \\
&= \det(I + sP - sQ_n) \det(I - s^2 Q_n a^{-1} Q a P) \\
&= (1 + s)^{nN} \det(I - s^2 Q_n a^{-1} Q a Q_n).
\end{aligned}$$

$\square$

At this point we have proved formula (6) for $n \geq 0$ and thus formula (5) and Theorem 1.2. We are left with switching from (6) to (7) and passing to negative n's.

Proposition 3.3. *If $a \in \Phi_l(K_1^1)$ and $n \geq 0$, then*

$$f_{-n}(s) = (1-s)^{nN} \det(I - s^2(I - Q_{-n})a^{-1}Pa(I - Q_{-n})). \tag{10}$$

Proof. We repeat the argument of the preceding proof, but now we work with the left canonical Wiener-Hopf factorization $a = v_+v_-$. It results that

$$\begin{aligned}
f_{-n}(s) &= \det(I + sP - saQ_{-n}a^{-1}) \\
&= \det(I - sQ + sa(I - Q_{-n})a^{-1}) \\
&= \det(I - sQ + sv_+v_-(I - Q_{-n})v_-^{-1}v_+^{-1}) \\
&= \det(I - sv_+^{-1}Qv_+ + sv_-(I - Q_{-n})v_-^{-1}) \\
&= \det(I - sPv_+^{-1}Qv_+Q - sQ + sv_-(I - Q_{-n})v_-^{-1}) \\
&= \det(I + (-sQ + sQv_-(I - Q_{-n})v_-^{-1})(I + sPv_+^{-1}Qv_+Q)) \\
&= \det(I - sQ + sQv_-(I - Q_{-n})v_-^{-1}Q + s^2Qv_-(I - Q_{-n})v_-^{-1}Pv_+^{-1}Pv_+Q) \\
&= \det(I - sQ + s(I - Q_{-n}) - s^2Qv_-^{-1}Qv_-(I - Q_{-n})v_-^{-1}Pv_+^{-1}Pv_+Qv_-Q) \\
&= \det(I - sQ + s(I - Q_{-n}) - s^2(I - Q_{-n})v_-^{-1}v_+^{-1}Pv_+v_-Q) \\
&= \det(I - sQ + s(I - Q_{-n}) - s^2(I - Q_{-n})a^{-1}PaQ) \\
&= \det(I - sQ + s(I - Q_{-n}))\det(I - s^2(I - Q_{-n})a^{-1}PaQ) \\
&= (1-s)^{nN}\det(I - s^2(I - Q_{-n})a^{-1}Pa(I - Q_{-n})).
\end{aligned}$$

□

Lemma 3.4. *If a and a^{-1} are in K_1^1 and n is in $\mathbf{Z}$, then*

$$f_n(-s)f_n(s) = \det(I - s^2(I - Q_n)a^{-1}Pa(I - Q_n))\det(I - s^2Q_na^{-1}QaQ_n).$$

Proof. We have

$$\begin{aligned}
&(I - sP + saQ_na^{-1})(I + sP - saQ_na^{-1}) \\
&= I - s^2P + s^2PaQ_na^{-1}P - s^2QaQ_na^{-1}Q \\
&= I - s^2Pa(I - Q_n)a^{-1}P - s^2QaQ_na^{-1}Q.
\end{aligned}$$

Taking determinants we obtain that

$$\begin{aligned}
f_n(-s)f_n(s) &= \det(I - s^2Pa(I - Q_n)a^{-1}P)\det(I - s^2QaQ_na^{-1}Q) \\
&= \det(I - s^2(I - Q_n)a^{-1}Pa(I - Q_n))\det(I - s^2Q_na^{-1}QaQ_n).
\end{aligned}$$

□

Proposition 3.5. *If $a \in \Phi_l(K_1^1)$ and $n \geq 0$, then*

$$(1+s)^{nN}f_{-n}(s) = \det(I - s^2Q_{-n}a^{-1}QaQ_{-n}), \tag{11}$$

and if $a \in \Phi_r(K_1^1)$ and $n \geq 0$, then

$$f_n(s) = (1-s)^{nN}\det(I - s^2(I - Q_n)a^{-1}Pa(I - Q_n)). \tag{12}$$

Proof. Proposition 3.3 and Lemma 3.4 give

$$f_{-n}(s)(1+s)^{nN}\det(I-s^2(I-Q_{-n})a^{-1}Pa(I-Q_{-n}))=f_{-n}(s)f_{-n}(-s)$$
$$=\det(I-s^2(I-Q_{-n})a^{-1}Pa(I-Q_{-n}))\det(I-s^2Q_{-n}a^{-1}QaQ_{-n}).$$

Since $\det(I-s^2(I-Q_{-n})a^{-1}Pa(I-Q_{-n}))\neq 0$ for sufficiently small s, we get (11) for these s and then by analytic continuation for all s. Analogously, using Propositions 3.2 and 3.3 we get

$$f_n(s)(1-s)^{nN}\det(I-s^2Q_na^{-1}QaQ_n)=f_n(s)f_n(-s)$$
$$=\det(I-s^2(I-Q_n)a^{-1}Pa(I-Q_n))\det(I-s^2Q_na^{-1}QaQ_n),$$

which implies (12). □

Obviously, Theorem 1.3 is the union of Proposition 3.2, Lemma 3.4, and Proposition 3.5.

4. Non-invertible operators

The hypothesis of Theorem 1.3 is that a be in $\Phi_r(K_1^1)\cap\Phi_l(K_1^1)$, which is equivalent to the invertibility of both $T(a)$ and $T(a^{-1})$. The theorem of this section, which is also from [1], relaxes this hypothesis essentially: we only require that $T(a)$ be Fredholm (which automatically implies that $T(a^{-1})$ is also Fredholm). Notice that if a is continuous (and matrix functions in K_1^1 are continuous) then $T(a)$ is a Fredholm operator if and only if $\det a$ has no zeros on $\mathbf{T}$. In that case the index of $T(a)$ is minus the winding number of $\det a$, $\operatorname{Ind}T(a)=-\operatorname{wind}\det a$.

Lemma 4.1. *If $a\in K_1^1$ and $T(a)$ is Fredholm of index zero, then* (6) *and* (7) *are valid.*

Proof. A theorem by Widom [6] tells us that there exist a trigonometric polynomial φ and a number $\varrho>0$ such that $T(a+\varepsilon\varphi)$ is invertible for all complex numbers ε satisfying $0<|\varepsilon|<\varrho$. Since $T(a+\varepsilon\varphi)$ is invertible, we conclude that $a+\varepsilon\varphi\in\Phi_r(K_1^1)$. Thus, (9) and (12) are true with a replaced by $a+\varepsilon\varphi$. From the proof of Lemma 3.1 we see that

$$L(a+\varepsilon\varphi)Q_nL((a+\varepsilon\varphi)^{-1})\to L(a)Q_nL(a^{-1}),$$
$$Q_nL((a+\varepsilon\varphi)^{-1})QL(a+\varepsilon\varphi)Q_n\to Q_nL(a^{-1})QL(a)Q_n$$

in the trace norm as $\varepsilon\to 0$. This gives (9) and (12). The proof of formulas (10) and (11) is analogous. □

Lemma 4.2. *If the scalar-valued function $a\in K_1^1$ has no zeros on the unit circle and winding number w about the origin, then for all $n\in\mathbf{Z}$,*

$$\det(I+sP-sL(a)Q_nL(a^{-1}))$$
$$=(1+s)^{n+w}\det(I-s^2Q_nL(a^{-1})QL(a)Q_n)\quad(s\neq -1),\tag{13}$$
$$\det(I+sP-sL(a)Q_nL(a^{-1}))$$
$$=(1-s)^{-n-w}\det(I-s^2(I-Q_n)L(a^{-1})PL(a)(I-Q_n))\quad(s\neq 1).\tag{14}$$

Proof. Recall that χ_w is defined by $\chi_w(t) = t^w$. We can write $a = \chi_w b$ with wind $b = 0$. The key observation is that $\chi_w Q_n \chi_{-w} = Q_{n+w}$. Consequently,

$$\begin{aligned}
&\det(I + sP - saQ_n a^{-1}) \\
&\quad = \det(I + sP - sb\chi_w Q_n \chi_{-w} b^{-1}) \\
&\quad = \det(I + sP - sbQ_{n+w} b^{-1}) \\
&\quad = (1+s)^{n+w} \det(I - s^2 Q_{n+w} b^{-1} Q b Q_{n+w}) \quad \text{(by Theorem 1.3)} \\
&\quad = (1+s)^{n+w} \det(I - s^2 \chi_w Q_n \chi_{-w} b^{-1} Q b \chi_w Q_n \chi_{-w}) \\
&\quad = (1+s)^{n+w} \det(I - s^2 Q_n a^{-1} Q a Q_n),
\end{aligned}$$

which is (13). Analogously one can derive (14) from Theorem 1.3. □

Theorem 4.3. (Baik-Deift-Rains) *Let* a *be an* $N \times N$ *matrix function in* K_1^1 *and suppose* $\det a$ *has no zeros on* $\mathbf{T}$. *Put* $w = \text{wind} \det a$. *Then for all* $n \in \mathbf{Z}$,

$$\begin{aligned}
&\det(I + sP - sL(a)Q_n L(a^{-1})) \\
&\quad = (1+s)^{nN+w} \det(I - s^2 Q_n L(a^{-1}) Q L(a) Q_n) \quad (s \neq -1), \qquad (15) \\
&\det(I + sP - sL(a)Q_n L(a^{-1})) \\
&\quad = (1-s)^{-nN-w} \det(I - s^2 (I - Q_n) L(a^{-1}) P L(a) (I - Q_n)) \quad (s \neq 1). \qquad (16)
\end{aligned}$$

Proof (after Percy Deift). We extend a to an $(N+1) \times (N+1)$ matrix function c by adding the $N+1, N+1$ entry χ_{-w}:

$$c = \begin{pmatrix} a & 0 \\ 0 & \chi_{-w} \end{pmatrix}.$$

Since $T(c)$ is Fredholm of index zero, we deduce from Lemma 4.1 that

$$\det(I + sP - scQ_n c^{-1}) = (1+s)^{n(N+1)} \det(I - s^2 Q_n c^{-1} Q c Q_n). \qquad (17)$$

Obviously,

$$\begin{aligned}
&\det(I + sP - scQ_n c^{-1}) \\
&\quad = \det(I + sP - saQ_n a^{-1}) \det(I + sP - s\chi_{-w} Q_n \chi_w), \qquad (18) \\
&\det(I - s^2 Q_n c^{-1} Q c Q_n) \\
&\quad = \det(I - s^2 Q_n a^{-1} Q a Q_n) \det(I - s^2 Q_n \chi_w Q \chi_{-w} Q_n). \qquad (19)
\end{aligned}$$

Lemma 4.2 implies that

$$\det(I + sP - s\chi_{-w} Q_n \chi_w) = (1+s)^{n-w} \det(I - s^2 Q_n \chi_w Q \chi_{-w} Q_n) \qquad (20)$$

(which, by the way, can also be verified straightforwardly in the particular case at hand). Combining (17), (18), (19), (20) we arrive at (15). The proof of (16) is analogous. □

Acknowledgement. I am grateful to Estelle Basor, Percy Deift, and Harold Widom for encouraging discussions and valuable comments.

References

[1] J. Baik, P. Deift, and E.M. Rains, A Fredholm determinant identity and the convergence of moments for random Young tableaux. *Comm. Math. Phys.*, **223** (2001), 627–672.

[2] E.L. Basor and H. Widom, On a Toeplitz determinant identity of Borodin and Okounkov, *Integral Equations Operator Theory,* **37** (2000), 397–401.

[3] A. Borodin and A. Okounkov, A Fredholm determinant formula for Toeplitz determinants, *Integral Equations Operator Theory,* **37** (2000), 386–396.

[4] A. Böttcher, One more proof of the Borodin-Okounkov formula for Toeplitz determinants, *Integral Equations Operator Theory,* **41** (2001), 123–125.

[5] A. Böttcher and B. Silbermann, *Analysis of Toeplitz Operators,* Springer-Verlag, Berlin, Heidelberg, New York 1990.

[6] H. Widom, Perturbing Fredholm operators to obtain invertible operators. *J. Funct. Analysis,* **20** (1975), 26–31.

[7] H. Widom, Asymptotic behavior of block Toeplitz matrices and determinants II. *Adv. Math.,* **21** (1976), 1–29.

A. Böttcher
Fakultät für Mathematik
TU Chemnitz
D-09107 Chemnitz, Germany
e-mail: aboettch@mathematik.tu-chemnitz.de

Received: 1 September 2001

Operator Theory:
Advances and Applications, Vol. 135, 101–106

On the Distance of a Large Toeplitz Band Matrix to the Nearest Singular Matrix

A. Böttcher, S. Grudsky and A. Kozak

Dedicated to Bernd Silbermann on His Sixtieth Birthday

Abstract. If the symbol of an infinite Toeplitz band matrix has nonzero winding number, then the distances of the $n \times n$ truncations of this matrix to the nearest singular matrices go to zero with exponential speed as $n \to \infty$. The purpose of this note is to show that if we measure the distance to the nearest singular matrix within the matrices of the same Toeplitz band structure as the original matrix, then things are less dramatic: this distance stays away from zero if only the origin does not belong to the Schmidt-Spitzer set.

1. Introduction and main results

Given a complex-valued continuous function c on the complex unit circle $\mathbf{T}$, we denote by $\{c_k\}_{k=-\infty}^{\infty}$ the sequence of its Fourier coefficients and by $T_n(c) = (c_{j-k})_{j,k=1}^{n}$ and $T(c) = (c_{j-k})_{j,k=1}^{\infty}$ the $n \times n$ Toeplitz matrix and the infinite Toeplitz matrix generated by c, respectively. The function c is referred to as the symbol of the Toeplitz matrices under consideration. We think of $T(c)$ as an operator on $\ell^2(\mathbf{N})$ and of $T_n(c)$ as an operator on $\mathbf{C}^n$ with the ℓ^2 norm. Fix two integers r, s such that $r \le 0 \le s$ and let $\mathcal{P}_{r,s}$ stand for the set of all trigonometric polynomials b of the form $b(t) = \sum_{k=r}^{s} b_k t^k$ $(t \in \mathbf{T})$.

The distance $d(A_n)$ of an $n \times n$ matrix A_n to the nearest singular matrix is defined as the infimum of the set of all $\varepsilon > 0$ for which the so-called ε-pseudospectrum

$$\mathrm{sp}_\varepsilon A_n = \bigcup_{\|K_n\| \le \varepsilon} \mathrm{sp}\,(A_n + K_n)$$

contains the origin; here sp denotes the spectrum, K_n stands for an $n \times n$ matrix, and $\|\cdot\|$ is the operator (= spectral) norm on the Hilbert space $\mathbf{C}^n$. It is well known (and easily seen) that $d(A_n)$ equals the smallest singular value of A_n. In case A_n is invertible, the smallest singular value of A_n is equal to $\|A_n^{-1}\|^{-1}$.

Now let $b \in \mathcal{P}_{r,s}$ and suppose b has no zeros on $\mathbf{T}$, that is, $0 \notin b(\mathbf{T})$. Let $\operatorname{wind} b$ denote the winding number of b about the origin. If $\operatorname{wind} b = 0$, then $T(b)$ is invertible (see, e.g., [3], [5]) and one can show (see [1] or [3]) that, as n goes to ∞, $d(T_n(b))$ converges to $\|T^{-1}(b)\|^{-1}$, which is a finite and nonzero number. On the other hand, if $\operatorname{wind} b \neq 0$, then $d(T_n(b))$ approaches zero with exponential speed (see [2], [11]). However, when working with Toeplitz matrices in practice, it is often reasonable not to admit general perturbations to the matrix, but only perturbations of the same structure as the matrix itself. For several approaches to the problem of studying the distance to the nearest singular matrix within specific (and not necessarily Toeplitz) structures we refer to [4], [6], [7], [9], [10], [12], [13], [14], for example. We here proceed as follows. For $b \in \mathcal{P}_{r,s}$, we define the "Toeplitz" ε-pseudospectrum by

$$\operatorname{sp}_\varepsilon^\tau T_n(b) = \bigcup_{\varphi \in \mathcal{P}_{r,s},\, \|\varphi\|_\infty \leq \varepsilon} \operatorname{sp} T_n(b+\varphi)$$

and, accordingly, the "Toeplitz" distance to the nearest singular matrix by

$$d_\tau(T_n(b)) = \inf\{\varepsilon > 0 : 0 \in \operatorname{sp}_\varepsilon^\tau T_n(b)\};$$

the τ is for Toeplitz.

The spectrum of the Toeplitz operator $T(b)$ is the union of $b(\mathbf{T})$ and all points in $\mathbf{C} \setminus b(\mathbf{T})$ that are encircled by $b(\mathbf{T})$ with nonzero winding number (see, e.g., [3], [5]). Schmidt and Spitzer [15] proved that if $b \in \mathcal{P}_{r,s}$, then, as $n \to \infty$, the spectrum of $T_n(b)$ converges in the Hausdorff metric to some limiting set $\Lambda(b)$, which is given by

$$\Lambda(b) = \bigcap_{\varrho > 0} \operatorname{sp} T(b_\varrho), \quad b_\varrho(t) := \sum_{k=r}^{s} b_k \varrho^k t^k \quad (t \in \mathbf{T}). \tag{1}$$

For instance, if $b \in \mathcal{P}_{-1,1}$, that is, if $T(b)$ is tridiagonal, then $\operatorname{sp} T(b)$ is the boundary and the interior of an ellipse, while $\Lambda(b)$ is the line segment between the foci of this ellipse. Thirteen more examples of sets $\Lambda(b)$ are on pages 155 to 157 of the text [3]. For $b \in \mathcal{P}_{r,s}$, we define

$$\Lambda_\varepsilon^\tau(b) = \bigcup_{\varphi \in \mathcal{P}_{r,s},\, \|\varphi\|_\infty \leq \varepsilon} \Lambda(b+\varphi).$$

Here are our main results.

Theorem 1.1. *If $b \in \mathcal{P}_{r,s}$, then the sets $\operatorname{sp}_\varepsilon^\tau T_n(b)$ converge to $\Lambda_\varepsilon^\tau(b)$ in the Hausdorff metric as $n \to \infty$ and*

$$\lim_{n\to\infty} d_\tau(T_n(b)) = \inf\{\varepsilon > 0 : 0 \in \Lambda_\varepsilon^\tau(b)\}. \tag{2}$$

We always have

$$\operatorname{dist}(0, \operatorname{sp} T(b)) \leq \lim_{n\to\infty} d_\tau(T_n(b)) \leq \operatorname{dist}(0, \Lambda(b)), \tag{3}$$

and this estimate is sharp.

Theorem 1.2. *Let $b \in \mathcal{P}_{r,s}$. If $0 \in \Lambda(b)$, then $d_\tau(T_n(b)) \to 0$ as $n \to \infty$. However, if $0 \notin \Lambda(b)$, then the distances $d_\tau(T_n(b))$ converge to some positive number as n goes to ∞.*

Thus, when measuring the distance of $T_n(b-\lambda)$ to the nearest singular matrix not within the set of all matrices but only within the set of the Toeplitz matrices of the same structure, then it is no longer all points λ in $\operatorname{sp} T(b)$ that are critical, but only the points λ in the Schmidt-Spitzer set $\Lambda(b)$.

2. Proofs

Let $\{E_n\}_{n=1}^{\infty}$ be a sequence of compact subsets of $\mathbf{C}$. We denote by $\liminf E_n$ the set of all points $\lambda \in \mathbf{C}$ that are the limit of some sequence $\{\lambda_n\}_{n=1}^{\infty}$ with $\lambda_n \in E_n$, and we denote by $\limsup \operatorname{sp}_\varepsilon^\tau T_n(b)$ the set of all points $\lambda \in \mathbf{C}$ that are a partial limit of some sequence $\{\lambda_n\}_{n=1}^{\infty}$ with $\lambda_n \in E_n$. It is well known (see, e.g., [8, Sections 3.1.1 and 3.1.2]) that E_n converges to some compact subset E of $\mathbf{C}$ in the Hausdorff metric if and only if $\liminf E_n = \limsup E_n = E$. Put $U_\sigma(\lambda_0) = \{\lambda \in \mathbf{C} : |\lambda - \lambda_0| < \sigma\}$.

Lemma 2.1. *Let $b \in \mathcal{P}_{r,s}$. If $\lambda_0 \notin \Lambda(b)$, then there exist $n_0 \in \mathbf{N}$, $\sigma > 0$, $\delta > 0$ such that $U_\sigma(\lambda_0) \cap \operatorname{sp} T_n(b+\varphi) = \emptyset$ whenever $\varphi \in \mathcal{P}_{r,s}$, $\|\varphi\|_\infty \le \delta$, and $n \ge n_0$.*

Proof. From (1) we infer that there is a $\varrho \in (0,\infty)$ such that $\lambda_0 \notin \operatorname{sp} T(b_\varrho)$. Hence, the finite section method is applicable to $T(b_\varrho - \lambda_0)$, which means that there exist $n_0 \in \mathbf{N}$ and $M \in (0,\infty)$ such that $\|T_n^{-1}(b_\varrho - \lambda_0)\| \le M$ for all $n \ge n_0$ (see, e.g., [3], [5]). Put $\sigma = 1/(4M)$ and suppose that $\lambda \in U_\sigma(\lambda_0)$ and $\varphi \in \mathcal{P}_{r,s}$. We have

$$T_n((b+\varphi)_\varrho - \lambda) = T_n(b_\varrho - \lambda_0) + T_n(\varphi_\varrho) + (\lambda_0 - \lambda) I_n$$

and $\|T_n^{-1}(b_\varrho - \lambda_0)\|^{-1} \ge 1/M > 1/(2M)$. Since $|\lambda_0 - \lambda| < 1/(4M)$ and since $\|T_n(\varphi_\varrho)\| \le \|\varphi_\varrho\|_\infty < 1/(4M)$ if only $\|\varphi\|_\infty \le \delta$ for some sufficiently small number $\delta > 0$, it follows that $T_n((b+\varphi)_\varrho - \lambda)$ is invertible for all $\lambda \in U_\sigma(\lambda_0)$ and all $\varphi \in \mathcal{P}_{r,s}$ with $\|\varphi\|_\infty \le \delta$. As the invertiblity of $T_n(b+\varphi-\lambda)$ is equivalent to the invertibility of $T_n((b+\varphi)_\varrho - \lambda)$ (notice that

$$\begin{aligned} &T_n((b+\varphi)_\varrho - \lambda) \\ &= \operatorname{diag}(1, \varrho, \ldots, \varrho^{n-1})\, T_n(b+\varphi-\lambda)\, \operatorname{diag}(1, \varrho^{-1}, \ldots, \varrho^{-(n-1)}), \end{aligned}$$

so that the two matrices are similar), we arrive at the assertion. □

Theorem 2.2. *If $b \in \mathcal{P}_{r,s}$, then the sets $\operatorname{sp}_\varepsilon^\tau T_n(b)$ converge to $\Lambda_\varepsilon^\tau(b)$ in the Hausdorff metric.*

Proof. Let $\lambda \in \limsup \operatorname{sp}_\varepsilon^\tau T_n(b)$. Then there are functions $\varphi_{n_k} \in \mathcal{P}_{r,s}$ and points $\lambda_{n_k} \in \operatorname{sp} T_{n_k}(b + \varphi_{n_k})$ such that $\|\varphi_{n_k}\|_\infty \le \varepsilon$ and $\lambda_{n_k} \to \lambda$. As the unit ball of $\mathbf{C}^{r+s+1}$ is compact, the sequence $\{\varphi_{n_k}\}$ has a subsequence $\{\varphi_{n_{k_\ell}}\}$ converging to some $\varphi \in \mathcal{P}_{r,s}$ with $\|\varphi\|_\infty \le \varepsilon$. We claim that $\lambda \in \Lambda(b+\varphi)$. Indeed, if $\lambda \notin \Lambda(b+\varphi)$ then, by Lemma 2.1, $U_\sigma(\lambda) \cap \operatorname{sp} T_{n_{k_\ell}}(b+\varphi_{n_{k_\ell}}) = \emptyset$ for some $\sigma > 0$ and all sufficiently large n_{k_ℓ}, which is impossible because $\lambda_{n_{k_\ell}} \to \lambda$. Thus, we have proved that λ is in $\Lambda_\varepsilon^\tau(b)$.

Now let $\lambda \in \Lambda(b+\varphi)$ for some $\varphi \in \mathcal{P}_{r,s}$ with $\|\varphi\|_\infty \le \varepsilon$. Then there are λ_n in $\operatorname{sp} T_n(b+\varphi)$ such that $\lambda_n \to \lambda$, which shows that λ is in $\liminf \operatorname{sp}_\varepsilon^\tau T_n(b)$. □

Theorem 2.3. *If $b \in \mathcal{P}_{r,s}$, then* $\lim d_\tau(T_n(b)) = \inf\{\varepsilon > 0 : 0 \in \Lambda_\varepsilon^\tau(b)\}$.

Proof. Clearly,

$$\inf\{\varepsilon > 0 : 0 \in \Lambda_\varepsilon^\tau(b)\} = \inf\{\|f\|_\infty : f \in \mathcal{P}_{r,s}, 0 \in \Lambda(b-f)\} =: \varrho.$$

Let $f \in \mathcal{P}_{r,s}$, $0 \in \Lambda(b-f)$, $\|f\|_\infty < \varrho + \varepsilon$. There are $\lambda_n \in \operatorname{sp} T_n(b-f)$ such that $\lambda_n \to 0$. Since $0 \in \operatorname{sp} T_n(b - f - \lambda_n)$, we have $d_\tau(T_n(b)) \le \|f + \lambda_n\|_\infty$, whence $\limsup d_\tau(T_n(b)) \le \|f\|_\infty < \varrho + \varepsilon$. As $\varepsilon > 0$ can be chosen arbitrarily, it follows that $\limsup d_\tau(T_n(b)) \le \varrho$.

Let $\sigma := \liminf d_\tau(T_n(b))$. Given any $\varepsilon > 0$, there exist $f_{n_k} \in \mathcal{P}_{r,s}$ such that $0 \in \operatorname{sp} T_{n_k}(b - f_{n_k})$ and $\|f_{n_k}\|_\infty < \sigma + \varepsilon$. The functions f_{n_k} converge in L^∞ to some $f \in \mathcal{P}_{r,s}$ satisfying $\|f\|_\infty \le \sigma + \varepsilon$. Assume $0 \notin \Lambda(b-f)$. Then, by Lemma 2.1, $U_\sigma(0) \cap \operatorname{sp} T_{n_k}(b - f_{n_k}) = \emptyset$ for all sufficiently large n_k, which is impossible. Hence $0 \in \Lambda(b-f)$. We arrive at the conclusion that $\varrho \le \sigma + \varepsilon$, and as $\varepsilon > 0$ is an arbitrary number, it follows that $\varrho \le \sigma$. This completes the proof. □

Theorems 2.2 and 2.3 prove the first assertion of Theorem 1.1 and equality (2). Estimates (3) can be easily verified: since $\Lambda(b-\delta) = \Lambda(b) - \delta$ for every $\delta \in \mathbf{C}$, it follows that always $\lim d_\tau(T_n(b)) \le \operatorname{dist}(0, \Lambda(b))$, and as $\Lambda(b+\varphi) \subset \operatorname{sp} T(b+\varphi)$ and the union of the spectra $\operatorname{sp} T(b+\varphi)$ over all $\varphi \in \mathcal{P}_{r,s}$ with $\|\varphi\|_\infty \le \varepsilon$ coincides with $\operatorname{sp} T(b) + \{\lambda \in \mathbf{C} : |\lambda| \le \varepsilon\}$, we see that $\operatorname{dist}(0, \operatorname{sp} T(b)) \le \lim d_\tau(T_n(b))$. Finally, the following two examples show that estimate (3) is sharp.

Example 2.4. Let $b \in \mathcal{P}_{0,1}$ be given by $b(t) = b_0 + b_1 t$ with $|b_0| > |b_1| > 0$. We put $\overline{\mathbf{D}} = \{\lambda \in \mathbf{C} : |\lambda| \le 1\}$. Then

$$\begin{aligned} &\operatorname{sp} T(b) = b_0 + b_1 \overline{\mathbf{D}}, \quad \Lambda(b) = \{b_0\} \\ &\operatorname{dist}(0, \operatorname{sp} T(b)) = |b_0| - |b_1|, \quad \operatorname{dist}(0, \Lambda(b)) = |b_0| \\ &\Lambda_\varepsilon^\tau(b) = |b_0| + \varepsilon \overline{\mathbf{D}}, \end{aligned}$$

and hence $\operatorname{dist}(0, \operatorname{sp} T(b)) < \lim d_\tau(T_n(b)) = \operatorname{dist}(0, \Lambda(b))$. □

Example 2.5. Define $b \in \mathcal{P}_{-1,1}$ by $b(t) = 4 + 2t + t^{-1}$. The range $b(\mathbf{T})$ is the ellipse $\{4 + 3\cos\theta + i \sin\theta : \theta \in [0, 2\pi)\}$. Thus, $\operatorname{dist}(0, \operatorname{sp} T(b)) = 1$. If $c(t) = c_{-1}t^{-1} + c_0 + c_1 t$, then the line segment between the foci of the ellipse $c(\mathbf{T})$ is

$$\Lambda(c) = [c_0 - 2\sqrt{c_1 c_{-1}}, c_0 + 2\sqrt{c_1 c_{-1}}\,]. \tag{4}$$

It follows that in the case at hand $\Lambda(b) = [4 - 2\sqrt{2}, 4 + 2\sqrt{2}\,]$, whence $\operatorname{dist}(0, \Lambda(b)) = 4 - 2\sqrt{2} > 1$. Put $\varphi(t) = t$. Then $\varphi \in \mathcal{P}_{-1,1}$ and from (4) we infer that $\Lambda(b+\varphi) = [4 - 2\sqrt{2 \cdot 2}, 4 + 2\sqrt{2 \cdot 2}\,] = [0, 8]$, which implies that $\lim d_\tau(T_n(b)) \le \|\varphi\|_\infty = 1$. Consequently, $\operatorname{dist}(0, \operatorname{sp} T(b)) = \lim d_\tau(T_n(b)) < \operatorname{dist}(0, \Lambda(b))$. □

We now turn to the proof of Theorem 1.2. If $0 \in \Lambda(b)$, then $d_\tau(T_n(b)) \to 0$ due to (3). To tackle the case where $0 \notin \Lambda(b)$, we need one more lemma. Let $\mathbf{D}$ be the open unit disk.

Lemma 2.6. *The function $b \mapsto \Lambda(b)$ is upper-semi continuous on $\mathcal{P}_{r,s}$, that is, given $b \in \mathcal{P}_{r,s}$ and $\varepsilon > 0$, there is a $\delta > 0$ such that $\Lambda(b+\varphi) \subset \Lambda(b) + \varepsilon\overline{\mathbf{D}}$ whenever $\varphi \in \mathcal{P}_{r,s}$ and $\|\varphi\|_\infty \le \delta$.*

Proof. The case where b vanishes identically is trivial. We may therefore without loss of generality assume that $\varepsilon \in (0, \|b\|_\infty)$. Let K be the compact set

$$K = \{\lambda \in \mathbf{C} : |\lambda| \le 2\|b\|_\infty, \lambda \notin \Lambda(b) + \varepsilon\mathbf{D}\}.$$

By virtue of Lemma 2.1, for each $\lambda \in K$ there are $n \in \mathbf{N}$, $\sigma > 0$, $\delta > 0$ such that $U_\sigma(\lambda) \cap \operatorname{sp} T_k(b+\varphi) = \emptyset$ for all $\varphi \in \mathcal{P}_{r,s}$ with $\|\varphi\|_\infty \le \delta$ and all $k \ge n$. Since K is compact, we can find finitely many $\lambda_j \in K$, $n_j \in \mathbf{N}$, $\sigma_j > 0$, $\delta_j > 0$ $(j = 1, \dots, m)$ such that

$$K \subset \bigcup_{j=1}^{m} U_{\sigma_j}(\lambda_j), \quad U_{\sigma_j}(\lambda_j) \cap \operatorname{sp} T_k(b+\varphi) = \emptyset$$

for $\varphi \in \mathcal{P}_{r,s}$, $\|\varphi\|_\infty \le \delta_j$, $k \ge n_j$. Put

$$n_0 = \max n_j, \quad \delta = \min\{\delta_1, \dots, \delta_m, \|b\|_\infty\}.$$

If $\lambda \in K$, then $\lambda \in U_{\sigma_j}(\lambda_j)$ for some j and hence $T_k(b+\varphi-\lambda)$ is invertible for $\varphi \in \mathcal{P}_{r,s}$, $\|\varphi\|_\infty \le \delta$, $k \ge n_0$. If $\lambda > 2\|b\|_\infty$, then $T_k(b+\varphi-\lambda)$ is invertible for all k and all $\varphi \in \mathcal{P}_{r,s}$ with $\|\varphi\|_\infty \le \delta$ because

$$\|T_k(b+\varphi)\| \le \|b\|_\infty + \|\varphi\|_\infty \le 2\|b\|_\infty < |\lambda|.$$

Thus, we have shown that if $\lambda \notin \Lambda(b)+\varepsilon\mathbf{D}$, then $T_k(b+\varphi-\lambda)$ is invertible for $k \ge n_0$ and $\varphi \in \mathcal{P}_{r,s}$, $\|\varphi\|_\infty \le \delta$. Consequently, $\operatorname{sp} T_k(b+\varphi) \subset \Lambda(b) + \varepsilon\overline{\mathbf{D}}$ for all $k \ge n_0$ and all $\varphi \in \mathcal{P}_{r,s}$ satisfying $\|\varphi\|_\infty \le \delta$. This implies that $\Lambda(b+\varphi) \subset \Lambda(b) + \varepsilon\overline{\mathbf{D}}$ for $\varphi \in \mathcal{P}_{r,s}$ with $\|\varphi\|_\infty \le \delta$. □

We can now complete the proof of Theorem 1.2. If $0 \notin \Lambda(b)$, then there is an $\varepsilon > 0$ such that $0 \notin \Lambda(b) + \varepsilon\overline{\mathbf{D}}$. Hence, by Lemma 2.6, $0 \notin \Lambda_\delta^\tau(b)$ for some $\delta > 0$. From (2) we finally conclude that $\lim d_\tau(T_n(b)) \ge \delta$.

References

[1] A. Böttcher, Pseudospectra and singular values of large convolution operators. *J. Integral Equations Appl.*, **6** (1994), 267–301.

[2] A. Böttcher and S. Grudsky, Toeplitz band matrices with exponentially growing condition numbers. *Electronic J. Linear Algebra,* **5** (1999), 104–125.

[3] A. Böttcher and B. Silbermann, *Introduction to Large Truncated Toeplitz Matrices.* Universitext, Springer-Verlag, New York 1999.

[4] J. Demmel, The componentwise distance to the nearest singular matrix. *SIAM J. Matrix Anal. Appl.*, **13** (1992), 10–19.

[5] I. Gohberg and I.A. Feldman, *Convolution Equations and Projection Methods for Their Solution.* Amer. Math. Soc., Providence, RI 1974.

[6] I. Gohberg and I. Koltracht, Mixed, componentwise, and structured condition numbers. *SIAM J. Matrix Anal. Appl.*, **14** (1993), 688–704.

[7] I. Gohberg and I. Koltracht, Structured condition numbers for linear matrix structures. In: Linear Algebra for Signal Processing (Minneapolis, 1992), pp. 17–26, *IMA Vol. Math. Appl.*, **69**, Springer-Verlag, New York 1995.

[8] R. Hagen, S. Roch and B. Silbermann, *C^*-Algebras and Numerical Analysis.* Monographs and Textbooks in Pure and Applied Mathematics, Vol. 236, Marcel Dekker, New York 2001.

[9] D.J. Higham and N.J. Higham, Backward error and condition of structured linear systems. *SIAM J. Matrix Anal. Appl.*, **13** (1992), 162–175.

[10] N.J. Higham, *Accuracy and Stability of Numerical Algorithms.* SIAM, Philadelphia, PA 1996.

[11] L. Reichel and L.N. Trefethen, Eigenvalues and pseudo-eigenvalues of Toeplitz matrices. *Linear Algebra Appl.*, **162** (1992), 153–185.

[12] S.M. Rump, Estimation of the sensitivity of linear and nonlinear algebraic problems. *Linear Algebra Appl.*, **153** (1991), 1–34.

[13] S.M. Rump, Bounds for the componentwise distance to the nearest singular matrix. *SIAM J. Matrix Anal. Appl.*, **18** (1997), 83–103.

[14] S.M. Rump, Structured perturbations and symmetric matrices. *Linear Algebra Appl.*, **278** (1998), 121–132.

[15] P. Schmidt and F. Spitzer, The Toeplitz matrices of an arbitrary Laurent polynomial. *Math. Scand.*, **8** (1960), 15–38.

A. Böttcher
Fakultät für Mathematik
TU Chemnitz
D-09107 Chemnitz, Germany
e-mail: `aboettch@mathematik.tu-chemnitz.de`

S. Grudsky and A. Kozak
Faculty of Mechanics and Mathematics
Rostov-on-Don State University
B. Sadovaya, 105
344711 Rostov-on-Don, Russia
e-mail: `grudsky@aaanet.ru`
e-mail: `kozak@aanet.ru`

Received: 24 October 2001

Operator Theory:
Advances and Applications, Vol. 135, 107–144

Singular Integral Equations on Piecewise Smooth Curves in Spaces of Smooth Functions

L.P. Castro, R. Duduchava and F.-O. Speck

To Bernd Silbermann on the occasion of his sixtieth birthday

Abstract. We prove the boundedness of the Cauchy singular integral operator in modified weighted Sobolev $\mathbb{KW}^m_p(\Gamma,\rho)$, Hölder-Zygmund $\mathbb{KZ}^0_\mu(\Gamma,\rho)$, Bessel potential $\mathbb{KH}^s_p(\Gamma,\rho)$ and Besov $\mathbb{KB}^s_{p,q}(\Gamma,\rho)$ spaces under the assumption that the smoothness parameters m,μ,s are large. The underlying contour Γ is piecewise smooth with angular points and even with cusps. We obtain Fredholm criteria and an index formula for singular integral equations with piecewise smooth coefficients and complex conjugation in these spaces provided the underlying contour has angular points but no cusps. The Fredholm property and the index turn out to be independent of the integer parts of the smoothness parameters m,μ,s. The results are applied to an oblique derivative problem (the Poincaré problem) in plane domains with angular points and peaks on the boundary.

Introduction

When considering a Cauchy singular integral equation with complex conjugation

$$A\varphi(t) \equiv a(t)\varphi(t) + \frac{b(t)}{\pi i}\int_\Gamma \frac{\varphi(\tau)d\tau}{\tau - t} + \frac{c(t)}{\pi i}\int_\Gamma \frac{\overline{\varphi(\tau)d\tau}}{\overline{\tau} - \overline{t}} = f(t)\,, \quad t\in\Gamma \tag{0.1}$$

on a piecewise smooth contour Γ (see § 1 below) we are restricted in the choice of the spaces where we can solve equation (0.1). Namely, the operator A in equation (0.1) is not bounded in important spaces of smooth functions: in the usual weighted Sobolev $\mathbb{W}^m_p(\Gamma,\rho)$, Hölder-Zygmund $\mathbb{Z}_\mu(\Gamma,\rho)$, Bessel potential $\mathbb{H}^s_p(\Gamma,\rho)$ and Besov $\mathbb{B}^s_{p,q}(\Gamma,\rho)$ spaces for large values of the smoothness parameters $m = 2,3,\ldots$, $\mu > 1$ and $|s| > 1+1/p$. These spaces cannot be even defined properly (i.e., independently of the choice of a parametrization) if Γ has knots, such as angular points or cusps.

This article was written during the second author's visit to Instituto Superior Técnico, U.T.L., and Universidade de Aveiro, Portugal, in May–July 2001. The work was supported by "Fundação para a Ciência e a Tecnologia" through "Centro de Matemática Aplicada" and "UI&D Matemática e Aplicações", respectively.

Even if Γ is sufficiently smooth and the spaces $\mathbb{W}^m_p(\Gamma,\rho)$, $\mathbb{Z}_\mu(\Gamma,\rho)$ etc. can be defined properly, the problem arises again when we take piecewise smooth coefficients $a(t)$, $b(t)$, $c(t)$ with jumps at the knots (for conciseness we relate discontinuity points to the knots of Γ as well).

On the other hand, especially in applications and numerical analysis, it is important to establish additional smoothness properties for the solutions at least outside the knots when the right-hand side f is sufficiently smooth.

We suggest the introduction of weighted spaces $\mathbb{KW}^m_p(\Gamma,\rho)$, $\mathbb{KZ}_\mu(\Gamma,\rho)$, $\mathbb{KH}^s_p(\Gamma,\rho)$, and $\mathbb{KB}^s_{p,q}(\Gamma,\rho)$ with the help of "Fuchs"-derivatives

$$\vartheta(t)\partial_t\varphi(t) := \vartheta(t)\frac{\partial\varphi(t)}{\partial t}, \quad \text{where} \quad \vartheta(t) := \prod_{t_j\in\mathcal{T}_\Gamma}(t-t_j) \tag{0.2}$$

and $\mathcal{T}_\Gamma$ is the collection of knots of Γ, instead of the usual derivatives $\partial_t\varphi(t)$ (see Lemmata 1.2, 1.3, and 2.4). It turns out that the operator A in (0.1) with piecewise smooth coefficients $a,b,c\in\mathbb{PC}^{\mathrm{m}}(\Gamma,\mathcal{T}_\Gamma)$ (and even with $a,b,c\in\mathbb{KPC}^{\mathrm{m}}(\Gamma,\mathcal{T}_\Gamma)$; see § 1 for the definitions) is bounded in the modified spaces $\mathbb{KW}^m_p(\Gamma,\rho)$, $\mathbb{KZ}_\mu(\Gamma,\rho)$, $\mathbb{KH}^s_p(\Gamma,\rho)$, and $\mathbb{KB}^s_{p,q}(\Gamma,\rho)$ provided the smoothness parameters m,μ and s are sufficiently large (see Lemmata 1.2, 1.3, 2.4 and Theorem 3.1). Moreover, the operator defined by (0.1) has one and the same kernel and cokernel in the spaces $\mathbb{KW}^{\widetilde{m}}_p(\Gamma,\rho)$, $\mathbb{KZ}_{\widetilde{\mu}}(\Gamma,\rho)$ and $\mathbb{KH}^{\widetilde{s}}_p(\Gamma,\rho)$, $\mathbb{KB}^{\widetilde{s}}_{p,q}(\Gamma,\rho)$ whatever the integer parts of the smoothness parameters $\widetilde{m}=0,\dots,m$, $0<\widetilde{\mu}\le\mu$, and $|\widetilde{s}|\le s$ are (see Theorem 3.2 and Remarks 3.4, 3.5).

The results on the Fredholm properties and also those on the boundedness of the operator A in the usual (non-modified) weighted Bessel potential and Besov spaces $\mathbb{H}^s_p(\Gamma,\rho)$ and $\mathbb{B}^s_{p,q}(\Gamma,\rho)$ for small s, $1/p-1<s<1/p$, when the multiplication by piecewise continuous functions represents a bounded operator (see Theorem 2.3), are new.

Although the space $\mathbb{KW}^m_p(\Gamma,\rho)$ coincides with $\mathbb{KH}^m_p(\Gamma,\rho)$ for any nonnegative integer m, we formulate the results for the modified Sobolev space $\mathbb{KW}^m_p(\Gamma,\rho)$ because these spaces are more common in applications and the proofs are simpler.

It is well known that the Bessel potential spaces are as natural in the theory of pseudodifferential operators as the Sobolev spaces are in the theory of partial differential operators. The norm in $\mathbb{H}^s_p(\mathbb{R}^2)$ is especially simple for even $s=2m, m=0,1,2,\dots$:

$$\|f\,|\,\mathbb{H}^{2m}_p(\mathbb{R}^2)\| = \|(I-\Delta)^m f\,|\mathbb{L}_p(\mathbb{R}^2)\|, \qquad \Delta=\frac{\partial^2}{\partial x_1^2}+\frac{\partial^2}{\partial x_2^2}.$$

But in the theory of boundary value problems we cannot confine ourselves to the Bessel potential spaces since the traces of functions $\Phi\in\mathbb{H}^s_p(\Omega^\pm)$ on the boundary belong to the Besov spaces $\mathbb{B}^{s-\frac{1}{p}}_{p,p}(\Gamma)$, provided the boundary Γ is sufficiently smooth and $s>1/p$.

The Besov spaces can be considered as the integral analogue of the Zygmund spaces.

In favor of the Hölder-Zygmund spaces we remark that many important operators, including singular integral operators, are unbounded in the spaces $C^m(\Gamma)$ and in the Hölder spaces $H_m(\Gamma)$ but that they are bounded in $\mathbb{Z}_m(\mathbb{R}^n)$ for every integer $m \in \mathbb{N} := \{1, 2, \ldots\}$. The Hölder-Zygmund spaces are the natural extensions of the scale of Hölder spaces to integer values of the smoothness exponent and have an important interpolation property (see § 2).

In § 4 we apply the obtained results to the oblique derivative problem for the Laplacian in domains with piecewise smooth boundary.

The space $\mathcal{L}^1_{p,\beta}(\Gamma) := \mathbb{KW}^1_p(\Gamma, \mathcal{T}_\Gamma, |t - t_1|^{\beta-1})$ (i.e., the particular case where $m = 1$, $\mathcal{T}_\Gamma = \{t_1\}$ and $\rho(t) := |t - t_1|^{\beta-1}$) was applied in [MS1] to the investigation of boundary integral equations. The anisotropic Bessel potential spaces $\mathbb{H}^{(s,\nu),m}_p(\mathcal{M})$, similar to $\mathbb{KW}^m_p(\Gamma)$, were introduced in [CD1] for the multi-dimensional case in which $\mathcal{M} = \mathbb{R}^n_+$ or $\mathcal{M}$ is a manifold with smooth boundary. In [CD1] the boundedness of a certain class of pseudodifferential operators was proved and a Fredholm criterion for them was established. The spaces $L^{p,m}(\mathbb{R}^+)$ and $X^{p,m}_\rho(\mathbb{R}^+)$, also similar to $\mathbb{KW}^m_p(\mathbb{R}^+, \{0\})$, were used by J. Elschner in a spline approximation method for convolution equations (see [El1] and [PS1, Ch. 5]).

Some results of §§ 1-2 were already announced in [DS2].

1. Weighted Sobolev and Hölder-Zygmund spaces

Let Γ be a piecewise smooth curve that consists of a finite union of smooth arcs which have in common at most endpoints, called knots:

$$\Gamma = \bigcup_{j=1}^{\ell} \Gamma_j, \quad \Gamma_j := [t_j, t_{j+1}] = \overset{\frown}{t_j t_{j+1}}, \quad j = 1, \ldots, n, \quad t_{n+1} = t_1.$$

Let $\mathcal{T}_\Gamma := \{t_j\}$ denote the collection of all different knots (i.e., all different endpoints of smooth arcs) of Γ. The curve may contain cusps, i.e., the angles between some arcs are allowed to be 0.

The closed arcs Γ_j, $j = 1, \ldots, n$, between the knots are sufficiently smooth: any parametrizations

$$\omega_j \; : \mathcal{I} := [0,1] \longrightarrow \Gamma_j, \qquad \omega_j(0) = t_j, \quad \omega_j(1) = t_{j+1} \tag{1.1}$$

are μ-smooth, $\omega_j \in \mathbb{Z}_\mu(\mathcal{I})$, $\mu \geq 1$, $j = 1, \ldots, n$, where $\mathbb{Z}_\mu(\mathcal{I})$ denotes the Hölder-Zygmund space (see below).

Let us suppose, until formula (1.7), that Γ is either a single smooth arc or a single smooth closed contour and that the natural parametrization of Γ with the

$\ldots, n.$ h parameter $0 < s \leq \ell$,

$$[0,\ell] \longrightarrow \Gamma, \qquad s \longmapsto t(s), \quad t(0) = t(\ell), \tag{1.2}$$

$t(\cdot) \in \mathbb{Z}_\mu([0,\ell])$.

For $0 < \nu \leq \mu \leq 1$, the Hölder-Zygmund space $\mathbb{Z}_\nu(\Gamma)$ is defined as the space of functions with finite norm

$$\|\psi \mid \mathbb{Z}_\nu(\Gamma)\| := \sup_{t\in\Gamma} |\psi(t)| + \sup_{\substack{0<s\leq\ell\\ h>0}} \frac{|\Delta_h^2 \psi(t(s))|}{h^\nu}, \tag{1.3}$$

$$\Delta_h\varphi(s) := \varphi(s+h) - \varphi(s), \qquad \Delta_h^2\varphi(s) := \varphi(s+h) - 2\varphi(s) + \varphi(s-h).$$

Note that for $\nu = 1$ the definition of $\mathbb{Z}_1(\Gamma)$ requires that Γ be smooth, $t(\cdot) \in C^1([0,\ell])$, and that Γ have no angular points: each knot $t_j \in \mathcal{T}_\Gamma$ is an endpoint of a separate arc.

If Γ is μ-smooth and

$$\mu = m + \nu \geq 1, \quad 0 < \nu \leq 1, \quad m \in \mathbb{N}_0 := \{0, 1, \ldots\}, \tag{1.4}$$

then angular points on Γ are absent and the Hölder-Zygmund space $\mathbb{Z}_\mu(\Gamma)$ is defined as a collection of all functions $\psi(t)$ which have finite norm

$$\|\psi \mid \mathbb{Z}_\mu(\Gamma)\| := \sum_{k=0}^{m} \sup_{t\in\Gamma} |\psi^{(k)}(t)| + \|\psi^{(m)}|\mathbb{Z}_\nu(\Gamma)\|, \tag{1.5}$$

$$\varphi^{(k)}(x) := \partial_x^k \varphi(x) = \frac{\partial^k \varphi(x)}{\partial x^k}.$$

The Hölder space $H_\mu(\Gamma)$ is defined as a collection of all functions $\psi(t)$ which have finite norm

$$\|\psi \mid H_\mu(\Gamma)\| := \sum_{k=0}^{m} \sup_{t\in\Gamma} |\psi^{(k)}(t)| + \sup_{\substack{0<s\leq\ell\\ h>0}} \frac{|\Delta_h \psi^{(m)}(t(s))|}{h^\nu}. \tag{1.6}$$

If $\mu = m + \nu \notin \mathbb{N} = \{1, 2, \ldots\}$ is not an integer, $0 < \nu < 1$, the norms in (1.5) and in (1.6) are equivalent and the spaces coincide: $\mathbb{Z}_\mu(\Gamma) = H_\mu(\Gamma)$ for $\mu \notin \mathbb{N}$ (see [St1]). Note that for an integer $m = 1, 2, \ldots$ the spaces $H_m(\Gamma)$ and $\mathbb{Z}_m(\Gamma)$ differ essentially from each other and from the space $C^m(\Gamma)$ of smooth functions with the natural norm

$$\|\psi \mid C^m(\Gamma)\| := \sum_{k=0}^{m} \sup_{t\in\Gamma} |\psi^{(k)}(t)|$$

(see [St1]) and that we have the following proper embedding instead:

$$C^m(\Gamma) \subset H_m(\Gamma) \subset \mathbb{Z}_m(\Gamma). \tag{1.7}$$

For an arbitrary piecewise smooth curve Γ, we denote by $\mathbb{PX}(\Gamma, \mathcal{T}_\Gamma)$ the space of piecewise smooth functions with jump discontinuities at the knots $t_j \in \mathcal{T}_\Gamma$:

$$\mathbb{PX}(\Gamma, \mathcal{T}_\Gamma) := \{g \in \mathbb{X}(\Gamma_j) \;:\; j = 1, \ldots, n\},$$

where $\mathbb{X}(\Gamma_j)$ denotes one of the spaces $C^m(\Gamma_j)$, $H_\mu(\Gamma_j)$, or $\mathbb{Z}_\mu(\Gamma_j)$, $j =$
For $m = 0$ we use the notation $\mathbb{PC}(\Gamma, \mathcal{T}_\Gamma)$ instead of $\mathbb{PC}^0(\Gamma, \mathcal{T}_\Gamma)$.

The Sobolev space $\mathbb{W}_p^m(\mathcal{I})$ on the unit interval is defined as

$$\mathbb{W}_p^m(\mathcal{I}) := \{\varphi \in \mathbb{L}_p(\mathcal{I}) \;:\; \partial^k\varphi \in \mathbb{L}_p(\mathcal{I}),\;\; k = 0, \dots, m\}$$

and is endowed with the norm

$$\|\varphi \mid \mathbb{W}_p^m(\mathcal{I})\| := \left(\sum_{k=0}^{m} \|\partial_x^k\varphi \mid \mathbb{L}_p(\mathcal{I})\|^p\right)^{\frac{1}{p}} = \left(\sum_{k=0}^{m} \int_0^1 |\partial_x^k\varphi(x)|^p dx\right)^{\frac{1}{p}} .$$

With the help of the parametrization (1.1) we define the Sobolev space $\mathbb{W}_p^m(\Gamma_j)$ on the smooth arcs Γ_j for $1 < p < \infty$ and $m \le \mu$ as the space of all functions φ for which $\varphi(\omega_j(x))$ is in $\mathbb{W}_p^m(\mathcal{I})$. For the entire piecewise smooth curve Γ the space $\mathbb{W}_p^m(\Gamma)$ can be defined only for $m = 0, 1$. In fact, for any parametrization

$$\omega \;:\; \mathcal{R} := [0, R] \longrightarrow \Gamma \tag{1.8}$$

of the entire curve Γ (cf. (1.2)) the derivative $(\partial_t\varphi)(\omega(x)) = [\omega'(x)]^{-1}\partial_x\varphi(\omega(x))$ involves a piecewise continuous factor $[\omega']^{-1} \in \mathbb{PC}(\mathcal{R}, \mathcal{T}_{\mathcal{R}})$, where

$$\mathcal{T}_{\mathcal{R}} := \{x_j \in \mathcal{R} \;:\; \omega(x_j) = t_j \in \mathcal{T}_\Gamma\}$$

is the set of all "knots" of $\mathcal{R}$. Therefore, the second derivative $(\partial_t^2\varphi)(\omega(x))$ is not defined properly because the second derivative $\partial_x^2\omega(x)$ of the parametrization, which participates as a factor, may contain delta functions:

$$\partial_x^2\omega(x) = \omega_0^{(2)}(x) - \sum_{j=1}^{n}[\omega'(x_j + 0) - \omega'(x_j - 0)]\delta_j(x)\,, \tag{1.9}$$

$$\begin{aligned}
&\omega_0^{(2)} \in \mathbb{PC}^{m-2}(\Gamma, \mathcal{T}_\Gamma), \qquad \omega_0^{(2)}(x_j \pm 0) = \omega^{(2)}(x_j \pm 0) = (\partial_x^2\omega)(x_j \pm 0)\,,\\
&\langle\delta_j, \psi\rangle := \psi(x_j)\,, \qquad \psi \in \mathbb{C}(\mathcal{R})\,, \qquad \forall\,\omega(x_j) = t_j \in \mathcal{T}_\Gamma\,.
\end{aligned}$$

To prove (1.9) we represent $\omega'(x)$ in the form

$$\omega'(x) = \omega_0'(x) - \sum_{j=1}^{n}[\omega'(x_j + 0) - \omega'(x_j - 0)]\chi_+(x - x_j)\,, \tag{1.10}$$

where $\omega_0'(x)$ is continuous $\omega_0' \in \mathbb{C}(\mathcal{R}) \cap \mathbb{PC}^{m-1}(\mathcal{R}, \mathcal{T}_{\mathcal{R}})$, $\partial\omega_0'(x_j \pm 0) = \omega'(x_j \pm 0)$ and $\chi_+(x)$ is the Heaviside function: $\chi_+(x) = 0$ for $x < 0$, $\chi_+(x) = 1$ for $x > 0$. It is easy to ascertain that the functions $\omega'(x)$ and $\omega_0'(x)$ differ by a piecewise constant function with jumps at x_j, $\omega(x_j) = t_j \in \mathcal{T}_\Gamma$ and, therefore, their derivatives coincide:

$$(\partial_x\omega')(x) = (\partial_x\omega_0')(x) = \omega^{(2)}(x) \quad \forall\, x \ne x_1, \dots, x_n$$

and even

$$(\partial_x\omega')(x_j \pm 0) = (\partial_x\omega_0')(x_j \pm 0) = \omega^{(2)}(x_j \pm 0) \quad \forall\, j = 1, \dots, n\,.$$

Since $\chi_+' = \delta$ in the sense of distributions, from (1.10) we derive (1.9).

By the same reason a multiplication operator

$$gI \,:\, \mathbb{W}_p^m(\Gamma) \longrightarrow \mathbb{W}_p^m(\Gamma)\,, \qquad g \in \mathbb{PC}^m(\Gamma, \mathcal{T}_\Gamma) \tag{1.11}$$

is bounded only for $m = 0$ (i.e., in the Lebesgue space $\mathbb{L}_p(\Gamma)$ only).

In order to treat finally singular integral equations for spaces of smooth functions in an efficient way, we need the boundedness (j) of differential operators, (jj) of multiplication operators (both were discussed before), and (jjj) of the Cauchy singular integral operator (see Theorem 3.1 below).

To guarantee all three listed space properties, we suggest to consider a special Sobolev space $\mathbb{KW}_p^m(\Gamma, \rho)$ with a power weight

$$\rho(t) := \prod_{j=1}^{n} (t - t_j)^{\alpha_j}\,, \qquad \alpha_j \in \mathbb{C}\,, \quad 1 < p < \infty \tag{1.12}$$

defined as follows:

$$\mathbb{KW}_p^m(\Gamma, \rho) := \left\{\varphi \in \mathbb{L}_p(\Gamma, \rho) \;:\; \partial^k \varphi \in \mathbb{L}_p(\Gamma, \rho^{(k)}),\ k = 0, \ldots, m\right\},$$

$$\rho^{(k)}(t) := \prod_{j=1}^{n} (t - t_j)^{\alpha_j + k}\,. \tag{1.13}$$

The space is endowed with a natural norm,

$$\|\varphi \,|\, \mathbb{KW}_p^m(\Gamma,\rho)\| := \left(\sum_{k=0}^{m} \|\partial_t^k \varphi \,|\, \mathbb{L}_p(\Gamma, \rho^{(k)})\|^p\right)^{\frac{1}{p}} := \left(\sum_{k=0}^{m} \int_\Gamma |\rho^{(k)}(t)\partial_t^k \varphi(t)|^p |dt|\right)^{\frac{1}{p}}, \tag{1.14}$$

which makes it a Banach space. It can be verified straightforwardly that the derivatives

$$\vartheta^k(t)\partial_t^k \varphi(t) \quad \text{and} \quad \partial_t^k \vartheta^k(t)\varphi(t)$$

(see (0.2) for $\vartheta(t)$) exist in the usual sense and that the following norms are equivalent to the original norm in (1.14):

$$\|\varphi \mid \mathbb{KW}_p^m(\Gamma, \rho)\|_1 := \left(\sum_{k=0}^{m} \|(\vartheta\partial_t)^k \varphi \mid \mathbb{L}_p(\Gamma, \rho)\|^p\right)^{\frac{1}{p}},$$

$$\|\varphi \mid \mathbb{KW}_p^m(\Gamma, \rho)\|_2 := \left(\sum_{k=0}^{m} \|\partial_t^k \vartheta^k \varphi \mid \mathbb{L}_p(\Gamma, \rho)\|^p\right)^{\frac{1}{p}}. \tag{1.15}$$

Let $\mu = m + \nu$ be as in (1.4), $\vartheta(t)$ be as in (0.2), and

$$\mathbb{KPZ}_\mu(\Gamma, \mathcal{T}_\Gamma) := \left\{g \in \mathbb{PZ}_\nu(\Gamma) \;:\; \vartheta^k \partial^k g \in \mathbb{PZ}_\nu(\Gamma, \mathcal{T}_\Gamma) \quad \forall j = 1, \ldots, m\right\}. \tag{1.16}$$

Let us prove that $\mathbb{PZ}_\mu(\Gamma, \mathcal{T}_\Gamma) \subset \mathbb{KPZ}_\mu(\Gamma, \mathcal{T}_\Gamma)$. In fact, from the definition of the δ-function and the above considerations it is clear that

$$(t - t_j)\delta_j(t) = 0\,, \quad j = 1, \ldots, n \tag{1.17}$$

(see (1.9)). Therefore, $\lim_{t\to t_j} \vartheta(t)g'(t) = 0$. Thus, dealing with "Fuchs" derivatives of $g \in \mathbb{KW}_p^m(\Gamma, \rho)$, we can ignore the δ-functions and take $\vartheta^k \partial_t^k g \in \mathbb{PZ}_{m-k}(\Gamma, \mathcal{T}_\Gamma) \subset \mathbb{PZ}_\nu(\Gamma, \mathcal{T}_\Gamma)$ for all $k = 1, \ldots, m$.

Moreover, $\mathbb{KPZ}_\mu(\Gamma, \mathcal{T}_\Gamma)$ is a Banach algebra. In fact, for arbitrary $g, h \in \mathbb{KPZ}_\mu(\Gamma, \mathcal{T}_\Gamma)$ we have

$$\vartheta^k \partial_t^k (gh) = \sum_{j=0}^{k} \begin{pmatrix} j \\ k \end{pmatrix} \vartheta^{k-j} (\partial_t^{k-j} g) \vartheta^j (\partial_t^j h) \in \mathbb{PZ}_\mu(\Gamma, \mathcal{T}_\Gamma) \tag{1.18}$$

for all $k = 0, 1, \ldots, m$, which implies that $g \cdot h \in \mathbb{KPZ}_\mu(\Gamma, \mathcal{T}_\Gamma)$.

The space $\mathbb{KPZ}_\mu(\Gamma, \mathcal{T}_\Gamma)$ is essentially larger than $\mathbb{PZ}_\mu(\Gamma, \mathcal{T}_\Gamma)$: the first contains, e.g., the functions $g_1(t) + (t - t_j)^\gamma g_2(t)$ with a complex γ such that $\operatorname{Re}\gamma > 0$ and $g_1, g_2 \in \mathbb{PZ}_\mu(\Gamma, \mathcal{T}_\Gamma)$, which are absent in the second one.

Lemma 1.1. *The space $\mathbb{KPZ}_\mu(\Gamma, \mathcal{T}_\Gamma)$, defined by (1.16), and the space $\mathbb{KPC}^m(\Gamma, \mathcal{T}_\Gamma)$, defined similarly, are Banach algebras and the embeddings*

$$\mathbb{PZ}_\mu(\Gamma, \mathcal{T}_\Gamma) \subset \mathbb{KPZ}_\mu(\Gamma, \mathcal{T}_\Gamma), \quad \mathbb{PC}^m(\Gamma, \mathcal{T}_\Gamma) \subset \mathbb{KPC}^m(\Gamma, \mathcal{T}_\Gamma)$$

are proper.

As usual, for a negative $m = -1, -2, \ldots$ the space $\mathbb{KW}_p^m(\Gamma, \rho)$ is defined as the dual space to $\mathbb{KW}_{p'}^{-m}(\Gamma, \rho^{-1})$, where $p' := p/(p-1)$.

Lemma 1.2. *Let Γ be piecewise μ-smooth, $m = 0, \pm1, \pm2, \ldots$, and $|m| < \mu$.*

The space $\mathbb{KW}_p^m(\Gamma, \rho)$ is defined correctly and is independent of the choice of parametrizations $\omega_j : \mathcal{I} \to \Gamma_j$, $j = 1, \ldots, n$ of the arcs Γ_j (see (1.1)).

The multiplication operator gI is bounded in $\mathbb{KW}_p^m(\Gamma, \rho)$ for arbitrary $g \in \mathbb{KPC}^{|m|}(\Gamma, \mathcal{T}_\Gamma)$.

Proof. We have to consider the case $m = 0, 1, \ldots$ only. For a negative $m = -1, -2 \ldots$ both assertions follow by duality. In fact, it suffices to prove that the dual space is correctly defined and that the dual operator to gI is $\overline{g}I$.

The equality

$$\vartheta^k \partial_t^k (g\varphi) = \sum_{j=0}^{k} \begin{pmatrix} j \\ k \end{pmatrix} \vartheta^{k-j} (\partial_t^{k-j} g) \vartheta^j (\partial_t^j \varphi) \in \mathbb{W}_p^{m-k}(\Gamma, \rho), \quad k = 0, 1, \ldots, m,$$

which is similar to (1.18), immediately implies that the multiplication operator gI is bounded in the space $\mathbb{KW}_p^m(\Gamma, \mathcal{T}_\Gamma)$.

From the equalities

$$\begin{aligned} \vartheta \partial_x \varphi(\omega) &= \vartheta (\partial_t \varphi)(\omega) \partial_x \omega, \\ \vartheta^2 \partial_x^2 \varphi(\omega) &= \vartheta^2 (\partial_t^2 \varphi)(\omega)(\partial_x \omega)^2 + \vartheta (\partial_t \varphi)(\omega) \vartheta \partial_x^2 \omega \end{aligned} \tag{1.19}$$

and similar formulas for higher derivatives and from the boundedness of the multiplication operators proved in the first part of the lemma it follows that the transformation operator

$$\omega_* : \mathbb{K}\mathbb{W}^m_p(\Gamma, \rho) \longrightarrow \mathbb{K}\mathbb{W}^m_p(\mathcal{I}, \rho_0), \qquad \omega_*\varphi(x) := \varphi(\omega(x)), \tag{1.20}$$
$$\omega(1-0) = \omega(0+0) = t_1, \ \ \omega(x_j) = t_j, \ \ j = 2, \dots, n,$$
$$\rho_0(x) := x^{\alpha_1}(x-1)^{\alpha_1}\prod_{j=2}^{n}(x-x_j)^{\alpha_j}$$

is a homeomorphism. Therefore the space $\mathbb{K}\mathbb{W}^m_p(\Gamma, \rho)$ is independent of the choice of the parametrization of Γ. □

Let us consider the weighted Hölder-Zygmund space

$$\mathbb{Z}^0_\mu(\Gamma, \rho) := \{\varphi_0 := \rho\varphi \in \mathbb{Z}_\mu(\Gamma) \ : \ \varphi_0(t_j) = 0, \ \ k = 0, \dots, m\}, \quad 0 < \mu \le 1,$$

which is endowed with a natural norm (cf. (1.3)):

$$\|\varphi \mid \mathbb{Z}^0_\mu(\Gamma, \rho)\| = \|\rho\varphi \mid \mathbb{Z}_\mu(\Gamma)\|. \tag{1.21}$$

For $0 < \mu < 1$, the weighted Hölder-Zygmund space $\mathbb{Z}^0_\mu(\Gamma, \rho)$ coincides with the weighted Hölder space $H^0_\mu(\Gamma, \rho)$ considered in [Du1, Du2, Du3]. As for the spaces $H^0_1(\Gamma, \rho)$ and $\mathbb{Z}^0_1(\Gamma, \rho)$, they are essentially different (see [St1]).

To give a straightforward definition of the Hölder-Zygmund spaces $\mathbb{Z}^0_\mu(\Gamma, \rho)$ for $\mu = m + \nu \ge 1$ we need a μ-smooth contour Γ. For a piecewise smooth Γ we suggest the following modification of the Hölder-Zygmund space without weight:

$$\mathbb{K}\mathbb{Z}_\mu(\Gamma) := \{\varphi \in \mathbb{Z}_\nu(\Gamma) \ : \ \vartheta^k\partial^k\varphi \in \mathbb{Z}_\nu(\Gamma), \ \ k = 1, \dots, m\},$$

provided $0 < \nu < 1$ (i.e., $\mu \notin \mathbb{N}$). For a weighted space we set

$$\mathbb{K}\mathbb{Z}^0_\mu(\Gamma, \rho) := \{\rho\varphi \in \mathbb{Z}_\nu(\Gamma) \ : \ \varphi_k := \rho^{(k)}\partial^k\varphi \in \mathbb{Z}_\nu(\Gamma), \tag{1.22}$$
$$\varphi_k(t_j) = 0, \ \ k = 0, \dots, m, \ j = 0, \dots, n\}.$$

We endow these spaces with the following natural norms (cf. (1.3) and (1.21)):

$$\|\varphi \mid \mathbb{K}\mathbb{Z}_\mu(\Gamma)\| := \sum_{k=0}^{m-1}\sup_{t\in\Gamma}|\vartheta^k(t)\partial^k_t\varphi(t)| + \|\vartheta^m\partial^m\varphi \mid \mathbb{Z}_\nu(\Gamma)\|,$$
$$\|\varphi \mid \mathbb{K}\mathbb{Z}^0_\mu(\Gamma, \rho)\| := \sum_{k=0}^{m-1}\sup_{t\in\Gamma}|\varphi_k(t)| + \|\varphi_m \mid \mathbb{Z}_\nu(\Gamma)\|. \tag{1.23}$$

Equivalent norms can be written down as in (1.15).

Lemma 1.3. *The spaces $\mathbb{K}\mathbb{Z}_\mu(\Gamma)$ for $\mu \notin \mathbb{N}$ and $\mathbb{K}\mathbb{Z}^0_\mu(\Gamma, \rho)$ for arbitrary $\mu > 0$ are correctly defined and are independent of the choice of the parametrizations $\omega_j : \mathcal{I} \to \Gamma_j$, $j = 1, \dots, n$ of the curve Γ (see (1.1)).*

The multiplication operator gI is bounded in the space $\mathbb{K}\mathbb{Z}^0_\mu(\Gamma, \rho)$ for arbitrary $g \in \mathbb{K}\mathbb{P}\mathbb{Z}_\mu(\Gamma, \mathcal{T}_\Gamma)$.

Proof. The proof follows word for word the proof of the preceding Lemma 1.2 with obvious modifications. □

2. Weighted Bessel potential and Besov spaces

It is possible to define new spaces by interpolation. Without going into the details of interpolation theory (we refer the reader to [Tr1] for that) let us note that interpolation assigns to a pair of Banach spaces $\mathbb{X}_0$, $\mathbb{X}_1$ embedded in a bigger Banach space, $\mathbb{X}_0, \mathbb{X}_1 \subset \mathbb{X}$ (an interpolation pair), a new space $\mathbb{X}_\vartheta := [\mathbb{X}_0, \mathbb{X}_1]_\vartheta$, $0 \leq \vartheta \leq 1$ (an interpolated space), with a comfortable interpolation property, stated in the next lemma. This lemma summarizes results on interpolation of operators by different methods exposed, e.g., in [Tr1, §§ 1.10.1, 1.16.4].

Lemma 2.1. (Interpolation Property). *If the operator*

$$A \;:\; \mathbb{X}_0 \longrightarrow \mathbb{Y}_0$$
$$\;:\; \mathbb{X}_1 \longrightarrow \mathbb{Y}_1$$

is bounded in both pairs, then A is bounded between pairs of interpolated spaces

$$A : \mathbb{X}_\vartheta := [\mathbb{X}_0, \mathbb{X}_1]_\vartheta \longrightarrow \mathbb{Y}_\vartheta := [\mathbb{Y}_0, \mathbb{Y}_1]_\vartheta$$

for all $0 \leq \vartheta \leq 1$ and, for some positive constant C_ϑ,

$$\|A|\mathbb{X}_\vartheta \to \mathbb{Y}_\vartheta\| \leq C_\vartheta \|A|\mathbb{X}_0 \to \mathbb{Y}_0\|^{1-\vartheta} \|A|\mathbb{X}_1 \to \mathbb{Y}_1\|^{\vartheta}.$$

Moreover, if $\mathbb{Y}_0 = \mathbb{Y}_1$ or $\mathbb{X}_0 = \mathbb{X}_1$ and the operator $A : \mathbb{X}_k \longrightarrow \mathbb{Y}_k$ is compact for $k = 0$ or for $k = 1$, then $A : \mathbb{X}_\vartheta \longrightarrow \mathbb{Y}_\vartheta$ is compact for all[1] $0 < \vartheta < 1$.

The Bessel potential space $\mathbb{H}^s_p(\Gamma)$, $s \geq 0$, $1 \leq p \leq \infty$, where

$$s = m + \vartheta, \qquad m \in \mathbb{N}, \quad 0 < \vartheta \leq 1, \tag{2.1}$$

can be defined as the result of complex interpolation of Sobolev spaces (cf. [Tr1, § 1.9, § 2.3]),

$$\mathbb{H}^s_p(\Gamma) = (\mathbb{W}^m_p(\Gamma), \mathbb{W}^{m+1}_p(\Gamma))_\vartheta, \tag{2.2}$$

while the Besov space $\mathbb{B}^s_{p,q}(\Gamma)$ is the result of real interpolation (cf. [Tr1, § 1.3, § 2.3]),

$$\mathbb{B}^s_{p,q}(\Gamma) = (\mathbb{W}^m_p(\Gamma), \mathbb{W}^{m+1}_p(\Gamma))_{\vartheta,q}, \quad 1 \leq q \leq \infty. \tag{2.3}$$

Similar definitions are valid for the spaces $\mathbb{H}^s_p(\Omega^\pm)$, $\mathbb{B}^s_{p,q}(\Omega^\pm)$.

We note that the spaces can also be defined rigorously if Γ is only s-smooth. Therefore, for piecewise smooth Γ we can take $s \leq 1$ (or, even, $s \leq 1 + 1/p$).

For the definition of equivalent norms in Bessel potential and Besov spaces we need some standard definitions and notations.

$\mathcal{S}(\mathbb{R}^n)$ denotes the space of rapidly decreasing smooth functions and $\mathcal{S}'(\mathbb{R}^n)$ is the space of tempered distributions, i.e., the space of continuous linear functionals on $\mathcal{S}(\mathbb{R}^n)$.

[1] The first result on interpolation of compact operators was obtained, to our knowledge, by M. Krasnosel'skij [Kr1] in 1960. This reference is missing in H. Triebel's fundamental monograph [Tr1, § 1.16.4]; he only cites papers devoted to the subject since 1964.

The direct and the inverse Fourier transforms $\mathcal{F}$ and $\mathcal{F}^{-1}$ are defined as follows:

$$\begin{aligned} \mathcal{F}\varphi(\xi) &:= \int_{\mathbb{R}^n} e^{i\xi x}\varphi(x)\,dx, \qquad \xi \in \mathbb{R}^n, \\ \mathcal{F}^{-1}\psi(x) &= (2\pi)^{-n}\int_{\mathbb{R}^n} e^{-ix\xi}\psi(\xi)\,d\xi, \qquad x \in \mathbb{R}^n. \end{aligned} \tag{2.4}$$

For the Euclidean space $\mathbb{R}^n$, the Bessel potential space $\mathbb{H}^s_p(\mathbb{R}^n)$ ($s \in \mathbb{R}$, $1 \leq p \leq \infty$) is the subset of $\mathcal{S}'(\mathbb{R}^n)$ ($s \in \mathbb{R}$, $1 < p < \infty$) consisting of the elements f with finite norm

$$\|f\big|\mathbb{H}^s_p(\mathbb{R}^n)\| := \|\mathcal{F}^{-1}\langle\xi\rangle^s\mathcal{F}f\,|\,\mathbb{L}_p(\mathbb{R}^n)\|,$$

where

$$\langle x\rangle = \left(1+|x|^2\right)^{\frac{1}{2}}, \quad x \in \mathbb{R}^n. \tag{2.5}$$

The Besov space $\mathbb{B}^s_{p,q}(\mathbb{R}^n)$ ($s = m + \vartheta > 0$, $1 \leq p, q \leq \infty$) is equipped with the norm

$$\|f\big|\mathbb{B}^s_{p,q}(\mathbb{R}^n)\| := \|f\big|\mathbb{W}^m_p(\mathbb{R}^n)\| + \sum_{|\alpha|=m}\left(\int_{\mathbb{R}^n}\frac{\|\Delta^2_h\partial^\alpha f\big|L_p(\mathbb{R}^n)\|^q}{h^{\vartheta q}}\frac{dh}{h^n}\right)^{1/q}$$

for $1 \leq q < \infty$ and with the obvious "esssup" modification (instead of the integration) when $q = \infty$.

For domains $\Omega^\pm \subset \mathbb{R}^2$, the spaces $\mathbb{H}^s_p(\Omega^\pm)$ and $\mathbb{B}^s_{p,q}(\Omega^\pm)$ are defined as the restrictions of $\mathbb{H}^s_p(\mathbb{R}^2)$ and $\mathbb{B}^s_{p,q}(\mathbb{R}^2)$ to $\Omega^\pm$, while $\widetilde{\mathbb{H}}^s_p(\Omega^\pm)$ and $\widetilde{\mathbb{B}}^s_{p,q}(\Omega^\pm)$ are subspaces of the corresponding spaces on $\mathbb{R}^2$ and consist of functions which are supported in $\overline{\Omega^\pm}$.

The space $\mathbb{W}^s_p(\Omega^\pm) = \mathbb{B}^s_{p,p}(\Omega^\pm)$ is also known as the Sobolev-Slobodetskij space.

Similarly we define the spaces $\mathbb{H}^s_p(\mathcal{I})$, $\mathbb{B}^s_{p,q}(\mathcal{I})$ (as the restrictions of $\mathbb{H}^s_p(\mathbb{R})$, $\mathbb{B}^s_{p,q}(\mathbb{R})$ to the interval $\mathcal{I} \subset \mathbb{R}$) and $\widetilde{\mathbb{H}}^s_p(\mathcal{I})$, $\widetilde{\mathbb{B}}^s_{p,q}(\mathcal{I})$ (as the subspaces of $\mathbb{H}^s_p(\mathbb{R})$, $\mathbb{B}^s_{p,q}(\mathbb{R})$ of functions which are supported in the closed interval $\overline{\mathcal{I}} = [0,1]$).

Using the parametrizations (1.1), (1.2) we define the spaces $\mathbb{H}^s_p(\Gamma_j), \ldots, \widetilde{\mathbb{B}}^s_{p,q}(\Gamma_j)$ in a standard way.

The weighted spaces $\mathbb{H}^s_p(\Gamma,\rho)$ and $\mathbb{B}^s_{p,q}(\Gamma,\rho)$ consist of all functions $\varphi(t)$ for which $\rho\varphi \in \mathbb{H}^s_p(\Gamma)$ and $\rho\varphi \in \mathbb{B}^s_{p,q}(\Gamma)$, respectively.

Since the spaces $\mathbb{H}^s_p(\Gamma,\rho)$ and $\widetilde{\mathbb{H}}^{-s}_{p'}(\Gamma,\rho^{-1})$ ($s \in \mathbb{R}$, $p' := p/(p-1)$) are dual (adjoint), see [Tr1], it is natural to define the spaces $\mathbb{B}^s_{p,q}(\Gamma,\rho)$ and $\widetilde{\mathbb{B}}^s_{p,q}(\Gamma,\rho)$ for a negative $s < 0$ as the dual spaces for $\widetilde{\mathbb{B}}^{-s}_{p',q'}(\Gamma,\rho^{-1})$ and $\mathbb{B}^{-s}_{p',q}(\Gamma,\rho^{-1})$, respectively.

The notation $\mathbb{H}^s_p$, ignoring the weight function and the domain of definition, will be used if the subsequent proposition is valid for any weight and any of the domains $\mathbb{R}^n$, $\Omega^\pm$, Γ_j and Γ. Moreover, when writing $\mathbb{X}(\Omega^\pm,\rho)$ we mean any of the spaces $\mathbb{H}^s_p(\Omega^\pm,\rho)$, $\mathbb{B}^s_{p,q}(\Omega^\pm,\rho)$, $\mathbb{W}^s_p(\Omega^\pm,\rho)$, $\mathbb{Z}_s(\Omega^\pm,\rho)$ while $\widetilde{\mathbb{X}}(\Omega^\pm,\rho)$ stands for the "tilde-spaces".

The Besov space $\mathbb{B}^s_{\infty,\infty} = \mathbb{W}^s_\infty$ and the Hölder-Zygmund space $\mathbb{Z}_s$ are isomorphic and coincide.

The equality $\mathbb{H}^s_2 = \mathbb{B}^s_{2,2}$ holds for all $s \in \mathbb{R}$ and, in particular, $\mathbb{H}^s_2(\mathbb{R}^n) = \mathbb{W}^s_2(\mathbb{R}^n)$ for all $s \geq 0$.

For a non-negative integer $m \in \mathbb{N}_0$ the Bessel potential space $\mathbb{H}^m_p$ and the Sobolev space $\mathbb{W}^m_p$ can be identified.

Theorem 2.2. (Interpolation Theorem; [Tr1, §§ 2.4, 3.3]). *Let*

$$s_0, s_1 \in \mathbb{R}, \quad 0 < \vartheta < 1, \quad 1 \leq p_0, p_1, \nu, q_0, q_1 \leq \infty,$$
$$\frac{1}{p} = \frac{1-\vartheta}{p_0} + \frac{\vartheta}{p_1}, \quad \frac{1}{q} = \frac{1-\vartheta}{q_0} + \frac{\vartheta}{q_1}, \quad s = (1-\vartheta)s_0 + \vartheta s_1.$$

For the real $(\cdot,\cdot)_{\vartheta,p}$, complex $(\cdot,\cdot)_\vartheta$, and modified complex $[\cdot,\cdot]_\vartheta$ interpolation functors we have the following:

i. $(\mathbb{H}^s_{p_0}, \mathbb{H}^s_{p_1})_{\vartheta,p} = \mathbb{H}^s_p$ *provided* $1 < p_0,\ p_1 < \infty$;

ii. $[\mathbb{H}^{s_0}_{p_0}, \mathbb{H}^{s_1}_{p_1}]_\vartheta = (\mathbb{H}^{s_0}_{p_0}, \mathbb{H}^{s_1}_{p_1})_\vartheta = \mathbb{H}^s_p$ *provided* $1 < p_0,\ p_1 < \infty$;

iii. $(\mathbb{H}^{s_0}_r, \mathbb{H}^{s_1}_r)_{\vartheta,\nu} = \mathbb{B}^s_{r,\nu}$ *provided* $s_0 \neq s_1,\ 1 < r < \infty$;

iv. $(\mathbb{B}^{s_0}_{r,q_0}, \mathbb{B}^{s_1}_{r,q_1})_{\vartheta,\nu} = \mathbb{B}^s_{r,\nu}$ *provided* $s_0 \neq s_1,\ 1 \leq r \leq \infty$;

v. $(\mathbb{B}^{s_0}_{p_0,q_0}, \mathbb{B}^{s_1}_{p_1,q_1})_\vartheta = \mathbb{B}^s_{p,q}$; *if in addition,* $1 < p_0,\ p_1 < \infty$ *and either* $q_0 \neq \infty$ *or* $q_1 \neq \infty$, *then* $[\mathbb{B}^{s_0}_{p_0,q_0}, \mathbb{B}^{s_1}_{p_1,q_1}]_\vartheta = \mathbb{B}^s_{p,q}$;

vi. $(\mathbb{Z}_{t_0}, \mathbb{Z}_{t_1})_{\vartheta,\infty} = (\mathbb{Z}_{t_0}, \mathbb{Z}_{t_1})_\vartheta = \mathbb{Z}_t$ *provided* $t_0, t_1 > 0,\ 0 < \vartheta < 1,\ t = (1-\vartheta)t_0 + \vartheta t_1$.

The next theorem addresses the boundedness of a multiplication operator and of the Cauchy singular integral operator, which are of a special importance for the theory of the equations in study (see §§ 3–4).

Theorem 2.3.

i. *If Γ is sufficiently smooth and if*

$$s \in \mathbb{R}, \quad 1 < p < \infty, \quad \textit{and} \quad \mu \begin{cases} = |s| & \textit{for} \quad s = 0, \pm 1, \dots, \\ > |s| & \textit{for} \quad s \neq 0, \pm 1, \dots, \end{cases} \tag{2.6}$$

then the multiplication operator gI with $g \in H_\mu(\Gamma)$ is bounded in the space $\mathbb{H}^s_p(\Gamma, \rho)$.

ii. *If Γ is sufficiently smooth and if*

$$s \in \mathbb{R}, \quad 1 \leq p, q \leq \infty, \quad \textit{and} \quad \mu > |s|, \tag{2.7}$$

then the multiplication operator gI with $g \in H_\mu(\Gamma)$ is bounded in the space $\mathbb{B}^s_{p,q}(\Gamma, \rho)$.

iii. *The spaces*

$$\mathbb{H}^\nu_p(\mathbb{R}^+, x^\alpha),\ \widetilde{\mathbb{H}}^\nu_p(\mathbb{R}^+, x^\alpha)\ \textit{and the spaces}\ \mathbb{B}^\nu_{p,q}(\mathbb{R}^+, x^\alpha),\ \widetilde{\mathbb{B}}^\nu_{p,q}(\mathbb{R}^+, x^\alpha)$$

are pairwise isomorphic and can be identified (extending the restricted functions by 0) provided

$$\frac{1}{p} - 1 < \nu < \frac{1}{p}, \qquad 1 < p < \infty. \tag{2.8}$$

iv. *If Γ is piecewise smooth and $\mu > |\nu|$, then the multiplication operator gI with $g \in \mathbb{PZ}_\mu(\Gamma, \mathcal{T}_\Gamma)$ is bounded in the spaces $\mathbb{H}^\nu_p(\Gamma, \rho)$ and $\mathbb{B}^\nu_{p,q}(\Gamma, \rho)$ provided condition* (2.8) *holds.*

Proof. For the proofs of assertion i we refer to [Tr1, § 3] (see also [MSh1]). Assertion ii follows from assertion i and Theorem 2.2.iii. Assertion iii is a slight modification of the corresponding assertion from [Tr1, § 2.10] (see also [MSh1]). Namely, it suffices to prove the equivalent boundedness of the multiplication operator $\chi_+ I$ by a characteristic function of the half-line in $\mathbb{H}^\nu_p(\mathbb{R}, x^\alpha)$ and in $\mathbb{B}^\nu_{p,q}(\mathbb{R}, x^\alpha)$. By the definition of the weighted spaces, this is equivalent to the boundedness of the operator $x^{-\alpha}\chi_+ x^\alpha I = \chi_+ I$ in the unweighted spaces $\mathbb{H}^\nu_p(\mathbb{R})$ and $\mathbb{B}^\nu_{p,q}(\mathbb{R})$, which is proved in [Tr1, § 2.10].

To prove assertion iv we recall the definition of the weighted Bessel potential space on a piecewise smooth contour. Due to assertion iii, under condition (2.8) the space $\mathbb{H}^\nu_p(\Gamma, \rho)$ can be identified with the space

$$\widetilde{\mathbb{H}}^\nu_p(\Gamma, \rho) := \{\varphi \ :\ \rho\varphi \in \widetilde{\mathbb{H}}^\nu_p(\Gamma_j)\ \forall\, j = 1, \ldots, n\}. \tag{2.9}$$

On the other hand, any function $g \in \mathbb{PZ}_\mu(\Gamma, \mathcal{T}_\Gamma)$ can be represented as a finite sum

$$g(t) = \sum_{j=1}^{n} g_j(t)\chi_j(t), \qquad g_j \in \mathbb{Z}_\mu(\Gamma), \quad j = 1, \ldots, n,$$

where $\chi_j(t)$ is the characteristic function of the smooth arc $\Gamma_j \subset \Gamma$, and due to assertion i it suffices to consider only the case $g(t) = \chi_j(t)$.

We are left to prove that $\chi_j I$ is bounded in $\widetilde{\mathbb{H}}^\nu_p(\Gamma, \rho)$ or, as in assertion iii, in $\widetilde{\mathbb{H}}^\nu_p(\Gamma)$, which is a simple exercise.

For the space $\mathbb{B}^\nu_{p,q}(\Gamma, \rho)$ the result can be proved similarly or obtained from the proved assertion by interpolation. □

From Theorem 2.3.iii we easily find that if Γ is piecewise smooth, then the spaces $\mathbb{H}^s_p(\Gamma, \rho)$ and $\mathbb{B}^s_{p,q}(\Gamma, \rho)$ can be defined rigorously, provided conditions (2.8) hold. As we have already noted above, the definition cannot be extended to large $|s|$. We must impose restrictions on s to ensure the boundedness of a multiplication operator by a piecewise smooth function and of the Cauchy singular integral operator (see Theorems 2.3 and 3.1).

Therefore we suggest to consider the following spaces: if

$$s = m + \nu, \qquad m \in \mathbb{N}_0 \tag{2.10}$$

and (2.8) holds, we define

$$\mathbb{KH}_p^s(\Gamma,\rho) := \left\{\varphi \in \mathbb{H}_p^\nu(\Gamma,\rho) \ : \ \vartheta^k\partial^k\varphi \in \mathbb{H}_p^\nu(\Gamma,\rho), \ k=0,\dots,m\right\}, \qquad (2.11)$$

$$\mathbb{KB}_{p,q}^s(\Gamma,\rho) := \left\{\varphi \in \mathbb{B}_{p,q}^\nu(\Gamma,\rho) \ : \ \vartheta^k\partial^k\varphi \in \mathbb{B}_{p,q}^\nu(\Gamma,\rho), \ k=0,\dots,m\right\} \qquad (2.12)$$

(see (0.2) for $\vartheta(t)$). The spaces are endowed with natural norms:

$$\|\varphi \mid \mathbb{KH}_p^s(\Gamma,\rho)\| := \left(\sum_{k=0}^m \|\vartheta^k\partial_t^k\varphi \mid \mathbb{H}_p^\nu(\Gamma,\rho)\|^p\right)^{\frac{1}{p}},$$

$$\|\varphi \mid \mathbb{KB}_{p,q}^s(\Gamma,\rho)\| := \left(\sum_{k=0}^m \|\vartheta^k\partial_t^k\varphi \mid \mathbb{B}_{p,q}^\nu(\Gamma,\rho)\|^p\right)^{\frac{1}{p}},$$

which make them be Banach spaces. It can be verified straightforwardly that equivalent norms can be defined analogously as in (1.15).

For a negative $s<0$, the spaces $\mathbb{KH}_p^s(\Gamma,\rho)$ and $\mathbb{KB}_{p,q}^s(\Gamma,\rho)$ are defined as the dual spaces to $\mathbb{KH}_{p'}^{-s}(\Gamma,\rho^{-1})$ and $\mathbb{KB}_{p',q'}^{-s}(\Gamma,\rho^{-1})$, respectively, where $p' := p/(p-1)$, $q' := q/(q-1)$.

Lemma 2.4. *Let Γ be piecewise μ-smooth and $|s|<\mu$.*

The spaces $\mathbb{KH}_p^s(\Gamma,\rho)$ and $\mathbb{KB}_{p,q}^s(\Gamma,\rho)$ are defined correctly and are independent of the choice of the parametrizations $\omega_j : \mathcal{I} \to \Gamma_j$, $j=1,\dots,n$ of the curve Γ (see (1.1)).

The multiplication operator gI is bounded in $\mathbb{KH}_p^s(\Gamma,\rho)$ and in $\mathbb{KB}_{p,q}^s(\Gamma,\rho)$ for arbitrary $g \in \mathbb{KPZ}_\mu(\Gamma,\mathcal{T}_\Gamma)$.

Proof. The proof is similar to the proof of Lemma 1.2. □

3. Singular integral equations in weighted spaces

If Γ is an m-smooth closed contour and $m \in \mathbb{N}$, then the Cauchy singular integral operator

$$S_\Gamma\varphi(t) := \frac{1}{\pi i}\int_\Gamma \frac{\varphi(\tau)d\tau}{\tau - t}, \qquad t \in \Gamma \qquad (3.1)$$

is bounded in the spaces $\mathbb{H}_p^s(\Gamma)$, $\mathbb{B}_{p,q}^s(\Gamma)$, and $\mathbb{Z}_s(\Gamma)$ provided

$$1<p,q<\infty, \quad s\in\mathbb{R}, \quad s \le m. \qquad (3.2)$$

In fact, let us recall that S_Γ is bounded in the Hölder spaces $H_\nu(\Gamma) = \mathbb{Z}_\nu(\Gamma)$ for $0<\nu<1$ (the Privalov Theorem; see [Du1, Du2, Mu1, St1]) and in the Lebesgue spaces $L_p(\Gamma)$ (the Riesz Theorem; see [GK1, Kh1, St1]). Since

$$(\partial^j S_\Gamma\varphi)(t) = (S_\Gamma\partial^j\varphi)(t) \qquad \forall j \in \mathbb{N}, \qquad (3.3)$$

we get

$$\begin{aligned}\|S_\Gamma\varphi|\mathbb{X}^{k+\nu}(\Gamma)\| &= \sum_{j=0}^{k}\|\partial^j S_\Gamma\varphi|\mathbb{X}^\nu(\Gamma)\| = \sum_{j=0}^{k}\|S_\Gamma\partial^j\varphi|\mathbb{X}^\nu(\Gamma)\| \\ &\leq C\sum_{j=0}^{k}\|\partial^j\varphi|\mathbb{X}^\nu(\Gamma)\| = C\|\varphi|\mathbb{X}^{k+\nu}(\Gamma)\|\,, \quad k\in\mathbb{N}.\end{aligned}$$

This implies the boundedness in the Sobolev spaces $\mathbb{X}^k(\Gamma) := \mathbb{W}^k_p(\Gamma) = \mathbb{H}^k_p(\Gamma)$ for all integers $k = 0,\dots,m$ and in the Hölder-Zygmund spaces $\mathbb{Z}_\mu(\Gamma)$ for all non-integers $\mu = k+\nu$, $0<\nu<1$. Due to the Interpolation Theorems 2.2.i, 2.2.iii and 2.2.vi, S_Γ is bounded in the spaces $\mathbb{H}^s_p(\Gamma)$, $\mathbb{B}^s_{p,q}(\Gamma)$, and $\mathbb{Z}_\mu(\Gamma)$ if conditions (3.2) hold.

If Γ is piecewise smooth, the boundedness result (3.1)–(3.2) does not hold any more, especially for the weighted spaces. Instead we can prove the following.

Theorem 3.1. *The Cauchy singular integral operator with a weight,*

$$S_{\Gamma,w}\varphi(t) := \frac{1}{\pi i}\int_\Gamma \frac{w(t)}{w(\tau)}\frac{\varphi(\tau)d\tau}{\tau - t}\,, \qquad w(t) := \prod_{j=1}^{n}(t-t_j)^{\beta_j}\,, \tag{3.4}$$

is bounded in the following spaces:

i. *in the modified weighted Sobolev space $\mathbb{KW}^m_p(\Gamma,\rho)$, $m\in\mathbb{N}_0$ provided*

$$-\frac{1}{p} < \alpha_j+\beta_j < 1-\frac{1}{p}\,, \quad j=1,\dots,n\,, \quad 1<p<\infty\,; \tag{3.5}$$

ii. *in the modified weighted Hölder-Zygmund space $\mathbb{KZ}^0_\mu(\Gamma,\rho)$ provided*

$$\begin{aligned}&\mu = m+\nu, \quad m\in\mathbb{N}_0\,, \quad 0<\nu\leq 1\,,\\ &0<\alpha_j+\beta_j-\nu<1\,, \quad j=1,\dots,n\,;\end{aligned} \tag{3.6}$$

iii. *in the modified weighted Bessel potential $\mathbb{KH}^s_p(\Gamma,\rho)$ and Besov $\mathbb{KB}^s_{p,q}(\Gamma,\rho)$ spaces, with $s = m+\nu$ for any integer m, provided that* (2.8) *holds and*

$$-\frac{1}{p} < \alpha_j+\beta_j-\nu\,,\ \alpha_j+\beta_j < 1-\frac{1}{p}\,, \quad j=1,\dots,n\,, \quad 1<p,q<\infty\,. \tag{3.7}$$

Proof. First let us prove the assertions for $m=0$: the operator

$$S_{\Gamma,w}\ :\ \mathbb{X}(\Gamma,\rho)\longrightarrow\mathbb{X}(\Gamma,\rho) \tag{3.8}$$

is bounded provided

$$\begin{aligned}\mathbb{X}(\Gamma,\rho) &= \mathbb{L}_p(\Gamma,\rho) &&\text{and (3.5) holds}\,,\\ \mathbb{X}(\Gamma,\rho) &= \mathbb{Z}_\nu(\Gamma,\rho) &&\text{and (3.6) holds}\,,\\ \mathbb{X}(\Gamma,\rho) &= \mathbb{H}^\nu_p(\Gamma,\rho) &&\text{and (3.7) holds}\,,\\ \mathbb{X}(\Gamma,\rho) &= \mathbb{B}^\nu_{p,q}(\Gamma,\rho) &&\text{and (3.7) holds}\,.\end{aligned} \tag{3.9}$$

In fact, for the weighted Lebesgue space $\mathbb{X}(\Gamma,\rho) = \mathbb{L}_p(\Gamma,\rho)$ the boundednes result (3.8) is well known (and first proved in [Kh1], see also [GK1] and [BK1]).

For the weighted Hölder space $\mathbb{X}(\Gamma,\rho) = H_\nu(\Gamma,\rho)$, the boundedness result (3.8) is proved in [Du1, Du2] (see also [GK1]) under the constraint $0 < \nu < 1$. In the case $\nu = 1$, the boundedness result can be proved similarly using the boundedness of S_Γ in $\mathbb{Z}_1(\Gamma)$ (see (3.1)–(3.2)).

The operator $S_{\Gamma,w}$ is bounded in the space $\mathbb{H}_p^\nu(\Gamma,\rho) = \widetilde{\mathbb{H}}_p^\nu(\Gamma,\rho)$ if and only if the operators $\chi_j S_{\Gamma,w\rho}\chi_k$ are bounded in $\mathbb{H}_p^\nu(\Gamma)$, where $\chi_j(t)$ is the characteristic function of the smooth arc $\Gamma_j \subset \Gamma$, $j = 1,\dots,n$. By rectification of the smooth arcs we easily derive the required boundedness property from the boundedness of the operators

$$S^{(\alpha)}_{\mathbb{R}^+}\varphi(x) := \frac{1}{\pi i}\int_0^\infty \left(\frac{x}{y}\right)^\alpha \frac{\varphi(y)dy}{y-x}, \qquad N^{(\alpha)}_{\mathbb{R}^+,\gamma}\varphi(x) := \frac{1}{\pi i}\int_0^\infty \left(\frac{x}{y}\right)^\alpha \frac{\varphi(y)dy}{y-e^{i\gamma}x},$$

$$S^{(\alpha)}_{\mathbb{R}^+},\ N^{(\alpha)}_{\mathbb{R}^+,\gamma}\ :\ \widetilde{\mathbb{H}}_p^\nu(\mathbb{R}^+) = \mathbb{H}_p^\nu(\mathbb{R}^+) \longrightarrow \mathbb{H}_p^\nu(\mathbb{R}^+), \tag{3.10}$$

$$\frac{1}{p}-1 < \nu < \frac{1}{p}, \quad -\frac{1}{p} < \alpha-\nu,\ \alpha < 1-\frac{1}{p}, \quad 0 < |\gamma| < \pi .$$

To prove the boundedness of the operators in (3.10) we will first consider the operator

$$A := -i\sin\pi\alpha\cos\pi\alpha + \sin^2\pi\alpha\, S_{\mathbb{R}^+}\ :\ \mathbb{L}_p(\mathbb{R}^+) \longrightarrow \mathbb{L}_p(\mathbb{R}^+). \tag{3.11}$$

This operator represents a Mellin convolution:

$$A = \mathfrak{M}^0_{\mathcal{A}_p}, \qquad \mathcal{A}_p(\xi) := -i\sin\pi\alpha\cos\pi\alpha + \sin^2\pi\alpha\,\coth\pi\left(\frac{i}{p}+\xi\right), \tag{3.12}$$

where

$$\mathfrak{M}^0_a\varphi(x) := \mathfrak{M}^{-1}_{\xi\to x}\left\{a(\xi)\mathfrak{M}_{y\to\xi}[\varphi(y)]\right\}(x), \quad x\in\mathbb{R}^+,\ \xi\in\mathbb{R}, \tag{3.13}$$

and $\mathfrak{M}^{\pm1} := \mathfrak{M}^{\pm1}_{\xi\to x}$ are the Mellin transforms

$$\begin{aligned} \mathfrak{M}\varphi(\xi) &:= \int_0^\infty y^{\frac{1}{p}-i\xi}\varphi(y)\frac{dy}{y}, \quad \xi\in\mathbb{R}, \\ \mathfrak{M}^{-1}\psi(x) &= (2\pi)^{-1}\int_{-\infty}^\infty x^{i\xi-\frac{1}{p}}\psi(\xi)\,d\xi, \quad x\in\mathbb{R}^+ \end{aligned} \tag{3.14}$$

(see [Du6, Du9, Du5, DLS1]). Therefore the operator $A = \mathfrak{M}^0_{\mathcal{A}_p}$ in (3.11) is invertible if and only if

$$\mathcal{A}_p(\xi) \neq 0 \iff -\frac{1}{p} < \alpha < 1-\frac{1}{p} \tag{3.15}$$

(note that $\mathcal{A}_p(\xi) = 0$ only for $\xi = 0$).

Let us prove that the Mellin convolution operator

$$B = S^{(\alpha)}_{\mathbb{R}^+} - i\cot\pi\alpha I = \mathfrak{M}^0_{\mathcal{B}_p}, \tag{3.16}$$

$$\mathcal{B}_p(\xi) := \coth\pi\left(\frac{i}{p} + i\alpha + \xi\right) - i\cot\pi\alpha$$

is the inverse of A. Indeed,

$$\mathcal{B}_p(\xi) := \coth\pi\left(\frac{i}{p} + i\alpha + \xi\right) - i\cot\pi\alpha = -i\left[\cot\pi\left(\frac{1}{p} + \alpha - i\xi\right) - \cot\pi\alpha\right]$$

$$= i\frac{1+\cot^2\pi\alpha}{\cot\pi(1/p - i\xi) + \cot\pi\alpha} = i\frac{\sin^{-2}\pi\alpha}{\cot\pi(1/p - i\xi) + \cot\pi\alpha} = \mathcal{A}_p^{-1}(\xi)$$

and hence we get $AB = BA = \mathfrak{M}^0_{\mathcal{A}_p\mathcal{A}_p^{-1}} = \mathfrak{M}^0_1 = I$.

The operator A in (3.11) can also be regarded as a Fourier convolution operator:

$$A := -i\sin\pi\alpha\cos\pi\alpha + \sin^2\pi\alpha\, S_{\mathbb{R}^+} = W_{a_p} \;:\; \widetilde{\mathbb{H}}^\nu_p(\Gamma) = \mathbb{H}^\nu_p(\mathbb{R}^+) \longrightarrow \mathbb{H}^\nu_p(\mathbb{R}^+),$$

$$a_p(\xi) = -i\sin\pi\alpha\cos\pi\alpha - \sin^2\pi\alpha\ \mathrm{sign}\,\xi \tag{3.17}$$

(see [Du6, Lemma 1.35]). Here

$$W^0_a\varphi(x) := \mathcal{F}^{-1}_{\xi\to x}\left\{a(\xi)\mathcal{F}_{y\to\xi}[\varphi(y)]\right\}(x), \quad x\in\mathbb{R} \tag{3.18}$$

is a convolution on $\mathbb{R}$ ($\mathcal{F}^{\pm1} = \mathcal{F}^{\pm1}_{\xi\to x}$ are the Fourier transformations; see (2.4)).

$$W_a := r_+W^0_a \;:\; \widetilde{\mathbb{X}}(\mathbb{R}^+) \longrightarrow \mathbb{X}(\mathbb{R}^+)$$

is the restriction of W^0_a to the positive half-line (r_+ denotes the restriction of a function from $\mathbb{R}$ to $\mathbb{R}^+$).

It is well known that the operators W^0_a and $\mathfrak{M}^0_a$ are isomorphic:

$$\mathfrak{M}^0_a = \mathbf{Z}^{-1}W^0_a\mathbf{Z},$$

$$\mathbf{Z} \;:\; \mathbb{L}_p(\mathbb{R}^+) \longrightarrow \mathbb{L}_p(\mathbb{R}), \qquad \mathbf{Z}\varphi(x) = e^{-\frac{1}{p}x}\varphi(e^{-x}),$$

$$\mathbf{Z}^{-1} \;:\; \mathbb{L}_p(\mathbb{R}) \longrightarrow \mathbb{L}_p(\mathbb{R}^+), \quad \mathbf{Z}^{-1}\psi(t) = t^{-\frac{1}{p}}\psi(-\log t)$$

(see [Du6, § 8]).

Next we apply a lifting procedure to the operator A in (3.17). For this we recall that the Bessel potential operators

$$\Lambda^r_+ := W^0_{\lambda^r_+} \;:\; \widetilde{\mathbb{H}}^s_p(\mathbb{R}^+) \longrightarrow \widetilde{\mathbb{H}}^{s-r}_p(\mathbb{R}^+), \quad \lambda^r_+(\xi) = (\xi + i)^r,$$

$$r_+\Lambda^r_-\ell := W_{\lambda^r_-}\ell \;:\; \mathbb{H}^s_p(\mathbb{R}^+) \longrightarrow \mathbb{H}^{s-r}_p(\mathbb{R}^+), \quad \lambda^r_-(\xi) = (\xi - i)^r, \tag{3.19}$$

where ℓ denotes an arbitrary extension from $\mathbb{R}^+$ to $\mathbb{R}$, arrange isomorphisms of the indicated spaces for arbitrary $s, r \in \mathbb{R}$ (see, e.g., [Du5, DS1, Es1, St1]) and that

$$r_+\Lambda^r_-\ell W_a\varphi = W_{a\lambda^r_-}\varphi, \qquad W_a r_+\Lambda^r_+\varphi = W_{a\lambda^r_+}\varphi, \qquad \chi_+\Lambda^r_+\varphi = \Lambda^r_+\varphi \tag{3.20}$$

for $\varphi \in \widetilde{\mathbb{H}}^s_p(\mathbb{R}^+)$. In particular, we have the isomorphisms

$$\begin{aligned} \Lambda_+^{-\nu} &: \mathbb{L}_p(\mathbb{R}^+) \longrightarrow \widetilde{\mathbb{H}}^\nu_p(\mathbb{R}^+), \\ r_+\Lambda_-^\nu \ell &: \mathbb{H}^\nu_p(\mathbb{R}^+) \longrightarrow r_+\mathbb{L}_p(\mathbb{R}), \end{aligned} \tag{3.21}$$

and since the spaces $r_+\mathbb{L}_p(\mathbb{R})$ and $\mathbb{L}_p(\mathbb{R}^+)$ can be identified (extending functions $\varphi \in \mathbb{L}_p(\mathbb{R}^+)$ by 0), we can lift the operator $A = W_{a_p}$ in (3.17). As a result we get the equivalent operator

$$r_+\Lambda_-^\nu \ell W_{a_p}\Lambda_+^{-\nu} = W_{\lambda_-^\nu a_p \lambda_+^{-\nu}} = W_{a_{p,\nu}}, \qquad a_{p,\nu}(\xi) = a_p(\xi)\left(\frac{\xi - i}{\xi + i}\right)^\nu. \tag{3.22}$$

Since

$$\begin{aligned} a_{p,\nu}(+\infty) &= \frac{2e^{-i\pi\alpha}\sin^2\pi\alpha}{e^{i\pi\alpha} + e^{-i\pi\alpha}}, \\ a_{p,\nu}(-\infty) &= \frac{2e^{i\pi(\alpha-2\nu)}\sin^2\pi\alpha}{e^{i\pi\alpha} + e^{-i\pi\alpha}}, \\ a_{p,\nu}(0-0) &= \frac{2e^{i\pi\alpha}\sin^2\pi\alpha}{e^{i\pi\alpha} + e^{-i\pi\alpha}}, \\ a_{p,\nu}(0+0) &= \frac{2e^{-i\pi\alpha}\sin^2\pi\alpha}{e^{i\pi\alpha} + e^{-i\pi\alpha}}, \end{aligned} \tag{3.23}$$

we get

$$\begin{aligned} \frac{1}{2\pi}\arg\frac{a_{p,\nu}(+\infty)}{a_{p,\nu}(-\infty)} &= \frac{1}{2\pi}\arg e^{2\pi(\nu-\alpha)i} = \nu - \alpha, \\ \frac{1}{2\pi}\arg\frac{a_{p,\nu}(0+0)}{a_{p,\nu}(0-0)} &= \frac{1}{2\pi}\arg e^{-2\pi\alpha i} = -\alpha, \end{aligned}$$

and the conditions in the third line of (3.10) ensure the invertibility of the lifted convolution operator (3.22) in $\mathbb{L}_p(\mathbb{R}^+)$ (see [Du6, Lemma 4.1, Theorem 4.2]). Therefore, conditions (3.15) guarantee the invertibility of the operator $A = W_{a_p}$ in (3.17), and since B in (3.16) is its inverse, B is a bounded operator in $\mathbb{H}^\nu_p(\mathbb{R}^+)$. Thus, $S^{(-\gamma)}_{\mathbb{R}^+} = B - i\cot\pi\gamma I$ is a bounded operator in $\mathbb{H}^\nu_p(\mathbb{R}^+)$.

To prove the boundedness of the operator $N^{(\alpha)}_{\mathbb{R}^+,\gamma}$ (see (3.10)) it is equivalent to consider the operator $N_{\mathbb{R}^+,\gamma} := N^{(0)}_{\mathbb{R}^+,\gamma}$ in $\mathbb{H}^\nu_p(\mathbb{R}^+, x^\alpha)$, due to the definition of the weighted spaces.

Let us apply the trick described in [Sc1]: there exists a sufficiently small and negative $\beta < 0$ such that

$$-\frac{1}{p} < \alpha + \beta - \nu, \ \alpha + \beta < 1 - \frac{1}{p}. \tag{3.24}$$

As already proved, these conditions ensure the boundedness of the operator $S^{(\beta+i\vartheta)}_{\mathbb{R}^+}$ in $\mathbb{H}^\nu_p(\mathbb{R}^+, x^\alpha)$, because $\|S^{(\beta+i\vartheta)}_{\mathbb{R}^+}\| = \|S^{(\beta)}_{\mathbb{R}^+}\|$ for all $\vartheta \in \mathbb{R}$. So the operator

$$R_{\mathbb{R}^+,\gamma} := \frac{e^{-i\gamma}-1}{2}\int_{-\infty}^{\infty}\frac{e^{i(\pi-\gamma)(\beta+i\vartheta)}}{\sin\pi(\beta+i\vartheta)}S^{(\beta+i\vartheta)}_{\mathbb{R}^+}\,d\vartheta$$

is also bounded in $\mathbb{H}^\nu_p(\mathbb{R}^+, x^\alpha)$ because the integral is absolutely convergent (cf. the last inequality in (3.10)).

Let $0 < t < 1$. By changing the order of integration and applying the residue theorem in the right complex half-plane $\mathrm{Re}\,\pi(\beta+i\vartheta) > 0$, where the integrand has simple poles at $\beta = 0, 1, \dots$, one finds that

$$R_{\mathbb{R}^+,\gamma}\varphi(x) = \frac{1}{\pi i}\int_0^\infty K\left(\frac{x}{y}\right)\frac{\varphi(y)dy}{y-x},$$

$$K(t) = \frac{e^{-i\gamma}-1}{2}\int_{-\infty}^{\infty}\frac{t^{\beta+i\vartheta}e^{i(\pi-\gamma)(\beta+i\vartheta)}}{\sin\pi(\beta+i\vartheta)}d\vartheta$$

$$= i(e^{-i\gamma}-1)\int_{-\infty}^{\infty}\frac{t^{\beta+i\vartheta}e^{i(\pi-\gamma)(\beta+i\vartheta)}}{e^{i\pi(\beta+i\vartheta)}-e^{-i\pi(\beta+i\vartheta)}}d\vartheta = \pi i\frac{e^{-i\gamma}-1}{2}\sum_{k=0}^{\infty}\frac{2it^k e^{i(\pi-\gamma)k}}{(-1)^{k+1}\pi}$$

$$= (1-e^{-i\gamma})\sum_{k=0}^{\infty}t^k e^{-i\gamma k} = \frac{1-e^{-i\gamma}}{1-te^{-i\gamma}}.$$

If $1 < t < \infty$, one can apply the residue theorem in the left complex half-plane $\mathrm{Re}\,\pi(\beta+i\vartheta) < 0$, where the integrand has simple poles at $\beta = -1, -2, \dots$. Similarly to the foregoing case we get

$$\begin{aligned} K(t) &= \frac{e^{-i\gamma}-1}{2}\int_{-\infty}^{\infty}\frac{t^{\beta+i\vartheta}e^{i(\pi-\gamma)(\beta+i\vartheta)}}{\sin\pi(\beta+i\vartheta)}d\vartheta = (e^{-i\gamma}-1)\sum_{k=1}^{\infty}t^{-k}e^{i\gamma k} \\ &= -\frac{t^{-1}e^{i\gamma}(1-e^{-i\gamma})}{1-t^{-1}e^{i\gamma}} = \frac{1-e^{-i\gamma}}{1-te^{-i\gamma}}. \end{aligned}$$

Therefore,

$$R_{\mathbb{R}^+,\gamma}\varphi(x) = \frac{1}{\pi i}\int_0^\infty\frac{1-e^{-i\gamma}}{1-\frac{x}{y}e^{-i\gamma}}\frac{\varphi(y)dy}{y-x} \tag{3.25}$$

and hence, by virtue of the equality $R_{\mathbb{R}^+,\gamma} = N_{\mathbb{R}^+,\gamma} - e^{-i\gamma}S_{\mathbb{R}^+}$, the boundedness of $N_{\mathbb{R}^+,\gamma}$ in the space $\mathbb{H}^\nu_p(\mathbb{R}^+, x^\alpha)$ follows.

To prove that the operator $S_{\Gamma,w}$ is bounded in the modified weighted Besov space $\widetilde{\mathbb{B}}^\nu_{p,q}(\Gamma,\rho) = \mathbb{B}^\nu_{p,q}(\Gamma,\rho)$ we can employ the interpolation method: if conditions (3.7) hold, the operator S_{Γ,w_ρ} is bounded not only in $\mathbb{H}^\nu_p(\Gamma)$, but also in $\mathbb{H}^{\nu\pm\varepsilon}_p(\Gamma,\rho)$

for a small $\varepsilon > 0$. Due to the Interpolation Theorem 2.2.iii, $S_{\Gamma,w\rho}$ is bounded in $\mathbb{B}^{\nu}_{p,q}(\Gamma)$ and, therefore, $S_{\Gamma,w}$ is bounded in $\mathbb{B}^{\nu}_{p,q}(\Gamma,\rho)$.

Now we will prove the boundedness result for a positive integer $m \in \mathbb{N}$. The assertions can be reformulated as follows: the operator

$$S_\Gamma \,:\, \mathbb{X}^m(\Gamma, w\rho) \longrightarrow \mathbb{X}^m(\Gamma, w\rho) \tag{3.26}$$

is bounded provided

$$\begin{aligned} &\mathbb{X}^m(\Gamma, w\rho) = \mathbb{K}\mathbb{W}^m_p(\Gamma, w\rho) && \text{and (3.5) holds},\\ &\mathbb{X}^m(\Gamma, w\rho) = \mathbb{K}\mathbb{Z}^0_{m+\nu}(\Gamma, w\rho) && \text{and (3.6) holds},\\ &\mathbb{X}^m(\Gamma, w\rho) = \mathbb{K}\mathbb{H}^{m+\nu}_p(\Gamma, w\rho) && \text{and (3.7) holds},\\ &\mathbb{X}^m(\Gamma, w\rho) = \mathbb{K}\mathbb{B}^{m+\nu}_{p,q}(\Gamma, w\rho) && \text{and (3.7) holds}. \end{aligned} \tag{3.27}$$

The following is easy to verify (cf. (3.3)):

$$(\vartheta^k \partial^k_t)(S_\Gamma\varphi)(t) = \vartheta^k(S_\Gamma \partial^k_\tau \varphi)(t) = S_\Gamma(\vartheta^k \partial^k_\tau \varphi)(t) + \sum_{j,r=0}^{k} (B_{j,r}\varphi)(t)\,, \tag{3.28}$$

where the functionals

$$\begin{aligned} &B_{j,k} \,:\, \mathbb{X}^m(\Gamma, w\rho) \longrightarrow \mathbb{C}\,, \qquad j = 1, \dots, k\,,\\ &B_{j,k}\varphi := (-1)^j \frac{t^{k-j}}{\pi i} \int_\Gamma \tau^{j-1} \partial^k_\tau \varphi(\tau) d\tau\,, \end{aligned}$$

are bounded (note that even if Γ contains open arcs, partial integration in (3.28) does not generate any summands at the boundary points, because these summands are eliminated by the factor $\vartheta^k(t)$).

In fact, from the corresponding conditions (3.5)–(3.7) we conclude that

$$\partial^k_\tau \varphi \in \mathbb{X}^{m-k}(\Gamma, w\rho) \subset \mathbb{L}_1(\Gamma)$$

and that the embedding is continuous, i.e.,

$$\|\partial^k_\tau\varphi|\mathbb{L}_1(\Gamma)\| \le C'_k \|\partial^k_\tau\varphi|\mathbb{X}^{m-k}(\Gamma, w\rho)\| \le C_k \|\varphi|\mathbb{X}^m(\Gamma, w\rho)\|\,.$$

Thus,

$$|B_{j,k}\varphi| \le C_j \|\varphi|\mathbb{X}^m(\Gamma, \rho)\| \qquad \forall\, j = 1, \dots, m\,. \tag{3.29}$$

Since the singular integral operator

$$S_\Gamma \,:\, \mathbb{X}(\Gamma, w\rho) := \mathbb{X}^0(\Gamma, w\rho) \longrightarrow \mathbb{X}(\Gamma, w\rho) \tag{3.30}$$

is bounded (see the first part of the proof) and, by definition, $\vartheta^k\partial_t^k\varphi\in\mathbb{X}^{m-k}(\Gamma,w\rho)$, from (3.28)–(3.30) we get

$$\begin{aligned}\|S_\Gamma\varphi \mid \mathbb{X}^m(\Gamma,w\rho)\| &= \sum_{k=0}^m \|\vartheta^k\partial_t^k S_\Gamma\varphi \mid \mathbb{X}(\Gamma,w\rho)\| \\ &\le \sum_{k=0}^m \|S_\Gamma\vartheta^k\partial_t^k\varphi \mid \mathbb{X}(\Gamma,w\rho)\| + \sum_{k=0}^m\sum_{j=0}^k |B_{j,k}\varphi| \\ &\le C\sum_{k=0}^m \|\vartheta^k\partial_t^k\varphi \mid \mathbb{X}(\Gamma,w\rho)\| = C\|\varphi \mid \mathbb{X}^m(\Gamma,w\rho)\| . \end{aligned} \tag{3.31}$$

The boundedness of $S_{\Gamma,w}$ in the modified spaces $\mathbb{KX}^m(\Gamma,\rho)$ for negative $m=-1,-2,\dots$ (excluding the Hölder-Zygmund $\mathbb{KZ}^0_\mu(\Gamma,\rho)$ spaces, which are defined only for positive $\mu>0$) follows by duality.

In fact, let the operator $S_{\Gamma,w}$ be bounded in the modified weighted spaces $\mathbb{KW}^m_p(\Gamma,\rho)$, in $\mathbb{KH}^m_p(\Gamma,\rho)$, or in $\mathbb{KB}^m_{p,q}(\Gamma,\rho)$ and suppose that the conditions (3.5)–(3.7) hold. The dual operator, $S_{\Gamma,w^{-1}}$, to $S_{\Gamma,w}$ is defined by the bilinear form

$$(\varphi, S_{\Gamma,w^{-1}}\psi) := (S_{\Gamma,w}\varphi,\psi)\,, \qquad (\varphi,\psi) := \int_\Gamma \varphi(\tau)\psi(\tau)d\tau\,, \tag{3.32}$$

and is thus bounded in the dual spaces $\mathbb{KW}^{-m}_{p'}(\Gamma,\rho^{-1})$, $\mathbb{KH}^{-m}_{p'}(\Gamma,\rho^{-1})$, and $\mathbb{KB}^{-m}_{p',q'}(\Gamma,\rho^{-1})$, respectively, because $-m\in\mathbb{N}$ is already positive and the parameters p',q', $-\alpha_j,-\beta_j$ satisfy the corresponding conditions (3.5)–(3.7). □

Theorem 3.1 enables us to establish a Fredholm criterion and an index formula for a singular integral operator with complex conjugation:

$$A\varphi := a\varphi + bS_\Gamma\varphi + cVS_\Gamma V\varphi = f\,, \quad V\varphi(t) := \overline{\varphi(t)}\,, \tag{3.33}$$

$$a,b,c\in\mathbb{KPC}^{m+1}(\Gamma,\mathcal{T}_\Gamma)$$

$$\text{for}\quad \varphi,\ f\in\mathbb{X}^m(\Gamma,\rho)=\mathbb{KW}^m_p(\Gamma,\rho),\ \mathbb{KH}^{m+\nu}_p(\Gamma,\rho)\,,\ \mathbb{KB}^{m+\nu}_{p,q}(\Gamma,\rho)\,,$$

$$a,b,c\in\mathbb{KPZ}_{m+\nu}(\Gamma,\mathcal{T}_\Gamma)\quad\text{for}\quad \varphi,\ f\in\mathbb{X}^m(\Gamma,\rho)=\mathbb{KZ}^0_{m+\nu}(\Gamma,\rho)\,,$$

where $\mathbb{X}^m(\Gamma,\rho)$ is defined by (3.27).

Although the coefficients of the operator A are $N\times N$ matrix functions and equation (3.33) is considered in weighted N-vector spaces, we use the same notation for spaces and classes of functions as in the scalar case $N=1$ for the sake of simplicity.

Also for conciseness, we assume that Γ is a closed piecewise smooth curve with smooth arcs Γ_{j-1}, Γ_j, having in common the knot t_j where they meet under the interior angle $\pi\gamma_j$ (measured from the bounded domain Ω^+ enclosed by Γ). Therefore, $0\le\gamma_j\le 2$, $j=1,\dots,n$, while the values $\gamma_j=0,2$ correspond to a cusp at t_j. This assumption simplifies the symbol of operator (3.33). In the general case the symbol can be written down in a similar but more complicated form (see [Du3, Du4, Du5, DLS1, RS1]).

When Γ has no cusps, $0 < \gamma_j < 2$, the symbol of the operator A in the space $\mathbb{X}^m(\Gamma, \rho)$ is defined as follows:

$$\mathcal{A}_{\mathbb{X}^m(\Gamma,\rho)}(t,\xi) := \widetilde{a}(t) + \widetilde{b}(t)\mathcal{S}_{\mathbb{X}^m(\Gamma,\rho)}(t,\xi) + \widetilde{c}(t)\overline{\mathcal{S}_{\mathbb{X}^m(\Gamma,\rho)}(t,-\xi)}\,, \tag{3.34}$$

where

$$\mathcal{S}_{\mathbb{X}^m(\Gamma,\rho)}(t,\xi) := \begin{bmatrix} \coth \pi(i\beta_t + \xi) & -\dfrac{e^{\pi(\gamma_t - 1)(i\beta_t+\xi)}}{\sinh \pi(i\beta_t + \xi)} \\ \dfrac{e^{\pi(1-\gamma_t)(i\beta_t+\xi)}}{\sinh \pi(i\beta_t + \xi)} & -\coth \pi(i\beta_t + \xi) \end{bmatrix}, \quad \xi \in \mathbb{R}, \tag{3.35}$$

$$\beta_t := \begin{cases} 1/p + \alpha_j - \nu & \text{if } t \in \Gamma, \quad \mathbb{X}^m(\Gamma,\rho) = \mathbb{KH}_p^{m+\nu}(\Gamma,\rho), \\ & \qquad\qquad\qquad \mathbb{KB}_{p,q}^{m+\nu}(\Gamma,\rho), \\ 1/p & \text{if } t \neq t_1, \dots, t_n, \quad \mathbb{X}^m(\Gamma,\rho) = \mathbb{KW}_p^m(\Gamma,\rho), \\ 1/2 & \text{if } t \neq t_1, \dots, t_n, \quad \mathbb{X}^m(\Gamma,\rho) = \mathbb{KZ}_{m+\nu}^0(\Gamma,\rho), \\ 1/p + \alpha_j & \text{if } t = t_j, \quad \mathbb{X}^m(\Gamma,\rho) = \mathbb{KW}_p^m(\Gamma,\rho), \\ \alpha_j - \nu & \text{if } t = t_j, \quad \mathbb{X}^m(\Gamma,\rho) = \mathbb{KZ}_{m+\nu}^0(\Gamma,\rho), \end{cases} \tag{3.36}$$

$$\widetilde{d}(t) := \begin{bmatrix} d(t+0) & 0 \\ 0 & d(t-0) \end{bmatrix}, \qquad d \in \mathbb{PC}(\Gamma, \mathcal{T}_\Gamma), \quad t \in \Gamma,$$

$$\gamma_t := \begin{cases} 1 & \text{if} \quad t \neq t_1, \dots, t_n, \\ \gamma_j & \text{if} \quad t = t_j. \end{cases} \tag{3.37}$$

Let us note that the symbol would be a full matrix function if the corresponding operator contains the terms VS_Γ, VaI, aV, or $S_\Gamma V$ (see Remark 3.5).

Due to assumptions (3.5)–(3.7) and (3.27) we have $0 < \beta_t < 1$ for all $t \in \Gamma$ and the symbol $\mathcal{A}_{\mathbb{X}^m(\Gamma,\rho)}(t,\xi)$ represents a piecewise continuous uniformly bounded function of the variables $(t,\xi) \in \Gamma \times \mathbb{R}$.

Theorem 3.2. *Let Γ have no cusps, i.e., $0 < \gamma_j < 2$, $j = 1, \dots, n$ and let $\mathbb{X}^m(\Gamma,\rho)$ be defined by* (3.27). *Then equation* (3.33) *is Fredholm in the space $\mathbb{X}^m(\Gamma,\rho)$ if and only if*

$$\inf_{t\in\Gamma,\ \xi\in\mathbb{R}} \left|\det \mathcal{A}_{\mathbb{X}^m(\Gamma,\rho)}(t,\xi)\right| > 0\,. \tag{3.38}$$

If condition (3.38) *holds, then*

$$\operatorname{Ind} A = -\frac{1}{2\pi}\left\{ \left[\arg\det \mathcal{A}_{\mathbb{X}^m(\Gamma,\rho)}(t,+\infty)\right]_\Gamma + \sum_{j=1}^n \left[\arg\det \mathcal{A}_{\mathbb{X}^m(\Gamma,\rho)}(t_j,\xi)\right]_\mathbb{R} \right\}.$$

If, in particular, $c = 0$ and the operator $A = aI + bS_\Gamma$ has scalar coefficients ($N = 1$), then A is invertible in $\mathbb{X}^m(\Gamma,\rho)$ from the left or the right in dependence on whether $\operatorname{Ind} A \leq 0$ *or* $\operatorname{Ind} A \geq 0$, *respectively.*

Proof. If a singular integral operator is bounded in the space $\mathbb{X}^0(\Gamma,\rho)$, it is bounded in $\mathbb{X}^m(\Gamma,\rho)$ (see Theorem 3.1). This is also valid for any inverse operator and any regularizer to the canonical operator $A = aI + bS_\Gamma$. The same is true if $\Gamma = \mathbb{R}$ and $\rho(x) \equiv 1$, or $\Gamma = \mathbb{R}^+$ and $\rho(x) = x^\alpha$.

A similar simultaneous boundedness property, for all values of the parameter $m \in \mathbb{N}_0$, holds also for Mellin convolution operators $\mathfrak{M}_g^0$ in the spaces $\mathbb{X}^m(\mathbb{R}^+, x^\alpha)$ (for general boundedness of Mellin convolution operators we refer to J. Elschner's results in [El1] and in [PS1, Ch. 5]).

Thus, it suffices to prove the theorem for $m = 0$. For this case we apply quasi-localization (see [Du5, DLS1, Si1, Sp1, Ra1]). Note that localization in the weighted Hölder space is a special case (see [Po1, Sc1]). Let us expose here a short description of the approach. If $\mathcal{L}(\mathbb{X})$ denotes the algebra of all bounded operators in a Banach space $\mathbb{X}$ and $\mathfrak{S}(\mathbb{X}) \subset \mathcal{L}(\mathbb{X})$ is the ideal of all compact operators, then in the quotient algebra $\mathcal{L}(\mathbb{X})/\mathfrak{S}(\mathbb{X})$ (the Calkin algebra) the essential norm of an operator,

$$|||B|\mathcal{L}(\mathbb{X})||| := \inf_{T\in\mathfrak{S}(\mathbb{X})} \|B+T|\mathcal{L}(\mathbb{X})\|\,, \tag{3.39}$$

defines a norm of the coset which contains this operator.

K. Kuratowski introduced the measure of non-compactness $||\mathbb{Y}||_d$ (the Kuratowski measure) of a bounded set $\mathbb{Y} \subset \mathbb{X}$ as the minimal value of all numbers ε for which $\mathbb{Y}$ can be covered by an ε-net of a finite number of elements. The Kuratowski measure of the image of the unit sphere under an operator B,

$$||B|\mathcal{L}(\mathbb{X})||_d := ||B\mathcal{B}_{\mathbb{X}}(0,1)||\,, \quad \mathcal{B}_{\mathbb{X}}(0,1) := \{x \in \mathbb{X}\ :\ \|x\| = 1\}, \tag{3.40}$$

is called the measure of non-compactness of the operator B (see [AKPRS1]). Obviously, $\|B|\mathcal{L}(\mathbb{X})\|_d \le |||B|\mathcal{L}(\mathbb{X})|||$, while the equality $||B|\mathcal{L}(\mathbb{X})||_d = |||B|\mathcal{L}(\mathbb{X})|||$ holds for $\mathbb{X} = \mathbb{L}_p$ and does not hold for the Hölder-Zygmund spaces, where we have the inequality $|||B|\mathcal{L}(\mathbb{X})||| \le C\|B|\mathcal{L}(\mathbb{X})\|_d$ with some constant C independent of the operator B (see [AKPRS1, Po1]).

In [Po1], R. Pöltz proved that the Kuratowski measure of a multiplication operator gI, $g \in H_\nu(\Gamma)$, in the weighted Hölder space $H_\nu^0(\Gamma,\rho)$ coincides with the supremum-norm,

$$|||gI|\mathcal{L}(H_\nu^0(\Gamma,\rho))||| \le C\|gI|\mathcal{L}(H_\nu^0(\Gamma,\rho))\|_d\,, \tag{3.41}$$

$$\|gI|\mathcal{L}(H_\nu^0(\Gamma,\rho))\|_d = \|gI|C(\Gamma)\| := \sup_{t\in\Gamma}|g(t)|\,,$$

although for the usual norm we have the inequality

$$\|gI|\mathcal{L}(H_\nu^0(\Gamma,\rho))\| \le \|g|\mathbb{P}H_\nu(\Gamma,\mathcal{T}_\Gamma)\| := \max_{j=1,\dots,n} \|g|H_\nu(\Gamma_j)\|\,.$$

Property (3.21) enables a localization of coefficients similar to the case of the space $\mathbb{L}_p(\Gamma,\rho)$, exposed in [DLS1] (see also [Du7, RS1]).

A local quasi-equivalent representative of A at a point $t_0 \in \Gamma$ has the form

$$\begin{aligned} A_{t_0} &:= \widetilde{a}(t_0)I + \widetilde{b}(t_0)\begin{bmatrix} S_{\mathbb{R}^+} & -N_{\mathbb{R}^+,-\gamma_{t_0}} \\ N_{\mathbb{R}^+,\gamma_{t_0}} & -S_{\mathbb{R}^+} \end{bmatrix} \\ &\quad + \widetilde{c}(t_0)V\begin{bmatrix} S_{\mathbb{R}^+} & -N_{\mathbb{R}^+,-\gamma_{t_0}} \\ N_{\mathbb{R}^+,\gamma_{t_0}} & -S_{\mathbb{R}^+} \end{bmatrix}V \\ &= \widetilde{a}(t_0)I + \widetilde{b}(t_0)\begin{bmatrix} S_{\mathbb{R}^+} & -N_{\mathbb{R}^+,-\gamma_{t_0}} \\ N_{\mathbb{R}^+,\gamma_{t_0}} & -S_{\mathbb{R}^+} \end{bmatrix} - \widetilde{c}(t_0)\begin{bmatrix} S_{\mathbb{R}^+} & -N_{\mathbb{R}^+,\gamma_{t_0}} \\ N_{\mathbb{R}^+,-\gamma_{t_0}} & -S_{\mathbb{R}^+} \end{bmatrix}, \end{aligned}$$

$$S_{\mathbb{R}^+}\varphi(x) := \frac{1}{\pi i}\int_0^\infty \frac{\varphi(y)dy}{y-x}, \quad N_{\mathbb{R}^+,\gamma}\varphi(x) := \frac{1}{\pi i}\int_0^\infty \frac{\varphi(y)dy}{y-e^{i\gamma}x}, \quad x \in \mathbb{R}^+ \tag{3.42}$$

(see, e.g., [DLS1]). We prove that the local quasi-equivalent representative A_{t_0} is locally invertible in the space $\mathbb{L}_p(\mathbb{R}^+, x^{\alpha_{t_0}})$ for the Sobolev space, in $H_\nu(\mathbb{R}^+, x^{\alpha_{t_0}})$ for the weighted Hölder space, and in $\mathbb{H}_p^\nu(\mathbb{R}^+, x^{\alpha_{t_0}})$ for the weighted Bessel potential space. Let us consider these cases separately and, afterwards, the case of Besov spaces.

I. *The weighted Lebesgue space* $\mathbb{L}_p(\mathbb{R}^+, x^{\alpha_{t_0}})$.
Since the operator A_{t_0} is dilation invariant, i.e.,

$$\mathcal{D}_\lambda A_{t_0} = A_{t_0}\mathcal{D}_\lambda, \qquad \mathcal{D}_\lambda\varphi(x) := \varphi(\lambda x), \quad \forall\, \lambda,\, x \in \mathbb{R}^+, \tag{3.43}$$

A_{t_0} is locally invertible at $0 \in \mathbb{R}^+$ if and only if it is invertible in $\mathbb{L}_p(\mathbb{R}^+, x^{\alpha_{t_0}})$ (see [Du8]).

Next we replace A_{t_0} by the equivalent operator

$$\begin{aligned} A^0_{t_0} &:= x^{\alpha_{t_0}+\frac{1}{p}}A_{t_0}x^{-\alpha_{t_0}-\frac{1}{p}}I = \widetilde{a}(t_0)I + \widetilde{b}(t_0)\mathfrak{M}^0_{S_{\mathfrak{X}^m(\Gamma,\rho)}(t_0,\cdot)} + \widetilde{c}(t_0)\mathfrak{M}^0_{\overline{S_{\mathfrak{X}^m(\Gamma,\rho)}(t_0,\cdot)}} \\ &= \mathfrak{M}^0_{\mathcal{A}_{\mathfrak{X}^m(\Gamma,\rho)}(t_0,\cdot)} \end{aligned} \tag{3.44}$$

in the weighted space $\mathbb{L}_p(\mathbb{R}^+, x^{-1})$ (see [Du5, DLS1] and [RS1]). We recall that the Mellin convolution operator $\mathfrak{M}^0_g$ is invertible in $\mathbb{L}_p(\mathbb{R}^+, x^{-1})$ if and only if the symbol $g(\xi)$, which belongs to the L_p-multiplier class $\mathcal{M}_p(\mathbb{R})$, does not vanish, $\inf|g(\xi)| \neq 0$ for $\xi \in \mathbb{R}$, and that the inverse is $\mathfrak{M}^0_{g^{-1}}$ (see [Du6, DLS1, RS1]). The proof is completed in a standard way by application of the local principle: condition (3.38) is necessary and sufficient for A to have the Fredholm property in $\mathbb{L}_p(\Gamma, \rho)$. The index formula is proved by a standard homotopy argument (see [DLS1]).

II. *The weighted Bessel potential space* $\mathbb{H}_p^\nu(\mathbb{R}^+, x^{\alpha_{t_0}})$, $1/p - 1 < \nu - \alpha_{t_0} < 1/p$. We apply the lifting procedure (see (3.19)–(3.22)). As a result we get an equivalent operator,

$$\Lambda_-^\nu A_{t_0} \Lambda_+^{-\nu} = \widetilde{a}(t_0)I + \widetilde{b}(t_0) \begin{bmatrix} S_{\mathbb{R}^+}^\nu & -N_{\mathbb{R}^+,-\gamma_{t_0}}^\nu \\ N_{\mathbb{R}^+,\gamma_{t_0}}^\nu & -S_{\mathbb{R}^+}^\nu \end{bmatrix} - \widetilde{c}(t_0) \begin{bmatrix} S_{\mathbb{R}^+}^\nu & -N_{\mathbb{R}^+,\gamma_{t_0}}^\nu \\ N_{\mathbb{R}^+,-\gamma_{t_0}}^\nu & -S_{\mathbb{R}^+}^\nu \end{bmatrix}, \tag{3.45}$$

$$S_{\mathbb{R}^+}^\nu := \Lambda_-^\nu S_{\mathbb{R}^+} \Lambda_+^{-\nu}, \qquad N_{\mathbb{R}^+,\gamma_{t_0}}^\nu := \Lambda_-^\nu N_{\mathbb{R}^+,\gamma_{t_0}} \Lambda_+^{-\nu}.$$

It is known that all Mellin convolution operators (e.g., $N_{\mathbb{R}^+,\gamma_{t_0}}$) belong to the Banach algebra generated by the Fourier convolution operators W_a with symbols discontinuous at infinity[2], i.e., $a(-\infty) \neq a(+\infty)$. Moreover, the algebra $\mathfrak{M}(L_p(\mathbb{R}^+, x^{\alpha_{t_0}}))$ of all Mellin convolution operators in $L_p(\mathbb{R}^+, x^{\alpha_{t_0}})$ is generated by only two operators: by the identity I and by the Cauchy singular integral operator,[3] which is, at the same time, a convolution operator:

$$S_{\mathbb{R}^+} = \mathfrak{M}_{g_p}^0 = W_{-\operatorname{sign}\xi}, \qquad g_p(\xi) := \coth \pi \left(\frac{i}{p} + \xi \right), \quad \xi \in \mathbb{R} \tag{3.46}$$

(see (3.22) and [Du9, Lemma 2.2]).

Since we need only the local invertibility of the lifted convolution operator

$$\Lambda_-^\nu W_a \Lambda_+^{-\nu} = W_{a\varkappa_\nu}, \quad \varkappa_\nu(\xi) := \left(\frac{\xi - i}{\xi + i} \right)^\nu, \quad \xi \in \mathbb{R} \tag{3.47}$$

at $0 \in \mathbb{R}^+$, we can work with any local representative and its symbol. Since the operator gW_h with $g(\pm\infty) = h(\pm\infty) = 0$ is compact in $L_p(\mathbb{R})$ (see, e.g., [Du9]), and $\varkappa_\nu(-\infty) = e^{-2\pi i\nu}$, $\varkappa_\nu(+\infty) = 1$, it is easy to ascertain the local equivalence

$$W_{a\varkappa_\nu} \overset{0}{\sim} W_{a_\nu}, \qquad a_\nu = e^{-2\pi i\nu} a_- \chi_-(\xi) + a_+ \chi_+(\xi), \tag{3.48}$$

where $\chi_\pm(\xi)$ are the characteristic functions of the half-lines $\mathbb{R}^\pm$ and $a(\pm\infty) = a_\pm$.

---

[2]We have already proved, by a different method, that all four entries of the lifted matrix operator $\Lambda_-^\nu A_{t_0} \Lambda_+^{-\nu}$ in (3.45) belong to the algebra generated by $S_{\mathbb{R}^+}$ and I (see (3.10)–(3.16) and (3.24)–(3.25) above).

[3]This result is proved in [Du9, Lemma 2.2] for $p = 2$, but extends easily to arbitrary $1 < p < \infty$ because the closed sub-algebra $\mathfrak{M}(L_p(\mathbb{R}^+, x^{\alpha_{t_0}})) \subset \mathfrak{M}(L_2(\mathbb{R}^+, x^{\alpha_{t_0}}))$ is generated by the same operators I and $S_{\mathbb{R}^+}$.

The symbol $\mathcal{W}_{a_\nu}(\xi)$ of W_{a_ν} is

$$\mathcal{W}_{a_\nu}(\xi) = e^{-2\pi i\nu} a_- \left[1 + \coth \pi\Xi\right] + a_+ \left[1 - \coth \pi\Xi\right] \tag{3.49}$$
$$= \mathcal{G}_p(\xi) \left\{a_- \left[1 + \coth \pi\left(\Xi - \nu i\right)\right] + a_+ \left[1 - \coth \pi\left(\Xi - \nu i\right)\right]\right\} ,$$

$$\mathcal{G}_p(\xi) := \frac{e^{-\pi\nu i} \sinh \pi\left(\Xi - \nu i\right)}{\sinh \pi\Xi} \neq 0 , \quad \xi \in \mathbb{R} , \qquad \mathcal{G}_p(\pm\infty) = e^{-\pi\nu i} , \tag{3.50}$$
$$\Xi := \frac{i}{p} + i\alpha_{t_0} + \xi ,$$

and (3.50) holds since $0 < 1/p + \alpha_{t_0}$ and $1/p + \alpha_{t_0} - \nu < 1$.

As can be seen from (3.49), (3.50), to write down the symbol of the lifted operator $\Lambda_-^\nu B \Lambda_+^{-\nu}$, where B belongs to the Banach algebra generated by convolutions, we should detach the common non-vanishing factor $\mathcal{G}_p(\xi)$ and replace $1/p$ by $1/p - \nu$ in the definition of the symbol. This can also be interpreted as considering operators in the weighted space $\mathbb{L}_p(\mathbb{R}^+, x^{\alpha_{t_0} - \nu})$ instead of $\mathbb{L}_p(\mathbb{R}^+, x^{\alpha_{t_0}})$. The same holds for all four entries of the lifted operator $\Lambda_-^\nu A_{t_0} \Lambda_+^{-\nu}$ in (3.45), and the symbol of this operator acquires the form described in (3.34)–(3.37) as the symbol of A in the weighted Bessel potential space $\mathbb{KH}_p^\nu(\Gamma, \rho)$. The index formula is proved by a standard homotopy argument (see [DLS1]).

III. *The weighted Hölder space* $H_\nu^0(\mathbb{R}^+, x^{\alpha_{t_0}})$, $0 < \nu < 1$, $\nu < \alpha_{t_0} < \nu + 1$. The weighted Hölder space on the half-line is defined as follows:

$$H_\nu^0(\mathbb{R}^+, x^{\alpha_{t_0}}) := \left\{\varphi_0 = x^{\alpha_{t_0}} \varphi \in H_\nu(\mathbb{R}^+) \; : \; \varphi_0(0) = 0\right\} ,$$

$$\|\psi \mid H_\nu(\mathbb{R}^+)\| := \sup_{x\in\mathbb{R}^+} |\psi(x)| + \sup_{\substack{x_1, x_2 \in \mathbb{R}^+ \\ x_1 \neq x_2}} \frac{|\psi(x_2) - \psi(x_1)|}{\left|\dfrac{x_2}{x_2 + i} - \dfrac{x_1}{x_1 + i}\right|^\nu} . \tag{3.51}$$

Absolutely similar to the case of weighted L_p-spaces we prove that the operators $S_{\mathbb{R}^+}^{(\alpha)}$ and $N_{\mathbb{R}^+,\gamma}^{(\alpha)}$ (see (3.10)) belong to the Banach algebra of operators in the space $H_\nu^0(\mathbb{R}^+, x^{\alpha_{t_0}})$ generated by the two operators $S_{\mathbb{R}}^+$ and I provided the conditions

$$0 < \alpha + \alpha_{t_0} - \nu, \; \alpha_{t_0} < 1 , \quad 0 < |\gamma| < \pi \tag{3.52}$$

hold (see (3.10)–(3.16) and (3.24)–(3.25)).

The symbol of the singular integral operator $S_{\mathbb{R}_+}$ (see (3.42)) in the space $H_\nu^0(\mathbb{R}^+, x^{\alpha_{t_0}})$, given in [Du3, Du4], can be rewritten in the equivalent form

$$S_{H_\nu^0(\mathbb{R}^+, x^{\alpha_{t_0}})} := \coth \pi(i(\alpha_{t_0} - \nu + i\xi)) , \qquad \xi \in \mathbb{R} , \tag{3.53}$$

where $t = t_0$, $\beta_{t_0} = \alpha_{t_0} - \nu$ (cf. the diagonal terms in (3.35); this corresponds to the symbol of $S_{\mathbb{R}^+}$ in $L_p(\mathbb{R}^+)$ with $p = (\alpha_{t_0} - \nu)^{-1}$). It is easy to ascertain that the symbols of the entries $N_{\mathbb{R}^+, \mp\gamma_{t_0}}$ in (3.45) are exactly those which are inserted as the off-diagonal terms at $t = t_0$ in the symbol of $S_{\mathbb{Z}_\mu^0(\Gamma,\rho)}$ in (3.35). Obviously, $VS_{\mathbb{R}^+}V = -S_{\mathbb{R}^+}$ and $VN_{\mathbb{R}^+,\gamma}V = -N_{\mathbb{R}^+,-\gamma}$. Therefore we can easily write the symbol $\mathcal{A}_{t_0}(\xi)$ of the operator A_{t_0} in $H^0_\nu(\mathbb{R}^+, x^{\alpha_{t_0}})$, which coincides with the symbol $\mathcal{A}_{\mathbb{Z}^0_\mu(\Gamma,\rho)}(t_0,\xi)$ in (3.34). This accomplishes the proof for the Hölder-Zygmund space $\mathbb{Z}^0_\mu(\Gamma,\rho)$ because the condition $\inf|\det \mathcal{A}_{\mathbb{Z}^0_\mu(\Gamma,\rho)}(t_0,\xi)| \neq 0$ provides the criterion of the invertibility of the local operator A_{t_0} in the local space $H^0_\nu(\mathbb{R}^+, x^{\alpha_{t_0}})$ for all $t_0 \in \Gamma$. By the local principle, this coincides with the Fredholm criterion of A in $\mathbb{H}^0_\nu(\Gamma,\rho)$ and, by the above considerations, in $\mathbb{Z}^0_\mu(\Gamma,\rho)$. The index formula is proved by a standard homotopy argument (see [DLS1]).

IV. *The weighted Besov space* $\mathbb{B}^\nu_{p,q}(\Gamma,\rho)$.
Let condition (3.38) for the space $\mathbb{X}^m(\Gamma,\rho) = \mathbb{B}^\mu_{p,q}(\Gamma,\rho)$, $\mu = m+\nu$, hold. Then, as already proved, A is Fredholm in $\mathbb{H}^\nu_p(\Gamma,\rho)$. Hence A has a regularizer R: $AR = I+T_r$, $RA = I+T_\ell$ in $\mathbb{H}^\nu_p(\Gamma,\rho)$, where T_r, T_ℓ are compact operators. There exists a sufficiently small $\varepsilon > 0$ such that A has a regularizer in $\mathbb{H}^{\nu\pm\varepsilon}_p(\Gamma,\rho)$. This implies that R and the operators T_r, T_ℓ are all bounded in the spaces $\mathbb{H}^{\nu\pm\varepsilon}_p(\Gamma,\rho)$ and, due to the Interpolation Theorem 2.2.iii, R, T_r and T_ℓ are all bounded in the space $\mathbb{B}^\nu_{p,q}(\Gamma,\rho)$. Moreover, due to the interpolation property of compact operators (see Lemma 2.1), T_r, T_ℓ are both compact in $\mathbb{B}^\nu_{p,q}(\Gamma,\rho)$. Therefore, A is Fredholm in the space $\mathbb{B}^\nu_{p,q}(\Gamma,\rho)$.

If appropriate conditions hold and A is invertible in $\mathbb{H}^\nu_p(\Gamma,\rho)$, then A is invertible in $\mathbb{H}^{\nu\pm\varepsilon}_p(\Gamma,\rho)$ for a sufficiently small ε and the inverse A^{-1} is bounded in $\mathbb{B}^\nu_{p,q}(\Gamma,\rho)$ (see the foregoing case). Therefore the operator A is invertible in $\mathbb{B}^\nu_{p,q}(\Gamma,\rho)$. The index formula is proved, again, by a standard homotopy argument (see [DLS1]). □

Corollary 3.3. *Let*

$$A_0 = a_0 I + a_1 S_\Gamma = (a_0+a_1)(P_+ + GP_-), \quad P_\pm := \frac{1}{2}(I \pm S_\Gamma), \quad G := \frac{a_0 - a_1}{a_0 + a_1}.$$

Then condition (3.38) *holds if and only if the following two conditions are satisfied:*

i. $\inf_{t\in\Gamma} |a_0(t) \pm a_1(t)| > 0$;

ii′. $-2\pi\beta_{t_j} < \arg\dfrac{G(t_j-0)}{G(t_j+0)} < 2\pi(1-\beta_{t_j})$, $j = 1,\dots,n$, *where* β_{t_j} *is defined by* (3.36).

Furthermore, condition ii′ *is equivalent to the following:*

ii″. *$G(t)$ has the representation*

$$G(t) = G_0(t)\prod_{j=1}^n (t - z_0)^{\nu_j}_{t_j}, \quad G_0 \in C(\Gamma),$$

$$z_0 \in \Omega^+, \quad -\beta_{t_j} < \nu_j < 1 - \beta_{t_j}, \quad j = 1,\dots,n$$

and $(t-z_0)^{\nu_j}_{t_j}$ *is taken as a branch of* $(t-z_0)^{\nu_j}$ *which has a jump only at the point* $t_j \in \Gamma$.

If conditions i *and* ii′ (*or* i *and* ii″) *hold, then* $\operatorname{Ind} A_0 = \operatorname{ind} G$.

Remark 3.4. From Theorem 3.2 we find that the Fredholm properties and the index of the operator A (see (3.33)) in the space $\mathbb{X}^m(\Gamma,\rho)$ are independent of the smoothness parameter $m \in \mathbb{N}_0$. This means that if equation (3.33) has a solution $\varphi \in \mathbb{X}^0(\Gamma,\rho)$ for a given $f \in \mathbb{X}^m(\Gamma,\rho)$, then automatically $\varphi \in \mathbb{X}^m(\Gamma,\rho)$.

Remark 3.5. Equations more general than (3.33), such as

$$\widetilde{A}\varphi := a\varphi + bV\varphi + cS_\Gamma\varphi + dVS_\Gamma\varphi + eS_\Gamma V\varphi + gVS_\Gamma V\varphi = f\,, \tag{3.54}$$

are linear in the space $\mathbb{X}^m(\Gamma,\rho)$ over the field of the real numbers $\mathbb{R}$. After "doubling" the equation by adding the composition $V\widetilde{A}\varphi = Vf$ and introducing new vector-functions $\Phi := (\varphi, V\varphi)$, $F := (f, Vf)$, we get the equivalent equation

$$\begin{bmatrix} a & b \\ \overline{b} & \overline{a} \end{bmatrix}\Phi + \begin{bmatrix} c & e \\ \overline{d} & \overline{g} \end{bmatrix} S_\Gamma\Phi + \begin{bmatrix} g & d \\ \overline{e} & \overline{c} \end{bmatrix} VS_\Gamma V\Phi = F\,, \tag{3.55}$$

which is linear (the same as in (3.33)) and can be treated in the space $\mathbb{X}^m(\Gamma,\rho)$ over the field of complex numbers $\mathbb{C}$ (see [DL1, Li1]). We will only indicate the symbol of the operator $\widetilde{A}$ because the Fredholm properties and the index are defined by the symbol as in Theorem 3.2 (note that we do not need to double the size of the symbol of the operator $\widetilde{A}$ as this was done for the operator A in order to characterize the Fredholm property and the index of A). Namely,

$$\begin{aligned} \mathcal{A}_{\mathbb{X}^m(\Gamma,\rho)}(t,\xi) &:= \widetilde{a}(t) + \widetilde{b}(t)\mathcal{V} + \widetilde{c}(t)\mathcal{S}_{\mathbb{X}^m(\Gamma,\rho)}(t,\xi) \\ &+\widetilde{d}(t)\mathcal{V}\mathcal{S}_{\mathbb{X}^m(\Gamma,\rho)}(t,\xi) + \widetilde{e}(t)\mathcal{S}_{\mathbb{X}^m(\Gamma,\rho)}(t,\xi)\mathcal{V} + \widetilde{g}\mathcal{V}\mathcal{S}_{\mathbb{X}^m(\Gamma,\rho)}(t,\xi)\mathcal{V}\,, \end{aligned} \tag{3.56}$$

where, in addition to (3.34)–(3.37), we have to indicate the symbol $\mathcal{V} = \mathcal{V}_{\mathbb{X}^m(\Gamma,\rho)}$ of the complex conjugate operator:

$$\mathcal{V} := \begin{bmatrix} 0 & 1 \\ 1 & 0 \end{bmatrix},$$

which is independent of the point $t \in \Gamma$ and the space $\mathbb{X}^m(\Gamma,\rho)$.

Note that if $\mathcal{B}_{\mathbb{X}^m(\Gamma,\rho)}(t,\xi)$ is the symbol of B, the symbol for the operator VBV is

$$(\mathcal{V}\mathcal{B}\mathcal{V})_{\mathbb{X}^m(\Gamma,\rho)}(t,\xi) = \overline{\mathcal{B}(t,-\xi)}$$

(see [DLS1, § 1]). Therefore, $\mathcal{V}\mathcal{S}_{\mathbb{X}^m(\Gamma,\rho)}(t,\xi)\mathcal{V} = \overline{\mathcal{S}_{\mathbb{X}^m(\Gamma,\rho)}(t,-\xi)}$ (cf. (3.35)).

Remark 3.6. The readers familiar with [Du3, Du5, DLS1] will find differences in writing the symbol of the operators A (cf. (3.34)–(3.37)): the symbol of the operator A_0 defined in [Du3, Du5, DLS1] has a block-diagonal form

$$\mathcal{A}_{\mathbb{X}^m(\Gamma,\rho)}(t,\xi) = \begin{bmatrix} (\mathcal{A}_0)_{\mathbb{X}^m(\Gamma,\rho)}(t,\xi) & 0 \\ 0 & \overline{(\mathcal{A}_0)_{\mathbb{X}^m(\Gamma,\rho)}(t,-\xi)} \end{bmatrix}.$$

It turns out that it is sufficient to consider only the first block as a symbol of A_0. It is obvious that this does not influence the Fredholm criterion

$$\inf \det (\mathcal{A}_0)_{\mathbb{X}^m(\Gamma,\rho)}(t,\xi) \neq 0,$$

while for the index formula with $\mathcal{A}_{\mathbb{X}^m(\Gamma,\rho)}(t,\xi)$ we need to add the factor $1/2$.

4. Application: an oblique derivative problem for the Laplacian in a domain with a piecewise smooth boundary

Throughout this section Γ is a closed, oriented, simple (i.e., without self-intersections), piecewise Ljapunov curve in the complex plane $\mathbb{C}$, which borders a bounded domain Ω^+ as well as an unbounded domain Ω^- and has knots at $\mathcal{T}_\Gamma := \{t_1, \dots, t_n\} \subset \Gamma$. The boundary $\Gamma = \partial\Omega^+ = \partial\Omega^-$ consists of n arcs $\Gamma_j := [t_j, t_{j+1}] = \overset{\frown}{t_j t_{j+1}}$, $j = 1, \dots, n$, which are μ-smooth (see § 1) and oriented, with μ as in (1.4). Let $\pi\gamma_j$ be the interior angle with respect to Γ between Γ_{j-1} and Γ_j at the knot $t_j \in \mathcal{T}_\Gamma$ $(0 \leq \gamma_j \leq 2,\ \ j = 1, \dots, n)$. When $\gamma_j = 0$ or $\gamma_j = 2$, the domain Ω^+ has an outward or an inward peak, respectively, or, what is the same, the boundary curve Γ has a cusp. As usual, $\vec{\nu}(t) := (\nu_1(t), \nu_2(t))$ denotes the outer unit normal vector to Ω^+ at the point $t \in \Gamma \setminus \mathcal{T}_\Gamma$.

The main objective of the present section is to study the following boundary value problem (BVP): find a harmonic function

$$\Delta u(x) = 0\,, \quad x \in \Omega^\pm \tag{4.1}$$

with given oblique derivative (also known as the Poincaré problem)

$$(\partial_{\vec{\ell}(t)} u)^\pm(t) + c(t)u^\pm(t) = f(t)\,, \qquad t \in \Gamma, \tag{4.2}$$

$$\partial_{\vec{\ell}(t)} := \ell_1(t)\partial_{t_1} + \ell_2(t)\partial_{t_2},$$

where the coefficients are piecewise smooth such that

$$\operatorname{Im} \ell_1(t) \equiv \operatorname{Im} \ell_2(t) \equiv \operatorname{Im} c(t) \equiv 0\,, \qquad \ell_1, \ell_2, c \in \mathbb{KPC}^m(\Gamma, \mathcal{T}_\Gamma)\,, \tag{4.3}$$

and the space $\mathbb{KPC}^m(\Gamma, \mathcal{T}_{\Gamma_1})$ of piecewise m-smooth functions is defined similarly to (1.16).

It is common to write the oblique derivative boundary condition (4.2) in the form

$$a(t)(\partial_{\vec{\nu}(t)} u)^\pm(t) + b(t)(\partial_{\vec{s}(t)} u)^\pm(t) + c(t)u^\pm(t) = f(t)\,, \qquad t \in \Gamma\,, \tag{4.4}$$

where

$$a(t) = \ell_1(t) \cos\vartheta_t + \ell_2(t) \sin\vartheta_t\,, \qquad b(t) = -\ell_1(t) \sin\vartheta_t + \ell_2(t) \cos\vartheta_t\,, \tag{4.5}$$

and

$$\partial_{\vec{\nu}(t)} := \cos\vartheta_t \partial_{t_1} + \sin\vartheta_t \partial_{t_2}\,, \qquad \partial_{\vec{s}(t)} := -\sin\vartheta_t \partial_{t_1} + \cos\vartheta_t \partial_{t_2} \tag{4.6}$$

are the normal and the tangential derivatives, respectively, i.e., the derivatives with respect to the outer unit normal vector and the positively directed tangent vector at $t \in \Gamma$,

$$\vec{\nu}(t) := (\cos\vartheta_t, \sin\vartheta_t), \qquad \vec{s}(t) := (-\sin\vartheta_t, \cos\vartheta_t). \tag{4.7}$$

As usual, ϑ_t denotes the inclination of the outer unit normal vector with respect to the abscissa axis (see, e.g., [Mu1, § 74] and the recent book [Pa1]).

In the particular case where the oblique derivative vector $\vec{\ell}(t) = (\ell_1(t), \ell_2(t))$ coincides with the outer unit normal vector $\vec{\ell}(t) = \vec{\nu}(t)$ and $c(t) \equiv 0$, we get the Neumann BVP. If $\vec{\ell}(t) \equiv 0$ and $c(t) \equiv 1$ we have the Dirichlet BVP.

It is known that the usual function spaces, $\mathbb{W}^1_2(\Omega^\pm)$ for the solutions and $\mathbb{W}_2^{-\frac{1}{2}}(\Gamma)$ for the right-hand sides, cannot ensure solvability and uniqueness of solutions of BVPs in domains with outward peaks (see [DS1]). To describe suitable function spaces for the solutions and boundary data we recall the modified Smirnov-Sobolev space $\mathcal{KW}^m_p(\overline{\Omega^\pm}, \rho)$. This space consists of all functions in $\Omega^\pm$ which have finite norm

$$\|\psi|\mathcal{KW}^m_p(\overline{\Omega^\pm}, \rho)\| := \sup_{0<r<1} \|\psi|\mathbb{KW}^m_p(\Gamma^{(r)}, \rho)\|,$$

where $\Gamma^{(r)} := \{z = \omega(r\zeta) \,:\, |\zeta| = 1\}$ are the images of the concentric circles of radius r under the conformal mapping of the unit disk $\mathcal{D}_1$ onto the domain $\Omega^\pm$,

$$\omega \,:\, \mathcal{D}_1 \longrightarrow \Omega^\pm. \tag{4.8}$$

The weight function $\rho(t)$ is defined by (1.12) and we assume that the following conditions hold:

$$1 < p < \infty, \quad m = 0, \pm 1, \dots, , \quad -\frac{1}{p} < \alpha_j < 1 - \frac{1}{p}, \quad j = 1, \dots, n. \tag{4.9}$$

An equivalent definition of the modified Smirnov-Sobolev spaces $\mathcal{KW}^m_p(\overline{\Omega^\pm}, \rho)$ is the following: $\Phi \in \mathcal{KW}^m_p(\overline{\Omega^\pm}, \rho)$ if and only if $\Phi(z)$ is represented by the Cauchy integral in the form

$$\begin{aligned} &\Phi(z) = c_0 + C_\Gamma\varphi(z), \quad c_0 = \text{const}, \quad \varphi \in \mathbb{KW}^m_p(\Gamma, \rho), \\ &C_\Gamma\varphi(z) := \frac{1}{2\pi i}\int_\Gamma \frac{\varphi(\tau)d\tau}{\tau - z}, \quad z \in \Omega^\pm, \end{aligned} \tag{4.10}$$

and for a compact Ω^+ one can take $c_0 = 0$ (cf. [Pv1]).

We know that a function $\Phi \in \mathbb{W}^m_p(\overline{\Omega^\pm}, \rho)$ in general has traces $\Phi^\pm$ on the boundary Γ only for $m \geq 1$, see [Tr1]; the same is true for the modified spaces $\Phi \in \mathbb{KW}^m_p(\overline{\Omega^\pm}, \rho)$. In contrast to this fact, a function from the modified Smirnov-Sobolev space $\Phi \in \mathcal{KW}^m_p(\Gamma, \rho)$, represented by the Cauchy integral in (4.10), has the traces

$$\Phi^\pm(t) = c_0 \pm \frac{1}{2}\varphi(t) + \frac{1}{2}S_\Gamma\varphi(t), \qquad \Phi^\pm \in \mathbb{KW}^m_p(\Gamma, \rho) \tag{4.11}$$

for arbitrary $m = 0, \pm 1, \dots$, $1 < p < \infty$ provided $\rho(t)$ is defined in (1.12) and conditions (4.9) hold. For a negative $m = -1, -2, \dots$, the space $\mathbb{KW}_p^m(\Omega^\pm, \rho)$ is defined as the dual space to $\mathbb{KW}_{p'}^{-m}(\Omega^\pm, \rho^{-1})$, where $p' := p/(p-1)$.

The Sokhotski-Plemelj [Mu1] formulae (4.11) are well known for Hölder continuous (see [Mu1]) and Lebesgue integrable functions (see [GK1]) and for functions $\varphi \in \mathbb{KW}_p^m(\Gamma, \rho)$, which follows from Theorem 3.1.i since the mentioned spaces are dense in $\mathbb{KW}_p^m(\Gamma, \rho)$ under the asserted conditions.

If $\vec{\ell}(t) \not\equiv 0$ we take the right-hand side of (4.2) in the modified Sobolev space $\mathbb{KW}_p^{m-1}(\Gamma, \rho)$ and look for the solutions in the corresponding Smirnov-Sobolev space $\mathcal{KW}_{p,loc}^m(\overline{\Omega^\pm}, \rho)$,

$$\begin{aligned} &f \in \mathbb{KW}_p^{m-1}(\Gamma, \rho), && u \in \mathcal{KW}_p^m(\overline{\Omega^+}, \rho) \quad \text{for} \quad \Omega^+, \\ &u \in \mathcal{KW}_{p,loc}^m(\overline{\Omega^-}, \rho), && u(x) = \mathcal{O}(1) \quad \text{as} \quad |x| \to \infty \quad \text{for} \quad \Omega^-. \end{aligned} \tag{4.12}$$

We suppose that $\rho(t)$ is defined by (1.12) and that conditions (4.9) hold.

If $\vec{\ell}(t) \equiv 0$ we get the Dirichlet problem and replace (4.12) by

$$\begin{aligned} &f \in \mathbb{KW}_p^m(\Gamma, \rho), && u \in \mathcal{KW}_p^m(\overline{\Omega^+}, \rho) \quad \text{for} \quad \Omega^+, \\ &u \in \mathcal{KW}_{p,loc}^m(\overline{\Omega^-}, \rho), && u(x) = \mathcal{O}(1) \quad \text{as} \quad |x| \to \infty \quad \text{for} \quad \Omega^-. \end{aligned} \tag{4.13}$$

If the domain $\Omega^\pm$ has no outward peak, conditions (4.12) can be replaced by the following equivalent conditions, which are simpler:

$$\begin{aligned} &f \in \mathbb{KW}_p^{m-1}(\Gamma, \rho), && u \in \mathbb{KW}_p^m(\overline{\Omega^+}, \rho) \quad \text{for} \quad \Omega^+, \\ &u \in \mathbb{KW}_{p,loc}^m(\overline{\Omega^-}, \rho), && u(x) = \mathcal{O}(1) \quad \text{as} \quad |x| \to \infty \quad \text{for} \quad \Omega^-, \end{aligned}$$

and similarly for (4.13) (see [DSi1]).

Theorem 4.1. *Let Γ be piecewise $\mathbb{C}^m$-smooth, let the weight function $\rho(t)$ be defined by* (1.12) *and let conditions* (4.9) *be satisfied. Let $\ell_1, \ell_2, c \in \mathbb{KPC}^m(\Gamma)$ (see* (4.3)*) and introduce $G(t) := \ell_1(t) + i\ell_2(t)$.*

Further, let $\mathbb{T} := \{\zeta \in \mathbb{C} \; : \; |\zeta| = 1\}$ be the unit circle and $\mathcal{T}_\mathbb{T} := \{\zeta_j \; : \; \omega(\zeta_j) = t_j\,, \; j = 1, \dots, n\}$ (see (4.3)*) be the pre-image of all knots (the angular points and peaks) of Γ under the conformal mapping $\omega(z)$ of* (4.8)*.*

The oblique derivative problem (4.1), (4.4) *(or* (4.1), (4.2)*) is Fredholm if and only if one of the following conditions* **A** *or* **B** *is satisfied.*

A. $\inf_{t \in \Gamma} |G(t)| \neq 0$, *conditions* (4.12) *hold, and the following singular integral equation on the unit circle is Fredholm:*

$$P_\mathbb{T}^+ \varphi(\zeta) + F(\zeta) P_\mathbb{T}^- \varphi(\zeta) = f_0(\zeta)\,, \qquad \zeta \in \mathbb{T}\,, \tag{4.14}$$

where

$$P^{\pm}_{\mathbb{T}} := \frac{1}{2}(I \pm S_{\mathbb{T}}), \qquad f_0, \varphi \in \mathbb{KW}^{m-1}(\mathbb{T}, \mathcal{T}_{\mathbb{T}}),$$

$$F(\zeta) := \rho(\omega(\zeta))\overline{G(\omega(\zeta))}[G(\omega(\zeta))\overline{\rho(\omega(\zeta))}]^{-1}[\omega'(\zeta)]^{\frac{1}{p}}[\overline{\omega'(\zeta)}]^{-\frac{1}{p}}, \qquad (4.15)$$

$$f_0(\zeta) := 2[G(\omega(\zeta))]^{-1}\rho(\omega(\zeta))[\omega'(\zeta)]^{\frac{1}{p}}f(\omega(\zeta)), \quad \zeta \in \mathbb{T}. \qquad (4.16)$$

B. $G(t) \equiv 0$, $\inf_{t\in\Gamma} |c(t)| \neq 0$, *conditions* (4.13) *hold and the following singular integral equation on the unit circle is Fredholm:*

$$P^{+}_{\mathbb{T}}\psi(\zeta) + F(\zeta)P^{-}_{\mathbb{T}}\psi(\zeta) + \frac{F(\zeta)-1}{2}K_0\psi(\zeta) = f_0(\zeta), \qquad (4.17)$$

$$F(\zeta) := \rho(\omega(\zeta))[\overline{\rho(\omega(\zeta))}]^{-1}[\omega'(\zeta)]^{\frac{1}{p}}[\overline{\omega'(\zeta)}]^{-\frac{1}{p}}, \qquad f_0, \psi \in \mathbb{KW}^m_p(\mathbb{T}, \mathcal{T}_{\mathbb{T}}),$$

$$f_0(\zeta) := 2[c(\omega(\zeta))]^{-1}\rho(\omega(\zeta))[\omega'(\zeta)]^{\frac{1}{p}}f(\omega(\zeta)), \quad \zeta \in \mathbb{T}.$$

If one of these two conditions is satisfied we have furthermore in the corresponding case:

A. *The indices of the BVP* (4.1), (4.4) *and of the integral equation* (4.14) *are equal.*

The coefficient $F(\zeta)$ *in* (4.14) *is piecewise smooth, i.e.,* F *is in* $\mathbb{KPC}^m(\mathbb{T}, \mathcal{T}_{\mathbb{T}})$ (*see Remark* 4.2).

If $c(t) \equiv 0$, *the BVP* (4.1), (4.4) *and the modified* (*with the help of the one-dimensional operator* K_0) *integral equations*

$$\begin{cases} P^{+}_{\mathbb{T}}\varphi(\zeta) + F(\zeta)P^{-}_{\mathbb{T}}\varphi(\zeta) = f_0(\zeta), \\ K_0\varphi(\zeta) := \dfrac{1}{2\pi}\displaystyle\int_{-\pi}^{\pi} \varphi^{-}(e^{i\vartheta})d\vartheta = 0, \qquad \text{for} \quad \Omega^{-}, \end{cases} \qquad (4.18)$$

$$P^{+}_{\mathbb{T}}\varphi(\zeta) + F(\zeta)P^{-}_{\mathbb{T}}\varphi(\zeta) + \frac{F(\zeta)-1}{2}K_0\varphi(\zeta) = f_0(\zeta), \quad \text{for} \quad \Omega^{+}$$

are equivalent in the sense that there is a one-to-one correspondence between their solutions.

B. *The indices of the BVP* (4.1), (4.4) *and of the integral equation* (4.17) *are equal and, moreover, they are equivalent in the sense that there is a one-to-one correspondence between their solutions.*

The coefficient $F(\zeta)$ *in* (4.17) *is piecewise smooth, that is,* F *is in* $\mathbb{KPC}^m(\mathbb{T}, \mathcal{T}_{\mathbb{T}})$ (*see Remark* 4.2).

Proof. The oblique derivative problem (4.1), (4.4) (or (4.1), (4.2)) can also be written as follows (see [Mu1, §§74,75]):

$$\operatorname{Re}\,[G(t)(\Psi')^{\pm}(t) + c(t)\Psi^{\pm}(t)] = f(t), \quad t \in \Gamma,$$

$$u(x) = \operatorname{Re}\Psi(x), \quad \Psi \in \mathcal{KW}^m_p(\overline{\Omega^{\pm}}, \rho), \quad x \in \Omega^{\pm}, \qquad (4.19)$$

$$G(t) = \ell_1(t) + i\ell_2(t) = e^{i\vartheta_t}a(t) + ie^{i\vartheta_t}b(t) = e^{i\vartheta_t}a(t) + e^{i\frac{\pi}{2}+i\vartheta_t}b(t)$$

(see (4.5)). Indeed, since

$$\Psi = u + iv \in \mathcal{KW}_p^m(\overline{\Omega^\pm}, \rho), \quad \Psi' := \frac{\partial u}{\partial x} - i\frac{\partial u}{\partial y} \in \mathcal{KW}_p^{m-1}(\overline{\Omega^\pm}, \rho),$$

with the help of (4.5) and (4.6) we get

$$\begin{aligned}\mathrm{Re}\left[G(t)(\Psi')^\pm(t) + c(t)\Psi^\pm(t)\right] &= \ell_1(t)(\partial_{t_1}u)^\pm(t) + \ell_2(t)(\partial_{t_2}u)^\pm(t) + c(t)u^\pm(t)\\ &= a(t)(\partial_{\vec{\nu}(t)}u)^\pm(t) + b(t)(\partial_{\vec{s}(t)}u)^\pm(t) + c(t)u^\pm(t)\end{aligned}$$

and (4.19) follows.

The case **B.** Thus, we suppose $G(t) \equiv 0$ and follow the scheme of [DSi1, Theorem 1.16]. The analytic function defined by

$$\Phi(z) := \begin{cases} \rho(\omega(z))[\omega'(z)]^{\frac{1}{p}}\Psi(\omega(z)) & \text{for} \quad |z| < 1, \\ \overline{\rho\left(\omega\left(\frac{1}{\overline{z}}\right)\right)}\,\overline{\left[\omega'\left(\frac{1}{\overline{z}}\right)\right]^{\frac{1}{p}}}\,\overline{\Psi\left(\omega\left(\frac{1}{\overline{z}}\right)\right)} & \text{for} \quad |z| > 1 \end{cases} \tag{4.20}$$

belongs to the space $\mathcal{KW}_p^m(\overline{\mathcal{D}_1}, \mathcal{T}_\mathbb{T})$. This can be verified straightforwardly with the help of the following property of the conformal mapping ω:

$$\prod_{\zeta_j \in \Theta} (z - \zeta_j)^k \partial_z^k \omega \in C(\overline{\mathcal{D}_1}) \tag{4.21}$$

for all $k = 1, \ldots, m$, where $m \in \mathbb{N}$. Notice that property (4.21) was already proved in [DSi2, Theorem 5.1].

For the analytic function $\Phi(z)$ in (4.20) the boundary condition (4.19) acquires the form

$$\mathrm{Re}\,[c(\omega(\zeta))\Psi^\pm(\omega(\zeta))] = \frac{c(\omega(\zeta))}{2}\left[\frac{\Phi^+(\zeta)}{\rho(\omega(\zeta))[\omega'(\zeta)]^{\frac{1}{p}}} - \frac{\Phi^-(\zeta)}{\overline{\rho(\omega(\zeta))}\,\overline{[\omega'(\zeta)]^{\frac{1}{p}}}}\right] = f(\omega(\zeta)),$$

which can also be written as follows:

$$\Phi^+(\zeta) - F(\zeta)\Phi^-(\zeta) = f_0(\zeta), \quad \zeta \in \mathbb{T}, \tag{4.22}$$

with $F(\zeta)$ and $f_0(\zeta)$ defined by (4.17). It is easy to verify by having recourse to (4.21) that $f_0 \in \mathbb{KW}_p^m(\mathbb{T}, \mathcal{T}_\mathbb{T})$.

Since $\Phi \in \mathcal{KW}_p^m(\overline{\mathcal{D}_1}, \mathcal{T}_\mathbb{T})$, it has a representation of the form

$$\Phi(z) = -\frac{i}{2}K_0\psi + C_\mathbb{T} i\psi(z) = -\frac{i}{4\pi}\int_{-\pi}^{\pi}\psi(e^{i\vartheta})d\vartheta + \frac{1}{2\pi}\int_{|\tau|=1}\frac{\psi(\tau)d\tau}{\tau - z} \tag{4.23}$$

for all $|z| \neq 1$ with a density $i\psi$, $\psi \in \mathbb{KW}_p^m(\mathbb{T}, \mathcal{T}_\mathbb{T})$. If we apply the Sokhotski-Plemelj formulae for the boundary values of Φ (see (4.11)) we obtain (for a density ψ)

$$\Phi^\pm(\zeta) = -\frac{1}{2}K_0\psi \pm \frac{1}{2}\left[\psi(\zeta) \pm S_\mathbb{T}\psi(\zeta)\right] = -\frac{1}{2}K_0\psi \pm P_\mathbb{T}^\pm\psi(\zeta), \quad \zeta \in \mathbb{T},$$

and inserting this into (4.22) we get (4.17) for the density $\psi \in \mathbb{KW}_p^m(\mathbb{T}, \mathcal{T}_\mathbb{T})$.

Let us remind that we need only the real-valued solution $\psi = \operatorname{Re}\psi$ of (4.17). To this end let us verify that if $\psi \in \mathbb{K}\mathbb{W}^m_p(\mathbb{T}, \mathcal{T}_\mathbb{T})$ is a solution, then $\overline{\psi}$ is a solution as well. In fact, applying the relations

$$\overline{\zeta} = \frac{1}{\zeta}, \quad |\zeta| = 1, \quad \overline{\tau} = \frac{1}{\tau}, \quad d\overline{\tau} = -\frac{d\tau}{\tau^2}, \quad \frac{d\tau}{\tau} = id\vartheta \text{ for } \tau = e^{i\vartheta}, \quad -\pi < \vartheta < \pi$$

we find that

$$\begin{aligned}
&\overline{F(\zeta)} = F^{-1}(\zeta), \quad \overline{f_0(\zeta)} = F^{-1}(\zeta) f_0(\zeta) \quad \text{since } \overline{f} = f, \\
&\overline{P^{\pm}_\mathbb{T}\psi(\zeta)} = \frac{1}{2}\overline{\psi(\zeta)} \mp \frac{1}{2\pi i}\int\limits_{|\tau|=1} \frac{\overline{\psi(\tau)d\tau}}{\overline{\tau} - \overline{\zeta}} = \frac{1}{2}\overline{\psi}(\zeta) \mp \frac{1}{2\pi i}\int\limits_{|\tau|=1} \frac{\zeta}{\tau}\frac{\overline{\psi(\tau)}d\tau}{\tau - \zeta} \\
&\quad = P^{\mp}_\mathbb{T}\overline{\psi}(\zeta) \pm \frac{1}{2\pi i}\int\limits_{|\tau|=1} \overline{\psi(\tau)}\frac{d\tau}{\tau} = P^{\mp}_\mathbb{T}\overline{\psi}(\zeta) \pm K_0\overline{\psi}.
\end{aligned} \tag{4.24}$$

Now, if $\psi_0 \in \mathbb{K}\mathbb{W}^m_p(\mathbb{T}, \mathcal{T}_\mathbb{T})$ is a solution of equation (4.17), taking the complex conjugate and invoking (4.24) we get the same equality for $\overline{\psi}_0$:

$$P^+_\mathbb{T}\overline{\psi}_0(\zeta) + F(\zeta)P^-_\mathbb{T}\overline{\psi}_0(\zeta) + \frac{F(\zeta) - 1}{2}K_0\overline{\psi}_0 = f_0(\zeta), \quad \zeta \in \mathbb{T}.$$

Therefore, the real-valued function $\psi := \operatorname{Re}\psi = \frac{1}{2}(\psi_0 + \overline{\psi}_0)$ is a solution that we look for.

With a solution $\psi = \operatorname{Re}\psi$ of (4.17) at hand we find $\Phi(z)$ from (4.22), but the latter has the following symmetry property:

$$\Phi_*(z) := \overline{\Phi\left(\frac{1}{\overline{z}}\right)} = \Phi(z), \quad z \in \Omega^+ \cup \Omega^-,$$

as it follows from the definition (4.20). This property can be verified similarly to (4.24):

$$\begin{aligned}
\Phi_*(z) &= \overline{\Phi\left(\frac{1}{\overline{z}}\right)} = \frac{i}{2}K_0\psi + \frac{1}{2\pi}\int\limits_{|\tau|=1} \frac{\psi(\tau)\overline{d\tau}}{\overline{\tau} - \frac{1}{z}} = \frac{i}{2}K_0\psi + \frac{1}{2\pi}\int\limits_{|\tau|=1} \frac{z}{\tau}\frac{\psi(\tau)d\tau}{\tau - z} \\
&= -\frac{i}{2}K_0\psi + \frac{1}{2\pi}\int\limits_{|\tau|=1} \frac{\psi(\tau)d\tau}{\tau - z} = -\frac{i}{2}K_0\psi + iC_\mathbb{T}\psi(z) = \Phi(z).
\end{aligned} \tag{4.25}$$

Inserting $\Phi(z)$ in (4.20) we find first $\Psi(z)$ and afterwards $u = \operatorname{Re}\Psi$.

Conversely, if $\psi(\zeta)$ is a solution of (4.17) we easily ascertain that $\Psi(z)$ defined by (4.23) and (4.20) solves the BVP (4.19) and $u(z) = \operatorname{Re}\Psi(z)$ solves the Dirichlet BVP (4.1), (4.2), (4.13) with $\vec{\ell} \equiv 0$.

The case **A.** In this case we can ignore $c(t)$ (take $c(t) \equiv 0$) because, after equivalent reduction, the corresponding summand in the integral equation has a weakly singular kernel (the corresponding operator is compact) and has no influence on the Fredholm property and the index of the equation. In the rest of the proof we follow the scheme of [DSi1, Theorem 1.17].

The analytic function

$$\Phi(z) := \begin{cases} \rho(\omega(z))[\omega'(z)]^{\frac{1}{p}}\Psi'(\omega(z)) & \text{for} \quad |z| < 1\,, \\ \overline{\rho\left(\omega\left(\frac{1}{\overline{z}}\right)\right)}\,\overline{\left[\omega'\left(\frac{1}{\overline{z}}\right)\right]^{\frac{1}{p}}}\,\overline{\Psi'\left(\omega\left(\frac{1}{\overline{z}}\right)\right)} & \text{for} \quad |z| > 1\,, \end{cases} \tag{4.26}$$

belongs to the space $\mathcal{KW}_p^{m-1}(\overline{\mathcal{D}_1}, \mathcal{T}_{\mathbb{T}})$. This can be verified straightforwardly with the help of (4.21).

For the analytic function $\Phi(z)$ in (4.26) we get the following BVP:

$$\Phi^+(\zeta) - F(\zeta)\Phi^-(\zeta) = f_0(\zeta)\,, \quad \zeta \in \mathbb{T}\,, \tag{4.27}$$

where $f_0(\zeta)$ and $F(\zeta)$ are defined in (4.15)–(4.16). It is easy to see, by applying (4.21), that $f_0 \in \mathbb{KW}_p^{m-1}(\mathbb{T}, \mathcal{T}_{\mathbb{T}})$.

Since $\Phi \in \mathcal{KW}_p^{m-1}(\overline{\mathcal{D}_1}, \mathcal{T}_{\mathbb{T}})$, it has a representation by the Cauchy integral

$$\Phi(z) = -\frac{i}{2}K_0\varphi + C_{\mathbb{T}}i\varphi(z) = -\frac{i}{4\pi}\int_{-\pi}^{\pi}\varphi(e^{i\vartheta})d\vartheta + \frac{1}{2\pi}\int_{|\tau|=1}\frac{\varphi(\tau)d\tau}{\tau - z} \tag{4.28}$$

for all $|z| \neq 1$ with the density $i\varphi$, $\varphi \in \mathbb{KW}_p^{m-1}(\mathbb{T}, \mathcal{T}_{\mathbb{T}})$. If we apply the Sokhotski-Plemelj formulae for the boundary values (see (4.11)) we get equation (4.18).

Note that for the domain Ω^- we have to require in addition (see the condition in (4.18)) that

$$K_0\varphi = \frac{1}{2\pi}\int_{-\pi}^{\pi}\varphi(e^{i\vartheta})d\vartheta = 0\,.$$

To justify this we remind that $\Psi \in \mathcal{KW}_p^m(\overline{\Omega^\pm}, \rho)$ and that the derivative must vanish at infinity, i.e., $\Psi'(\infty) = 0$ (see (4.12)); therefore (see (4.26), (4.28))

$$\int_{-\pi}^{\pi}\varphi(e^{i\vartheta})d\vartheta = 2\pi\Phi(0) = 2\pi\rho(\omega(0))[\omega'(0)]^{\frac{1}{p}}\Psi'(\omega(0)) = 0$$

because $\omega(0) = \infty$.

Since we need only real-valued solutions $\varphi = \operatorname{Re}\varphi$ of (4.18), we verify by analogy to (4.24) that, together with φ_0, equations (4.18) have the solution $\overline{\varphi_0}$. Therefore the real-valued solution $\varphi := \operatorname{Re}\varphi_0 = \frac{1}{2}(\varphi_0 + \overline{\varphi_0})$ is the one we look for.

The function $\Phi(z)$ in (4.28) must have the symmetry property $\Phi_*(z) = \Phi(z)$ (cf. (4.25) and (4.26)). This can also be verified with the help of properties similar to (4.24) (see (4.25)).

Conversely, if $\varphi = \operatorname{Re}\varphi$ is a real-valued solution of (4.18), then the function $\Phi(z)$ defined by (4.26) solves the BVP (4.27), which implies that $u(x) := \operatorname{Re}\,\Phi(z)$ solves the BVP (4.1), (4.2) and (4.12) with $c(t) \equiv 0$. □

Remark 4.2. The coefficient $F(\zeta)$ in (4.14) and in (4.17) is piecewise smooth, namely $F \in \mathbb{KPC}^m(\mathbb{T}, \mathcal{T}_{\mathbb{T}})$. Moreover, we can indicate the jumps at the knots:

$$\frac{F(\zeta_j - 0)}{F(\zeta_j + 0)} = \frac{G(t_j+0)}{G(t_j-0)} \overline{\left[\frac{G(t_j-0)}{G(t_j+0)}\right]} \frac{\rho(\omega(\zeta_j-0))}{\rho(\omega(\zeta_j+0))} \overline{\left[\frac{\rho(\omega(\zeta_j+0))}{\rho(\omega(\zeta_j-0))}\right]}$$

$$\times \left[\frac{\omega'(\zeta_j-0)}{\omega'(\zeta_j+0)}\right]^{\frac{1}{p}} \overline{\left[\frac{\omega'(\zeta_j+0)}{\omega'(\zeta_j-0)}\right]^{\frac{1}{p}}}$$

$$= \exp\left\{2i[\arg G(t_j+0) - \arg G(t_j-0)] - 2\pi i\left(\frac{1}{p} + \alpha_j\right)(1-\gamma_j)\right\} \tag{4.29}$$

$$= \exp\left\{2i[\arg G(t_j+0) - \arg G(t_j-0)] - \frac{2\pi i}{p} - 2\pi i\alpha_j + 2\pi i\left(\frac{1}{p} + \alpha_j\right)\gamma_j\right\},$$

where $\pi\gamma_j$ is the interior angle at the knot $t_j \in \mathcal{T}_\Gamma$ and α_j is the exponent of the weight at the same t_j. In fact,

$$\frac{\rho(\omega(\zeta_j-0))}{\rho(\omega(\zeta_j+0))} = \lim_{\varepsilon\to 0}\left[\frac{\omega(e^{-i\varepsilon}\zeta_j) - t_j}{\omega(e^{i\varepsilon}\zeta_j) - t_j}\right]^{\alpha_j} = \lim_{\varepsilon\to 0}\left[\frac{\dfrac{\omega(e^{-i\varepsilon}\zeta_j) - \omega(\zeta_j)}{e^{-i\varepsilon}\zeta_j - \zeta_j}}{\dfrac{\omega(e^{i\varepsilon}\zeta_j) - \omega(\zeta_j)}{e^{i\varepsilon}\zeta_j - \zeta_j}}\right]^{\alpha_j}$$

$$= \left[\frac{\omega'(\zeta_j-0)}{\omega'(\zeta_j+0)}\right]^{\alpha_j} = \exp[-2\pi i\alpha_j(1-\gamma_j)].$$

From (4.29) and Corollary 3.3 it is clear that even if $G(t)$ is continuous at one of the outward peaks,

$$G(t_j - 0) = G(t_j + 0) \quad \text{when} \quad \gamma_j = 0,$$

then the corresponding singular integral operator in (4.14) and (4.17) is not Fredholm (moreover, is not normally solvable, i.e., has non-closed image).

Due to Theorem 4.1 we are able to apply Theorem 3.2 to the oblique derivative problem (4.1), (4.4) (or to (4.1), (4.2); cf. [DSi1]).

Acknowledgement. The authors would like to thank Albrecht Böttcher for various helpful suggestions during the preparation of the manuscript.

References

[AKPRS1] R. Akhmerov, M. Kamenskij, A. Potapov, R. Rodkina and B. Sadovskij, *Measures of Noncompactness and Condensing Operators* (Transl. from the Russian), Operator Theory: Advances and Applications **55**, Birkhäuser, Basel 1992.

[BK1] A. Böttcher and Yu.I. Karlovich, *Carleson Curves, Muckenhoupt Weights, and Toeplitz Operators*, Progress in Mathematics **154**, Birkhäuser Verlag, Basel 1997.

[CD1] O. Chkadua and R. Duduchava, Pseudodifferential equations on manifolds with boundary: Fredholm property and asymptotic, *Mathematische Nachrichten* **222** (2001), 79–139.

[Du1] R. Duduchava, On singular integral operators in Hölder spaces with weights, *Sov. Math. Dokl.* **11** (1970), 304–308; translation from *Dokl. Akad. Nauk SSSR* **191** (1970), 16–19.

[Du2] R. Duduchava, On the boundedness of the singular integral operator in Hölder spaces with weights, *Matematicheskie Issledovania* **5**:1 (1970), 56–76, Kishinev, Stiintsa (Russian).

[Du3] R. Duduchava, The algebra of one-dimensional singular integral operators in spaces of Hölder functions with weight, *Trudy Tbiliskogo Matematicheskogo Instituta Akademii Nauk Gruzinskoi SSR* **41** (1973), 19–52 (Russian).

[Du4] R. Duduchava, On singular integral operators on piecewise smooth lines; in: *Research Notes in Mathematics* **8**, *Functional Theoretic Methods in Differential Equations*, Pitman, London, pp. 109–131, 1976.

[Du5] R. Duduchava, On bisingular integral operators with discontinuous coefficients, *Mathematics USSR, Sbornik* **30** (1976), 515–537.

[Du6] R. Duduchava, *Integral Equations with Fixed Singularities*, Teubner, Leipzig 1979.

[Du7] R. Duduchava, On general singular integral operators of the plane theory of elasticity, *Rendiconti Sem. Mat. Univers. e Politecn. Torino* **42**:3 (1984), 15–41.

[Du8] R. Duduchava, On multidimensional singular integral operators, I–II, *Journal of Operator Theory* **11** (1984), 41–76, 199–214.

[Du9] R. Duduchava, On algebras generated by convolutions and discontinuous functions, *Integral Equations and Operator Theory* **10** (1987), 505–530.

[DL1] R. Duduchava and T. Latsabidze, On the index of singular integral equations with complex conjugate functions on piecewise smooth lines, *Trudy Tbilisskogo Matematicheskogo Instituta Akademii Nauk Gruzinskoi SSR* **76** (1985), 40–59 (Russian).

[DLS1] R. Duduchava, T. Latsabidze and A. Saginashvili, Singular integral operators with the complex conjugation on curves with cusps, *Integral Equations and Operator Theory* **22** (1995), 1–36.

[DSi1] R. Duduchava and B. Silbermann, Boundary value problems in domains with peaks, *Memoirs on Differential Equations and Mathematical Physics* **21** (2000), 1–121.

[DSi2] R. Duduchava and B. Silbermann, The Cisotti formulae and conformal mapping of domains with peaks (to appear).

[DS1] R. Duduchava and F.-O. Speck, Pseudo-differential operators on compact manifolds with Lipschitz boundary, *Mathematische Nachrichten* **160** (1993), 149–191.

[DS2] R. Duduchava and F.-O. Speck, Singular integral equations in special weighted spaces, *Georgian Mathematical Journal* **7** (2000), 633–642.

[El1] J. Elschner, On spline approximation for a class of non-compact integral equations, *Mathematische Nachrichten* **146** (1990), 271–321.

[Es1] G. Eskin, *Boundary Value Problems for Elliptic Pseudodifferential Equations* AMS, Providence, Rhode Island 1981.

[GK1] I. Gohberg and N. Krupnik, *One-Dimensional Linear Singular Integral Equations*, **I–II**, Operator Theory: Advances and Applications **53–54**, Birkhäuser Verlag, Basel 1992.

[Kh1] B. Khvedelidze, Linear discontinuous boundary value problems of function theory, singular integral equations and their applications, *Trudy Tbilisskogo Matematicheskogo Instituta Akademii Nauk Gruzinskoi SSR* **23** (1957), 3–158 (Russian).

[Kr1] M. Krasnosel'skij, On a theorem of M. Riesz, *Sov. Math. Dokl.* **1** (1960), 229–231; translation from *Dokl. Akad. Nauk SSSR* **131** (1960), 246–248.

[Li1] G. Litvinchuk, *Solvability Theory of Boundary Value Problems and Singular Integral Equations With Shift*, Mathematics and its Applications **523**, Kluwer Academic Publishers, Dordrecht 2000.

[MSh1] V. Maz'ya and T. Shaposhnikova, *Theory of Multipliers in Spaces of Differentiable Functions*, Monographs and Studies in Mathematics **23**, Pitman, Boston 1985.

[MS1] V. Maz'ya and A. Solov'ev, L_p-theory of boundary integral equations on a contour with outward peak, *Integral Equations and Operator Theory* **32** (1998), 75–100.

[Mu1] N. Muskhelishvili, *Singular Integral Equations*, Nordhoff, Groningen 1953 (The last Russian edition: Nauka, Moscow 1968).

[Pa1] B. Paneah, *The Oblique Derivative Problem. The Poincaré-problem*, Mathematical Topics **17**, Wiley-VCH, Weinheim 2000.

[Pv1] I. Privalov, *Randeigenschaften analytischer Funktionen*, Deutscher Verlag der Wissenschaften, Berlin 1956 (Russian original: GITTL, Moskva 1950).

[Po1] R. Pöltz, Operators of local type in spaces of Hölder functions, Semin. Anal. 1986/87, Weierstrass Institute, Berlin 1987, 107–122 (Russian).

[PS1] S. Prößdorf and B. Silbermann, *Numerical Analysis for Integral and Related Operator Equations*, Operator Theory: Advances and Applications **52**, Birkhäuser Verlag, Basel 1991.

[Ra1] V. Rabinovich, Pseudodifferential equations in unbounded regions, *Sov. Math. Dokl.* **12** (1971), 452–456 (translation from *Dokl. Akad. Nauk SSSR* **197** (1971), 284–287.

[RS1] S. Roch and B. Silbermann, The Calkin image of algebras of singular integral operators, *Integral Equations and Operator Theory* **12** (1989), 854–897.

[Sc1] H. Schulze, On singular integral operators on weighted Hölder spaces, *Wiss. Z. Techn. Univ. Chemnitz* **33**:1 (1991), 37–47.

[Si1] I. Simonenko, A new general method of investigating linear operator equations of the type of singular integral equations, I, II *Izv. Akad. Nauk SSSR, Ser. Mat.* **29** (1965), 567–586, 757–782 (Russian).

[Sp1] F.-O. Speck, *On Generalized Convolution Operators and a Class of Integro-Differential Equations*, PhD dissertation, TH Darmstadt, 1974 (German).

[St1] E. Stein, *Singular Integrals and Differentiability Properties of Functions*, Princeton Univ. Press, Princeton 1970.

[Tr1] H. Triebel, *Interpolation Theory, Function Spaces, Differential Operators*, 2nd edition, Johann Ambrosius Barth Verlag, Heidelberg and Leipzig 1995.

L.P. Castro
Departamento de Matemática,
Universidade de Aveiro,
Campus Universitário,
3810-193 Aveiro, Portugal
e-mail: lcastro@mat.ua.pt

R. Duduchava
A. Razmadze Mathematical Institute,
Academy of Sciences of Georgia,
1, M. Alexidze str.,
Tbilisi 93, Georgia
e-mail: duduch@rmi.acnet.ge

F.-O. Speck
Departamento de Matemática,
Instituto Superior Técnico, U.T.L.,
Avenida Rovisco Pais,
1049-001 Lisboa, Portugal
e-mail: fspeck@math.ist.utl.pt

Received: 4 October 2001

Operator Theory:
Advances and Applications, Vol. 135, 145–160

Spline Approximation Methods for the Biharmonic Dirichlet Problem on Non-Smooth Domains

Victor D. Didenko and Bernd Silbermann

Abstract. For the approximate solution of the biharmonic Dirichlet problem we propose and study a boundary element method based on the integral equation of Muskhelishvili. Such an approach has a number of advantages, for instance, this equation does not have any critical geometry and in the case of smooth boundaries the method always converges. If the boundary has corner points, then the convergence of the method depends on the invertibility of some operators from a Toeplitz algebra.

1. Introduction

Let D be a two-dimensional domain bounded by a simple closed piecewise smooth contour Γ and let $\bar{D} := D \cup \Gamma$. In the present work we analyse spline Galerkin approximation methods for the biharmonic Dirichlet problem

$$\begin{cases} \Delta^2 \mathbf{U}|_D &= 0, \\ \mathbf{U}|_\Gamma &= f_1, \\ \left.\frac{\partial \mathbf{U}}{\partial \mathrm{n}}\right|_\Gamma &= f_2. \end{cases} \tag{1}$$

Here $\frac{\partial \mathbf{U}}{\partial \mathrm{n}}(t)$ is the outward normal derivative to Γ at the point $t \in \Gamma$, which is defined everywhere except the corner points of Γ. We assume that the function $\mathbf{U}$ is sought in $W_p^1(\overline{D}) \cap W_p^4(D)$. The notation $W_p^k(X)$ is used for the Sobolev space of k-times differentiable functions on X the derivatives of which belong to the corresponding space $L_p(X)$. The biharmonic problem (1) has been intensively studied by different authors. The reader can consult [17] for results concerning its solvability and solution properties.

Problem (1) arises in different branches of applied mathematics. For example, the behaviour of plane "slow" viscous flows, deflection of plates, elastic equilibrium of solids as well as a number of other problems can be modelled by means of the biharmonic equation [8, 9, 22, 23, 24, 25]. It is therefore no wonder that this problem has been attracting great attention by numerical analysts. It suffices to say that one of the most powerful approximation procedures, namely, the Galerkin method was discovered while considering a special case of problem (1). Among the variety of approaches to the numerical treatment of problem (1) one can distinguish the so-called boundary element methods [2, 4, 11, 12, 19]. They allow us to reduce the dimension of the initial problem and, as a result, to reduce the computation cost drastically. The authors of the afore-mentioned papers usually use different modifications of the integral equation proposed by C. Christiansen and P. Hougaard [5, 6, 16] or an integral equation of first kind [18]. Although such approaches are widely used, they have some drawbacks. Thus, for some boundaries called "critical" the corresponding integral operators become non-invertible and corrections are to be done before one can start with the construction of approximation methods. In addition, the analysis of the stability of the approximation methods proposed has been accomplished for smooth boundaries only, though conditions for the invertibility of the corresponding integral operators are available for piecewise smooth contours as well [10].

On the other hand, there is a very nice "complex" approach to the problem (1). It takes its beginning from the works of N.I. Muskhelishvili [23] and leads to integral equations without critical geometry. However, in a strange way the Muskhelishvili equation remains an almost unknown quantity in numerical analysis. The most common approach to the approximate solution of this equation is based on trigonometric Fourier expansions and was proposed by N.I. Muskhelishvili himself in the middle of thirties (cf. [3, 7, 21]). Since then in numerical analysis there has been developed a lot of new powerful approximation methods, but they have not been implemented and studied in the case of the Muskhelishvili equation. We mention the papers [26, 27], which deal with spline approximations of solutions of the Muskhelishvili equation but do not contain any stability analysis.

It is notable that the integral operators in the Muskhelishvili equation can be "locally" represented as elements of an algebra of Mellin operators with conjugation. The stability of approximation methods for Mellin convolutions was studied in [20], so we could apply some of those results to the operators appearing in the case of the Muskhelishvili equation. It should be also mentioned that from the practical point of view the convergence of the methods considered has to be proved in spaces of differentiable functions. However, the technique used here is well adapted to the norms of L_p spaces. Therefore, we first show the stability in the spaces L_p and then, using [28], we obtain some results for Sobolev norms.

This paper is organized as follows. First, we reduce the initial problem (1) to a boundary value problem for two analytic functions in the domain D. One of the functions is a solution of the Muskhelishvili equation. However, the corresponding integral operator is not invertible on the space we work with. Nevertheless, it can be corrected in such a way that a newly obtained operator is invertible and the solution of the associated integral equation is simultaneously a solution of the Muskhelishvili equation. Afterwards we study the stability of spline Galerkin method for the auxiliary integral equation and obtain an approximate solution for the equation of Muskhelishvili.

2. Auxiliary results

This section contains results which are needed to construct an approximate solution to problem (1). Some of these results are known. In slightly modified form, they can be found in the literature dealing with problems of elasticity theory. However, we include them to make the paper self-contained. On the other hand, below we will mention auxiliary results concerning the invertibility of the integral operators related to our problem. Their proofs will be published in another paper.

Let z denote the complex coordinate of a point of $D \cup \Gamma$, so $z = x + iy$, and let $c_1, c_2, \ldots, c_l$ be the corner points of Γ. By ω_j we denote the angle between the corresponding semi-tangents at the point c_j. Hence,

$$\omega_j \in (0, 2\pi), \quad \omega_j \neq \pi, \quad j = 1, 2, \ldots, l.$$

For real numbers p and α_j $(j = 1, 2, \ldots, l)$ satisfying $p > 1$ and

$$0 < \alpha_j + \frac{1}{p} < 1, \quad j = 1, 2, \ldots, l, \tag{2}$$

we introduce the weight function

$$\rho = \rho(t) = \prod_{j=1}^{l} |t - c_j|^{\alpha_j}, \quad t \in \Gamma,$$

and by $L_p(\Gamma, \rho)$ we denote the set of all Lebesgue measurable functions f such that

$$||f|| := \left(\int_\Gamma |f(t)\rho(t)|^p |dt| \right)^{1/p} < \infty.$$

Then, $W_p^1(\Gamma, \rho)$ refers to the Sobolev space of all functions the derivatives of which belong to $L_p(\Gamma, \rho)$.

Let $\alpha = \alpha(t), t \in \Gamma$ be the angle between the real axis and the outward normal n to Γ at the point t. By s we denote the unit vector defined by the requirement that the angle between s and the real axis is $\alpha - \pi/2$.

Proposition 2.1. *Let* $(f_1, f_2) \in W_p^1(\Gamma, \rho) \times L_p(\Gamma, \rho)$. *If* $\mathbf{U} = \mathbf{U}(x, y)$ *is a solution of problem* (1), *then it can be represented in the form*

$$\mathbf{U}(x, y) = \operatorname{Re}\left[\bar{z}\psi(z) + \chi(z)\right], \quad z = x + iy, \tag{3}$$

where ψ *and* χ *are the functions analytic in* D *that satisfy the boundary condition*

$$\overline{\psi(t)} + \bar{t}\psi'(t) + \chi'(t) = e^{-i\alpha}\left(f_2 + i\frac{\partial f_1}{\partial s}\right), \quad t \in \Gamma. \tag{4}$$

Proof. It is well known [23] that any biharmonic function can be represented in the form (3) with some analytic functions ψ and χ. So our task now is to find these functions in such a way that they will satisfy the boundary conditions (1).

Since f_1 belongs to the Sobolev space $W^1_p(\Gamma,\rho)$ and since $\mathbf{U}$ was supposed to be differentiable on $\overline{D}$, one can write

$$f_2 + i\frac{\partial f_1}{\partial s} = \frac{\partial \mathbf{U}}{\partial \mathrm{n}} + i\frac{\partial \mathbf{U}}{\partial s}$$

$$= (\cos\alpha + i\sin\alpha)\left(\frac{\partial \mathbf{U}}{\partial x} - i\frac{\partial \mathbf{U}}{\partial y}\right) = e^{i\alpha}\overline{(\mathbf{U}_x + i\mathbf{U}_y)}. \tag{5}$$

On the other hand, if we represent the function ψ of (3) in the form $\psi = u + iv$, then

$$\mathbf{U}(x,y) = xu(x,y) + yv(x,y) + \operatorname{Re}\chi(x,y).$$

Immediate calculations and the Cauchy-Riemann equations lead to the formula

$$\mathbf{U}_x(z) + i\mathbf{U}_y(z)$$

$$= u(x,y) + xu_x(x,y) + yv_x(x,y) + \operatorname{Re}\chi_x(x,y) +$$

$$+ i(xu_y(x,y) + v(x,y) + yv_y(x,y) + \operatorname{Re}\chi_y(x,y))$$

$$= \psi(z) + z\overline{\psi'(z)} + \overline{\chi'(z)}. \tag{6}$$

Comparing (5) and (6) we obtain that the functions ψ and χ of (3) satisfy the boundary condition (4). □

Assume for a moment that the boundary values $\psi = \psi(t), \chi = \chi(t), t \in \Gamma$ of the functions ψ and χ have been found. Then by the Cauchy's integral formula,

$$\psi(z) = \frac{1}{2\pi i}\int_\Gamma \frac{\psi(t)dt}{t-z}, \quad z \in D,$$

$$\chi(z) = \frac{1}{2\pi i}\int_\Gamma \frac{\chi(t)dt}{t-z}, \quad z \in D, \tag{7}$$

one will be able to represent the biharmonic function $\mathbf{U}$ inside of the domain D with the help of (3). Later on we will see that the function $\psi = \psi(t), t \in \Gamma$ is the solution of an integral equation, so if ψ is known, then

$$\chi'(t) = f(t) - \overline{\psi(t)} - \bar{t}\psi'(t), \quad t \in \Gamma, \tag{8}$$

where f denotes the right-hand side of (4).

It is clear that equation (8) allows us to restore the function χ. Really, let $t = t(s)$ be a 1-periodic parameterization of Γ, and let $s_1, s_2, \ldots, s_l$ be those points on [0,1) which correspond to the corner points of Γ, i.e.,

$$t(s_j) = c_j,\ j = 1, 2, \ldots, l.$$

The function $t = t(s)$ is continuously differentiable on (s_j, s_{j+1}) because Γ was supposed to be a piecewise smooth curve. We set $s_{l+1} = s_1 + 1$, let Γ_k be the subarc of Γ which joins the points t_k and t_{k+1}, and let $s \in [s_k, s_{k+1}]$. Using (8) we can write

$$\chi'(t(s))t'(s) = f(t(s))t'(s) - \overline{\psi(t(s))}t'(s) - \overline{t(s)}\psi'(t(s))t'(s)$$

for any $s \in (s_k, s_{k+1})$ and hence introduce the functions

$$\begin{aligned}\zeta_k(t(s)) &= \int_{s_k}^{s} f(t(s))t'(s)ds - \int_{s_k}^{s} \overline{\psi(t(s))}t'(s)ds \\ &+ \int_{s_k}^{s} \frac{d\overline{t(s)}}{ds}\psi(t(s))ds - \overline{t(s)}\psi(t(s)),\ k = 1, 2, \ldots, l. \end{aligned} \qquad (9)$$

Then the function χ may be represented in the form

$$\chi(t) = \begin{cases} \zeta_1(t) + C & \text{if} \quad t \in \Gamma_1 \\ \zeta_2(t) + (\zeta_1(c_2) - \zeta_2(c_2)) + C & \text{if} \quad t \in \Gamma_2 \\ \ldots\ldots\ldots\ldots\ldots & \\ \zeta_l(t) + (\zeta_{l-1}(c_l) - \zeta_l(c_l)) + \ldots + (\zeta_1(c_2) - \zeta_2(c_2)) + C & \text{if} \quad t \in \Gamma_l \end{cases} \qquad (10)$$

The indefinite constant C must be chosen to satisfy the first of the boundary conditions (1).

Therefore, if we would have found some approximations $\{\psi_n\}$ of the function ψ such that

$$||\psi_n - \psi||_{L_p(\Gamma,\rho)} \to 0 \quad \text{as } n \to \infty,$$

then we might construct "good" approximations χ_n of χ replacing ψ by ψ_n in (9)–(10) and obtain

$$||\chi_n - \chi||_{L_p(\Gamma,\rho)} \to 0 \quad \text{as } n \to \infty,$$

with the same rate of convergence. So, our task now is to find approximations of the functions ψ. To achieve the goal we will use the Muskhelishvili equation. Let us recall [14, 15] that if ψ is a solution of (4) and belongs to $W_p^1(\Gamma, \rho)$, then it satisfies the equation

$$R\psi(t) \equiv -\overline{\psi(t)} - \frac{1}{2\pi i}\int_\Gamma \overline{\psi(\tau)}d\log\frac{\overline{\tau} - \overline{t}}{\tau - t} - \frac{1}{2\pi i}\int_\Gamma \psi(\tau)d\frac{\overline{\tau} - \overline{t}}{\tau - t} = f_0(t),\ t \in \Gamma, \quad (11)$$

where

$$f_0(t) = -\frac{1}{2}f(t) + \frac{1}{2\pi i}\int_\Gamma \frac{f(\tau)d\tau}{\tau - t}$$

and f is the function of the right-hand side of (4). Here we can mention the two properties of the integral operator R which probably prevent us from applying approximation methods to the equation (11). First, the operator R has been investigated in the spaces $W_p^1(\Gamma, \rho)$ (see [14, 15]), but at present there are no efficient tools for studying the stability of approximation methods in spaces $W_p^1(\Gamma, \rho)$ when

Γ possesses corner points. Secondly, R is not invertible, neither as an operator acting on $W_p^1(\Gamma, \rho)$ nor as one on $L_p(\Gamma, \rho)$. Fortunately, things can be corrected in an appropriate way.

We consider the following finite-dimensional operator T :

$$(T\psi)(t) = \frac{1}{2\pi i} \int_\Gamma \frac{\psi(\tau)d\tau}{\tau} + \frac{1}{t} \frac{1}{2\pi i} \int_\Gamma \left(\frac{\psi(\tau)}{\tau^2} d\tau + \frac{\overline{\psi(\tau)d\tau}}{\overline{\tau^2}} \right).$$

Theorem 2.2. *Let Γ be a piecewise smooth curve. There exist $\delta < 1/2$ and $\delta' > 1/2$ such that if*

$$\delta < \min_{1 \le j \le l} \left\{ \frac{1}{p} + \alpha_j \right\} \le \max_{1 \le j \le l} \left\{ \frac{1}{p} + \alpha_j \right\} < \delta', \tag{12}$$

then the operator

$$R_1 = R + T : L_p(\Gamma, \rho) \to L_p(\Gamma, \rho)$$

is invertible. If the weight ρ satisfies inequality (12), $f \in W_p^1(\Gamma, \rho)$, *and*

$$\mathrm{Re} \int_\Gamma f(t)dt = 0, \tag{13}$$

then the solution of the equation

$$R_1 \psi = f_0 \tag{14}$$

is simultaneously a solution of the Muskhelishvili equation (11) *and of the boundary problem* (4).

For a proof of this result see [13].

Corollary 2.3. *The operator $R_1 : L_2(\Gamma) \to L_2(\Gamma)$ is invertible.*

3. Approximate solution of the Muskhelishvili equation on special contours

In this section we consider the stability of a spline Galerkin method for the Muskhelishvili operator on special curves. The corresponding results concern the so-called local models. In the sequel they will be used to obtain necessary and sufficient conditions for the stability of the spline Galerkin method on Γ.

Let $\beta \in [0, 2\pi)$ and $\omega \in (0, \pi) \cup (\pi, 2\pi)$ be real numbers. We denote by $\Gamma_{\beta,\omega}$ the infinite angle

$$\Gamma_{\beta,\omega} := \Gamma_1 \cup \Gamma_2,$$

where the ray $\Gamma_1 := e^{i(\beta+\omega)}\mathbb{R}^+$ is directed to 0 but $\Gamma_2 := e^{i\beta}\mathbb{R}^+$ is directed away from 0. Then $L_p(\Gamma_{\beta,\omega}, \alpha), 0 < 1/p + \alpha$, is the space of all Lebesgue measurable functions f equipped with the norm

$$||f||_{p,\alpha,\omega} = \left(\int_{\Gamma_{\beta,\omega}} |f(t)|^p |t|^{\alpha p} |dt| \right)^{1/p}.$$

Note that in the case $\omega = \pi$ all the following computations are trivial and will be omitted.

Let R_ω denote the Muskhelishvili operator on $L_p(\Gamma_{\beta,\omega}, \alpha)$:

$$R_\omega x(t) \equiv -\overline{x(t)} - \frac{1}{2\pi i}\int_{\Gamma_{\beta,\omega}} \overline{x(\tau)} d \log\frac{\overline{\tau} - \overline{t}}{\tau - t} - \frac{1}{2\pi i}\int_{\Gamma_{\beta,\omega}} x(\tau) d\frac{\overline{\tau} - \overline{t}}{\tau - t}. \tag{15}$$

First of all we construct spline spaces on $\Gamma_{\beta,\omega}$. Let $\chi_{[0,1)} = \chi_{[0,1)}(s), s \in \mathbb{R}$, denote the characteristic function of the interval [0,1). For any natural number m, one may introduce the function

$$\varphi^m(s) := (\varphi^0 * \varphi^{m-1})(s), \quad s \in \mathbb{R},$$

with

$$\varphi^0(s) = \chi_{[0,1)}(s), \quad s \in \mathbb{R},$$

and with $(f * g)$ denoting the convolution of the functions f and g, i.e.,

$$(f * g)(s) := \int_{\mathbb{R}} f(s - x)g(x)dx.$$

From now on we fix $m \in \mathbb{N}$ and set

$$\varphi(s) = \varphi^m(s), \quad s \in \mathbb{R}.$$

As a next step we fix a natural number n and, for each $k \in \mathbb{N}$, we define the function $\varphi_{kn} = \varphi_{kn}(s)$ by

$$\varphi_{kn}(s) := \varphi(ns - k), \quad s \in \mathbb{R}. \tag{16}$$

Lemma 3.1. *Let* φ *denote the function defined by* (16). *Then*

1. $\operatorname{supp}\{\varphi\} \subset [0, m+1]$;
2. *for every* $s \geq 0$ *one has*

$$\varphi(-s + m + 1) = \varphi(s). \tag{17}$$

Proof. Both assertions of this lemma can be proved by induction. We show the second one only. Thus, if $m = 1$, then

$$\varphi(s) = \begin{cases} s & \text{if } 0 \leq s < 1, \\ 2 - s & \text{if } 1 \leq s < 2, \\ 0 & \text{otherwise}, \end{cases}$$

and (17) is obvious. Suppose that equality (17) is satisfied for $k = m$ and consider the case $k = m+1$. One has

$$\begin{aligned}\varphi^{m+1}(-s+m+2) &= \int_{\mathbb{R}} \chi_{[0,1)}(-s+m+2-x)\varphi^m(x)dx \\ &= \int_{\mathbb{R}} \chi_{[0,1)}(-s+1+u)\varphi^m(-u+m+1)du \\ &= \int_{\mathbb{R}} \chi_{[0,1)}(-s+1+u)\varphi^m(u)du \\ &= \int_{\mathbb{R}} \chi_{[0,1)}(s-u)\varphi^m(u)du = \varphi^{m+1}(s),\end{aligned}$$

and the proof is complete. □

Now we are able to introduce spline spaces on $\Gamma_{\beta,\omega}$. Namely, we denote by $S_n^{\beta,\omega}$ the smallest closed subspace of $L_p(\Gamma_{\beta,\omega},\alpha)$ which contains all the functions

$$\widetilde{\phi}_k^{(n)}(t) := \begin{cases} \begin{cases} \varphi_{kn}(s) & \text{if} \quad t = e^{i\beta}s \\ 0 & \text{otherwise} \end{cases} & k \geq 0, \\ \\ \begin{cases} \varphi_{k-m,n}(s) & \text{if} \quad t = e^{i(\beta+\omega)}s \\ 0 & \text{otherwise} \end{cases} & k < 0. \end{cases} \tag{18}$$

Let us consider the semi-linear form

$$\langle f, g\rangle := \int_{\Gamma_{\beta,\omega}} f(t)\overline{g(t)}|dt|,$$

where $f \in L_p(\Gamma_{\beta,\omega},\alpha)$, $g \in L_q(\Gamma_{\beta,\omega},-\alpha)$ and $1/p+1/q=1$. The Galerkin projection operators $\widetilde{L}_n$ from $L_p(\Gamma_{\beta,\omega},\alpha)$ onto $S_n^{\beta,\omega}$ can be defined by

$$\langle \widetilde{L}_n f, \phi_{kn}\rangle := \langle f, \phi_{kn}\rangle \quad \text{for} \quad f \in L_p(\Gamma_{\beta,\omega},\alpha) \quad \text{and} \quad \phi_{kn} \in S_n^{\beta,\omega}. \tag{19}$$

It is known [20] that the operators $\widetilde{L}_n$, $(n = 1, 2, \ldots)$ are well-defined and that the sequence $\{\widetilde{L}_n\}$ converges strongly to the identity operator I as n tends to ∞.

Definition 3.2. *An operator sequence* $\{A_n\}, A_n : \operatorname{Im}\widetilde{L}_n \to \operatorname{Im}\widetilde{L}_n$, *is said to be stable if there exists an* n_0 *such that for all* $n \geq n_0$ *the operators* $A_n : \operatorname{Im}\widetilde{L}_n \to \operatorname{Im}\widetilde{L}_n$ *are invertible and*

$$\sup_{n\geq n_0} ||A_n^{-1}\widetilde{L}_n|| < \infty.$$

Our task now is to establish stability conditions for the sequence of the Galerkin operators $\{\widetilde{L}_n R_\omega \widetilde{L}_n\}$ in the case where R_ω is the Muskhelishvili operator (15). To proceed with this problem we have to recall some notions. We denote by M and M^{-1} the direct and inverse Mellin transforms, respectively:

$$(Mf)(z) = \int_0^{+\infty} x^{1/p+\alpha-zi-1} f(x)dx, \quad z \in \mathbb{R},$$

$$(M^{-1}f)(x) = \frac{1}{2\pi}\int_{-\infty}^{+\infty} x^{zi-1/p-\alpha} f(z)dz, \quad x \in \mathbb{R}^+.$$

Let $L_p(\mathbb{R}^+, \alpha), p > 1, 0 < 1/p+\alpha < 1$, refer to the space of all Lebesgue measurable functions on the half-line $\mathbb{R}^+$ equipped with the norm

$$||f||_{p,\alpha} := \left(\int_{\mathbb{R}^+} |f(t)|^p t^{\alpha p} dt\right)^{1/p}.$$

We will also consider the set $L_p^2(\mathbb{R}^+, \alpha)$ of all pairs $(f_1, f_2)^T, f_1, f_2 \in L_p(\mathbb{R}^+, \alpha)$. The norm in $L_p^2(\mathbb{R}^+, \alpha)$ is defined by

$$||(f_1, f_2)^T|| = \left(||f_1||_{p,\alpha}^p + ||f_2||_{p,\alpha}^p\right)^{1/p}.$$

It is known (see, e.g., [20]) that under some restrictions on the function b the rule

$$(\mathcal{M}(b)f)(\sigma) = ((M^{-1}bM)f)(\sigma)$$

defines a bounded linear operator on $L_p(\mathbb{R}^+, \alpha)$. The operator $\mathcal{M}$ is called the Mellin operator with the symbol b, and it can be represented in the integral form

$$(\mathcal{M}(b)f)(\sigma) = \int_{\mathbb{R}^+} k\left(\frac{\sigma}{s}\right) f(s)\frac{ds}{s}, \quad k = M^{-1}b.$$

For any Banach space X, we denote by $\mathcal{L}(X)$ ($\mathcal{L}_{add}(X)$) the space of all bounded linear (additive) operators on X.

Following [13] we introduce a mapping $\eta : L_p(\Gamma_{\beta,\omega}, \alpha) \to L_p^2(\mathbb{R}^+, \alpha)$ by

$$\eta(f) = (\eta_1(f), \eta_2(f))^T \tag{20}$$

where

$$\begin{aligned} \eta_1(f)(s) &= f(se^{i(\beta+\omega)}), \quad s \in \mathbb{R}^+, \\ \eta_2(f)(s) &= f(se^{i\beta}), \quad s \in \mathbb{R}^+. \end{aligned}$$

Obviously, the mapping $\eta : L_p(\Gamma_{\beta,\omega}, \alpha) \to L_p^2(\mathbb{R}^+, \alpha)$ is invertible and $A \to \eta A \eta^{-1}$ is an isometric algebra isomorphism of $\mathcal{L}(L_p(\Gamma_{\beta,\omega}, \alpha))$ onto $\mathcal{L}(L_p^2(\mathbb{R}^+, \alpha))$.

Lemma 3.3. *Let S_n be the smallest closed subspace of $L_p(\mathbb{R}^+, \alpha)$ which contains all functions $\varphi_{kn} = \varphi_{kn}(s), s \in \mathbb{R}^+$ of (16) with $k \geq 0$ and let $L_n : L_p(\mathbb{R}^+, \alpha) \to S_n$ denote the Galerkin projection onto S_n. Then for every $n \in \mathbb{N}$ the operator $\widetilde{L}_n \in \mathcal{L}(L_p(\Gamma_{\beta,\omega}, \alpha))$ is isometrically isomorphic to the operator* diag $(L_n, L_n) \in \mathcal{L}(L_p^2(\mathbb{R}^+, \alpha))$.

Proof. Immediate calculations and Lemma 3.1 show that

$$\eta(\widetilde{L}_n)\eta^{-1} = \operatorname{diag}(L_n, L_n).$$

□

Note that in the sequel any diagonal operator of the form diag $(T, T, \dots, T)$ will be written as T, so we write L_n instead of diag (L_n, L_n).

Lemma 3.4. *The operator $R_\omega \in \mathcal{L}_{add}(L_p(\Gamma_{\beta,\omega}, \alpha))$ is isometrically isomorphic to the operator $A_\omega \in \mathcal{L}_{add}(L_p^2(\mathbb{R}^+, \alpha))$ defined by*

$$A_\omega = \mathcal{A} + \mathcal{B}\widetilde{V}, \tag{21}$$

where the operators $\mathcal{A}, \mathcal{B}$ and $\widetilde{V}$ are given by

$$\mathcal{A} = \begin{pmatrix} 0 & e^{-i2\beta}\mathcal{M}_\omega \\ -e^{-i2(\beta+\omega)}\mathcal{M}_{2\pi-\omega} & 0 \end{pmatrix},$$

$$\mathcal{B} = \begin{pmatrix} -I & \frac{1}{2}[\mathcal{N}_\omega - \mathcal{N}_{2\pi-\omega}] \\ \frac{1}{2}[\mathcal{N}_\omega - \mathcal{N}_{2\pi-\omega}] & -I \end{pmatrix},$$

and

$$\widetilde{V}(f_1, f_2) = (\overline{f_1}, \overline{f_2}),$$

and where $\mathcal{M}_\nu, \mathcal{N}_\nu, \nu \in (0, 2\pi)$ are the Mellin operators

$$(\mathcal{M}_\nu(\varphi))(\sigma) = \frac{1}{\pi}\int_0^{+\infty} \left(\frac{\sigma}{s}\right) \frac{\sin\nu}{(1-(\sigma/s)e^{i\nu})^2} \frac{\varphi(s)}{s} ds$$

and

$$(\mathcal{N}_\nu(\varphi))(\sigma) = \frac{1}{2\pi i}\int_0^{+\infty} \frac{\varphi(s)ds}{s - \sigma e^{i\nu}}.$$

A proof of the last lemma can be found in [13].

The next result immediately follows from the Lemmas 3.3 and 3.4.

Proposition 3.5. *The sequence $\{\widetilde{L}_n R_\omega \widetilde{L}_n\}$ is stable if and only if so is the sequence $L_n A_\omega L_n$.*

Now we can make further simplification. Namely, we consider the space $l_{p,\alpha}$ of all sequences $\{\xi_j\}_{j=0}^{+\infty}$ of complex numbers ξ_j such that

$$||\{\xi_j\}||_{p,\alpha}^p = \sum_{j=0}^{+\infty}(1+j)^{\alpha p}|\xi_j|^p < \infty,$$

and we define the operators $E_n : l_{p,\alpha} \to S_n$ and $E_{-n} : S_n \to l_{p,\alpha}$ by

$$E_n : \{\xi_j\} \to \sum_{j=0}^{+\infty} \xi_j \varphi_{jn}(t),$$

$$E_{-n} : \sum_{j=0}^{+\infty} \xi_j \varphi_{jn}(t) \to \{\xi_j\}.$$

Proposition 3.6. ([1]) *The mappings $E_n : l_{p,\alpha} \to S_n$ and $E_{-n} : S_n \to l_{p,\alpha}$ are bounded linear operators, and there are constants $C_1 > 0$ and $C_2 > 0$ such that*

$$\|\sum_{j=0}^{+\infty} \xi_j \varphi_{jn}\|_{L_p(\mathbb{R}^+,\alpha)} \leq C_1 n^{-(1/p+\alpha)} \|\{\xi_j\}\|_{l_{p,\alpha}}, \tag{22}$$

$$\|\{\xi_j\}\| \leq C_2 n^{(1/p+\alpha)} \|\sum_{j=0}^{+\infty} \xi_j \varphi_{jn}\|_{L_p(\mathbb{R}^+,\alpha)}. \tag{23}$$

Let $l^2_{p,\alpha}$ denote the cartesian product of two copies of $l_{p,\alpha}$.

Lemma 3.7. *The sequence $\{\widetilde{L}_n R_\omega \widetilde{L}_n\}$ is stable if and only if the operator $B^1_\omega := E_{-1} L_1 A_\omega L_1 E_1 : l^2_{p,\alpha} \to l^2_{p,\alpha}$ is invertible.*

Proof. Let $\mathcal{M}(b)$ be a Mellin convolution operator and $k = M^{-1}(b)$ and let $G(b) = (a_{lq})^\infty_{l,q=0}$ be the matrix of the operator $E_{-n} L_n (M^{-1} b M) L_n E_n : l_{p,\alpha} \to l_{p,\alpha}$. Then

$$\begin{aligned} a_{lq} &= \int_{\mathbb{R}} (\mathcal{M}(b)\varphi_{qn})(s)\varphi_{ln}(\sigma)d\sigma \\ &= \int_{\mathbb{R}} \int_{\mathbb{R}^+} k\left(\frac{\sigma}{s}\right) \varphi(ns-q)\frac{ds}{s}\varphi(\sigma n - l)du \\ &= \int_{\mathbb{R}} \int_{\mathbb{R}^+} k\left(\frac{u+l}{t+q}\right) \varphi(t)\frac{dt}{t}\varphi(u)du. \end{aligned}$$

Thus, the entries of the matrix $G(b)$ are independent of n. Taking into account that the operator $B^n_\omega := E_{-n} L_n A_\omega L_n E_n$ admits the representation

$$B^n_\omega = \begin{pmatrix} 0 & G(b_1) \\ G(b_2) & 0 \end{pmatrix} + \begin{pmatrix} -I & G(b_3) \\ G(b_3) & -I \end{pmatrix} \begin{pmatrix} \overline{V} & 0 \\ 0 & \overline{V} \end{pmatrix} \tag{24}$$

where $\overline{V}(\{\xi_j\}) = \{\overline{\xi}_j\}$ and $b_1 = e^{-2i\beta} m_\omega(y)$, $b_2 = e^{-2i(\beta+\omega)} m_{2\pi-\omega}(y)$, $b_3 = k n_\omega(y), y = z + i(1/p+\alpha), z \in \mathbb{R}$, one obtains the claim. □

It is worth mentioning that studying invertibility of the operator B^1_ω is a very difficult problem. However, the conditions for the invertibility of the operator A_ω are simultaneously the conditions for the Fredholmness of the operator B^1_ω.

Corollary 3.8. *For every $\omega \in (0, 2\pi)$ there exists a real number $\delta'_\omega > 1/2$ such that for every $p > 1$ and every α satisfying the inequality*

$$0 < \frac{1}{p} + \alpha < \delta'_\omega \tag{25}$$

the operator $B^1_\omega : l^2_{p,\alpha} \to l^2_{p,\alpha}$ is Fredholm of index zero.

Proof. Indeed, let us consider the operator $G(b)$ again, and let R and S denote the real and imaginary parts of the function $k = M^{-1}b$, respectively. Then

$$a_{lq} = \int_{\mathbb{R}}\int_{\mathbb{R}+} R\left(\frac{u+l}{t+q}\right)\varphi(t)\frac{dt}{t}\varphi(u)du + i\int_{\mathbb{R}}\int_{\mathbb{R}+} S\left(\frac{u+l}{t+q}\right)\varphi(t)\frac{dt}{t}\varphi(u)du.$$

Using the generalized mean value theorem we find that there exist points $u_{lq}^1, u_{lq}^2 \in [0,m]$ and $t_{lq}^1, t_{lq}^2 \in [0,m]$ such that

$$a_{lq} = \left(\int_{\mathbb{R}+}\varphi(t)dt\right)^2 \left(R\left(\frac{u_{lq}^1+l}{t_{lq}^1+q}\right)\frac{1}{t_{lq}^1} + iS\left(\frac{u_{lq}^2+l}{t_{lq}^2+q}\right)\frac{1}{t_{lq}^2}\right).$$

Now by [20] (cf. Corollary 2.1 and Proposition 2.11 on page 65) there exists a compact operator K_1 such that

$$G(b) = \left(\int_{\mathbb{R}}\varphi(t)dt\right)^2 \left(\int_l^{l+1}\int_q^{q+1} k\left(\frac{t}{s}\right)\frac{ds}{s}dt\right)_{l,q=0}^{+\infty} + K_1.$$

The rest of the proof is as the proof of Proposition 3.6 of [13] □

4. Galerkin method for the Muskhelishvili equation

This section contains some results concerning the stability and convergence of the Galerkin method based on the splines of degree $m \geq 1$.

Let γ be a 1-periodic parameterization of Γ such that the corner points $c_j, j = 0,1,\dots,l-1$, are represented as follows:

$$c_j = \gamma(j/l), \quad j = 0,1,\dots,l-1.$$

We also assume that γ is twice continuously differentiable on each of the intervals $(j/l,(j+1)/l), j = 0,1,\dots,l-1$, and that there exist one-sided limits $\gamma'(j/l \pm 0)$ and $\gamma''(j/l \pm 0)$ and

$$|\gamma'(j/l+0)| = |\gamma'(j/l-0)|, \quad j = 0,1,\dots,l-1.$$

We choose $n = lr, r \in \mathbb{N}$ and set

$$\widetilde{\varphi}_{kn}(t) := \varphi_{kn}(s), \quad t = \gamma(s), \quad s \in [0,1), \tag{26}$$

with φ_{kn} defined by (16).

An approximate solution ψ_n of equation (11) is sought in the form

$$\psi_n = \sum{}' c_k \widetilde{\varphi}_{kn}, \tag{27}$$

where the sum $\sum'$ includes only those functions $\widetilde{\varphi}_{kn}$ the support of which is entirely contained in one of the arcs $[\gamma(j/l), \gamma((j+1)/l)), j = 0,1,\dots,l-1$. The smallest subspace of $L_p(\Gamma,\rho)$ containing all such functions $\widetilde{\varphi}_{kn}$ will be referred to as $S_n(\Gamma)$. The corresponding subset of indices $k\,(0 \leq k \leq n-1)$ such that $\widetilde{\varphi}_{kn} \in S_n(\Gamma)$ is denoted by A'.

To find the coefficients $c_k, k \in A'$ on the right-hand side of (27) we use the following system of algebraic equations:

$$(R_1\psi_n, \widetilde{\varphi}_{kn})_\Gamma = (f_0, \widetilde{\varphi}_{kn})_\Gamma, \quad k \in A', \tag{28}$$

where the operator R_1 was defined in the statement of Theorem 2.2 and

$$(x,y)_\Gamma = \sum_{j=0}^{l-1} \int_{j/l}^{(j+1)/l} x(\gamma(s))\overline{y(\gamma(s))}ds, \; x \in L_p(\Gamma,\rho), y \in L_q(\Gamma,\rho^{-1}), \; 1/p+1/q=1.$$

With each corner point c_r $(r = 0,1,\dots,l-1)$ of Γ we associate an operator $B_{\omega_r} = B_{\beta_r,\omega_r}$. These operators are defined similarly to the operator B_ω in (24), but the parameters ω and β are replaced by ω_r and β_r, respectively. We recall that ω_r is the angle between corresponding semi-tangents to Γ at the point c_r and that β_r is the angle between the right semi-tangent to Γ at the point c_r and the real axis.

Theorem 4.1. *Let $\alpha_r, r = 0,1,\dots,l-1$, and $p \in (1,\infty)$ satisfy inequality* (12). *Then:*

1. *The Galerkin method* (28) *is stable for the operator R_1 in $L_p(\Gamma,\rho)$ if and only if the operators $B_{\omega_r} \in \mathcal{L}_{add}(l^2_{p,\alpha_r}), r = 0,1,2,\dots,l-1$, are invertible.*
2. *Let the operators $B_{\omega_r} \in \mathcal{L}_{add}(l^2_{p,\alpha_r}), r = 0,1,2,\dots,l-1$, be invertible. If, in addition, $f \in W^1_p(\Gamma,\rho)$ and if f satisfies condition* (13), *then the approximate solutions* (27) *converge to a solution of Muskhelishvili equation* (11) *in the norm of $L_p(\Gamma,\rho)$ as $n \to \infty$.*

The proof of the first claim immediately follows from the corresponding results concerning the stability of approximation methods for singular integral operators on piecewise smooth curves (cf. [20], Chapter 5) and from the first part of Theorem 2.2. Afterwards, the second claim follows from the second part of Theorem 2.2 and standard results on the convergence of stable approximation methods [20, 28].

Remark 4.2. *If the boundary Γ is smooth, the first assertion in Theorem* 4.1 *disappears.*

Thus, the sequence $\{\psi_n\}$ constructed by using the Galerkin method (28) converges to an exact solution of (11) in the norm $L_p(\Gamma,\rho)$. What is more important is that in the case $p = 2$ one can guarantee the convergence of the sequence in the space $W^1_2(\Gamma)$.

Corollary 4.3. *Let the operators $B_{\omega_r}, r = 0,1,\dots,l-1$, be invertible in l^2_2. Then the Galerkin method* (11) *is stable in $W^1_2(\Gamma)$, and if f satisfies condition* (13), *then the sequence $\{\psi_n\}$ converges to a solution of equation* (11) *in the norm of $W^1_2(\Gamma)$.*

Proof. It follows from Corollary 2.3 and from the proof of Theorem 2.14 of [13] that the operator R_1 is simultaneously invertible in the spaces $L_2(\Gamma)$ and $W^1_2(\Gamma)$. Hence, to prove the stability of the method (28) in $W^1_2(\Gamma)$ one can use Theorem

1.37 of [28]. Let L_n denote the corresponding Galerkin projection onto the subspace $S_n(\Gamma)$. A slight modification of the proof of Theorem 2.7 of [28] shows that there exists a constant $c_1 > 0$ such that

$$||g - L_n g||_{L_2(\Gamma)} < c_1 n^{-1} ||g||_{W_2^1(\Gamma)}, \quad g \in W_2^1(\Gamma).$$

The latter inequality yields the estimate

$$||Ag - L_n A L_n g||_{L_2(\Gamma)} < c_2 n^{-1} ||g||_{W_2^1(\Gamma)}, \quad g \in W_2^1(\Gamma). \tag{29}$$

Note that the positive constants c_1, c_2 are independent of $g \in W_2^1(\Gamma)$ and n.

Taking into account the inverse properties of the splines ψ_n (recall that $n = lr, r \in \mathbb{N}$ and that the support of $\widetilde{\varphi}_{kn}$ is entirely contained in one of the arcs $(\gamma(j/l), \gamma((j+1)/l)), j = 0, 1, \dots, l-1$) and applying Theorem 1.37 of [28] one obtains the stability of the Galerkin method (28). This implies the convergence of $\{\psi_n\}$ in $W_2^1(\Gamma)$. Using Theorem 2.2 once more we get the result. □

Thus, the Galerkin method (28) can be used to find approximate solutions of the Muskhelishvili equation (11). This allows us to construct approximate solutions of the biharmonic problem (1) (see Section 2 for details).

Acknowledgement. The work of V.D. Didenko was supported in part by Universiti Brunei Darussalam under the Research Grant **UBD/PNC2/2/RG/1(12)**.

References

[1] C. de Boor, *A Practical Guide to Splines,* Springer-Verlag, New York, Heidelberg, Berlin **1978**.

[2] C.V. Camp and G.S. Gipson, *A boundary element method for viscous flows at low Reynold numbers,* Eng. Analysis with Boundary Elements **6** (1989), 144–151.

[3] R.H. Chan, T.K. DeLilo and M.A. Horn, *Superlinear convergence estimates for a conjugate gradient method for the biharmonic equation,* SIAM J. Sci. Comput. **19(1)** (1998), 139–147.

[4] C. Chang-jun and W. Rong, *Boundary integral equations and the boundary element method for buckling analysis of perforated plates,* Eng. Analysis with Boundary Elements **17** (1996), 54–68.

[5] S. Christiansen, *Derivation and analytical investigation of three direct boundary integral equations for the fundamental biharmonic problem,* J. Comput. Appl. Math. **91** (1998), 231–247.

[6] S. Christiansen and P. Hougaard, *An Investigation of a Pair of Integral Equations for the Biharmonic Problem,* J. Inst. Mat. Appl. **22** (1978), 15–27.

[7] J.M. Chuang and S.Z. Hu, *Numerical computation of Muskhelishvili's integral equation in plane elasticity,* J. Comput. Appl. Math. **66** (1996), 123–138.

[8] C. Constanda, *Sur le problème de Dirichlet dans déformation plane',* Comptes Rendus de l'Académie de Sciences Paris, Serie I, Mathematique **316** (1993), 1107–1109.

[9] C. Constanda, *On the Dirichlet problem for the two-dimensional biharmonic equation,* Math. Meth. Appl. Sc. **20** (1997), 885–890.

[10] M. Costabel and M. Dauge, *Invertibility of the biharmonic single layer potential operator,* Integr. Equat. Oper. Th. **24** (1996), 46–67.

[11] M. Costabel, I. Lusikka and J. Saranen. *Comparison of three boundary element approaches for the solution of the clamped plate problem,* Boundary Elements,Vol. IX, Edited by C. A. Brebbia, Springer-Verlag, New York **1989**, 19–34.

[12] M. Costabel and J. Saranen, *Boundary element analysis of a direct method for a biharmonic Dirichlet problem,* Operator Theory: Advances and Applications **41** (1989), 77–95.

[13] V.D. Didenko and B. Silbermann, *On stability of approximation methods for the Muskhelishvili equation,* Preprint, 2000.

[14] R.V. Duduchava, *On general singular integral operators of the plane theory of elasticity,* Rend. Politechn. Torino **42(3)** (1984), 15–41.

[15] R.V. Duduchava, *On general singular integral operators of the plane theory of elasticity,* Trudy Tbilissk. Matem. Inst. **82** (1986), 45–89 (Russian).

[16] B. Fuglege, *On a direct method of integral equations for solving the biharmonic Dirichlet problem,* ZAMM **61** (1981), 449–459.

[17] P. Grisvard, *Boundary Value Problems in Non-Smooth Domains,* Monographs and Studies in Math. Vol. 24, Pitman, Boston, London, Melbourne **1985**.

[18] G.C. Hsiao and R. MacCamy, *Solution of boundary value problems by integral equations of the first kind,* SIAM Rev. **15** (1973), 687–705.

[19] G.C. Hsiao, P. Kopp and W.L. Wendland, *A Galerkin collocation method for some integral equations of the first kind,* Computing **25** (1980), 89–130.

[20] R. Hagen, S. Roch and B. Silbermann, *Spectral Theory of Approximation Methods for Convolution Equations,* Operator Theory. Advances and Applications. Vol. 74, Birkhäuser Verlag, Basel, Boston, Stuttgart **1995.**

[21] A.I. Kalandiya, *Mathematical Methods of Two-Dimensional Elasticity,* Nauka, Moscow **1973** (Russian); *Engl. transl.:* Mir Publisher, Moscow **1975**.

[22] V.V. Meleshko, *Biharmonic problem in a rectangle,* Appl. Sci. Res. **48** (1998), 217–249.

[23] N.I. Muskhelishvili, *Fundamental Problems in the Theory of Elasticity,* Nauka, Moscow **1966** (Russian).

[24] N.I. Muskhelishvili, *Singular Integral Equations,* Nauka, Moscow **1968** (Russian).

[25] V.Z. Parton and P.I. Perlin, *Integral Equations of Elasticity Theory,* Nauka, Moscow **1977** (Russian).

[26] P.I. Perlin and Yu.N. Shalyukhin, *On the numerical solution of the integral equations of plane elasticity theory,* Izv. Akad. Nauk Kazah. SSR, Ser. fiz.-mat. **1** (1976), 86–88 (Russian).

[27] P.I. Perlin and Yu.N. Shalyukhin, *On the numerical solution of some plane problems of elasticity theory,* Prikl. Mekh. **15(4)** (1977), 83–86 (Russian).

[28] S. Prössdorf and B. Silbermann, *Numerical Analysis for Integral and Related Operator Equations,* Akademie-Verlag, Berlin **1991**; Birkhäuser Verlag, Basel **1991**.

V.D. Didenko
Department of Mathematics
Universiti Brunei Darussalam
Bandar Seri Begawan
BE1410 Brunei
e-mail: victor@fos.ubd.edu.bn

B. Silbermann
Faculty of Mathematics
University of Technology Chemnitz
09107 Chemnitz
Germany
e-mail: silbermn@mathematik.tu-chemnitz.de

Received: 12 November 2001

Operator Theory:
Advances and Applications, Vol. 135, 161–181

On Rank Invariance of Schwarz-Pick-Potapov Block Matrices of Matrix-Valued Carathéodory Functions

Bernd Fritzsche, Bernd Kirstein and Andreas Lasarow

Dedicated to Professor Bernd Silbermann with admiration and friendship on the occasion of his sixtieth birthday

Abstract. We derive statements on rank invariance of Schwarz-Pick-Potapov block matrices of matrix-valued Carathéodory functions. The rank of the Schwarz-Pick-Potapov block matrices of a matrix-valued Carathéodory function coincides with the rank of the real part of the corresponding section matrices of its Taylor coefficients.

0. Introduction

This paper deals with matrix-valued Carathéodory functions in the open unit disk $\mathbb{D} := \{z \in \mathbb{C} : |z| < 1\}$. By a $q \times q$ Carathéodory function $\Omega : \mathbb{D} \to \mathbb{C}^{q\times q}$ we mean a function which is holomorphic in $\mathbb{D}$ and which has nonnegative Hermitian real part $\frac{1}{2}(\Omega(w)+[\Omega(w)]^*)$ for each $w \in \mathbb{D}$. The treatment of Nevanlinna-Pick interpolation problems for matrix-valued Carathéodory functions has led to the consideration of certain block matrices which are now called Schwarz-Pick-Potapov block matrices. Namely, it has turned out that such a Nevanlinna-Pick problem has a solution if and only if the corresponding Schwarz-Pick-Potapov block matrix is nonnegative Hermitian. V.P. Potapov (see [P], [EP]) found several matrix inequalities which can be considered as far reaching generalizations of the classical lemma of H.A. Schwarz and its reformulation by G. Pick. These inequalities (which are now often called Schwarz-Pick-Potapov inequalities) form the basis of Potapov's "Method of Fundamental Matrix Inequality", which turned out to be a powerful tool to treat matrix versions of classical interpolation and moment problems (see, e.g., Kovalishina [K], Dubovoj [D], [DFK]).

In [FKL] the authors have gone first steps towards constructing a matrix generalization of the theory of orthogonal rational functions created by Bultheel/González-Vera/Hendriksen/Njåstad [BGHN]. There we obtained several statements on rank invariance of various Gramian matrices. These results are the starting point of this paper. From them we derive corresponding results on rank invariance of Schwarz-Pick-Potapov block matrices of matrix-valued Carathéodory functions. In particular, we will see that this rank concept can be treated both on the basis of Taylor coefficients and on the basis of the values of the underlying Carathéodory functions. Special attention will be paid to the full rank case.

Moreover, we will discuss matrix-valued Carathéodory functions with molecular Riesz-Herglotz measure. In a sense, the Schwarz-Pick-Potapov block matrices of these functions possess the highest degree of degeneracy.

1. Notation

Throughout this paper, let p and q belong to the set $\mathbb{N}$ of all positive integers. We will use $\mathbb{C}, \mathbb{R}, \mathbb{Z}$ and $\mathbb{N}_0$ to denote the sets of all complex numbers, of all real numbers, of all integers, and of all nonnegative integers, respectively. If $m \in \mathbb{N}_0$ and if $n \in \mathbb{N}_0$ or $n = +\infty$, then $\mathbb{N}_{m,n}$ stands for the set of all integers k which satisfy $m \leq k \leq n$. If $\mathfrak{X}$ is a nonempty set, we let $\mathfrak{X}^{p\times q}$ be the set of all $p \times q$ matrices each entry of which belongs to $\mathfrak{X}$. The notation $\mathbf{0}_{p\times q}$ stands for the null matrix that belongs to $\mathbb{C}^{p\times q}$, and the identity matrix which belongs to $\mathbb{C}^{q\times q}$ is designated by $\mathbf{I}_q$. If $\mathbf{A} \in \mathbb{C}^{q\times q}$, we let Re $\mathbf{A}$ and Im $\mathbf{A}$ be the real part of $\mathbf{A}$ and the imaginary part of $\mathbf{A}$: $\operatorname{Re} \mathbf{A} := \frac{1}{2}(\mathbf{A}+\mathbf{A}^*)$ and $\operatorname{Im} \mathbf{A} := \frac{1}{2i}(\mathbf{A}-\mathbf{A}^*)$. The set of all nonnegative Hermitian complex $q \times q$ matrices is denoted by $\mathbb{C}^{q\times q}_{\geq}$. The symbol $\mathbb{T}$ stands for the unit circle, $\mathbb{D}$ for its interior and $\mathbb{E}$ for its exterior with respect to the extended complex plane $\mathbb{C}_0 := \mathbb{C} \cup \{\infty\}$, i.e., $\mathbb{T} := \{z \in \mathbb{C} : |z| = 1\}$, $\mathbb{D} := \{z \in \mathbb{C} : |z| < 1\}$ and $\mathbb{E} := \mathbb{C}_0 \setminus \{\mathbb{D} \cup \mathbb{T}\}$. Let $\mathfrak{B}_1$ be the σ-algebra of all Borelian subsets of $\mathbb{C}$, and let $\mathfrak{B}_{\mathbb{T}} := \mathfrak{B}_1 \cap \mathbb{T}$.

If f is a $q \times q$ matrix-valued function defined on $\mathbb{D}$, then let $\hat{f} : \mathbb{C}_0 \setminus \mathbb{T} \to \mathbb{C}^{q\times q}$ be defined by

$$\hat{f}(w) := \begin{cases} f(w) & \text{if} \quad w \in \mathbb{D} \\ -\left(f\left(\frac{1}{\overline{w}}\right)\right)^* & \text{if} \quad w \in \mathbb{C} \setminus \{\mathbb{D} \cup \mathbb{T}\} \\ -(f(0))^* & \text{if} \quad w = \infty \end{cases}. \tag{1}$$

If $\frac{1}{0} := \infty$, then it is readily checked that

$$\hat{f}\left(\frac{1}{\overline{z}}\right) = -(\hat{f}(z))^* \tag{2}$$

holds for all $z \in \mathbb{C}_0 \setminus \mathbb{T}$. Let $m \in \mathbb{N}_0$, and let $n \in \mathbb{N}_0$ with $n \geq m$ or let $n = +\infty$. A sequence $(\alpha_j)_{j=m}^n$ of complex numbers is said to be a system of points in good position (with respect to $\mathbb{T}$) if $\alpha_j \overline{\alpha_k} \neq 1$ for all integers j and k which satisfy $m \leq j \leq k \leq n$. We will use $\mathcal{T}_{m,n}$ to denote all sequences $(\alpha_j)_{j=m}^n$ which form a system of points in good position (with respect to $\mathbb{T}$). The set of all sequences $(\alpha_j)_{j=m}^n$ which belong to $\mathcal{T}_{m,n}$ and which satisfy $\alpha_j \neq \alpha_k$ for all integers j and k with $m \leq j < k \leq n$ will be designated by $\mathcal{T}^{\#}_{m,n}$.

If Ω is a complex $p \times q$ matrix-valued function which is holomorphic in a neighborhood of the point $w = 0$, then, for each $n \in \mathbb{N}_0$, let $\mathbf{S}_n^{(\Omega)}$ be the block Toeplitz matrix given by

$$\mathbf{S}_n^{(\Omega)} := \begin{pmatrix} \mathbf{A}_0 & \mathbf{0}_{p\times q} & \mathbf{0}_{p\times q} & \dots & \mathbf{0}_{p\times q} \\ \mathbf{A}_1 & \mathbf{A}_0 & \mathbf{0}_{p\times q} & \dots & \mathbf{0}_{p\times q} \\ \mathbf{A}_2 & \mathbf{A}_1 & \mathbf{A}_0 & \dots & \mathbf{0}_{p\times q} \\ \vdots & \vdots & \vdots & & \vdots \\ \mathbf{A}_n & \mathbf{A}_{n-1} & \mathbf{A}_{n-2} & \dots & \mathbf{A}_0 \end{pmatrix} \tag{3}$$

where

$$\Omega(w) = \sum_{j=0}^{\infty} \mathbf{A}_j w^j \tag{4}$$

is the Taylor series representation of Ω.

2. The rank of block Pick matrices of Carathéodory type

A matrix-valued function $\Omega : \mathbb{D} \to \mathbb{C}^{q\times q}$ is called a $q \times q$ Carathéodory function (in $\mathbb{D}$) if Ω is holomorphic in $\mathbb{D}$ and the real part $\operatorname{Re} \Omega(w)$ of $\Omega(w)$ is nonnegative Hermitian for each $w \in \mathbb{D}$. We will use the notation $\mathcal{C}_q(\mathbb{D})$ to designate the set of all $q \times q$ Carathéodory functions (in $\mathbb{D}$). Let m be a nonnegative integer or let $m = +\infty$, let $(\alpha_j)_{j=0}^m$ be a sequence of pairwise different points belonging to $\mathbb{D}$, and let $(\mathbf{A}_j)_{j=0}^m$ be a sequence of complex $q \times q$ matrices. Results on matrix versions of classical Carathéodory-Nevanlinna-Pick interpolation (see, e.g., [DGK]) then show that there is a function $\Omega \in \mathcal{C}_q(\mathbb{D})$ such that $\Omega(\alpha_j) = \mathbf{A}_j$ for each $j \in \mathbb{N}_{0,m}$ if and only if, for every nonnegative integer with $n \leq m$, the matrix

$$\mathbf{R}_{\alpha,\mathbf{A},n} := \left(\frac{1}{1 - \alpha_j \overline{\alpha_k}} (\mathbf{A}_j + \mathbf{A}_k^*) \right)_{j,k=0}^{n}$$

is nonnegative Hermitian. Moreover, the matrix $\mathbf{R}_{\alpha,\mathbf{A},n}$ is nonnegative Hermitian if and only if the matrix

$$\mathbf{L}_{\alpha,\mathbf{A},n} := \left(\frac{1}{1 - \overline{\alpha_j} \alpha_k} (\mathbf{A}_j^* + \mathbf{A}_k) \right)_{j,k=0}^{n}$$

is nonnegative Hermitian. Furthermore, the following well-known theorem (see, e.g., [DFK, Theorem 2.3.1]) holds.

Theorem 2.1. *Let Ω be a complex $q \times q$ matrix-valued function defined on $\mathbb{D}$ which is holomorphic in $\mathbb{D}$, let n be a nonnegative integer, and let $\hat{\Omega}$ be given by* (1). *Then the following statements are equivalent:*

(i) *The matrix-valued function* Ω *belongs to the Carathéodory class* $\mathcal{C}_q(\mathbb{D})$.

(ii) *For all sequences* $(\alpha_j)_{j=0}^n \in \mathcal{T}_{0,n}$ *the matrix*

$$\mathbf{R}_n^{(\alpha,\Omega)} := \left(\frac{1}{1-\alpha_j\overline{\alpha_k}} \left(\hat{\Omega}(\alpha_j) + [\hat{\Omega}(\alpha_k)]^* \right) \right)_{j,k=0}^n \tag{5}$$

is nonnegative Hermitian.

(iii) *For all sequences* $(\alpha_j)_{j=0}^n \in \mathcal{T}_{0,n}$, *the matrix*

$$\mathbf{L}_n^{(\alpha,\Omega)} := \left(\frac{1}{1-\overline{\alpha_j}\alpha_k} \left(\left[\hat{\Omega}(\alpha_j)\right]^* + \hat{\Omega}(\alpha_k) \right) \right)_{j,k=0}^n \tag{6}$$

is nonnegative Hermitian.

The consideration of the extended function $\hat{\Omega}$ instead of Ω is one of the cornerstones of V.P. Potapov's "Method of Fundamental Matrix Inequality". In this way, his school obtained far reaching generalizations of the classical Schwarz-Pick inequalities (see [P], [EP], [D], and [K]). Observe that the matrices given in (5) and (6) are also called Schwarz-Pick-Potapov block matrices of the first kind.

The main result of this paper shows that the rank of these matrices is independent of the choice of the sequence $(\alpha_j)_{j=0}^n$ in $\mathcal{T}_{0,n}^{\#}$:

Theorem 2.2. *Let* $\Omega \in \mathcal{C}_q(\mathbb{D})$, *and let* n *be a nonnegative integer. Then for every choice of* $(\alpha_j)_{j=0}^n$ *in* $\mathcal{T}_{0,n}^{\#}$,

$$\operatorname{rank} \mathbf{R}_n^{(\alpha,\Omega)} = \operatorname{rank} \operatorname{Re} \mathbf{S}_n^{(\Omega)} \quad \text{and} \quad \operatorname{rank} \mathbf{L}_n^{(\alpha,\Omega)} = \operatorname{rank} \operatorname{Re} \mathbf{S}_n^{(\Omega)}.$$

One can see immediately that in the special case $n = 0$ Theorem 2.2 provides the well-known fact that, for each $\Omega \in \mathcal{C}_q(\mathbb{D})$, the equation $\operatorname{rank}(\operatorname{Re}\Omega(z)) = \operatorname{rank}(\operatorname{Re}\Omega(0))$ holds for each $z \in \mathbb{D}$.

Particular attention is turned to the so-called nondegenerate $q \times q$ Carathéodory functions. If $n \in \mathbb{N}_0$, then a function $\Omega \in \mathcal{C}_q(\mathbb{D})$ is said to be nondegenerate of order n (respectively, degenerate of order n) if the matrix Re $\mathbf{S}_n^{(\Omega)}$ is nonsingular (respectively, singular). Then one can easily see that Theorems 2.1 and 2.2 provide the following characterization of nondegenerate $q \times q$ Carathéodory functions.

Theorem 2.3. *Let* $\Omega \in \mathcal{C}_q(\mathbb{D})$ *and let* n *be a nonnegative integer. Then the following statements are equivalent:*

(i) *The* $q \times q$ *Carathéodory function* Ω *is nondegenerate of order* n.

(ii) *There is a sequence* $(\alpha_j)_{j=0}^n \in \mathcal{T}_{0,n}^{\#}$ *such that the matrix* $\mathbf{R}_n^{(\alpha,\Omega)}$ *is nonsingular.*

(iii) *There is a sequence* $(\alpha_j)_{j=0}^n \in \mathcal{T}_{0,n}^{\#}$ *such that the matrix* $\mathbf{L}_n^{(\alpha,\Omega)}$ *is nonsingular.*

(iv) *For each sequence* $(\alpha_j)_{j=0}^n \in \mathcal{T}_{0,n}^{\#}$, *the matrices* $\mathbf{R}_n^{(\alpha,\Omega)}$ *and* $\mathbf{L}_n^{(\alpha,\Omega)}$ *are both positive Hermitian.*

Our proof to Theorem 2.2 is based on a well-known connection between $q \times q$ Carathéodory functions and $q \times q$ nonnegative Hermitian-valued Borel measures on the unit circle $\mathbb{T}$. Let $\mathcal{M}^q_{\geq}(\mathbb{T}, \mathfrak{B}_{\mathbb{T}})$ be the set of all $q \times q$ nonnegative Hermitian-valued Borel measures on $\mathbb{T}$, i.e., the set of all countably additive mappings $F : \mathfrak{B}_{\mathbb{T}} \to \mathbb{C}^{q\times q}_{\geq}$. If $\Omega \in \mathcal{C}_q(\mathbb{D})$, then the matricial version of a famous theorem due to F. Riesz and G. Herglotz shows that there is a unique nonnegative Hermitian-valued measure $F \in \mathcal{M}^q_{\geq}(\mathbb{T}, \mathfrak{B}_{\mathbb{T}})$ such that

$$\Omega(w) - i \operatorname{Im} \Omega(0) = \int_{\mathbb{T}} \frac{z+w}{z-w} F(dz) \tag{7}$$

for all $w \in \mathbb{D}$. This unique $F \in \mathcal{M}^q_{\geq}(\mathbb{T}, \mathfrak{B}_{\mathbb{T}})$ is said to be the Riesz-Herglotz measure associated with Ω. In this case, even the identity

$$\hat{\Omega}(w) - i \operatorname{Im} \Omega(0) = \int_{\mathbb{T}} \frac{z+w}{z-w} F(dz) \tag{8}$$

holds for all $w \in \mathbb{C} \setminus \mathbb{T}$. If F is the Riesz-Herglotz measure associated with a given $q \times q$ Carathéodory function Ω, then Ω admits the representation

$$\Omega(w) - i \operatorname{Im} \Omega(0) = \mathbf{\Gamma}_0^{(F)} + 2\sum_{j=1}^{\infty} \mathbf{\Gamma}_j^{(F)} w^j, \quad w \in \mathbb{D},$$

where

$$\mathbf{\Gamma}_k^{(F)} := \int_{\mathbb{T}} z^{-k} F(dz), \quad k \in \mathbb{Z}, \tag{9}$$

are the Fourier coefficients of F, and we have

$$\operatorname{Re} \mathbf{S}_n^{(\Omega)} = \mathbf{T}_n^{(F)} \tag{10}$$

for each $n \in \mathbb{N}_0$, where $\mathbf{T}_n^{(F)}$ is the n-th block Toeplitz matrix associated with the sequence $(\mathbf{\Gamma}_k^{(F)})_{k\in\mathbb{Z}}$ of the Fourier coefficients of F:

$$\mathbf{T}_n^{(F)} := \left(\mathbf{\Gamma}_{j-k}^{(F)}\right)_{j,k=0}^{n}. \tag{11}$$

In particular we see that, for each $n \in \mathbb{N}_0$, the matrix $\operatorname{Re} \mathbf{S}_n^{(\Omega)}$ is nonnegative Hermitian. Furthermore, we obtain that a $q \times q$ Carathéodory function Ω is nondegenerate of order n if and only if $\operatorname{Re} \mathbf{S}_k^{(\Omega)}$ is positive Hermitian for all nonnegative integers k with $k \leq n$. Thus a $q \times q$ Carathéodory function is nondegenerate of order n if and only if Ω is nondegenerate of order m for each integer $m \in \mathbb{N}_{0,n}$. Observe that, conversely, if a measure $F \in \mathcal{M}^q_{\geq}(\mathbb{T}, \mathfrak{B}_{\mathbb{T}})$ is given, then the mapping $\Omega_F : \mathbb{D} \to \mathbb{C}^{q\times q}$ defined by

$$\Omega_F(w) := \int_{\mathbb{T}} \frac{z+w}{z-w} F(dz), \quad w \in \mathbb{D}, \tag{12}$$

belongs to $\mathcal{C}_q(\mathbb{D})$ and satisfies $\operatorname{Im} \Omega_F(0) = \mathbf{0}_{q\times q}$. In this case Ω_F is called the Riesz-Herglotz transform of F. Hence, if we consider a $q \times q$ Carathéodory function Ω (in $\mathbb{D}$), then $\Psi := \Omega - i \operatorname{Im} \Omega(0)$ is exactly the Riesz-Herglotz transform of the Riesz-Herglotz measure associated with Ω.

The main idea to prove Theorems 2.2 and 2.3 is to consider the so-called (α, n)-moment matrices of the Riesz-Herglotz measure associated with the given $q \times q$ Carathéodory function. Before we will explain this notion let us give some remarks on a particular class of rational matrix-valued functions. If $\eta \in \mathbb{C}$, we define $\rho_\eta : \mathbb{C} \to \mathbb{C}$ by

$$\rho_\eta(w) := 1 - \overline{\eta} w. \tag{13}$$

The notation $\mathcal{R}_{\alpha,0}$ stands for the set of all constant complex-valued functions defined on $\mathbb{C}_0$, and $\pi_{\alpha,0}$ designates the constant function defined on $\mathbb{C}_0$ with value 1. If $n \in \mathbb{N}$, and if $\alpha_1, \alpha_2, \ldots, \alpha_n$ are given complex numbers, then let

$$\pi_{\alpha,n} := \prod_{j=1}^{n} \rho_{\alpha_j}, \tag{14}$$

and let $\mathcal{R}_{\alpha,n}$ denote the set of all functions f which are meromorphic in $\mathbb{C}_0$ and which admit a representation $f = \frac{1}{\pi_{\alpha,n}} P$ with some complex polynomial P of degree not greater than n. Observe that if $(\alpha_j)_{j=1}^{\infty}$ is a given sequence of complex numbers, for each $n \in \mathbb{N}_0$, the class $\mathcal{R}_{\alpha,n}^{p\times q}$ can be considered as right $\mathbb{C}^{q\times q}$-submodule of the right $\mathbb{C}^{q\times q}$-module $\mathcal{R}_{\alpha}^{p\times q} := \bigcup_{k=0}^{\infty} \mathcal{R}_{\alpha,k}^{p\times q}$ (respectively, left $\mathbb{C}^{p\times p}$-submodule of the left $\mathbb{C}^{p\times p}$-module $\mathcal{R}_{\alpha}^{p\times q}$). Furthermore, one can easily see that, for each $n \in \mathbb{N}$, the set $\tilde{\mathcal{R}}_{\alpha,n}^{p\times q}$ of all $Z \in \mathcal{R}_{\alpha}^{p\times q}$ which can be represented via $Z = \frac{1}{\pi_{\alpha,n}} Q$ where Q is some complex $p \times q$ matrix polynomial of degree not greater than $n-1$ is both a right $\mathbb{C}^{q\times q}$-submodule of the right $\mathbb{C}^{q\times q}$-module $\mathcal{R}_{\alpha,n}^{p\times q}$ and a left $\mathbb{C}^{p\times p}$-submodule of the left $\mathbb{C}^{p\times p}$-module $\mathcal{R}_{\alpha,n}^{p\times q}$. We will see that the $\mathbb{C}^{q\times q}$-submodule $\tilde{\mathcal{R}}_{\alpha,n}^{q\times q}$ of the $\mathbb{C}^{q\times q}$-module $\mathcal{R}_{\alpha,n}^{q\times q}$ plays a key role in our following considerations.

Remark 2.4. Let $n \in \mathbb{N}$ and let $(\alpha_j)_{j=1}^{n} \in \mathcal{T}_{1,n}$. For each $j \in \mathbb{N}_{1,n-1}$, let $\beta_j := \alpha_{j+1}$. Then it is readily checked that the mapping $\varphi_{n-1} : \mathcal{R}_{\beta,n-1}^{q\times q} \to \tilde{\mathcal{R}}_{\alpha,n}^{q\times q}$ given by

$$\varphi_{n-1}(Y) := \frac{1}{\pi_{\alpha,1}} Y$$

is bijective, right-hand $\mathbb{C}^{q\times q}$-linear and left-hand $\mathbb{C}^{q\times q}$-linear.

Now we turn our attention to particular bases of the $\mathbb{C}^{q\times q}$-modules $\mathcal{R}_{\alpha,n}^{q\times q}$ and $\tilde{\mathcal{R}}_{\alpha,n}^{q\times q}$. If $\alpha \in \mathbb{C} \setminus (\mathbb{T} \cup \{0\})$, we define the function $b_\alpha : \mathbb{C}_0 \setminus \{\frac{1}{\overline{\alpha}}\} \to \mathbb{C}$ by

$$b_\alpha(w) := \begin{cases} \frac{\overline{\alpha}}{|\alpha|} \cdot \frac{\alpha - w}{1 - \overline{\alpha} w} & \text{for} \quad w \in \mathbb{C} \setminus \{\frac{1}{\overline{\alpha}}\} \\ \frac{1}{|\alpha|} & \text{for} \quad w = \infty \end{cases}.$$

Further, let $b_0 : \mathbb{C} \to \mathbb{C}$ be given by $b_0(w) := w$ for all $w \in \mathbb{C}$. Let $m \in \mathbb{N}$ or let $m = +\infty$, and let $(\alpha_j)_{j=1}^{m} \in \mathcal{T}_{1,m}$. For each $n \in \mathbb{N}_{1,m}$, let the matrix-valued function $B_{\alpha,n}^{(q)} : \mathbb{C}_0 \setminus \left(\bigcup_{j=1}^{n} \left\{ \frac{1}{\overline{\alpha_j}} \right\} \right) \to \mathbb{C}^{q\times q}$ be defined by

$$B_{\alpha,n}^{(q)}(w) := \left(\prod_{j=1}^{n} b_{\alpha_j}(w) \right) \mathbf{I}_q.$$

Moreover, let $B_{\alpha,0}^{(q)} : \mathbb{C}_0 \to \mathbb{C}^{q\times q}$ denote the constant matrix-valued function with value $\mathbf{I}_q$. It is readily checked that, for each $n \in \mathbb{N}_0$, the system $\{B_{\alpha,k}^{(q)} : k \in \mathbb{N}_{0,n}\}$ is both a basis of the right $\mathbb{C}^{q\times q}$-module $\mathcal{R}_{\alpha,n}^{q\times q}$ and a basis of the left $\mathbb{C}^{q\times q}$-module $\mathcal{R}_{\alpha,n}^{q\times q}$.

Lemma 2.5. *Let $n \in \mathbb{N}$ and let $(\alpha_j)_{j=1}^n \in \mathcal{T}_{1,n}$.*

(a) *For each $j \in \mathbb{N}_{1,n-1}$, let $\beta_j := \alpha_{j+1}$. Then $\left\{ \frac{1}{\pi_{\alpha,1}} B_{\beta,j}^{(q)} : j \in \mathbb{N}_{0,n-1} \right\}$ is both a basis of the right $\mathbb{C}^{q\times q}$-module $\tilde{\mathcal{R}}_{\alpha,n}^{q\times q}$ and a basis of the left $\mathbb{C}^{q\times q}$-module $\tilde{\mathcal{R}}_{\alpha,n}^{q\times q}$.*

(b) *If $\alpha_1, \alpha_2, \ldots, \alpha_n$ are pairwise different, then $\left\{ \frac{1}{\rho_{\alpha_j}} \mathbf{I}_q : j \in \mathbb{N}_{1,n} \right\}$ is both a basis of the right $\mathbb{C}^{q\times q}$-module $\tilde{\mathcal{R}}_{\alpha,n}^{q\times q}$ and a basis of the left $\mathbb{C}^{q\times q}$-module $\tilde{\mathcal{R}}_{\alpha,n}^{q\times q}$.*

Proof. (a) Since $\{B_{\beta,j}^{(q)} : j \in \mathbb{N}_{0,n-1}\}$ is both a basis of the right $\mathbb{C}^{q\times q}$-module $\mathcal{R}_{\beta,n-1}^{q\times q}$ and of the left $\mathbb{C}^{q\times q}$-module $\mathcal{R}_{\beta,n-1}^{q\times q}$, we easily infer from Remark 2.4 that the assertion stated in part (a) is true.

(b) One can easily check that $\left\{ \frac{1}{\rho_{\alpha_j}} \mathbf{I}_q : j \in \mathbb{N}_{1,n} \right\}$ is a right-hand (respectively, left-hand) $\mathbb{C}^{q\times q}$-linear independent system. Obviously, $\frac{1}{\rho_{\alpha_1}} \mathbf{I}_q = \frac{1}{\pi_{\alpha,1}} \mathbf{I}_q$. If $n > 1$, then, for each $j \in \mathbb{N}_{2,n}$, we have

$$\frac{1}{\rho_{\alpha_j}} \mathbf{I}_q = \frac{1}{\pi_{\alpha,n}} P_j, \quad \text{where} \quad P_j := \left(\prod_{\substack{m=1 \\ m\neq j}}^{n} \rho_{\alpha_m} \right) \mathbf{I}_q$$

is a $q \times q$ matrix polynomial of degree not greater than $n-1$. Thus we see that, for each $j \in \mathbb{N}_{1,n}$, the matrix-valued function $\frac{1}{\rho_{\alpha_j}} \mathbf{I}_q$ belongs to $\tilde{\mathcal{R}}_{\alpha,n}^{q\times q}$. In view of part (a), then it follows easily that the assertion stated in part (b) is true. □

A central theorem in [FKL] is concerned with right and left (α, n)-moment matrices of $q \times q$ nonnegative Hermitian-valued Borel measures on the unit circle $\mathbb{T}$. To state this it seems to be useful to introduce a further notation. If X is a matrix-valued function defined on a subset G of $\mathbb{C}_0$ with $\mathbb{T} \subseteq G$, then we will write $\underline{X}$ to denote the restriction of X onto $\mathbb{T}$. Let $F \in \mathcal{M}_{\geq}^{q}(\mathbb{T}, \mathfrak{B}_{\mathbb{T}})$, and let $(\alpha_j)_{j\in\mathbb{N}}$ be a system of points in good position with respect to $\mathbb{T}$. For every nonnegative integer

n, the matrix $\mathbf{G}_n^{(\alpha,F)} := (\mathbf{g}_{jk}^{(\alpha,F)})_{j,k=0}^n$ (respectively, $\mathbf{H}_n^{(\alpha,F)} = (\mathbf{h}_{jk}^{(\alpha,F)})_{j,k=0}^n$) given by

$$\mathbf{g}_{jk}^{(\alpha,F)} := \int_{\mathbb{T}} (\underline{B_{\alpha,j}^{(q)}})^* dF \underline{B_{\alpha,k}^{(q)}} \quad \left(\text{respectively}, \mathbf{h}_{jk}^{(\alpha,F)} := \int_{\mathbb{T}} \underline{B_{\alpha,j}^{(q)}} dF (\underline{B_{\alpha,k}^{(q)}})^* \right)$$

is called the right (respectively, left) (α,n)-moment matrix of F. It can be easily verified that if $n \in \mathbb{N}_0$ and if $\alpha_j = 0$ for each integer j with $1 \le j \le n$, the matrix $\mathbf{G}_n^{(\alpha,F)}$ coincides with the block Toeplitz matrix $\mathbf{T}_n^{(F)}$ of the Fourier coefficients of F (see (9) and (11)). In [FKL, Corollary 4.5] the following result is proved.

Theorem 2.6. *Let $F \in \mathcal{M}_{\geq}^q(\mathbb{T}, \mathfrak{B}_{\mathbb{T}})$ and let n be a positive integer. Then*

$$\operatorname{rank} \mathbf{G}_n^{(\alpha,F)} = \operatorname{rank} \mathbf{T}_n^{(F)} \quad \text{and} \quad \operatorname{rank} \mathbf{H}_n^{(\alpha,F)} = \operatorname{rank} \mathbf{T}_n^{(F)}$$

for all sequences $(\alpha_j)_{j=1}^n \in \mathcal{T}_{1,n}$.

In our subsequent considerations, Theorem 2.6 will play a key role. Now we compare the rank of special matrices constructed by integrals of distinguished rational matrix-valued functions with respect to a nonnegative Hermitian-valued Borel measure F on $\mathbb{T}$ with the rank of the block Toeplitz matrix $\mathbf{T}_n^{(F)}$.

Theorem 2.7. *Let $F \in \mathcal{M}_{\geq}^q(\mathbb{T}, \mathfrak{B}_{\mathbb{T}})$, let $n \in \mathbb{N}$, and let $(\alpha_j)_{j=1}^n \in \mathcal{T}_{1,n}$. Further, let $Y_0, Y_1, \ldots, Y_{n-1}$ be a basis of the right $\mathbb{C}^{q\times q}$-module $\tilde{\mathcal{R}}_{\alpha,n}^{q\times q}$ and let*

$$\mathbf{Y}_{n-1}^{(\alpha,F)} := \left(\int_{\mathbb{T}} \underline{Y_j^*} dF \underline{Y_k} \right)_{j,k=0}^{n-1}. \tag{15}$$

Then

$$\operatorname{rank} \mathbf{Y}_{n-1}^{(\alpha,F)} = \operatorname{rank} \mathbf{T}_{n-1}^{(F)}. \tag{16}$$

Proof. It can be easily seen that the mapping $F^{(\alpha,1)} : \mathfrak{B}_{\mathbb{T}} \to \mathbb{C}^{q\times q}$ defined by

$$F^{(\alpha,1)}(A) := \int_A \left(\underline{\frac{1}{\pi_{\alpha,1}}} \mathbf{I}_q \right)^* dF \left(\underline{\frac{1}{\pi_{\alpha,1}}} \mathbf{I}_q \right) \tag{17}$$

belongs to $\mathcal{M}_{\geq}^q(\mathbb{T}, \mathfrak{B}_{\mathbb{T}})$. In view of [FKL, Remark 4.7] and Theorem 2.6, we obtain

$$\operatorname{rank} \mathbf{G}_{n-1}^{(\beta, F^{(\alpha,1)})} = \operatorname{rank} \mathbf{G}_{n-1}^{(\alpha,F)} = \operatorname{rank} \mathbf{T}_{n-1}^{(F)} \tag{18}$$

where $\beta_j := \alpha_{j+1}$ for each $j \in \mathbb{N}_{1,n-1}$. On the other hand, a standard argument from the integration theory with respect to nonnegative Hermitian-valued measures (see [R] or also [FKL, Remark 1.1]) yields

$$\mathbf{G}_{n-1}^{(\beta, F^{(\alpha,1)})} = \left(\int_{\mathbb{T}} \left(\underline{\frac{1}{\pi_{\alpha,1}} B_{\beta,j}^{(q)}} \right)^* dF \left(\underline{\frac{1}{\pi_{\alpha,1}} B_{\beta,k}^{(q)}} \right) \right)_{j,k=0}^{n-1}.$$

Since we know from part (a) of Lemma 2.5 that $\left\{\frac{1}{\pi_{\alpha,1}}B_{\beta,j}^{(q)} : j \in \mathbb{N}_{0,n-1}\right\}$ is a basis of the right $\mathbb{C}^{q\times q}$-module $\tilde{\mathcal{R}}_{\alpha,n}^{q\times q}$, we can conclude that

$$\operatorname{rank} \mathbf{Y}_{n-1}^{(\alpha,F)} = \operatorname{rank} \mathbf{G}_{n-1}^{(\beta,F^{(\alpha,1)})}. \tag{19}$$

Comparison of (18) and (19) completes the proof. □

The following statement can be proved analogously.

Theorem 2.8. *Let $F \in \mathcal{M}_{\geq}^{q}(\mathbb{T},\mathfrak{B}_{\mathbb{T}})$, let $n \in \mathbb{N}$, and let $(\alpha_j)_{j=1}^{n} \in \mathcal{T}_{1,n}$. Further, let $Z_0, Z_1, \ldots, Z_{n-1}$ be a basis of the left $\mathbb{C}^{q\times q}$-module $\tilde{\mathcal{R}}_{\alpha,n}^{q\times q}$ and let*

$$\mathbf{Z}_{n-1}^{(\alpha,F)} := \left(\int_{\mathbb{T}} \underline{Z_j} dF \underline{Z_k}^*\right)_{j,k=0}^{n-1}.$$

Then

$$\operatorname{rank} \mathbf{Z}_{n-1}^{(\alpha,F)} = \operatorname{rank} \mathbf{T}_{n-1}^{(F)}.$$

Remark 2.9. An easy calculation shows that the identity

$$\frac{1}{(1-\beta_1\overline{z})(1-\overline{\beta_2}z)} = \frac{1}{2(1-\beta_1\overline{\beta_2})}\left(\frac{z+\beta_1}{z-\beta_1} + \overline{\left(\frac{z+\beta_2}{z-\beta_2}\right)}\right)$$

is satisfied for all $\beta_1, \beta_2 \in \mathcal{T}_{1,2}$ and all $z \in \mathbb{T}$.

Remark 2.10. Let $\Omega \in \mathcal{C}_q(\mathbb{D})$, let $n \in \mathbb{N}_0$ and let $(\alpha_j)_{j=0}^{n} \in \mathcal{T}_{0,n}$. In view of (8) and Remark 2.9, it is then readily checked that the matrix $\mathbf{R}_n^{(\alpha,\Omega)}$ given by (5) satisfies

$$\mathbf{R}_n^{(\alpha,\Omega)} = 2\mathbf{\Phi}_n^{(\alpha,F)} \tag{20}$$

where F is the Riesz-Herglotz measure associated with Ω, and where

$$\mathbf{\Phi}_n^{(\alpha,F)} := \left(\int_{\mathbb{T}} \left(\frac{1}{\underline{\rho_{\alpha_j}}}\mathbf{I}_q\right)^* dF \left(\frac{1}{\underline{\rho_{\alpha_k}}}\mathbf{I}_q\right)\right)_{j,k=0}^{n}. \tag{21}$$

Since the nonnegative Hermitian-valued measure F^T satisfies

$$\left(\mathbf{\Phi}_n^{(\alpha,F^T)}\right)^T = \left(\int_{\mathbb{T}} \left(\frac{1}{\underline{\rho_{\alpha_j}}}\mathbf{I}_q\right) dF \left(\frac{1}{\underline{\rho_{\alpha_k}}}\mathbf{I}_q\right)^*\right)_{j,k=0}^{n}, \tag{22}$$

and

$$\hat{\Omega}(w) - i\operatorname{Im}\Omega(0) = \left(\int_{\mathbb{T}} \frac{z+w}{z-w} F^T(dz)\right)^T$$

for all $w \in \mathbb{C} \setminus \mathbb{T}$, one can easily verify that

$$\mathbf{L}_n^{(\alpha,\Omega)} = 2\left(\mathbf{\Phi}_n^{(\alpha,F^T)}\right)^T \tag{23}$$

is true as well.

Observe that application of the formulas (20)–(23) yields a proof of Theorem 2.1. Now we prove Theorem 2.2.

Proof of Theorem 2.2. We consider an arbitrary sequence $(\alpha_j)_{j=0}^n \in \mathcal{T}_{0,n}^{\#}$. For each $j \in \mathbb{N}_{1,n+1}$, let $\beta_j := \alpha_{j-1}$. Then we know from part (b) of Lemma 2.5 that $\left\{\frac{1}{\rho_{\beta_j}}\mathbf{I}_q : j \in \mathbb{N}_{1,n+1}\right\}$ is both a basis of the right $\mathbb{C}^{q\times q}$-module $\tilde{\mathcal{R}}_{\beta,n+1}^{q\times q}$ and a basis of the left $\mathbb{C}^{q\times q}$-module $\tilde{\mathcal{R}}_{\beta,n+1}^{q\times q}$. Thus application of Theorem 2.7 gives $\operatorname{rank}\mathbf{\Phi}_n^{(\alpha,F)} = \operatorname{rank}\mathbf{T}_n^{(F)}$ where F denotes the Riesz-Herglotz measure associated with Ω. Hence we get $\operatorname{rank}\mathbf{R}_n^{(\alpha,\Omega)} = \operatorname{rank}\mathbf{T}_n^{(F)}$ from Remark 2.10. Using Theorem 2.8, (22) and (23) we can conclude similarly that $\operatorname{rank}\mathbf{L}_n^{(\alpha,\Omega)} = \operatorname{rank}\mathbf{T}_n^{(F)}$. In view of (10), the proof is complete. □

Observe that application of Theorem 2.2 provides us immediately with a proof of Theorem 2.3. As already mentioned above, from Theorem 2.2 one also obtains immediately the following well-known result (see, e.g., [DFK, Proposition 2.1.3]).

Corollary 2.11. *Let* $\Omega \in \mathcal{C}_q(\mathbb{D})$. *Then* $\operatorname{rank}(\operatorname{Re}\hat{\Omega}(z)) = \operatorname{rank}(\operatorname{Re}\Omega(0))$ *for each* $z \in \mathbb{C}_0 \setminus \mathbb{T}$.

Corollary 2.12. *Let* $n \in \mathbb{N}_0$ *and let* $(\alpha_j)_{j=0}^n$ *be a sequence of pairwise different points belonging to* $\mathbb{D}$. *If* $\mathbf{A}$ *is a positive Hermitian complex* $q \times q$ *matrix, then the matrix*

$$\mathbf{B} := \left(\frac{1}{1-\alpha_j\overline{\alpha_k}}\mathbf{A}\right)_{j,k=0}^n$$

is positive Hermitian as well.

Proof. Clearly, the constant function $\Omega : \mathbb{D} \to \mathbb{C}^{q\times q}$ with value $\frac{1}{2}\mathbf{A}$ belongs to $\mathcal{C}_q(\mathbb{D})$. Hence

$$\mathbf{R}_n^{(\alpha,\Omega)} = \mathbf{B}.$$

On the other hand, we easily see that Re $\mathbf{S}_n^{(\Omega)}$ coincides with the complex $(n+1)q \times (n+1)q$ matrix $\frac{1}{2}\operatorname{diag}(\mathbf{A},\mathbf{A},\dots,\mathbf{A})$. Thus the $q\times q$ Carathéodory function Ω is nondegenerate of order n and the assertion follows from Theorem 2.3. □

Corollary 2.13. *Let* $n \in \mathbb{N}$ *and let* $\Omega \in \mathcal{C}_q(\mathbb{D})$ *be such that the following two conditions are satisfied:*

(i) *There is a* $z \in \mathbb{C}_0 \setminus \mathbb{T}$ *such that the matrix* $\operatorname{Re}\hat{\Omega}(z)$ *is nonsingular.*

(ii) *There is a sequence* $(\alpha_j)_{j=0}^n \in \mathcal{T}_{0,n}^{\#}$ *such that* $\hat{\Omega}(\alpha_j) = \hat{\Omega}(\alpha_0)$ *for each* $\mathbb{N}_{1,n}$.

Then the $q \times q$ *Carathéodory function is nondegenerate of order* n.

Proof. Because of Theorem 2.3 it is sufficient to show that the matrix $\mathbf{R}_n^{(\alpha,\Omega)}$ given by (5) is nonsingular. We observe that condition (i) and Corollary 2.11 imply that the matrix $\operatorname{Re}\hat{\Omega}(w)$ is nonsingular for each $w \in \mathbb{C}_0 \setminus \mathbb{T}$. In particular, $\operatorname{Re}\Omega(w)$ is positive Hermitian for each $w \in \mathbb{D}$. Now we consider the case in which there is some $k \in \mathbb{N}_{0,n}$ such that $\alpha_k \in \mathbb{D}$. Then condition (ii) and (1) give

$$\hat{\Omega}(\alpha_0) = \hat{\Omega}(\alpha_k) = \Omega(\alpha_k).$$

In particular, the matrix $\mathrm{Re}\hat{\Omega}(\alpha_0)$ is nonnegative Hermitian. Assume that there is an $l \in \mathbb{N}_{0,n}$ such that $\alpha_l \notin \mathbb{D}$. Since $(\alpha_j)_{j=0}^n$ belongs to $\mathcal{T}_{0,n}^{\#}$ we then get $\alpha_l \in \mathbb{C} \setminus (\mathbb{D} \cup \mathbb{T})$. Thus $\frac{1}{\overline{\alpha}} \in \mathbb{D}$, and from (ii) and (1) we obtain that

$$-\hat{\Omega}(\alpha_0) = -\hat{\Omega}(\alpha_l) = \left[\Omega\left(\frac{1}{\overline{\alpha_l}}\right)\right]^*.$$

Hence the matrix $-\mathrm{Re}\,\hat{\Omega}(\alpha_0)$ is also nonnegative Hermitian. Consequently

$$\mathrm{Re}\,\hat{\Omega}(\alpha_0) = \mathbf{0}_{q\times q}.$$

However this is a contradiction to the fact that $\mathrm{Re}\,\hat{\Omega}(\alpha_0)$ is nonsingular. Therefore, for each $j \in \mathbb{N}_{0,n}$ we have $\alpha_j \in \mathbb{D}$ and, in view of (ii) and (1),

$$\mathrm{Re}\,\hat{\Omega}(\alpha_j) = \mathrm{Re}\,\hat{\Omega}(\alpha_0) = \mathrm{Re}\,\Omega(\alpha_0),$$

where $\mathrm{Re}\,\Omega(\alpha_0)$ is positive Hermitian. According to (5), now the equality

$$\mathbf{R}_n^{(\alpha,\Omega)} = \left(\frac{2}{1-\alpha_j\overline{\alpha_k}}\,\mathrm{Re}\,\Omega(\alpha_0)\right)_{j,k=0}^n$$

follows, and Corollary 2.12 shows that $\mathbf{R}_n^{(\alpha,\Omega)}$ is positive Hermitian. It remains to consider the case where $\alpha_j \notin \mathbb{D}$ for each $j \in \mathbb{N}_{0,n}$. Because $(\alpha_j)_{j=0}^n \in \mathcal{T}_{0,n}^{\#}$, we obtain from (ii) and (1) that

$$\mathrm{Re}\,\hat{\Omega}(\alpha_j) = \mathrm{Re}\,\hat{\Omega}(\alpha_0) = \mathrm{Re}\left(-\left[\Omega\left(\frac{1}{\overline{\alpha_0}}\right)\right]^*\right) = -\mathrm{Re}\,\Omega\left(\frac{1}{\overline{\alpha_0}}\right).$$

Thus it follows that

$$\begin{aligned}\mathbf{R}_n^{(\alpha,\Omega)} &= \left(-\frac{2}{1-\alpha_j\overline{\alpha_k}}\,\mathrm{Re}\,\Omega\left(\frac{1}{\overline{\alpha_0}}\right)\right)_{j,k=0}^n \\ &= \mathbf{D}_{0,n}^{(\alpha)}\cdot\left(\frac{2}{1-\frac{1}{\alpha_j}\frac{1}{\overline{\alpha_k}}}\,\mathrm{Re}\,\Omega\left(\frac{1}{\overline{\alpha_0}}\right)\right)_{j,k=0}^n\cdot(\mathbf{D}_{0,n}^{(\alpha)})^*\end{aligned}$$

where

$$\mathbf{D}_{0,n}^{(\alpha)} := \mathrm{diag}(\alpha_0\mathbf{I}_q, \alpha_1\mathbf{I}_q, \dots, \alpha_n\mathbf{I}_q). \tag{24}$$

Since the matrix $\mathrm{Re}\,\Omega\left(\frac{1}{\overline{\alpha_0}}\right)$ is positive Hermitian we can conclude from Corollary 2.12 that $\mathbf{R}_n^{(\alpha,\Omega)}$ is nonsingular. The proof is complete. □

If $\Omega \in \mathcal{C}_q(\mathbb{D})$ and if $n \in \mathbb{N}_0$, then it is clear from (10) that the $q\times q$ Carathéodory function Ω is nondegenerate of order n if and only if the Riesz-Herglotz measure F associated with Ω is nondegenerate of order n. Thus the characterizations of the set $\mathcal{M}_{\geq}^{q,n}(\mathbb{T},\mathfrak{B}_{\mathbb{T}})$ given in [FKL, Section 5] can be applied immediately to describe the set of all functions $\Omega \in \mathcal{C}_q(\mathbb{D})$ which are nondegenerate of order n.

Now we turn our attention to the Schwarz-Pick-Potapov block matrices of second kind for matricial Carathéodory functions. For this class of matrices, we will also verify some results on rank invariance.

Remark 2.14. Let $n \in \mathbb{N}_0$, let $m \in \mathbb{N}$, and let $(\alpha_j)_{j=0}^{n+m} \in \mathcal{T}_{0,n+m}$ be such that

$$\{\alpha_0, \alpha_1, \ldots, \alpha_n\} \cap \{\alpha_{n+1}, \alpha_{n+1}, \ldots, \alpha_{n+m}\} = \emptyset. \tag{25}$$

Let Ω be a complex $q \times q$ matrix-valued function defined on $\mathbb{D}$, and let

$$\tilde{\mathbf{R}}_{n,m}^{(\alpha,\Omega)} := \begin{pmatrix} \left(\frac{\hat{\Omega}(\alpha_j)+[\hat{\Omega}(\alpha_k)]^*}{1-\alpha_j\overline{\alpha_k}}\right)_{j,k=0,\ldots,n} & \left(\frac{\hat{\Omega}(\alpha_j)-\hat{\Omega}(\alpha_k)}{\alpha_j-\alpha_k}\right)_{\substack{j=0,\ldots,n\\k=n+1,\ldots,n+m}} \\ \left[\left(\frac{\hat{\Omega}(\alpha_j)-\hat{\Omega}(\alpha_k)}{\alpha_j-\alpha_k}\right)_{\substack{j=0,\ldots,n\\k=n+1,\ldots,n+m}}\right]^* & \left(\frac{[\hat{\Omega}(\alpha_j)]^*+\hat{\Omega}(\alpha_k)}{1-\overline{\alpha_j}\alpha_k}\right)_{j,k=n+1,\ldots,n+m} \end{pmatrix}. \tag{26}$$

(a) If $\alpha_k \neq 0$ for all $k \in \mathbb{N}_{n+1,m}$, then one can easily see that the matrices $\mathbf{D}_{n+1,n+m}^{(\alpha)} := \operatorname{diag}(\alpha_{n+1}\mathbf{I}_q, \alpha_{n+2}\mathbf{I}_q, \ldots, \alpha_{n+m}\mathbf{I}_q)$ and

$$\mathbf{V}_{n,m}^{(\alpha)} := \operatorname{diag}(\mathbf{I}_{(n+1)q}, -(\mathbf{D}_{n+1,n+m}^{(\alpha)})^{-1})$$

are nonsingular, that the sequence $(\tau_j)_{j=0}^{n+m}$ given by

$$\tau_j := \begin{cases} \alpha_j & \text{if } j \in \mathbb{N}_{0,n} \\ \frac{1}{\overline{\alpha_j}} & \text{if } j \in \mathbb{N}_{n+1,n+m} \end{cases}$$

belongs to $\mathcal{T}_{0,n+m}$, and, in view of (2), that the matrix $\tilde{\mathbf{R}}_{n,m}^{(\alpha,\Omega)}$ admits the representation

$$\tilde{\mathbf{R}}_{n,m}^{(\alpha,\Omega)} = (\mathbf{V}_{n,m}^{(\alpha)})^* \mathbf{R}_{n+m}^{(\tau,\Omega)} \mathbf{V}_{n,m}^{(\alpha)}. \tag{27}$$

(b) If $\alpha_k \neq 0$ for all $k \in \mathbb{N}_{0,n}$, then it is readily checked that the matrices $\mathbf{D}_{0,n}^{(\alpha)}$ given by (24) and

$$\mathbf{W}_{n,m}^{(\alpha)} := \operatorname{diag}(-(\mathbf{D}_{0,n}^{(\alpha)})^{-1}, \mathbf{I}_{mq})$$

are nonsingular, that the sequence $(\sigma_j)_{j=0}^{n+m}$ given by

$$\sigma_j := \begin{cases} \frac{1}{\overline{\alpha_j}} & \text{if } j \in \mathbb{N}_{0,n} \\ \alpha_j & \text{if } j \in \mathbb{N}_{n+1,n+m} \end{cases}$$

belongs to $\mathcal{T}_{0,n+m}$, and that the matrix $\tilde{\mathbf{R}}_{n,m}^{(\alpha,\Omega)}$ can be represented via

$$\tilde{\mathbf{R}}_{n,m}^{(\alpha,\Omega)} = \mathbf{W}_{n,m}^{(\alpha)} \mathbf{L}_{n+m}^{(\sigma,\Omega)} (\mathbf{W}_{m}^{(\alpha)})^*. \tag{28}$$

Now we give a slight generalization of a theorem which goes essentially back to Kovalishina [K] (see also [DFK, Theorem 2.3.2]).

Theorem 2.15. *Let Ω be a complex $q \times q$ matrix-valued function which is defined on $\mathbb{D}$ and which holomorphic in $\mathbb{D}$. Further, let n be a nonnegative integer and let m be a positive integer. Then the following statements are equivalent:*

(i) Ω *belongs to the Carathéodory class* $\mathcal{C}_q(\mathbb{D})$.
(ii) *For every sequence* $(\alpha_j)_{j=0}^{n+m} \in \mathcal{T}_{0,n+m}$ *which satisfies* (25), *the matrix* $\tilde{\mathbf{R}}_{n,m}^{(\alpha,\Omega)}$ *is nonnegative Hermitian.*

Proof. (i) $\Rightarrow$ (ii): Let $(\alpha_j)_{j=0}^{n+m} \in \mathcal{T}_{0,n+m}$ be such that (25) is fulfilled, and let the sequence $(\tau_j)_{j=0}^{n+m}$ (respectively, $(\sigma_j)_{j=0}^{n+m}$) be defined as in Remark 2.14. Using Theorem 2.1 we obtain that the matrix $\mathbf{R}_{n+m}^{(\tau,\Omega)}$ (respectively, $\mathbf{L}_{n+m}^{(\sigma,\Omega)}$) is nonnegative Hermitian. Thus we see from Remark 2.14 that (ii) holds.
(ii) $\Rightarrow$ (i): This implication is obvious. □

In Theorem 2.2 we observed a rank invariance of Schwarz-Pick-Potapov block matrices of the first kind. An analogous result is true for the matrices which characterize via Theorem 2.15 the class $\mathcal{C}_q(\mathbb{D})$.

Theorem 2.16. *Let* $\Omega \in \mathcal{C}_q(\mathbb{D})$, *let* n *be a nonnegative integer, and let* m *be a positive integer. Then*

$$\operatorname{rank} \tilde{\mathbf{R}}_{n,m}^{(\alpha,\Omega)} = \operatorname{rank} \operatorname{Re} \mathbf{S}_{n+m}^{(\Omega)} \tag{29}$$

is satisfied for all sequences $(\alpha_j)_{j=0}^{n+m} \in \mathcal{T}_{0,n+m}^{\#}$.

Proof. Use Remark 2.14 and Theorem 2.2. □

It is clear that, analogously to Theorem 2.2, Theorem 2.16 immediately gives a characterization of the k-th order nondegeneracy of a $q \times q$ Carathéodory functions. Furthermore, note that Corollary 2.13 can also be proved using Theorem 2.16 (and Theorem 2.3 and Corollary 2.11).

For each complex $q \times q$ matrix $\mathbf{A}$, the notation $\mathcal{R}(\mathbf{A})$ stands for the linear subspace of $\mathbb{C}^q$ which is generated by the columns of $\mathbf{A}$.

Corollary 2.17. *Let* $\Omega \in \mathcal{C}_q(\mathbb{D})$. *Then*

$$\mathcal{R}(\operatorname{Re} \hat{\Omega}(z)) = \mathcal{R}(\operatorname{Re} \Omega(0)) \tag{30}$$

for each $z \in \mathbb{C}_0 \setminus \mathbb{T}$.

Proof. In view of (1) it is sufficient to verify that (30) holds for all $z \in \mathbb{D}$. For each $z \in \mathbb{D}$, let $\mathcal{K}(z)$ denote the null space of the matrix $\operatorname{Re} \Omega(z)$. Obviously, there is a number $z_0 \in \mathbb{D}$ such that $\dim \mathcal{K}(z) \leq \dim \mathcal{K}(z_0)$ for all $z \in \mathbb{D}$. Let $z \in \mathbb{D} \setminus \{z_0\}$. Since the matrix $\operatorname{Re}\, \Omega(z)$ is nonnegative Hermitian it suffices to verify that

$$\mathcal{K}(z_0) \subseteq \mathcal{K}(z) \tag{31}$$

is satisfied. Let $g \in \mathcal{K}(z_0)$. For each $h \in \mathbb{C}^q$, Theorem 2.15 implies

$$\begin{aligned} \mathbf{0}_{2\times 2} \quad &\leq \quad \begin{pmatrix} g & \mathbf{0}_{q\times 1} \\ \mathbf{0}_{q\times 1} & h \end{pmatrix}^* \begin{pmatrix} \frac{\Omega(z_0)+[\Omega(z_0)]^*}{1-z_0\overline{z_0}} & \frac{\Omega(z_0)-\Omega(z)}{z_0-z} \\ \left(\frac{\Omega(z_0)-\Omega(z)}{z_0-z}\right)^* & \frac{[\Omega(z)]^*+\Omega(z)}{1-\overline{z}z} \end{pmatrix} \begin{pmatrix} g & \mathbf{0}_{q\times 1} \\ \mathbf{0}_{q\times 1} & h \end{pmatrix} \\ &= \quad \begin{pmatrix} 0 & g^* \frac{\Omega(z_0)-\Omega(z)}{z_0-z} h \\ h^* \left(\frac{\Omega(z_0)-\Omega(z)}{z_0-z}\right)^* g & h^* \frac{[\Omega(z)]^*+\Omega(z)}{1-\overline{z}z} h \end{pmatrix}. \end{aligned}$$

Consequently

$$g^* \frac{\Omega(z_0) - \Omega(z)}{z_0 - z} h = 0$$

for each $h \in \mathbb{C}^q$. Hence $g^*\Omega(z) = g^*\Omega(z_0)$. Thus

$$g^*(\operatorname{Re}\Omega(z))g = g^*(\operatorname{Re}\Omega(z_0))g = 0$$

follows. Since the matrix Re $\Omega(z)$ is nonnegative Hermitian, we obtain $g \in \mathcal{K}(z)$. Therefore (31) is checked, and the proof is complete. □

Obviously, we see from Corollary 2.17 that $\mathcal{R}(\mathbf{R}_0^{(\alpha,\Omega)}) = \mathcal{R}(\operatorname{Re}\mathbf{S}_0^{(\Omega)})$ for each $\Omega \in \mathcal{C}_q(\mathbb{D})$ and each $(\alpha_j)_{j=0}^0 \in \mathcal{T}_{0,0}^{\#}$. Consideration of the 1×1 Carathéodory function $\psi : \mathbb{D} \to \mathbb{C}$ defined by $\psi(w) := \frac{1+w}{1-w}$ shows that, in the case $n \in \mathbb{N}$, there is a sequence $(\alpha_j)_{j=0}^n \in \mathcal{T}_{0,n}^{\#}$ such that the linear spaces $\mathcal{R}(\mathbf{R}_n^{(\alpha,\Omega)})$ and $\mathcal{R}(\operatorname{Re}\mathbf{S}_n^{(\Omega)})$ do not coincide.

Let us finish this section with a remark on the inverse of an arbitrary $q \times q$ Carathéodory function which has a nonsingular value.

Remark 2.18. Let $\Omega \in \mathcal{C}_q(\mathbb{D})$ be such that the matrix $\Omega(w_0)$ is nonsingular for some $w_0 \in \mathbb{D}$. Then the matrix $\Omega(w)$ is nonsingular for all $w \in \mathbb{D}$, the matrix-valued function Ω^{-1} belongs to $\mathcal{C}_q(\mathbb{D})$ as well, and

$$\operatorname{rank}(\operatorname{Re}[\Omega^{-1}(w)]) = \operatorname{rank}(\operatorname{Re}[\Omega(w)]) = \operatorname{rank}(\operatorname{Re}[\Omega(w_0)])$$

for each $w \in \mathbb{D}$ (see, e.g., [DFK, Lemma 2.1.10]). Moreover, for all $n \in \mathbb{N}_0$,

$$\operatorname{Re}\mathbf{S}_n^{(\Omega^{-1})} = \operatorname{Re}[(\mathbf{S}_n^{(\Omega)})^{-1}] = (\mathbf{S}_n^{(\Omega)})^{-1}(\operatorname{Re}\mathbf{S}_n^{(\Omega)})(\mathbf{S}_n^{(\Omega)})^{-*}$$

and therefore

$$\operatorname{rank}(\operatorname{Re}\mathbf{S}_n^{(\Omega^{-1})}) = \operatorname{rank}(\operatorname{Re}\mathbf{S}_n^{(\Omega)}) \tag{32}$$

(see, e.g., [DFK, Lemma 1.1.21, Lemma 1.1.15]). This shows in particular that the $q \times q$ Carathéodory function Ω is nondegenerate of order n if and only if the $q \times q$ Carathéodory function Ω^{-1} is nondegenerate of order n.

3. On matricial Carathéodory functions associated with molecular nonnegative Hermitian-valued Borel measures

In this section, we will turn our attention to the matrix-valued Carathéodory functions the Riesz-Herglotz measure of which is molecular. Recall that a measure $F \in \mathcal{M}_{\geq}^q(\mathbb{T}, \mathfrak{B}_{\mathbb{T}})$ is called molecular if there is a finite subset B of $\mathbb{T}$ such that $F(\mathbb{T} \setminus B) = \mathbf{0}_{q\times q}$. Molecular nonnegative Hermitian-valued Borel measures have, in a sense, the highest degree of degeneracy (see Section 6 in [FKL]). The $q \times q$ Carathéodory functions with molecular Riesz-Herglotz measure are rational with poles on the unit circle. Now let us describe them in an explicit way.

Let $r \in \mathbb{N}$, let $(\mathbf{A}_j)_{j=1}^r$ be a sequence of nonnegative Hermitian complex $q \times q$ matrices, let $(z_j)_{j=1}^r$ be a sequence of unimodular complex numbers, and let $\mathbf{H}$ be a

Hermitian complex $q \times q$ matrix. Then it is readily checked that the matrix-valued function $\Psi : \mathbb{D} \to \mathbb{C}^{q\times q}$ defined by

$$\Psi(w) := \sum_{j=1}^{r} \frac{z_j + w}{z_j - w} \mathbf{A}_j + i\,\mathbf{H} \tag{33}$$

satisfies

$$\Psi(w) = \int_{\mathbb{T}} \frac{z + w}{z - w} F(dz) + i\,\mathbf{H}$$

where F is the molecular $q \times q$ nonnegative Hermitian-valued Borel measure on $\mathbb{T}$ given by

$$F := \sum_{j=1}^{r} \mathbf{A}_j \varepsilon_{z_j, \mathfrak{B}_{\mathbb{T}}}, \tag{34}$$

and where $\varepsilon_{z_j, B_{\mathbb{T}}}$ is the Dirac measure with unit mass located at z_j. In particular, Ψ belongs to $\mathcal{C}_q(\mathbb{D})$, and

$$\operatorname{Re} \mathbf{S}_n^{(\Psi)} = \left(\sum_{j=1}^{r} z_j^{k-l} \mathbf{A}_j \right)_{l,k=0}^{n}. \tag{35}$$

Let $n \in \mathbb{N}_0$, and let $(\alpha_j)_{j=0}^{n} \in \mathcal{T}_{0,n}^{\#}$. In view of Remark 2.9, (5) and (6), we have

$$\mathbf{R}_n^{(\alpha,\Psi)} = \left(\sum_{j=1}^{r} \frac{2}{(1 - \alpha_k \overline{z_j})(1 - \overline{\alpha_l} z_j)} \mathbf{A}_j \right)_{k,l=0,\ldots,n}, \tag{36}$$

and

$$\mathbf{L}_n^{(\alpha,\Psi)} = \left(\sum_{j=1}^{r} \frac{2}{(1 - \overline{\alpha_k} z_j)(1 - \alpha_l \overline{z_j})} \mathbf{A}_j \right)_{k,l=0,\ldots,n}. \tag{37}$$

Furthermore, if $n \in \mathbb{N}_0$, if $m \in \mathbb{N}$ and if $(\alpha_j)_{j=0}^{n+m} \in \mathcal{T}_{0,n+m}^{\#}$, then we see from (26) that

$$\tilde{\mathbf{R}}_{n,m}^{(\alpha,\Psi)} =$$

$$\begin{pmatrix} \left(\sum\limits_{j=1}^{r} \frac{2\mathbf{A}_j}{(1-\alpha_k \overline{z_j})(1-\overline{\alpha_l} z_j)} \right)_{k,l=0,\ldots,n} & \left(\sum\limits_{j=1}^{r} \frac{2 z_j \mathbf{A}_j}{(z_j - \alpha_k)(z_j - \alpha_l)} \right)_{\substack{k=0,\ldots,n \\ l=n+1,\ldots,n+m}} \\ \left[\left(\sum\limits_{j=1}^{r} \frac{2 z_j \mathbf{A}_j}{(z_j - \alpha_k)(z_j - \alpha_l)} \right)_{\substack{k=0,\ldots,n \\ l=n+1,\ldots,n+m}} \right]^* & \left(\sum\limits_{j=1}^{r} \frac{2\mathbf{A}_j}{(1-\overline{\alpha_k} z_j)(1-\alpha_l \overline{z_j})} \right)_{k,l=n+1,\ldots,n+m} \end{pmatrix}. \tag{38}$$

It is clear how the results stated in Theorems 2.2 and 2.16 can be specified for the matrices we encounter in (35), (36), (37) and (38).

One can give a more detailed estimation of the rank for the special case of matricial Carathéodory functions associated with molecular nonnegative Hermitian-valued Borel measures. Indeed, in view of (33) and (34), it follows easily from identity (10) and [FKL, Remark 3.9, Theorem 6.6, Corollary 6.8 and Theorem 6.11] that the following statements are true.

Remark 3.1. Let $r \in \mathbb{N}$, let $(\mathbf{A}_j)_{j=1}^r$ be a sequence of nonnegative Hermitian complex $q \times q$ matrices, let $(z_j)_{j=1}^r$ be a sequence from $\mathbb{T}$, let $\mathbf{H}$ be a Hermitian complex $q \times q$ matrix, and let $\Psi : \mathbb{D} \to \mathbb{C}^{q\times q}$ be defined by (33). For each $n \in \mathbb{N}_0$, and for each $(\alpha_j)_{j=0}^n \in \mathcal{T}_{0,n}^{\#}$, we then have

$$\max_{j\in\{1,\ldots,r\}} \operatorname{rank} \mathbf{A}_j \leq \operatorname{rank} \mathbf{R}_n^{(\alpha,\Psi)} \leq \sum_{j=1}^r \operatorname{rank} \mathbf{A}_j$$

and, in particular, if $n \geq r-1$, we have

$$\operatorname{rank} \mathbf{R}_n^{(\alpha,\Psi)} = \operatorname{rank} \operatorname{Re} \mathbf{S}_{r-1}^{(\Psi)}.$$

If each of the matrices $\mathbf{A}_1, \mathbf{A}_2, \ldots, \mathbf{A}_r$ is nonsingular and if the points $z_1, z_2, \ldots, z_r$ are pairwise different, then, for each $n \in \mathbb{N}_0$,

$$\operatorname{rank} \operatorname{Re} \mathbf{S}_n^{(\Psi)} = \begin{cases} (n+1)q & \text{if } \quad 0 \leq n \leq r-1 \\ rq & \text{if } \quad n \geq r \end{cases}.$$

Proposition 3.2. *Let Ψ be a complex-valued function defined on $\mathbb{D}$ and let $r \in \mathbb{N}$. Then the following statements are equivalent:*

(i) *Ψ is a 1×1 Carathéodory function which is degenerate of order r.*
(ii) *There are a sequence $(z_j)_{j=1}^r$ from $\mathbb{T}$, a sequence $(a_j)_{j=1}^r$ of nonnegative real numbers, and a real number h such that*

$$\Psi(w) = \sum_{j=1}^r \frac{z_j + w}{z_j - w}\, a_j + i\, h \tag{39}$$

for each $w \in \mathbb{D}$.

Proof. Using identity (10), [FKL, Proposition 6.4] and the considerations at the beginning of this section one can easily see that (i) is sufficient for (ii). On the other hand, from Theorem 2.2 and Remark 3.1 we obtain that (i) is necessary for (ii). □

If $\mathbf{A}$ is a complex $q \times q$ matrix, then tr $\mathbf{A}$ stands for the trace of $\mathbf{A}$.

Theorem 3.3. *Let $\Psi \in \mathcal{C}_q(\mathbb{D})$, let $r \in \mathbb{N}$, and let $(\alpha_j)_{j=0}^r \in \mathcal{T}_{0,r}^{\#}$. Further, let $(z_j)_{j=1}^r$ be a sequence of pairwise different unimodular complex numbers, and let $(A_j)_{j=1}^r$ be a sequence of nonnegative Hermitian complex $q \times q$ matrices.*

(a) *The following statements are equivalent:*
   (i) *There is a Hermitian complex $q \times q$ matrix $\mathbf{H}$ such that Ψ admits the representation*

$$\Psi(w) = \sum_{j=1}^r \frac{z_j + w}{z_j - w}\, \mathbf{A}_j + i\, \mathbf{H} \tag{40}$$

for each $w \in \mathbb{D}$.

(ii) *The matrix* $\mathbf{R}_r^{(\alpha,\Psi)}$ *given by* (5) *can be represented via* (36) *with* $r = n$.

(iii) *The matrix* $\mathbf{S}_r^{(\Psi)}$ *satisfies* (35) *with* $r = n$.

(b) *Let (ii) (respectively, (iii)) be satisfied. Then there is a unique* $q \times q$ *Carathéodory function* Ω *which satisfies* $\mathbf{R}_r^{(\alpha,\Omega)} = \mathbf{R}_r^{(\alpha,\Psi)}$ *(respectively,* $\operatorname{Re}\mathbf{S}_r^{(\Omega)} = \operatorname{Re}\mathbf{S}_r^{(\Psi)}$*) and* $\operatorname{Im}\Omega(0) = \mathbf{0}_{q\times q}$, *namely* $\Omega = \Psi - i\operatorname{Im}\Psi(0)$.

Proof. In this section we have already verified that (ii) is necessary for (i). Now assume that (ii) is fulfilled. The function $\varphi := \operatorname{tr}\Psi$ obviously belongs to $\mathcal{C}_1(\mathbb{D})$, and from (5) and (36) we get

$$\mathbf{R}_r^{(\alpha,\varphi)} = \left(\sum_{j=1}^{r} \frac{2}{(1-\alpha_k\overline{z_j})(1-\overline{\alpha_l}z_j)} \operatorname{tr}\mathbf{A}_j\right)_{k,l=0,\dots,r}. \tag{41}$$

The consideration at the beginning of this section shows that the function $\Theta : \mathbb{D} \to \mathbb{C}$ defined by

$$\Theta(w) := \sum_{j=1}^{r} \frac{z_j + w}{z_j - w} \operatorname{tr}\mathbf{A}_j, \quad w \in \mathbb{D},$$

belongs to $\mathcal{C}_1(\mathbb{D})$ and that, in view of (41), the identity $\mathbf{R}_r^{(\alpha,\Theta)} = \mathbf{R}_r^{(\alpha,\varphi)}$ holds. From Remark 3.1 we then see that the matrix $\mathbf{R}_r^{(\alpha,\varphi)}$ is singular. By virtue of Theorem 2.2, we thus infer that the matrix $\operatorname{Re}\mathbf{S}_r^{(\varphi)}$ is singular. According to Proposition 3.2, then the function φ admits the representation

$$\varphi(w) = \sum_{j=1}^{r} \frac{\zeta_j + w}{\zeta_j - w} a_j + i\,h, \quad w \in \mathbb{D},$$

where $(\zeta_j)_{j=1}^r$ is some sequence from $\mathbb{T}$, where $(a_j)_{j=1}^r$ is some sequence of nonnegative real numbers, and where h is some real number. Obviously,

$$\mu := \sum_{j=1}^{r} a_j \varepsilon_{\zeta_j,\mathfrak{B}_\mathbb{T}} \tag{42}$$

is the Riesz-Herglotz measure associated with φ. Now, in view of $\varphi = \operatorname{tr}\Psi$, it is readily checked that μ is exactly the trace measure of the Riesz-Herglotz measure G associated with Ψ. Because of (42), there are nonnegative Hermitian complex $q \times q$ matrices $\mathbf{B}_1, \mathbf{B}_2, \dots, \mathbf{B}_r$ such that

$$G = \sum_{j=1}^{r} \mathbf{B}_j \varepsilon_{\zeta_j,\mathfrak{B}_\mathbb{T}}.$$

Hence

$$\Psi(w) = \sum_{j=1}^{r} \frac{\zeta_j + w}{\zeta_j - w} \mathbf{B}_j + i\,\mathbf{H}$$

for each $w \in \mathbb{D}$, where $\mathbf{H}$ is some Hermitian complex $q \times q$ matrix. It can be checked easily that the mapping $\Phi : \mathbb{D} \to \mathbb{C}^{q\times q}$ given by

$$\Phi(w) := \sum_{j=1}^{r} \frac{z_j + w}{z_j - w} \mathbf{A}_j + i\,\mathbf{H}, \quad w \in \mathbb{D},$$

is a $q \times q$ Carathéodory function the associated Riesz-Herglotz measure of which is the F defined by (34). In view of the considerations at the beginning of this section and (ii), we therefore arrive at the equation

$$\mathbf{R}_r^{(\alpha,\Phi)} = \mathbf{R}_r^{(\alpha,\Psi)}. \tag{43}$$

The sequence $(\beta_j)_{j=1}^{r+1}$ given by $\beta_j := \alpha_{j-1}$ for each $j \in \mathbb{N}_{1,r+1}$ belongs to $\mathcal{T}_{1,r+1}^{\#}$. For all $X \in \tilde{\mathcal{R}}_{\beta,r+1}^{q\times q}$, we have

$$\int_{\mathbb{T}} \underline{X}^* dG \underline{X} = \sum_{j=1}^{r} [X(\zeta_j)]^* \mathbf{B}_j X(\zeta_j)$$

and

$$\int_{\mathbb{T}} \underline{X}^* dF \underline{X} = \sum_{j=1}^{r} [X(z_j)]^* \mathbf{A}_j X(z_j).$$

Using part (b) of Lemma 2.5, Remark 2.10 and (43), it follows that

$$\int_{\mathbb{T}} \underline{X}^* dG \underline{X} = \int_{\mathbb{T}} \underline{X}^* dF \underline{X}$$

for each $X \in \tilde{\mathcal{R}}_{\beta,r+1}^{q\times q}$. Therefore

$$\sum_{j=1}^{r} [X(\zeta_j)]^* \mathbf{B}_j X(\zeta_j) = \sum_{j=1}^{r} [X(z_j)]^* \mathbf{A}_j X(z_j)$$

for each $X \in \tilde{\mathcal{R}}_{\beta,r+1}^{q\times q}$. Now it is readily checked that $\zeta_j = z_j$ and $\mathbf{B}_j = \mathbf{A}_j$ for each $j \in \mathbb{N}_{1,r}$. Thus the equivalence of (i) and (ii) is verified. Analogously, one can show that (iii) is necessary and sufficient for (i) (compare also [FKL, Theorems 6.5 and 6.6]). The assertion stated in part (b) is clear. □

Corollary 3.4. *Let $r \in \mathbb{N}$, let $(z_j)_{j=1}^{r}$ be a sequence of pairwise different points belonging to the unit circle $\mathbb{T}$, let $(\mathbf{A}_j)_{j=1}^{r}$ be a sequence of nonnegative Hermitian complex $q \times q$ matrices, and let $\mathbf{H}$ be a Hermitian complex $q \times q$ matrix.*

(a) *Let $(\alpha_j)_{j=0}^{r}$ be a sequence from $\mathbb{D}$, and let*

$$\mathbf{\Psi}_k := \sum_{j=1}^{r} \frac{z_j + \alpha_k}{z_j - \alpha_k} \mathbf{A}_j + i\mathbf{H}$$

for each $k \in \mathbb{N}_{0,r}$. Then there is a unique $\Psi \in \mathcal{C}_q(\mathbb{D})$ such that $\Psi(\alpha_k) = \mathbf{\Psi}_k$ for each $k \in \mathbb{N}_{0,r}$, namely the Ψ given by (40).

(b) *Let* $\mathbf{\Gamma}_0 := \sum_{j=1}^r \mathbf{A}_j + i\mathbf{H}$ *and let* $\mathbf{\Gamma}_k := 2\sum_{j=1}^r z_j^{-k}\mathbf{A}_j$ *for each* $k \in \mathbb{N}_{1,r}$. *Then there is a unique function* $\Psi \in \mathcal{C}_q(\mathbb{D})$ *such that*

$$\frac{\Psi^{(k)}(0)}{k!} = \mathbf{\Gamma}_k$$

for each $k \in \mathbb{N}_{0,r}$, *namely again the* Ψ *given by* (40).

Proof. Use Theorem 3.3 and the consideration at the beginning of Section 3. □

Let us finish this paper with some corollaries on complex-valued Caratheodory functions.

Corollary 3.5. *Let* $\Psi \in \mathcal{C}_1(\mathbb{D})$, *let* $r \in \mathbb{N}$, *and let* $(\alpha_j)_{j=0}^r \in \mathcal{T}_{0,r}^{\#}$. *Then the following statements are equivalent:*

(i) Ψ *is degenerate of order* r.

(ii) *There are a sequence* $(z_j)_{j=1}^r$ *of unimodular complex numbers and a sequence* $(b_j)_{j=1}^r$ *of nonnegative real numbers such that*

$$\mathbf{R}_r^{(\alpha,\Psi)} = \left(\sum_{j=1}^r \frac{b_j}{(1-\alpha_k\overline{z_j})(1-\overline{\alpha_l}z_j)}\right)_{k,l=0,\dots,r}$$

(iii) *There are a sequence* $(\zeta_j)_{j=1}^r$ *of unimodular complex numbers and a sequence* $(c_j)_{j=1}^r$ *of nonnegative real numbers such*

$$\operatorname{Re}\mathbf{S}_r^{(\Psi)} = \left(\sum_{j=1}^r \zeta_j^{l-k}c_j\right)_{k,l=0}^r$$

Proof. Use Proposition 3.2 and Theorem 3.3. □

Corollary 3.6. *Let* $\Psi \in \mathcal{C}_1(\mathbb{D})$, *let* $r \in \mathbb{N}$, *and let* $(\alpha_j)_{j=0}^r \in \mathcal{T}_{0,r}^{\#}$. *Then the following statements are equivalent:*

(i) *There are a sequence* $(z_j)_{j=1}^r$ *of unimodular complex numbers, a sequence* $(a_j)_{j=1}^r$ *of nonnegative real numbers and a real number* h *such that* (39) *holds for all* $w \in \mathbb{D}$.

(ii) $\operatorname{rank}\mathbf{R}_r^{(\alpha,\Psi)} = \operatorname{rank}\operatorname{Re}\mathbf{S}_{r-1}^{(\Psi)}$

Proof. (i) ⇒ (ii): This implication follows immediately from Remark 3.1.
(ii) ⇒ (i): Use Proposition 3.2. □

Corollary 3.7. *Let* $m \in \mathbb{N}$, *and let* Ψ *be a* 1×1 *Carathéodory function which is degenerate of order* m *and which satisfies* $\operatorname{Re}\Psi(z) \neq 0$ *for some* $z \in \mathbb{D}$. *Let* $s := \operatorname{rank}\operatorname{Re}\mathbf{S}_m^{(\Psi)}$ *Then there are a sequence* $(z_j)_{j=1}^s$ *of pairwise different unimodular complex numbers, a sequence* $(a_j)_{j=1}^s$ *of positive real numbers, and a real number* h *such that* $\Psi(w) = \sum\limits_{j=1}^s \frac{z_j+w}{z_j-w}a_j + i\,h$ *for each* $w \in \mathbb{D}$.

Proof. First, observe that $1 \leq s \leq m$. By virtue of Proposition 3.2, there is an integer r with $1 \leq r \leq m$ such that Ψ can be represented via

$$\Psi(w) = \sum_{j=1}^{r} \frac{z_j + w}{z_j - w} a_j + i\, h$$

where $(z_j)_{j=1}^r$ is a sequence of pairwise different points from $\mathbb{T}$, where $(a_j)_{j=1}^r$ is a sequence of positive real numbers, and where h is some real number. In view of Remark 3.1, we finally get

$$r = \operatorname{rank} \operatorname{Re} \mathbf{S}_r^{(\Psi)} = \operatorname{rank} \operatorname{Re} \mathbf{S}_m^{(\Psi)} = s.$$

□

Corollary 3.8. *Let $r \in \mathbb{N}$, let $h \in \mathbb{R}$, let $(z_j)_{j=1}^r$ be a sequence of pairwise different points belonging to $\mathbb{T}$, and let $(a_j)_{j=1}^r$ be a sequence of nonnegative real numbers. Suppose that the function $\psi : \mathbb{D} \to \mathbb{C}$ given by*

$$\psi(w) := \sum_{j=1}^{r} \frac{z_j + w}{z_j - w} a_j + i\, h$$

satisfies $\psi(w_0) \neq 0$ for some $w_0 \in \mathbb{D}$. Then $\psi(w) \neq 0$ for all $w \in \mathbb{D}$, and there are a sequence $(v_j)_{j=1}^r$ of points belonging to $\mathbb{T}$, a sequence $(b_j)_{j=1}^r$ of nonnegative real numbers, and a real number g such that

$$\frac{1}{\psi(w)} = \sum_{j=1}^{r} \frac{v_j + w}{v_j - w} b_j + i\, g \tag{44}$$

for all $w \in \mathbb{D}$, where $g = 0$ if and only if $h = 0$. Moreover, if $a_j > 0$ for each $j \in \mathbb{N}_{1,r}$, then $(v_j)_{j=1}^r$ is a sequence of pairwise different points and $b_j > 0$ for each $j \in \mathbb{N}_{1,r}$.

Proof. The function ψ belongs to $\mathcal{C}_1(\mathbb{D})$, and, in view of Remark 2.18, the function $\frac{1}{\psi}$ belongs to $\mathcal{C}_1(\mathbb{D})$ as well. Furthermore, Remark 2.18 and Proposition 3.2 show that there are a sequence $(v_j)_{j=1}^r$ from $\mathbb{T}$, a sequence $(b_j)_{j=1}^r$ of nonnegative real numbers and a real number g such that (44) holds for all $w \in \mathbb{D}$. Obviously, $g = 0$ if and only if $h = 0$. From Remark 2.18 we also know that rank $(\operatorname{Re} \mathbf{S}_n^{(\psi)}) = \operatorname{rank} (\operatorname{Re} \mathbf{S}_n^{(\frac{1}{\psi})})$ for all $n \in \mathbb{N}_0$. Hence the rest of the assertion follows from Remark 3.1. □

Observe that Corollary 3.8 can be generalized to $q \times q$ Carathéodory functions of the type given by (33) for which $\Omega(w_0)$ is nonsingular for some $w_0 \in \mathbb{D}$. However, the representation of Ω^{-1} of the type (44) then has $qr+1$ terms in the sum, and the corresponding proof uses another technique.

The results stated in this paper can be used to prove similar statements on Schwarz-Pick-Potapov block matrices of matricial Schur functions and of functions which belong to the Potapov class $\mathcal{P}_{J_1,J_2}(\mathbb{D})$. We want to do this in future work.

References

[BGHN] A. Bultheel, P. González-Vera, E. Hendriksen and O. Njåstad, *Orthogonal Rational Functions*, Cambridge Mono. on Applied and Comput. Math. 5, Cambridge University Press, Cambridge 1999.

[DGK] P. Delsarte, Y. Genin and Y. Kamp, *The Nevanlinna Pick problem for matrix-valued functions*, SIAM J. Appl. Math. **36** (1979), 47–61.

[D] V.K. Dubovoj, *Indefinite metric in the interpolation problem of Schur for analytic matrix functions* (in Russian), Teor. Funktsii, Funktsional. Anal. i Prilozhen., Part I: **37** (1982), 14–26; Part II: **38** (1982), 32–39; Part III: **41** (1984), 55–64; Part IV: **42** (1984), 46–57; Part V: **45** (1986), 16–21; Part VI: **47** (1987), 112–119.

[DFK] V.K. Dubovoj, B. Fritzsche and B. Kirstein, *Matricial Version of the Classical Schur Problem*, Teubner-Texte zur Mathematik, Bd. 129, B.G. Teubner, Stuttgart-Leipzig 1992.

[EP] A.V. Efimov and V.P. Potapov, *J-expansive matrix-valued functions and their role in the analytic theory of electrical circuits* (in Russian), Uspekhi Mat. Nauk **28** (1973), No. 1, 65–130; Russian Math. Surveys **28** (1973), No. 1, 69–140.

[FKL] B. Fritzsche, B. Kirstein and A. Lasarow, *On rank invariance of moment matrices of nonnegative Hermitian-valued Borel measures on the unit circle*, submitted to: J. Comput. Appl. Math.

[K] I.V. Kovalishina, *Analytic theory of a class of interpolation problems* (in Russian), Izv. Akad. Nauk SSSR, Ser. Mat. **47** (1983), 455–497.

[P] V.P. Potapov, *General theorems on the structure and splitting-off of elementary factors of analytic matrix functions* (in Russian), Dokl. Akad. Nauk Armyan. SSR, Ser. Mat. **48** (1969), 257–262.

[R] M. Rosenberg, *The square integrability of matrix-valued functions with respect to a non-negative Hermitian measure*, Duke Math. J. **31** (1964), 291–298.

B. Fritzsche and B. Kirstein
Mathematisches Institut
Universität Leipzig
Augustusplatz 10/11
04109 Leipzig
e-mail: `fritzsche@mathematik.uni-leipzig.de`
e-mail: `wimath@mathematik.uni-leipzig.de`

A. Lasarow
Max-Planck-Institut
für Mathematik in den Naturwissenschaften
Inselstrasse 22
04103 Leipzig
e-mail: `lasarow@mis.mpg.de`

Received: 20 October 2001

Operator Theory:
Advances and Applications, Vol. 135, 183–191

Finite Section Method for Linear Ordinary Differential Equations Revisited

I. Gohberg, M.A. Kaashoek and F. van Schagen

To Bernd Silbermann on the occasion of his sixtieth birthday

Abstract. The sufficiency condition in the main theorem of [6] is shown to be necessary too. Also a new proof is given of the corresponding result for the time-invariant case.

1. Introduction

In this paper solutions of the initial value problem

$$\begin{cases} \dot{x}(t) - A(t)x(t) = f(t), & 0 \le t < \infty, \\ Lx(0) = 0, \end{cases} \tag{1}$$

are obtained as limits of equations on a finite interval of the following form

$$\begin{cases} \dot{x}(t) - A(t)x(t) = f(t), & 0 \le t \le \tau, \\ Lx(0) + R(\tau)x(\tau) = 0. \end{cases} \tag{2}$$

The main results of the paper complement those in [6]. To state the main theorems we first recall the conditions imposed on (1) and (2) in [6].

The main coefficient $A(t)$ in (1) (and (2)) is assumed to be a locally integrable $n \times n$ matrix function on $[0, \infty)$. The initial value coefficient L is an $n \times n$ matrix which is required to be an *exponential dichotomy* of

$$\dot{x}(t) - A(t)x(t) = 0, \qquad 0 \le t < \infty. \tag{3}$$

The latter means (see [2]) that L is a projection and that there exist positive real constants M and α such that

$$\begin{aligned} \|U(t)LU(s)^{-1}\| &\le M\mathrm{e}^{-\alpha(t-s)}, \quad 0 \le s \le t < \infty, \\ \|U(t)(I-L)U(s)^{-1}\| &\le M\mathrm{e}^{-\alpha(s-t)}, \quad 0 \le t \le s < \infty. \end{aligned} \tag{4}$$

Here $U(t)$ is the fundamental matrix of (3), i.e., $U(t)$ is absolutely continuous on finite intervals, $U(0)$ is the $n \times n$ identity matrix, and $(d/dt)U(t) = A(t)U(t)$ a.e. on $0 \le t < \infty$. The dichotomy condition on L implies (see [2] and [6]) that equation (1) has a unique solution x in $L_n^2[0, \infty)$ for each right-hand side f in $L_n^2[0, \infty)$. The boundary value coefficient $R(\tau)$ in (2) is an $n \times n$ matrix such that

$\operatorname{rank} L + \operatorname{rank} R(\tau) = n$. Furthermore, we require that there exists a number $\tau_\circ$ such that

$$\det\big(L + R(\tau)U(\tau)\big) \neq 0, \quad \tau \geq \tau_\circ.$$

The latter is equivalent to the condition (see [4]) that for $\tau \geq \tau_\circ$ the equation (2) has a unique solution for each right-hand side f in $L_n^2[0,\infty)$. In the sequel we refer to the conditions on (1) and (2) listed in this paragraph as our *standing assumptions* on (1) and (2).

Given $f \in L_n^2[0,\infty)$, our aim is to approximate the unique solution x of (1) by the solution x_τ of (2) for $\tau \to \infty$. We shall say that for the equation (1) the *finite section method with respect to the boundary value matrices* $\{R(\tau)\}$ *converges in* L^2 if for each f in $L_n^2[0,\infty)$ the unique solution x of (1) in $L_n^2[0,\infty)$ is obtained as the limit in L^2 of the unique solution x_τ $(\tau \geq \tau_\circ)$ of equation (2). Our main result is the following theorem.

Theorem 1.1. *Assume the standing assumptions on* (1) *and* (2) *are satisfied. Then the finite section method for* (1) *relative to the boundary value matrices* $\{R(\tau)\}$ *converges in* L^2 *if and only if*

$$\sup_{\tau \geq \tau_\circ} \|U(\tau)(L + R(\tau)U(\tau))^{-1} R(\tau)\| < \infty. \tag{5}$$

This result is specified further for the time-invariant case when $A(t)$ does not depend on t. In this case we also assume that R does not depend on t. In other words we consider the equations:

$$\begin{cases} \dot{x}(t) - Ax(t) = f(t), & 0 \leq t < \infty, \\ Lx(0) = 0, \end{cases} \tag{6}$$

and

$$\begin{cases} \dot{x}(t) - Ax(t) = f(t), & 0 \leq t < \tau, \\ Lx(0) + Rx(\tau) = 0. \end{cases} \tag{7}$$

In the time-invariant case when $A(t) = A$ the condition that L is an exponential dichotomy means that A has no eigenvalues on the imaginary axis and that L is a projection such that $\operatorname{Im} L = \operatorname{Im} L_A$, where L_A is the spectral projection of A corresponding to the eigenvalues of A in the left half-plane. For equation (7) our standing assumptions mean that $\operatorname{rank} L + \operatorname{rank} R = n$ and there exists a real number $\tau_\circ$ with the property that for each $\tau \geq \tau_\circ$ we have $\det(L + Re^{\tau A}) \neq 0$.

Theorem 1.2. *Assume the standing assumptions on* (6) *and* (7) *are satisfied, and let* L_A *be the spectral projection of* A *with respect to the left half-plane. Then for* (6) *the finite section method relative to the boundary value matrix* R *converges if and only if* $\operatorname{Ker} L_A \oplus \operatorname{Ker} R = \mathbb{C}^n$.

Theorem 1.2 with the additional assumption that R is a projection appears in [6]; in this paper we derive the theorem as a corollary of Theorem 1.1. The proof of Theorem 1.1 is based on an idea that was used earlier in [7] for the discrete version of Theorem 1.1. The "if part" of Theorem 1.1 has been proved by a different method in [6]. We view Theorems 1.1 and 1.2 as an addition to the projection method (see [1], [3]).

The paper consists of three sections. The second section contains the proof of Theorem 1.1, and the third that of Theorem 1.2.

2. Proof of Theorem 1.1

Throughout this section $U(t)$ is the fundamental matrix of the differential equation (3) and L is an exponential dichotomy of (3). A dichotomy L (assuming that it exists) is not unique. In fact, only the image of L is determined by (4). This is the contents of the next proposition (see [4, pp. 16, 17]).

Proposition 2.1. *Let L be an exponential dichotomy of* (3), *and let $U(t)$ be the fundamental matrix of* (3). *Then*

$$\operatorname{Im} L = \{x \in \mathbb{C}^n \mid U(t)x \in L_n^2[0,\infty)\}. \tag{8}$$

Moreover, if L' is a projection of $\mathbb{C}^n$ such that $\operatorname{Im} L = \operatorname{Im} L'$, *then L' is also an exponential dichotomy of* (3).

The proof of Theorem 1.1 will be based on a series of lemmas.

Lemma 2.2. *The (unique) solution $x(t)$ of* (1) *is given by $x = Tf$, where T is the integral operator on $L_n^2[0,\infty)$ defined by*

$$(Tf)(t) = \int_0^\infty \gamma(t,s) f(s)ds, \quad 0 \le t < \infty, \tag{9}$$

with

$$\gamma(t,s) = \begin{cases} U(t)LU(s)^{-1}, & 0 \le s < t < \infty, \\ -U(t)(I-L)U(s)^{-1}, & 0 \le t < s < \infty. \end{cases} \tag{10}$$

The operator T is bounded.

The above lemma may be obtained as a corollary from [5, Section I.2]. The lemma can also be proved by a straightforward computation verifying directly that $x = Tf$ is a solution of (1).

Lemma 2.3. *Assume that our standing assumptions on* (1) *and* (2) *are satisfied. Let $P_R(\tau) = \big(L + R(\tau)U(\tau)\big)^{-1} R(\tau)U(\tau)$. Then the (unique) solution of* (2) *is given by $x = T_{\tau,R} f$, where $T_{\tau,R}$ is the integral operator on $L_n^2[0,\tau)$ defined by*

$$(T_{\tau,R} f)(t) = \int_0^\tau \gamma_\tau(t,s) f(s)ds, \quad 0 \le t \le \tau, \tag{11}$$

with

$$\gamma_\tau(t,s) = \begin{cases} U(t)(I - P_R(\tau))U(s)^{-1}, & 0 \le s < t \le \tau, \\ -U(t)P_R(\tau)U(s)^{-1}, & 0 \le t < s \le \tau, \end{cases} \tag{12}$$

Furthermore, $P_R(\tau)$ is a projection with $\operatorname{Ker}(I - P_R(\tau)) = \operatorname{Ker} L$, *and hence $LP_R(\tau) = 0$ and $P_R(\tau)(I-L) = I - L$.*

Proof. First we prove that $P_R(\tau)$ has the properties mentioned in the final statement of the theorem, and that $\operatorname{Im} P_R(\tau) = \operatorname{Ker} R(\tau)U(\tau)$. If both $Lx = 0$ and $R(\tau)U(\tau)x = 0$, then $(L + R(\tau)U(\tau))x = 0$ and hence $x = 0$. Since also $\operatorname{rank} L + \operatorname{rank} R(\tau)U(\tau) = n$, this implies that $\operatorname{Ker} L \oplus \operatorname{Ker} R(\tau)U(\tau) = \mathbb{C}^n$. For $x \in \mathbb{C}^n$ write $x = y + z$ with $y \in \operatorname{Ker} L$ and $z \in \operatorname{Ker} R(\tau)U(\tau)$. Then

$$\big(L + R(\tau)U(\tau)\big)^{-1} R(\tau)U(\tau)x = \big(L + R(\tau)U(\tau)\big)^{-1} R(\tau)U(\tau)y = y$$

because $\big(L + R(\tau)U(\tau)\big)y = R(\tau)U(\tau)y$. Therefore $\big(L + R(\tau)U(\tau)\big)^{-1} R(\tau)U(\tau)$ is a projection along $\operatorname{Ker} L$ onto $\operatorname{Ker} R(\tau)U(\tau)$.

By direct verification it follows that $x = T_{\tau,R}f$ is the solution of (2). □

From the previous lemma it also follows that under our standing assumptions on (1) and (2) the boundary conditions in (2) can be separated and written as $Lx(0) = 0$ and $R(\tau)x(\tau) = 0$.

Lemma 2.4. *Assume that the standing assumptions on* (1) *and* (2) *are satisfied. Put* $P_R(\tau) = \big(L + R(\tau)U(\tau)\big)^{-1} R(\tau)U(\tau)$. *Then condition* (5) *is equivalent to*

$$\sup_{\tau \geq \tau_0} \|U(\tau)(I - L)P_R(\tau)LU(\tau)^{-1}\| < \infty. \tag{13}$$

Proof. Using the second statement of Lemma 2.3 one obtains

$$\begin{aligned} U(\tau)(L + R(\tau)U(\tau))^{-1}R(\tau) &= U(\tau)P_R(\tau)U(\tau)^{-1} \\ &= U(\tau)P_R(\tau)LU(\tau)^{-1} + U(\tau)P_R(\tau)(I - L)U(\tau)^{-1} \\ &= U(\tau)(I - L)P_R(\tau)LU(\tau)^{-1} + U(\tau)(I - L)U(\tau)^{-1} \end{aligned}$$

Since L is an exponential dichotomy, the term $U(\tau)(I - L)U(\tau)^{-1}$ is uniformly norm bounded in τ. This shows that (5) is bounded if and only if (13) is bounded. □

Proof of Theorem 1.1. Let

$$S(\tau) = (I - L)U(\tau)^{-1}. \tag{14}$$

Then $S(\tau)$ satisfies our standing assumptions because $\operatorname{rank} S(\tau) = \operatorname{rank}(I - L)$ and $L + S(\tau)U(\tau) = I$. Therefore the equation

$$\begin{cases} \dot{y}(t) - A(t)y(t) = f(t), & 0 \leq t \leq \tau, \\ Lx(0) + S(\tau)y(\tau) = 0, \end{cases}$$

is uniquely solvable. Note that $P_S(\tau) = (L + S(\tau)U(\tau))^{-1}S(\tau)U(\tau) = I - L$. Then Lemma 2.3 gives that

$$(T_{\tau,S}f)(t) = \int_0^\tau \gamma(t,s)f(s)ds, \quad 0 \leq t \leq \tau,$$

with γ given by (10). We view $L_n^2[0,\tau]$ as the subspace of $L_n^2[0,\infty)$ consisting of all $f \in L_n^2[0,\infty)$ such that $f(t) = 0$ a.e. on $t > \tau$, and we write Π_τ for the projection of

$L_n^2[0,\infty)$ onto $L_n^2[0,\tau]$. Then for each $f \in L_n^2[0,\infty)$ we have $T_{\tau,S}\Pi_\tau f = \Pi_\tau T \Pi_\tau f$, where T is the operator defined in Lemma 2.2. Hence we obtain

$$Tf - T_{\tau,S}\Pi_\tau f = (Tf - \Pi_\tau Tf) + \Pi_\tau T(f - \Pi_\tau f).$$

Since T is bounded and $\Pi_\tau g \to g \quad (\tau \to \infty)$ for each $g \in L_n^2[0,\infty)$, we conclude that $T_{\tau,S}\Pi_\tau f \to Tf \quad (\tau \to \infty)$, i.e., the finite section method for (1) relative to the boundary value matrices $\{S(\tau)\}$ converges in L^2.

Let $T_{\tau,R}$ be the operator given in Lemma 2.3. From the result in the previous paragraph we see that it remains to prove that

$$\lim_{\tau\to\infty} \|(T_{\tau,S} - T_{\tau,R})\Pi_\tau f\| = 0, \quad f \in L_n^2[0,\infty) \tag{15}$$

if and only if condition (5) holds true.

From Lemma 2.3 it follows that

$$(T_{\tau,S}\Pi_\tau f - T_{\tau,R}\Pi_\tau f)(t) = \int_0^\tau U(t)\big(P_R(\tau) - (I-L)\big)U(s)^{-1}f(s)\mathrm{d}s. \tag{16}$$

Again use Lemma 2.3 to obtain

$$\begin{aligned} P_R(\tau) - (I-L) &= P_R(\tau) - P_R(\tau)(I-L) = P_R(\tau)L = (I-L)P_R(\tau)L \\ &= (I-L)U(\tau)^{-1}U(\tau)(I-L)P_R(\tau)LU(\tau)^{-1}U(\tau)L. \end{aligned} \tag{17}$$

We define

$$Q_\tau = T_{\tau,S} - T_{\tau,R}. \tag{18}$$

Then $Q_\tau = \Gamma_\tau \Xi_\tau \Lambda_\tau$ with the operators Γ_τ, Ξ_τ, and Λ_τ being defined by

$$\begin{aligned} &\Gamma_\tau : \mathbb{C}^n \to L_n^2[0,\tau], && (\Gamma_\tau x)(t) = U(t)(I-L)U(\tau)^{-1}x; \\ &\Xi_\tau : \mathbb{C}^n \to \mathbb{C}^n, && \Xi_\tau = U(\tau)(I-L)P_R(\tau)LU(\tau)^{-1}; \\ &\Lambda_\tau : L_n^2[0,\tau] \to \mathbb{C}_n, && \Lambda_\tau f = \textstyle\int_0^\tau U(\tau)LU(s)^{-1}f(s)\mathrm{d}s. \end{aligned}$$

Since L is an exponential dichotomy, we have that $\sup_\tau \|\Gamma_\tau\| < \infty$. Moreover, for each f in $L_n^2[0,\infty)$ it follows that $\Lambda_\tau \Pi_\tau f \to 0$ if $\tau \to \infty$. Indeed, since L is an exponential dichotomy of (3), we have, with M and α as in (4)

$$\begin{aligned} \|\Lambda_\tau \Pi_\tau f\| &\le \int_0^\tau M\mathrm{e}^{-\alpha(\tau-s)}\|f(s)\|\mathrm{d}s \\ &= \int_0^{\frac{\tau}{2}} M\mathrm{e}^{-\alpha(\tau-s)}\|f(s)\|\mathrm{d}s + \int_{\frac{\tau}{2}}^\tau M\mathrm{e}^{-\alpha(\tau-s)}\|f(s)\|\mathrm{d}s. \end{aligned} \tag{19}$$

Now note that

$$\begin{aligned} \int_0^{\frac{\tau}{2}} M\mathrm{e}^{-\alpha(\tau-s)}\|f(s)\|\mathrm{d}s &\le \left(\int_0^{\frac{\tau}{2}} M^2\mathrm{e}^{-2\alpha(\tau-s)}\mathrm{d}s\right)^{\frac{1}{2}} \left(\int_0^{\frac{\tau}{2}} \|f(s)\|^2\mathrm{d}s\right)^{\frac{1}{2}} \\ &= \left(M^2\frac{\mathrm{e}^{-\alpha\tau} - \mathrm{e}^{-2\alpha\tau}}{2\alpha}\right)^{\frac{1}{2}} \left(\int_0^{\frac{\tau}{2}} \|f(s)\|^2\mathrm{d}s\right)^{\frac{1}{2}}. \end{aligned}$$

For $\tau \to \infty$ the first factor in the right-hand side converges to 0 and the second factor to $\|f\|$. For the second term of (19) we use that

$$\begin{aligned}\int_{\frac{\tau}{2}}^{\tau} M\mathrm{e}^{-\alpha(\tau-s)}\|f(s)\|\mathrm{d}s &\leq \left(\int_{\frac{\tau}{2}}^{\tau} M^2\mathrm{e}^{-2\alpha(\tau-s)}\mathrm{d}s\right)^{\frac{1}{2}}\left(\int_{\frac{\tau}{2}}^{\tau}\|f(s)\|^2\mathrm{d}s\right)^{\frac{1}{2}}\\ &= \left(M^2\frac{1-\mathrm{e}^{-\alpha\tau}}{2\alpha}\right)^{\frac{1}{2}}\left(\int_{\frac{\tau}{2}}^{\tau}\|f(s)\|^2\mathrm{d}s\right)^{\frac{1}{2}}\end{aligned}$$

For $\tau \to \infty$ in this product the first factor converges to $M/\sqrt{2\alpha}$ and the second to 0, because f is in $L_n^2[0,\infty)$. We proved $\Lambda_\tau\Pi_\tau f \to 0$ if $\tau \to \infty$.

Furthermore, Lemma 2.4 gives that $\sup_{\tau\geq\tau_0}\|\Xi_\tau\| < \infty$ is equivalent to condition (5). It follows that condition (5) implies (15), and we conclude that condition (5) is sufficient for the finite section method to converge with respect to the boundary value matrices $\{R(\tau)\}$.

Conversely, assume that (15) holds true for each right-hand side f in $L_n^2[0,\infty)$. This means that $Q_\tau\Pi_\tau f \to 0$, where Q_τ is defined by (18), for $\tau \to \infty$ and for each $f \in L_n^2[0,\infty)$. We will prove that condition (13) holds true, which in view of Lemma 2.4 will complete the proof. The principle of uniform boundedness implies that

$$m = \sup_\tau \|Q_\tau\| < \infty.$$

Hence, for each pair of functions f and g in $L_n^2[0,\infty)$ we have that

$$|\langle \Pi_\tau g, Q_\tau\Pi_\tau f\rangle| \leq m\|f\|\|g\|.$$

Recall that from (16) and (17) we have that

$$Q_\tau\Pi_\tau f = U(t)\big((I-L)P_R(\tau)L\big)\int_0^\tau U(s)^{-1}f(s)\mathrm{d}s.$$

Hence

$$\langle \Pi_\tau g, Q_\tau\Pi_\tau f\rangle = \int_0^\tau g(t)^*U(t)\mathrm{d}t\big((I-L)P_R(\tau)L\big)\int_0^\tau U(s)^{-1}f(s)\mathrm{d}s.$$

Now choose $g(t) = \frac{1}{\delta}\chi_{[\tau-\delta,\tau]}g_\circ$ and $f(t) = \frac{1}{\delta}\chi_{[\tau-\delta,\tau]}f_\circ$ with $f_\circ$ and $g_\circ$ vectors in $\mathbb{C}^n$ and $\chi_{[\tau-\delta,\tau]}$ the characteristic function of the interval $[\tau-\delta,\tau]$. It follows that

$$|g_\circ^*\int_{\tau-\delta}^{\tau}\frac{1}{\delta}U(t)\mathrm{d}t(I-L)P_R(\tau)L\int_{\tau-\delta}^{\tau}\frac{1}{\delta}U(s)^{-1}\mathrm{d}sf_\circ| \leq m\|g_\circ\|\|f_\circ\|.$$

By letting $\delta \to 0$ we obtain

$$|g_\circ^*U(\tau)(I-L)P_R(\tau)LU(\tau)^{-1}f_\circ| \leq m\|g_\circ\|\|f_\circ\|,$$

for each pair of vectors $f_\circ$ and $g_\circ$ in $\mathbb{C}^n$. Hence it follows that

$$\|U(\tau)(I-L)P_R(\tau)LU(\tau)^{-1}\| \leq m,$$

for each value of τ, which proves (13). □

3. Time-invariant case

Throughout this section A is an $n \times n$ matrix with no eigenvalues on the imaginary axis, and L_A is the spectral projection of A corresponding to the eigenvalues of A in the left half-plane. Furthermore, throughout this section L is a projection on $\mathbb{C}^n$ such that $\operatorname{Im} L = \operatorname{Im} L_A$.

Theorem 1.2 is an immediate corollary of Theorem 1.1 and the next proposition.

Proposition 3.1. *Assume that our standing assumptions on* (6) *and* (7) *are satisfied, and let L_A be the spectral projection of A corresponding to the eigenvalues in the left half-plane. Then the following are equivalent:*

(a) $\sup_{\tau \geq \tau_\circ} \|e^{\tau A}(L + Re^{\tau A})^{-1} R\| < \infty$;

(b) $\operatorname{Ker} L_A \oplus \operatorname{Ker} R = \mathbb{C}^n$.

Moreover, in that case $\lim_{\tau \to \infty} e^{\tau A}(L + Re^{\tau A})^{-1} R$ exists and is equal to the projection of $\mathbb{C}^n$ along $\operatorname{Ker} R$ *onto* $\operatorname{Ker} L_A$.

For the proof of Proposition 3.1 it will be convenient to use the following lemma.

Lemma 3.2. *Let $\mathbb{C}^r = \mathcal{X}_1 \oplus \mathcal{X}_2$ be a direct sum decomposition, and for $i = 1, 2$ let $\mathcal{M}_i$ be a subspace of $\mathbb{C}^r$ with $\dim \mathcal{M}_i = \dim \mathcal{X}_i$. Then there exist*

(a) *a surjective mapping $\begin{pmatrix} V_1 & V_2 \end{pmatrix} : \mathcal{X}_1 \oplus \mathcal{X}_2 \to \mathcal{X}_2$, and*

(b) *an injective mapping*

$$\begin{pmatrix} W_1 \\ W_2 \end{pmatrix} : \mathcal{X}_2 \to \mathcal{X}_1 \oplus \mathcal{X}_2,$$

such that $\mathcal{M}_1$ and $\mathcal{M}_2$ can be represented in the following way

$$\mathcal{M}_1 = \operatorname{Ker} \begin{pmatrix} V_1 & V_2 \end{pmatrix}, \qquad \mathcal{M}_2 = \operatorname{Im} \begin{pmatrix} W_1 \\ W_2 \end{pmatrix}.$$

Furthermore, $\mathbb{C}^r = \mathcal{M}_1 \oplus \mathcal{M}_2$ if and only if $V_1 W_1 + V_2 W_2$ is invertible, and in this case the projection Γ of $\mathbb{C}^r$ along $\mathcal{M}_1$ onto $\mathcal{M}_2$ is given by

$$\Gamma = \begin{pmatrix} W_1 \\ W_2 \end{pmatrix} (V_1 W_1 + V_2 W_2)^{-1} \begin{pmatrix} V_1 & V_2 \end{pmatrix}.$$

Proof. Since $\dim \mathcal{M}_2 = \dim \mathcal{X}_2$, there exists an invertible linear mapping $W_0 : \mathcal{X}_2 \to \mathcal{M}_2$ and hence there exists an injective linear mapping $W : \mathcal{X}_2 \to \mathcal{X}_1 \oplus \mathcal{X}_2$ such that $\operatorname{Im} W = \mathcal{M}_2$. Write $\mathbb{C}^r = \mathcal{M}_1 \oplus \mathcal{M}_1'$. Then $\dim \mathcal{M}_1' = \dim \mathcal{X}_2$. Hence there exists an invertible mapping $V_0 : \mathcal{M}_1' \to \mathcal{X}_2$. Define $V : \mathcal{M}_1 \oplus \mathcal{M}_1' \to \mathcal{X}_2$ by $V(m_1 + m_1') = V_0 m_1'$ for each pair $m_1 \in \mathcal{M}_1$ and $m_1' \in \mathcal{M}_1'$. Then $V : \mathcal{X}_1 \oplus \mathcal{X}_2 \to \mathcal{X}_2$ is surjective and $\operatorname{Ker} V = \mathcal{M}_1$.

Next notice that for $x \in \mathcal{X}_2$ we have $VWx = 0$ if and only if $Wx \in \operatorname{Ker} V$. Thus we see that $VWx = 0$ if and only if $Wx \in \operatorname{Ker} V \cap \operatorname{Im} W = \mathcal{M}_1 \cap \mathcal{M}_2$. Since W is injective, we may conclude that VW is invertible if and only if $\mathcal{M}_1 \cap \mathcal{M}_2 = \{0\}$, or equivalently $\mathbb{C}^r = \mathcal{M}_1 \oplus \mathcal{M}_2$. Here we use that $\dim \mathcal{M}_1 + \dim \mathcal{M}_2 = r$.

In case VW is invertible we define $\Gamma = W(VW)^{-1}V$. Then Γ is a projection and $\operatorname{Ker}\Gamma = \operatorname{Ker} V$ because W is injective, and $\operatorname{Im}\Gamma = \operatorname{Im} W$, because V is surjective. Finally we represent W and V as operator matrices with respect to the decomposition of $\mathbb{C}^r = \mathcal{X}_2 \oplus \mathcal{X}_1$ to obtain V_i and W_i $(i = 1, 2)$. □

Proof of Proposition 3.1. We begin with some preliminaries. Decompose the space $\mathbb{C}^n$ as $\mathbb{C}^n = \operatorname{Im} L_A \oplus \operatorname{Ker} L_A$. With respect to this decomposition we write L_A, L and A as operator matrices

$$L_A = \begin{pmatrix} I & 0 \\ 0 & 0 \end{pmatrix}, \quad L = \begin{pmatrix} I & E \\ 0 & 0 \end{pmatrix}, \quad A = \begin{pmatrix} A_1 & 0 \\ 0 & A_2 \end{pmatrix}. \tag{20}$$

Notice that A_1 has all its eigenvalues in the open left-hand plane and A_2 has all its eigenvalues in the open right-hand plane. From the operator matrix representations in (20) it follows that

$$\operatorname{Ker} L\mathrm{e}^{-\tau A} = \operatorname{Im} \begin{pmatrix} -\mathrm{e}^{\tau A_1} E \mathrm{e}^{-\tau A_2} \\ I \end{pmatrix}. \tag{21}$$

In particular, $\dim \operatorname{Ker} L\mathrm{e}^{-\tau A} = \dim \operatorname{Ker} L_A$.

From Lemma 2.3 we know that $\mathrm{e}^{\tau A}(L + R\mathrm{e}^{\tau A})^{-1}R = (L\mathrm{e}^{-\tau A} + R)^{-1}R$ is the projection along $\operatorname{Ker} R$ onto $\operatorname{Ker} L\mathrm{e}^{-\tau A}$. Hence,

$$\dim \operatorname{Ker} R = n - \dim \operatorname{Ker} L\mathrm{e}^{-\tau A} = \dim \operatorname{Im} L_A.$$

According to the first part of Lemma 3.2, this allows us to represent $\operatorname{Ker} R$ as $\operatorname{Ker} R = \operatorname{Ker} \begin{pmatrix} S & T \end{pmatrix}$, where $\begin{pmatrix} S & T \end{pmatrix}$ is a surjective mapping from $\operatorname{Im} L_A \oplus \operatorname{Ker} L_A$ into $\operatorname{Ker} L_A$. By combining this with (21), we can use the second part of Lemma 3.2 to show that

$$(L\mathrm{e}^{-\tau A} + R)^{-1}R = \begin{pmatrix} -\mathrm{e}^{\tau A_1} E \mathrm{e}^{-\tau A_2} \\ I \end{pmatrix} \left(T - S\mathrm{e}^{\tau A_1} E \mathrm{e}^{-\tau A_2}\right)^{-1} \begin{pmatrix} S & T \end{pmatrix}. \tag{22}$$

Next, since

$$\operatorname{Ker} R = \operatorname{Ker} \begin{pmatrix} S & T \end{pmatrix}, \qquad \operatorname{Ker} L_A = \operatorname{Im} \begin{pmatrix} 0 \\ I \end{pmatrix}, \tag{23}$$

Lemma 3.2 shows that (b) is equivalent to the invertibility of T.

We are now ready to prove the equivalence of (a) and (b).

(b) ⇒ (a). Assume (b) holds. Then T is invertible. Since

$$\lim_{\tau \to \infty} \mathrm{e}^{\tau A_1} E \mathrm{e}^{-\tau A_2} = 0, \tag{24}$$

the invertibility of T and (22) yield

$$\lim_{\tau \to \infty} (L\mathrm{e}^{-\tau A} + R)^{-1}R = \begin{pmatrix} 0 \\ I \end{pmatrix} T^{-1} \begin{pmatrix} S & T \end{pmatrix}, \tag{25}$$

which proves (a). Furthermore, from the second part of Lemma 3.2 and (23) we see that the right-hand side of (25) is precisely equal to the projection of $\mathbb{C}^n$ along $\operatorname{Ker} R$ onto $\operatorname{Ker} L_A$. Thus the final part of Proposition 3.1 is proved too.

(a) $\Rightarrow$ (b). Assume (a) holds. Property (a) implies that the operator matrix $\left(T - Se^{\tau A_1}Ee^{-\tau A_2}\right)^{-1} \begin{pmatrix} S & T \end{pmatrix}$ is uniformly bounded with respect to τ for $\tau \geq \tau_\circ$, because this operator matrix is precisely the second block row of the right-hand side of (22). Now recall that $\begin{pmatrix} S & T \end{pmatrix}$ is surjective. It follows that $\left(T - Se^{\tau A_1}Ee^{-\tau A_2}\right)^{-1}$ is uniformly bounded in τ for $\tau \geq \tau_\circ$. By (24) the latter can only happen when T is invertible. Indeed, assume $Tx = 0$. Then, using (24) and the fact that $\left(T - Se^{\tau A_1}Ee^{-\tau A_2}\right)^{-1}$ is bounded, we have

$$x = -\left(T - Se^{\tau A_1}Ee^{-\tau A_2}\right)^{-1} Se^{\tau A_1}Ee^{-\tau A_2}x \to 0 \quad (\tau \to \infty).$$

Hence $x = 0$ and T is invertible. But the invertibility of T is equivalent with (b), and therefore (b) is proved. □

References

[1] A. Böttcher and B. Silbermann, *Analysis of Toeplitz Operators*, Springer-Verlag, Berlin, Heidelberg, New York, 1990.

[2] W.A. Coppel, *Dichotomies in Stability Theory*, Lecture Notes in Mathematics, Springer-Verlag, Berlin, Heidelberg, New York, 1978.

[3] I. Gohberg and I.A. Feldman, *Convolution Equations and Projection Methods for Their Solution*, Amer. Math. Soc., Transl. Math. Monographs 41, Providence (RI), 1974 (Russian Original: Nauka, Moscow 1971).

[4] I. Gohberg and M.A. Kaashoek, Time varying systems with boundary conditions and integral operators, I. The transfer operator and its properties, *Integral Equations and Operator Theory,* **7** (1984), 325–391.

[5] I. Gohberg and M.A. Kaashoek, F. van Schagen, Non-compact integral operators with semi-separable kernels and their discrete analogues: inversion and Fredholm properties, *Integral Equations and Operator Theory,* **7** (1984), 642–703.

[6] I. Gohberg and M.A. Kaashoek, F. van Schagen, Finite Section Method for Linear Ordinary Differential Equations, *J. Differential Equations*, **163** (2000), 312–334.

[7] I. Gohberg and M.A. Kaashoek, F. van Schagen, Finite Section Method for Difference Equations, to appear

I. Gohberg
School of Mathematical Sciences
Raymond and Beverly Sackler Faculty of Exact Sciences
Tel Aviv University, Ramat Aviv, Israel
e-mail: `Gohberg@math.tau.ac.il`

M.A. Kaashoek and F. van Schagen
Department of Mathematics, Vrije Universiteit
De Boelelaan 1081a, 1081 HV Amsterdam, The Netherlands
e-mail: `kaash@cs.vu.nl` and `freek@cs.vu.nl`

Received: 20 September 2001

Operator Theory:
Advances and Applications, Vol. 135, 193–208

Fast Algorithms for Skewsymmetric Toeplitz Matrices

Georg Heinig and Karla Rost

Dedicated to Bernd Silbermann on the Occasion of His 60th Birthday

Abstract. In the paper we consider nonsingular $n \times n$ skewsymmetric Toeplitz matrices and develop fast algorithms for inversion, solution of linear systems, LU- and ZW-factorization that fully utilize the given symmetry properties. Skewsymmetry is a significant peculiarity, so that the properties and algorithms discussed here differ from those for symmetric Toeplitz matrices.

1. Introduction

This paper is dedicated to nonsingular skewsymmetric Toeplitz matrices, i.e. matrices of the form $T_n = [a_{i-j}]_{i,j=1}^n$ with $a_{-j} = -a_j$. We assume that the entries are from a field $\mathbb{F}$ of characteristic different from two. Symmetric Toeplitz matrices have found considerable attention in the literature. In particular many works address the question how the symmetry can be exploited to make algorithms for inversion and factorization more efficient. The first result in this direction is the split Levinson algorithm of P. Delsarte and Y. Genin [1], which saves about half of the number of multiplications. Later these authors presented also a Schur-type algorithm (see [2]). An interpretation of these algorithms as ZW- or bowtie factorization was given by C.J. Demeure in [3]. In the series of papers [13], [14], [15] A. Melman improved the split Levinson algorithm. Let us note that the best version of Melman's algorithms is closely related to an algorithm proposed in [6] and has the same computational complexity. Moreover, [6] contains also a Schur-type and a superfast algorithm. The main idea in [6] is to reduce the Toeplitz system to a system with a Chebyshev-Hankel coefficient matrix of half the size.

In contrast to symmetric Toeplitz matrices, skewsymmetric Toeplitz matrices have not yet found very much attention in the literature, although they appear in some applications. In [11] we observed that there are some essential differences between symmetric and skewsymmetric Toeplitz matrices. For example, the fundamental system of a symmetric Toeplitz matrix consists of a symmetric and a skewsymmetric vector, whereas the fundamental system of a skewsymmetric Toeplitz matrix consists of two symmetric vectors. Furthermore, the restriction of the operator generated by a skewsymmetric Toeplitz matrix to the subspace of all symmetric vectors completely defines the matrix in the skewsymmetric but not always in the symmetric case.

The aim of this paper is to show that for skewsymmetric Toeplitz matrices fast algorithms utilizing symmetry properties can be designed like for symmetric Toeplitz matrices. But we show also that in the skewsymmetric case these algorithms possess some features that are not completely analogous to the symmetric case. It is remarkable that in the skewsymmetric case the algorithms are simpler and seem to have more structure than in the symmetric case.

An obvious difference between skewsymmetric and symmetric Toeplitz matrices is that the former are never strongly nonsingular. In fact, all submatrices of odd order $T_{2k-1} = [\, a_{i-j} \,]_{i,j=1}^{2k-1}$ are singular, so that, firstly, the Gohberg-Semencul formula describing the inverse matrix in terms of its first column in the symmetric case never can be applied. We will show that in the skewsymmetric case T_n^{-1} can be described in terms of the two vectors spanning the (one-dimensional) kernels of T_{n-1} and any skewsymmetric $(n+1) \times (n+1)$ Toeplitz extension T_{n+1} of T_n. It follows from results of [11] that these two vectors are symmetric. We will here give an independent proof of this surprising fact.

Secondly, the classical Levinson-type and Schur-type algorithms cannot be applied because they work only for strongly nonsingular Toeplitz matrices. We here show that for skewsymmetric Toeplitz matrices there are nevertheless very efficient fast algorithms. In comparison with other fast algorithms, there are two sources of the efficiency of these algorithms. The first is the fact that almost all operations are carried out on symmetric vectors, which saves about half of the computational amount, while the second is that in each step the size of the matrix is increased by 2 rather than 1.

The Levinson-type algorithm presented here has the same complexity as Melman's algorithm in [15] and the algorithm in [6]. We also present a Schur-type algorithm. One of them needs only $9/4n^2 + O(n)$ arithmetic operations to solve a linear system of equations, so that this is the fastest algorithm not using FFT. Apparently, there is no symmetric counterpart of this algorithm with the same complexity. It can also be expected that, in analogy to the symmetric case, it is more stable in many cases. The Schur-type algorithm has two more advantages. The first is that it can be completely parallelized and has computational complexity $O(n)$ if n processors are available. Secondly, as pointed out in [6], it can be speeded up to a"superfast" $O(n \log^2 n)$-complexity algorithm in sequential processing.

A Schur-type algorithm can be used for triangular matrix factorization and a Levinson-type algorithm for triangular factorization of the inverse. In our situation of a skewsymmetric matrix we will have a generalized LU-factorization $T_n = LDL^T$, where L is lower triangular with units on the main diagonal and D is a block diagonal matrix with skewsymmetric 2×2 blocks on the diagonal, provided that all principal subsections of T_n of even order are nonsingular.

Besides the LU-factorization we will consider another kind of matrix factorizations, the ZW-factorization. This factorization is closely related to the "quadrant interlocking" or WZ-factorization, which was originally introduced and studied by D.J. Evans and his coworkers for the parallel solution of tridiagonal systems (see [16], [4] and references therein). While the LU-factorization of a matrix $[\,a_{ij}\,]_{i,j=1}^{n}$ relies on the principal submatrices $[\,a_{ij}\,]_{i,j=1}^{k}$ for $k = 1, \ldots, n$, the ZW-factorization relies on the central submatrices $[\,a_{ij}\,]_{i,j=l}^{n+1-l}$, $l = 1, \ldots, n/2$. The ZW-factorization has some advantages in comparison with the LU-factorization, since it makes use of the fact that skewsymmetric Toeplitz matrices are centro-skewsymmetric. This leads to a reduction of the computational complexity.

In the final Section 7 we discuss and compare different methods to solve linear systems with a skewsymmetric Toeplitz coefficient matrix which emerge from the algorithms presented before.

Let us finally mention that most of the algorithms described in this paper can been generalized to centro-skewsymmetric Toeplitz-plus-Hankel matrices. This set of problems will be discussed in the forthcoming paper [12].

2. Formula for the inverse

To begin with we recall some facts concerning inverses of Toeplitz matrices. It is well known that inverses of Toeplitz matrices are, in general, not Toeplitz matrices again but rather so-called Toeplitz Bezoutians. We give the definition in terms of the generating function of a matrix. An alternative definition in matrix language has the form of the Gohberg-Semencul formula [5].

If $A = [\,a_{ij}\,]_{i,j=1}^{n}$ is a matrix, then the generating function is, by definition, the bivariate polynomial $A(t,s) = \sum_{i,j=1}^{n} a_{ij} t^{i-1} s^{j-1}$. In the same spirit the polynomial $\mathbf{x}(t)$ is defined for a vector $\mathbf{x}$.

Let $\mathbf{p}, \mathbf{q} \in \mathbb{F}^{n+1}$ and let

$$J_n = \begin{bmatrix} 0 & & 1 \\ & \cdot^{\cdot^{\cdot}} & \\ 1 & & 0 \end{bmatrix}$$

be the $n \times n$ matrix of the counteridentity. Then the (Toeplitz) Bezoutian of $\mathbf{p}$ and $\mathbf{q}$ is defined as the $n \times n$ matrix $B = \mathrm{Bez}\,(\mathbf{p}, \mathbf{q})$ with the generating function

$$B(t,s) = \frac{\mathbf{p}(t)\widehat{\mathbf{q}}(s) - \mathbf{q}(t)\widehat{\mathbf{p}}(s)}{1 - ts},$$

where $\widehat{\mathbf{p}} = J_{n+1}\mathbf{p}$. The entries of the matrix B can be constructed recursively from $\mathbf{p}$ and $\mathbf{q}$ in $O(n^2)$ operations (see [7]). The recursion is closely related to Trench's representation of Toeplitz matrix inverses in [17].

For our purposes it is more important to mention that matrix-vector multiplication by Bezoutians can be carried out with a computational complexity of $O(n \log n)$ if FFT or fast trigonometric transformations are used (see [8], [9], [10]).

In what follows, let $\mathbf{e_k}$ denote the kth vector in the standard basis of $\mathbb{F}^n$.

The following is well known (see [7]).

Proposition 2.1. *The inverse of a nonsingular $n \times n$ Toeplitz matrix T_n admits the representation*

$$T_n^{-1} = \mathrm{Bez}\,(\mathbf{p}, \mathbf{q})$$

where

$$\mathbf{p} = \begin{bmatrix} \mathbf{p}' \\ 0 \end{bmatrix}, \ \mathbf{p}' = T_n^{-1}\mathbf{e}_1, \quad \mathbf{q} = \begin{bmatrix} \mathbf{q}' \\ 1 \end{bmatrix}, \ \mathbf{q}' = T_n^{-1}\mathbf{g}, \ \mathbf{g} = (-a_{i-n})_{i=0}^{n-1},$$

and $a_{-n} \in \mathbb{F}$ is arbitrary.

A disadvantage of this formula is that if T_n has a symmetry property, then this is not reflected in the formula. But this is desirable in order to design efficient algorithms exploiting the symmetry. The Gohberg-Semencul formula reflects the symmetry in the case of a symmetric Toeplitz matrix. But, as mentioned above, this formula can never be applied in the skewsymmetric case. We show now that, nevertheless, Proposition 2.1 takes a nice symmetric form in the skewsymmetric case.

From now on, let $T_n = [\, a_{i-j}\,]_{i,j=1}^{n}$ be a nonsingular skewsymmetric Toeplitz matrix and T_{n+1} any skewsymmetric $(n+1)\times(n+1)$ Toeplitz extension of T_n. Then n must be even, and T_{n-1} and T_{n+1} have kernel dimension equal to 1. Let $\mathbf{x}$ and $\mathbf{u}$ be nonzero vectors spanning $\ker T_{n-1}$ and $\ker T_{n+1}$, respectively. It follows from the relation $J_n T_n J_n = -T_n$ that $\mathbf{x}$ and $\mathbf{u}$ are either symmetric or skewsymmetric.

Recall that a vector $\mathbf{p} \in \mathbb{F}^n$ is called *symmetric* if $\mathbf{p} = J_n\mathbf{p}$ and *skewsymmetric* if $\mathbf{p} = -J_n\mathbf{p}$.

But we can show more. The following somehow surprising fact follows from the results of [11], but here we give an independent proof.

Proposition 2.2. *The vectors* $\mathbf{x}$ *and* $\mathbf{u}$ *are symmetric.*

Proof. Let f_j denote the jth row of T_{n+1}. Note that the middle row $f_{n/2+1}$ is skewsymmetric. We define $f_j^{\pm} = f_j \mp f_{n+2-j}$ for $j = 1, \ldots, n/2$. Then the f_j^+ are symmetric and the f_j^- are skewsymmetric. Moreover, the system $T_{n+1}\mathbf{v} = 0$ is equivalent to $f_j^{\pm}\mathbf{v} = 0$ for $j = 1, \ldots, n/2$ and $f_{n/2+1}\mathbf{v} = 0$. Since the dimension of the subspace of all symmetric row vectors of length $n+1$ equals $n/2+1$, there exists a symmetric vector $\mathbf{v} \neq 0$ such that $f_j^+\mathbf{v} = 0$ for $j = 1, \ldots, n/2$. For symmetric $\mathbf{v}$ we automatically have $f_j^-\mathbf{v} = 0$ and $f_{n/2+1}\mathbf{u} = 0$, which gives $T_{n+1}\mathbf{v} = 0$. Since the kernel dimension of T_{n+1} equals 1, we conclude that $\mathbf{u} = c\mathbf{v}$ for some $c \in \mathbb{F}$. Thus $\mathbf{u}$ and, analogously, $\mathbf{x}$ are symmetric. □

Observe that the number $[a_{n-1} \ \cdots \ a_1]\mathbf{x}$ cannot be zero, since otherwise T_n would be singular. Thus we can normalize the vector $\mathbf{x}$ by requiring that

$$[\, a_{n-1} \cdots \ a_1\,]\,\mathbf{x} = 1\,.$$

Then we have

$$T_n \begin{bmatrix} \mathbf{x} \\ 0 \end{bmatrix} = \mathbf{e}_n \quad \text{and} \quad T_n \begin{bmatrix} 0 \\ \mathbf{x} \end{bmatrix} = -\mathbf{e}_1. \tag{1}$$

The first component of $\mathbf{u}$ must be nonzero, since otherwise T_n would be singular. Thus we can normalize $\mathbf{u}$ so that its first component equals 1.

The following is an immediate consequence of Proposition 2.1.

Theorem 2.3. *The inverse of a nonsingular skewsymmetric Toeplitz matrix is given by*

$$T_n^{-1} = \operatorname{Bez}(\mathbf{u}, \widetilde{\mathbf{x}}),$$

where $\mathbf{x}$, $\mathbf{u}$ *are nonzero vectors spanning the nullspaces of* T_{n-1} *and* T_{n+1}, *respectively, with the normalizations made above, and* $\widetilde{\mathbf{x}} = \begin{bmatrix} 0 \\ \mathbf{x} \\ 0 \end{bmatrix}$.

3. Levinson-type algorithm

In this section we show how the two vectors $\mathbf{x}$ and $\mathbf{u}$ in Theorem 2.3 can be constructed in an efficient way utilizing their symmetry. For this we consider the principal subsections $T_k = [\,a_{i-j}\,]_{i,j=1}^{k}$ for $k = 1, \ldots, n$. From now on, we assume that all T_{2k} are nonsingular. Note that T_{2k} can also be interpreted as the central submatrix $[\,a_{i-j}\,]_{i,j=l}^{n+1-l}$, $n - 2l = 2k - 2$.

A matrix for which all these central submatrices are nonsingular is called *centro-nonsingular.* In other words, we now restrict ourselves to centro-nonsingular skewsymmetric Toeplitz matrices. As above, T_{n+1} will denote a skewsymmetric $(n+1) \times (n+1)$ Toeplitz extension of T_n.

Let $\mathbf{x}_k$ $(k = 1, \ldots, n/2)$ denote the vectors that are uniquely defined by

$$T_{2k-1}\mathbf{x}_k = \mathbf{0}, \quad [\,a_{2k-1} \;\cdots\; a_1\,]\,\mathbf{x}_k = 1\,. \tag{2}$$

In particular, $\mathbf{x} = \mathbf{x}_{n/2}$. For convenience we denote any nonzero vector from the nullspace of T_{n+1} by $\mathbf{x}_{n/2+1}$. Dividing it by its first component we will obtain $\mathbf{u}$. Furthermore, let r_k and r'_k be defined by

$$r_k = [\,a_{2k} \;\cdots\; a_2\,]\,\mathbf{x}_k, \quad r'_k = [\,a_{2k+1} \;\cdots\; a_3\,]\,\mathbf{x}_k.$$

Then we have

$$T_{2k+3}\begin{bmatrix} 0 \\ 0 \\ \mathbf{x}_{k-1} \\ 0 \\ 0 \end{bmatrix} = \begin{bmatrix} -r_{k-1} \\ -1 \\ \mathbf{0} \\ 1 \\ r_{k-1} \end{bmatrix}, \quad T_{2k+3}\begin{bmatrix} 0 \\ \mathbf{x}_k \\ 0 \end{bmatrix} = \begin{bmatrix} -1 \\ \mathbf{0} \\ 1 \end{bmatrix}$$

and

$$T_{2k+3}\begin{bmatrix} 0 \\ 0 \\ \mathbf{x}_k \end{bmatrix} = \begin{bmatrix} -r_k \\ -1 \\ \mathbf{0} \end{bmatrix}, \quad T_{2k+3}\begin{bmatrix} \mathbf{x}_k \\ 0 \\ 0 \end{bmatrix} = \begin{bmatrix} \mathbf{0} \\ 1 \\ r_k \end{bmatrix},$$

where $\mathbf{0}$ means a zero vector of appropriate length.

Theorem 3.1. *The vectors* $\mathbf{x}_k$ *satisfy the three-term recursion*

$$\mathbf{x}_{k+1} = \frac{1}{\alpha_k}\left(\begin{bmatrix} 0 \\ 0 \\ \mathbf{x}_k \end{bmatrix} + \begin{bmatrix} \mathbf{x}_k \\ 0 \\ 0 \end{bmatrix} - (r_k - r_{k-1})\begin{bmatrix} 0 \\ \mathbf{x}_k \\ 0 \end{bmatrix} - \begin{bmatrix} 0 \\ 0 \\ \mathbf{x}_{k-1} \\ 0 \\ 0 \end{bmatrix}\right),$$

where

$$\alpha_k = r'_k - r'_{k-1} - r_k(r_k - r_{k-1}).$$

Proof. The recursion follows from the relation

$$\begin{bmatrix} 1 & 1 & 0 \\ r_k & r_{k-1} & 1 \\ r'_k & r'_{k-1} & r_k \end{bmatrix}\begin{bmatrix} 1 \\ -1 \\ r_{k-1} - r_k \end{bmatrix} = \begin{bmatrix} 0 \\ 0 \\ \alpha_k \end{bmatrix}.$$

□

With the help of this theorem the generating vectors $\mathbf{x}$ and $\mathbf{u}$ for the inverse matrix can be computed recursively starting with the initial vectors

$$\mathbf{x}_1 = \frac{1}{a_1} \quad \text{and} \quad \mathbf{x}_2 = \frac{1}{\alpha_1}\begin{bmatrix} 1 \\ -\frac{a_2}{a_1} \\ 1 \end{bmatrix}, \quad \text{where} \quad \alpha_1 = a_1 + a_3 - \frac{a_2^2}{a_1}.$$

Besides the vectors $\mathbf{x}_k$ we consider the vectors

$$\mathbf{u}_k = \frac{1}{\xi_k}\mathbf{x}_k,$$

where ξ_k denotes the first component of $\mathbf{x}_k$ $(k = 1, \dots, n/2)$. Note that these numbers are nonzero, due to the centro-nonsingularity of T_n.

In analogy to the numbers r_k and r'_k we introduce the numbers

$$s_k = [\, a_{2k-1} \;\cdots\; a_1 \,]\, \mathbf{u}_k, \quad s'_k = [\, a_{2k} \;\cdots\; a_2 \,]\, \mathbf{u}_k.$$

(Note that $s_k = \frac{1}{\xi_k}$.)

Theorem 3.2. *The vectors* $\mathbf{u}_k$ *satisfy the three-term recursion*

$$\mathbf{u}_{k+1} = \begin{bmatrix} 0 \\ 0 \\ \mathbf{u}_k \end{bmatrix} + \begin{bmatrix} \mathbf{u}_k \\ 0 \\ 0 \end{bmatrix} - \gamma_k \begin{bmatrix} 0 \\ \mathbf{u}_k \\ 0 \end{bmatrix} - \beta_k \begin{bmatrix} 0 \\ 0 \\ \mathbf{u}_{k-1} \\ 0 \\ 0 \end{bmatrix},$$

where

$$\beta_k = \frac{s_k}{s_{k-1}}, \quad \gamma_k = \frac{s'_k}{s_k} - \frac{s'_{k-1}}{s_{k-1}}.$$

Complexity. Let us discuss the computational complexity of the algorithm resulting from Theorem 3.1. One step of the recursions requires the computation of two inner products of a general vector by a symmetric vector, which amounts to $2k + O(1)$ additions and $k + O(1)$ multiplications. Then we have the addition of three symmetric vectors and the multiplication by two scalars, which requires $\frac{3}{2}k$ additions and k multiplications. Since we make double steps in the recursion, we have $\frac{n}{2}$ steps. Altogether this results in $\frac{7}{8}n^2 + O(n)$ additions and $\frac{1}{2}n^2 + O(n)$ multiplications for the computation of the generating vectors $\mathbf{x}$ and $\mathbf{u}$. The amount for the recursion in Theorem 3.2 is in its principal term the same.

4. Schur-type algorithm

The recursions in Theorem 3.1 can be computed in parallel. However, the bottleneck here is the computation of the inner products for r_k and r'_k or s_k and s'_k. In order to achieve a complexity of $O(n)$ for n processors these numbers have to be precomputed. This can be done by a Schur-type algorithm, as shown below. We will see in the next sections that this algorithm also leads to factorizations of the matrix T_n.

For $k = 1, \dots, n/2$, we introduce the residuals

$$r_{j,k} = [\, a_{j+2k-2} \ \cdots \ a_j \,]\, \mathbf{x}_k, \tag{3}$$

where $j = -2k+2, \dots, 2l$ and $l = n/2 - k + 1$. In particular, we have $r_{j,k} = 0$ for $j < 1$, $r_{1,k} = 1$, $r_{2,k} = r_k$, and $r_{3,k} = r'_k$.

From Theorem 3.1 we conclude the following.

Theorem 4.1. *The residuals r_{jk} satisfy the recursion*

$$r_{j,k+1} = \frac{1}{\alpha_k}\left(r_{j+2,k} + r_{j,k} - r_{j+2,k-1} - (r_{2,k} - r_{2,k-1})r_{j+1,k}\right)$$

for $j = 1, \dots, n-2k$, where

$$\alpha_k = r_{3,k} - r_{3,k-1} + r_{2,k}(r_{2,k} - r_{2,k-1}).$$

The initialization for the recursions of the r_{jk} are obtained from the initialization of the $\mathbf{x}_k$:

$$r_{j,1} = \frac{a_j}{a_1} \quad (j = 2, \dots, n), \quad r_{j,2} = \frac{1}{\alpha_1}(a_j + a_{j+2} - \frac{a_2}{a_1}a_{j+1}) \quad (j = 2, \dots, n-2).$$

We present now the corresponding result of Theorem 4.1 for the vectors $\mathbf{u}_k$. For this purpose we introduce the residuals

$$s_{j,k} = [\, a_{j+2k-2} \ \cdots \ a_j \,]\, \mathbf{u}_k\,. \tag{4}$$

In particular, $s_{1,k} = s_k\,, s_{2,k} = s'_k$.

Theorem 4.2. *The coefficients $s_{j,k}$ in the recursion of Theorem 3.2 can be computed from*

$$s_{j,k+1} = s_{j+2,k} + s_{j,k} - \beta_k s_{j+2,k-1} - \gamma_k s_{j+1,k}$$

for $j = 1, \dots, n-2k+2$, where $\beta_k = \dfrac{s_{1,k}}{s_{1,k-1}}$, $\gamma_k = \dfrac{s_{2,k}}{s_{1,k}} - \dfrac{s_{2,k-1}}{s_{1,k-1}}$.

Complexity. The recursions in Theorems 4.1 and 4.2 can be combined with the recursions in Theorems 3.1 and 3.2. In this way the computation of inner products is replaced by the recursion of the residuals. This results in $\frac{9}{8}n^2 + O(n)$ additions and $\frac{3}{4}n^2 + O(n)$ multiplications. Consequently, this version is more expensive in sequential processing. However all computations are fully parallelizable. In particular, at a machine with n processors we need only $\frac{3}{2}n + O(1)$ additions and $n + O(1)$ multiplications.

5. LU-factorization

A nonsingular $n \times n$ skewsymmetric matrix $A_n = [a_{ij}]_{i,j=1}^n$ with the property that all submatrices $[a_{ij}]_{i,j=1}^{2k}$ $(k = 1, \ldots, n/2)$ are nonsingular admits a unique (generalized) LU-factorization

$$A_n = LDL^T, \tag{5}$$

where L is lower triangular with ones on the main diagonal and a zero entry just below the main diagonal in the columns with an odd number and D is a block diagonal matrix with diagonal blocks of the form

$$D_k = \begin{bmatrix} 0 & d_k \\ -d_k & 0 \end{bmatrix} \qquad (k = 1, \ldots, n/2).$$

Centro-nonsingular skewsymmetric Toeplitz matrices possess the property just mentioned. We show how the factors L and D can be constructed with the help of the Schur-type algorithm in Theorem 4.1.

Let $\mathbf{l}_j$ denote the jth column of the factor L in (5) when $A_n = T_n$. Since

$$T_n \begin{bmatrix} \mathbf{x}_k \\ 0 \end{bmatrix} = \begin{bmatrix} \mathbf{0} \\ \mathbf{r}_k \end{bmatrix},$$

with $\mathbf{r}_k = (r_{j,k})_{j=1}^{n-2k+1}$ and r_{jk} introduced in (3) and $\mathbf{x}_k \in \mathbb{F}^{2k-1}$, we have

$$\mathbf{l}_{2k} = \begin{bmatrix} \mathbf{0} \\ \mathbf{r}_k \end{bmatrix} \tag{6}$$

for $k = 1, \ldots, n/2$. Furthermore, $\mathbf{l}_1$ is the second column of T_n divided by $-a_1$. Thus it remains to construct the columns $\mathbf{l}_{2k-1}$ for $k = 2, \ldots, n/2$ and the block diagonal factor.

Let S_n denote the matrix of the forward shift in $\mathbb{F}^n$,

$$S_n = \begin{bmatrix} 0 & \cdots & & 0 \\ 1 & & & \\ & \ddots & & \vdots \\ 0 & & 1 & 0 \end{bmatrix}.$$

Then we have

$$T_n S_n \begin{bmatrix} \mathbf{x}_k \\ \mathbf{0} \end{bmatrix} = S_n \mathbf{l}_{2k} - \mathbf{e}_1 \,.$$

Hence

$$T_n S_n \left(\begin{bmatrix} \mathbf{x}_{k-1} \\ \mathbf{0} \end{bmatrix} - \begin{bmatrix} \mathbf{x}_k \\ \mathbf{0} \end{bmatrix} \right) = S_n(\mathbf{l}_{2k-2} - \mathbf{l}_{2k}),$$

from which we conclude

$$\mathbf{l}_{2k-1} = S_n(\mathbf{l}_{2k-2} - \mathbf{l}_{2k}) - r_{2,k-1}\mathbf{l}_{2k}. \tag{7}$$

The d_k in the block diagonal factor D are given by $d_k = \xi_k^{-1}$, where ξ_k denotes the first component of $\mathbf{x}_k$. According to Theorem 3.1 we have $\xi_{k+1} = \frac{1}{\alpha_k}\xi_k$. Hence

$$d_{k+1} = \alpha_k d_k \tag{8}$$

Theorem 5.1. *The LU-factorization* (5) *of a centro-nonsingular skewsymmetric Toeplitz matrix* $A_n = T_n$ *can be computed with the help of* (6), (7), *the recursion in Theorem* 4.1, *and* (8).

The recursion in Theorem 3.2 leads to a UL-factorization of T_n^{-1} :

$$T_n^{-1} = U\Delta U^T. \tag{9}$$

Here U is an upper triangular matrix with ones on the main diagonal and a zero entry just above the main diagonal in the columns with an even number, and Δ is a block diagonal matrix with diagonal blocks of the form

$$\Delta_k = \begin{bmatrix} 0 & \delta_k \\ -\delta_k & 0 \end{bmatrix}.$$

Let $\mathbf{v}_j$ denote the jth column of U. Then we have

$$\mathbf{v}_{2k-1} = \begin{bmatrix} \mathbf{u}_k \\ \mathbf{0} \end{bmatrix} \tag{10}$$

for $k = 1, \ldots, n/2$.

Furthermore,

$$\mathbf{v}_{2k} = S_n\left(\mathbf{v}_{2k-1} - \frac{s_k}{s_{k-1}}\mathbf{v}_{2k-3}\right) - \rho_k \mathbf{v}_{2k-1}, \tag{11}$$

where ρ_k is the last but one component of $\mathbf{u}_k$. Finally we observe that

$$\delta_k = \frac{1}{s_k}. \tag{12}$$

Theorem 5.2. *The UL-factorization* (9) *of the inverse of a centro-nonsingular skewsymmetric Toeplitz matrix* T_n *can be computed with the help of* (10), (11), *the recursion in Theorem* 3.2, *and* (12).

Complexity. In the following complexity estimations we neglect lower order terms. In order to find the LU-factorization of T_n we need $\frac{3}{4}n^2$ additions and $\frac{1}{2}n^2$ multiplications to compute the $\mathbf{l}_{2k}$ and additional $\frac{1}{2}n^2$ additions and $\frac{1}{4}n^2$ multiplications to compute the $\mathbf{l}_{2k-1}$.

In order to find the UL-factorization of T_n^{-1} we need $\frac{7}{8}n^2$ additions and $\frac{1}{2}n^2$ multiplications to compute the $\mathbf{v}_{2k-1}$ and additional $\frac{1}{2}n^2$ additions and $\frac{1}{2}n^2$ multiplications to compute the $\mathbf{v}_{2k}$.

6. ZW-factorization

We show that the algorithm described in the previous section can be used to compute a so-called ZW-factorization of T_n and that the ZW-factorization seems to be more appropriate for this kind of matrices than the LU-factorization. First let us recall some concepts.

A matrix $A = [\,a_{ij}\,]_{i,j=1}^{n}$ is called a *W-matrix* (or a bowtie matrix) if $a_{ij} = 0$ for all (i, j) for which $i > j$ and $i+j > n$ or $i < j$ and $i+j \leq n$. The matrix A will be called a *unit W-matrix* if in addition $a_{ii} = 1$ for $i = 1, \dots, n$ and $a_{i,n+1-i} = 0$ for $i \neq (n+1)/2$. The transpose of a W-matrix is called a *Z-matrix* (or hourglass matrix). A matrix which is both a Z- and a W-matrix will be called an *X-matrix.* All these names come from the shapes of the set of all possible positions for nonzero entries, which are as follows:

$$W = \begin{bmatrix} \bullet & & & & & \bullet \\ \bullet & \circ & & & \circ & \bullet \\ \bullet & \circ & \circ & \circ & \circ & \bullet \\ \bullet & \circ & \bullet & \bullet & \circ & \bullet \\ \bullet & \bullet & & & \bullet & \bullet \\ \bullet & & & & & \bullet \end{bmatrix}, \quad Z = \begin{bmatrix} \bullet & \bullet & \bullet & \bullet & \bullet & \bullet \\ & \circ & \circ & \circ & \bullet & \\ & & \circ & \bullet & & \\ & & \bullet & \circ & & \\ & \bullet & \circ & \circ & \circ & \\ \bullet & \bullet & \bullet & \bullet & \bullet & \bullet \end{bmatrix},$$

$$X = \begin{bmatrix} \bullet & & & & & \bullet \\ & \bullet & & & \bullet & \\ & & \bullet & \bullet & & \\ & & \bullet & \bullet & & \\ & \bullet & & & \bullet & \\ \bullet & & & & & \bullet \end{bmatrix}.$$

A unit Z- or W-matrix is obviously nonsingular and a linear system with such a coefficient matrix can be solved by back substitution with $n^2/2$ additions and $n^2/2$ multiplications.

A representation $A = ZW$ or $A = WZ$ in which Z is a nonsingular Z- and W is a nonsingular W-matrix is called a *ZW- or WZ-factorization of* A, respectively.

The following facts are easily checked.

- The inverse of a unit or nonsingular Z- or W-matrix as well as the inverse of a nonsingular X-matrix is again a matrix of this kind.
- The transpose of a (unit) Z-matrix is a (unit) W-matrix, and the transpose of a (unit) W-matrix is a (unit) Z-matrix.
- If Z is an $n \times n$ Z-matrix, then $J_n Z J_n$ is a Z-matrix again. If W is an $n \times n$ W-matrix, then $J_n W J_n$ is a W-matrix as well.
- If Z is a nonsingular Z matrix, then there exist unique nonsingular X-matrices X_1 and X_2 such that $Z = Z_1 X_1 = X_2 Z_2$, where Z_1 and Z_2 are unit Z-matrices.

- A matrix admits a ZW-factorization if and only if it is centro-nonsingular. The factorization can also be written as $A = ZXW$, where X is an X-matrix, Z is a unit Z-matrix and W is a unit W-matrix. The latter factorization is unique.
- If A is symmetric and centro-nonsingular, then A admits a unique factorization $A = ZXZ^T$, where Z is a unit Z-matrix and X is a symmetric X-matrix.

In this paper we are dealing with skewsymmetric matrices. In this case the matrix A admits a unique factorization $A = ZXZ^T$, where Z is again a unit Z-matrix and X is a skewsymmetric X-matrix, which is a skewsymmetric antidiagonal matrix.

A skewsymmetric Toeplitz matrix A is also centro-skewsymmetric, which means that $A = -J_nAJ_n$. In fact, since for a Toeplitz matrix $A = T_n$ the matrix $J_nT_nJ_n$ is just its transpose, T_n is skewsymmetric if and only if T_n is centro-skewsymmetric. Utilizing besides the skewsymmetry also the centro-skewsymmetry we obtain the following.

Proposition 6.1. *An $n \times n$ centro-nonsingular matrix A which is skewsymmetric and centro-skewsymmetric admits a unique factorization*

$$A = ZXZ^T, \tag{13}$$

where Z is a centro-symmetric unit Z-matrix and X is a skewsymmetric antidiagonal matrix.

We show how the recursions in Theorem 4.1 can be used to obtain a ZW-factorization of T_n. For this we define

$$\mathbf{x}_k^- = \begin{bmatrix} 0 \\ \mathbf{x}_k \end{bmatrix}, \quad \mathbf{x}_k^+ = \begin{bmatrix} \mathbf{x}_k \\ 0 \end{bmatrix}.$$

Note that $J_{2k}\mathbf{x}_k^+ = \mathbf{x}_k^-$, $T_{2k}\mathbf{x}_k^- = -\mathbf{e}_1$ and $T_{2k}\mathbf{x}_k^+ = \mathbf{e}_{2k}$. We arrange these vectors to an $n \times n$ W-matrix as follows:

$$V = \begin{bmatrix} & 0 & & \mathbf{0} & \mathbf{0} & & 0 & \\ -\mathbf{x}_m^- & -\mathbf{x}_{m-1}^- & \cdots & -\mathbf{x}_1^- & \mathbf{x}_1^+ & \cdots & \mathbf{x}_{m-1}^+ & \mathbf{x}_m^+ \\ & 0 & & \mathbf{0} & \mathbf{0} & & 0 & \end{bmatrix} \quad \left(m = \frac{n}{2}\right).$$

From the residuals (3) of the vectors $\mathbf{x}_k$ we form the vectors $\mathbf{r}_k = (r_{j,k})_{j=1}^l$ (The definition of $\mathbf{r}_k$ is here slightly different from that in Section 5) and $\mathbf{r}'_k = (r_{j-1,k})_{j=1}^l$, where $l = \frac{n}{2} - k + 1$. Then we have $T_nV = Z$, where

$$Z = \begin{bmatrix} \widehat{\mathbf{r}}_m & & \widehat{\mathbf{r}}_2 & \widehat{\mathbf{r}}_1 & -\widehat{\mathbf{r}}'_1 & -\widehat{\mathbf{r}}'_2 & & -\widehat{\mathbf{r}}'_m \\ \mathbf{0} & \cdots & \mathbf{0} & & & \mathbf{0} & \cdots & \mathbf{0} \\ -\mathbf{r}'_m & & -\mathbf{r}'_2 & -\mathbf{r}'_1 & \mathbf{r}_1 & \mathbf{r}_2 & & \mathbf{r}_m \end{bmatrix}$$

and $\widehat{\mathbf{r}_k} = J_l\mathbf{r}_k$. Note that $\mathbf{r}'_m = \widehat{\mathbf{r}}'_m = 0$. The matrix Z is a unit Z-matrix. It is actually the Z-factor in the ZW-factorization (13) of $A = T_n$.

It remains to find the antidiagonal factor X. Suppose that

$$X = \operatorname{diag}(-d_m, \dots, -d_1, d_1, \dots, d_m) J_n .$$

As for the LU-factorization, we have $d_k = \xi_k^{-1}$. Thus we have the same recursion as for the LU-factorization, namely

$$d_{k+1} = \alpha_k d_k \,. \tag{14}$$

Theorem 6.2. *The ZW-factorization $T_n = ZXZ^T$ of a skewsymmetric and centro-nonsingular Toeplitz matrix T_n can be computed using the recursion of Theorem 4.1 and* (14).

Let us note that the Z-factor of this ZW-factorization has a remarkable additional symmetry property: The jth column of it is skewsymmetric after cancelling its first component if $j \le m$, or its last component if $j > m$. We illustrate this for the case of a 6×6 matrix. In this case the Z-factor is of the form

$$Z = \begin{bmatrix} 1 & c & b & -a & -1 & 0 \\ 0 & 1 & a & -1 & 0 & 0 \\ 0 & 0 & 1 & 0 & 0 & 0 \\ 0 & 0 & 0 & 1 & 0 & 0 \\ 0 & 0 & -1 & a & 1 & 0 \\ 0 & -1 & -a & b & c & 1 \end{bmatrix} .$$

In the same way as in Section 6 the recursions in Theorem 3.2 lead to a WZ-factorization of T_n^{-1} in the form

$$T_n^{-1} = W \,\Xi\, W^T . \tag{15}$$

In fact, W is the centrosymmetric unit W-matrix

$$W = \left[\begin{array}{cccccccc} & 0 & & \mathbf{0} & \mathbf{0} & & 0 & \\ \mathbf{u}_m^+ & \mathbf{u}_{m-1}^+ & \cdots & \mathbf{u}_1^+ & \mathbf{u}_1^- & \cdots & \mathbf{u}_{m-1}^- & \mathbf{u}_m^- \\ & 0 & & \mathbf{0} & \mathbf{0} & & 0 & \end{array} \right] . \tag{16}$$

Furthermore,

$$\Xi = \operatorname{diag}(\,\xi_m, \dots, \xi_1, -\xi_1, \dots, -\xi_m\,)\, J_n \,, \tag{17}$$

where $\xi_k = \dfrac{1}{s_k}$.

Theorem 6.3. *The WZ-factorization* (15) *of the inverse of a skewsymmetric and centro-nonsingular Toeplitz matrix T_n can be computed using the recursion of Theorem* 3.2, (16) *and* (17).

Complexity. To find the ZW-factorization of T_n we need $\frac{3}{4}n^2$ additions and $\frac{1}{2}n^2$ multiplications, to find the WZ-factorization of T_n^{-1} we need $\frac{7}{8}n^2$ additions and $\frac{1}{2}n^2$ multiplications.

7. Solving linear systems

In this section we discuss and compare different possibilities to solve a linear system $T_n\mathbf{f} = \mathbf{b}$ with a centro-nonsingular skewsymmetric Toeplitz coefficient matrix T_n. We have two types of algorithms: Levinson-type and Schur-type. The Levinson-type algorithms are labeled with (a) and the Schur-type algorithms with (b). All Schur-type algorithms will have parallel complexity $O(n)$, Levinson-type algorithms only $O(n \log n)$.

(1) The first possibility is to compute the data in the inversion formulas of Theorem 2.3 and then to use this formula to compute $\mathbf{f} = T_n^{-1}\mathbf{b}$. The matrix-vector multiplication requires only $O(n \log n)$ operations if FFT or fast algorithms for trigonometric transformations are employed (see [8], [9], [10]).

The computation of the data in the inversion formula can be done using the recursion of Theorem 3.1 alone, which includes inner product calculations (Method (1a)), or the recursions of Theorem 3.1 together with those of Theorem 4.1 (Method (1b)). Let us reiterate that the latter is more expensive in sequential processing but that it is preferable in parallel processing, because it avoids inner products. Method (1b) can also be speeded up to a "superfast" $O(n \log^2 n)$-complexity algorithm, like shown in [6] for centro-symmetric Toeplitz-plus-Hankel matrices.

(2) The second possibility is to find first the generalized LU-factorization $T_n = LDL^T$ and then to use back substitution to solve the two resulting triangular systems (Method (2b)). We can also use the UL-factorization of T_n^{-1} and matrix-vector multiplication to obtain the solution (Method (2a)).

(3) A variant of this possibility is to compute the ZW-factorization $T_n = ZXZ^T$ and then to use back substitution to solve a system with a Z- and a system with a W-coefficient matrix (Method (3b)). This version is preferable, since the construction of the Z-factor requires less operations.

Another reduction of the computational amount can be achieved by taking advantage of the property that the outer factors in the ZW-factorization are centrosymmetric. We split the right-hand side into its symmetric and skewsymmetric parts, $\mathbf{b} = \mathbf{b}^+ + \mathbf{b}^-$. Then the solution of $T_n\mathbf{f}^{\mp} = \mathbf{b}^{\pm}$ reduces to the solution of $n/2 \times n/2$ triangular systems. In this way we need for the back substitution only $n^2/2 + O(n)$ additions and the same number of multiplications compared with $n^2 + O(n)$ if we do not utilize centro-symmetry.

We can also work with the WZ-factorization of T_n^{-1} and matrix-vector multiplication (Method (3a)). Here we can also utilize the centro-symmetry of the factors in order to reduce the amount of matrix-vector multiplication by 50%.

Method (3b) is believed to be more stable than Method (3a).

(4) A fourth possibility is to solve the system recursively not using the inversion formula. Let $\mathbf{b} = (b_j)_{j=1}^n$, $\mathbf{b}_k = (b_j)_{j=l+1}^{n-l} \in \mathbb{F}^{2k}$ for $k = 1, \ldots, n/2$, $l = n/2 - k$, and let $\mathbf{f}_k$ be the solutions of $T_{2k}\mathbf{f}_k = \mathbf{b}_k$. Then it is easy to check that

$$\mathbf{f}_{k+1} = \begin{bmatrix} 0 \\ \mathbf{f}_k \\ 0 \end{bmatrix} + (b_{n-l+1} - p_k) \begin{bmatrix} \mathbf{x}_{k+1} \\ 0 \end{bmatrix} - (b_l + q_k) \begin{bmatrix} 0 \\ \mathbf{x}_{k+1} \end{bmatrix}, \qquad (18)$$

where

$$p_k = [\, a_{2k} \ \dots \ a_1 \,]\, \mathbf{f}_k \,, \qquad q_k = [\, a_1 \ \dots \ a_{2k} \,]\, \mathbf{f_k} \,.$$

The recursion can now be combined with the recursion in Theorem 3.1.

Again we can reduce the number of operations if we split the right-hand side into the symmetric and skewsymmetric parts. Let $\mathbf{b} = \mathbf{b}^+ + \mathbf{b}^-$, where $\mathbf{b}^+$ is symmetric and $\mathbf{b}^-$ is skewsymmetric, and let $\mathbf{b}_k^\pm = (b_j^\pm)_{j=l+1}^{n-l}$ be accordingly defined. We consider the two systems $T_{2k}\mathbf{f}_k^\mp = \mathbf{b}_k^\pm$. Then $\mathbf{f}_k^+$ is symmetric and $\mathbf{f}_k^-$ is skewsymmetric, and $\mathbf{f}_k = \mathbf{f}_k^+ + \mathbf{f}_k^-$. From (18) we conclude

$$\mathbf{f}_{k+1}^+ = \begin{bmatrix} 0 \\ \mathbf{f}_k^+ \\ 0 \end{bmatrix} - (b_l^- + p_k^+)\left(\begin{bmatrix} \mathbf{x}_{k+1} \\ 0 \end{bmatrix} + \begin{bmatrix} 0 \\ \mathbf{x}_{k+1} \end{bmatrix}\right) \tag{19}$$

and

$$\mathbf{f}_{k+1}^- = \begin{bmatrix} 0 \\ \mathbf{f}_k^- \\ 0 \end{bmatrix} + (b_l^+ - p_k^-)\left(\begin{bmatrix} \mathbf{x}_{k+1} \\ 0 \end{bmatrix} - \begin{bmatrix} 0 \\ \mathbf{x}_{k+1} \end{bmatrix}\right), \tag{20}$$

where

$$p_k^\pm = [\, a_{2k} \ \dots \ a_1 \,]\, \mathbf{f}_k^\pm \,.$$

In the unsplit version (18) we have $8k + O(1)$ additions and $6k + O(1)$ multiplications in each step, in addition to the recursion of the $\mathbf{x}_k$. In the split version (19), (20) we can utilize the symmetry properties of the vectors to reduce the number of multiplications to $4k + O(1)$. We refer to this algorithm as Method (4a).

The numbers $p_k^\pm$ can also be precomputed. For this we introduce the numbers

$$p_{j,k}^\pm = [\, a_{2k+j-1} \ \dots \ a_j \,]\, \mathbf{f}_k^\pm \,.$$

We have $p_k^\pm = p_{1,k}^\pm$. From (19) and (20) we obtain the recursion

$$p_{j,k+1}^\pm = p_{j+1,k}^\pm \mp (b_l^\mp \pm p_{1,k}^\pm)(r_{j+1,k+1} \pm r_{j,k+1}) \,. \tag{21}$$

Method (4b) is the combination of (21), (19), (20) and the recursion given by Theorem 4.1.

We compare the complexity of these methods in sequential processing in the following table. According to this table, Method (1a) is the most efficient from the view point of sequential complexity. However, there are more criteria than complexity. One of them is stability. Taking also this into account we think that Method (3b) is of high practical value.

Method	Additions	Multiplications	Lower order terms
(1a)	$\frac{7}{8}n^2$	$\frac{1}{2}n^2$	$O(n\log n)$
(1b)	$\frac{9}{8}n^2$	$\frac{3}{4}n^2$	$O(n\log n)$
(2a)	$\frac{19}{8}n^2$	$2n^2$	$O(n)$
(2b)	$\frac{9}{4}n^2$	$\frac{7}{4}n^2$	$O(n)$
(3a)	$\frac{11}{8}n^2$	n^2	$O(n)$
(3b)	$\frac{5}{4}n^2$	n^2	$O(n)$
(4a)	$\frac{15}{8}n^2$	n^2	$O(n)$
(4b)	$\frac{15}{8}n^2$	n^2	$O(n)$

References

[1] P. Delsarte and Y. Genin, The split Levinson algorithm, *IEEE Transactions on Acoustics Speech, and Signal Processing* ASSP-**34** (1986), 470–477.

[2] P. Delsarte and Y. Genin, On the splitting of classical algorithms in linear prediction theory, *IEEE Transactions on Acoustics Speech, and Signal Processing* ASSP-**35** (1987), 645–653.

[3] C.J. Demeure, Bowtie factors of Toeplitz matrices by means of split algorithms, *IEEE Trans. Acoustics Speech, and Signal Processing*, ASSP-**37**, 10 (1989), 1601–1603.

[4] D.J. Evans and M. Hatzopoulos, A parallel linear systems solver, *Internat. J. Comput. Math.*, **7**, 3 (1979), 227–238.

[5] I. Gohberg and A.A. Semencul, On the inversion of finite Toeplitz matrices and their continous analogs (in Russian), *Matemat. Issledovanya*, **7**, 2 (1972), 201–223.

[6] G. Heinig, Chebyshev-Hankel matrices and the splitting approach for centrosymmetric Toeplitz-plus-Hankel matrices, *Linear Algebra Appl.*, **327**, 1–3 (2001), 181–196.

[7] G. Heinig and K. Rost, *Algebraic Methods for Toeplitz-like Matrices and Operators*, Birkhäuser Verlag, Basel, Boston, Stuttgart 1984.

[8] G. Heinig and K. Rost, DFT representations of Toeplitz-plus-Hankel Bezoutians with application to fast matrix-vector multiplication, *Linear Algebra Appl.*, **284** (1998), 157–175.

[9] G. Heinig and K. Rost, Hartley transform representations of inverses of real Toeplitz-plus-Hankel matrices, *Numer. Funct. Anal. and Optimiz.*, **21** (2000), 175–189.

[10] G. Heinig and K. Rost, Efficient inversion formulas for Toeplitz-plus-Hankel matrices using trigonometric transformations, In: V. Olshevsky (Ed.), *Structured Matrices in Mathematics, Computer Science, and Engineering*, vol.2, AMS-Series *Contemporary Mathematics* **281** (2001), 247–264.

[11] G. Heinig and K. Rost, Centro-symmetric and centro-skewsymmetric Toeplitz matrices and Bezoutians, *Linear Algebra Appl.*, **343–344** (2002), 195–209.

[12] G. Heinig and K. Rost, Fast algorithms for centro-symmetric and centro-skewsymmetric Toeplitz-plus-Hankel matrices, submitted.

[13] A. Melman, A symmetric algorithm for Toeplitz systems, *Linear Algebra Appl.*, **301** (1999), 145–152.

[14] A. Melman, The even-odd split Levinson Algorithm for Toeplitz systems, *SIAM J. Matrix Anal. Appl.*, **23**, 1 (2001), 256–270.

[15] A. Melman, A two-step even-odd split Levinson algorithm for Toeplitz systems, *Linear Algebra Appl.*, **338** (2001), 219–237.

[16] S. Chandra Sekhara Rao, Existence and uniqueness of WZ factorization, *Parallel Comp.*, **23**, 8 (1997), 1129–1139.

[17] W.F. Trench, An algorithm for the inversion of finite Toeplitz matrices, *J. Soc. Indust. Appl. Math.*, **12** (1964), 515–522.

G. Heinig
Department of Mathematics and Computer Sciences
P.O.Box 5969
Safat 1306, Kuwait
e-mail: georg@mcs.sci.kuniv.edu.kw

K. Rost
Department of Mathematics
Chemnitz University of Technology
D-09107 Chemnitz, Germany
e-mail: karla.rost@mathematik.tu-chemnitz.de

Received: 1 October 2001

Operator Theory:
Advances and Applications, Vol. 135, 209–233

On Collocation Methods for Nonlinear Cauchy Singular Integral Equations

Peter Junghanns, Katja Müller and Karla Rost

Dedicated to Bernd Silbermann on the Occasion of His 60th Birthday

Abstract. Collocation methods with respect to the Chebyshev nodes of first and of second kind together with both the modified and the classical Newton iteration method for a class of nonlinear Cauchy singular integral equations are investigated. The proof of the convergence of the Newton methods is based on the stability of the respective collocation methods applied to linear Cauchy singular integral equations, which is in turn proved by using Banach algebra techniques. The effective numerical realization of the methods and their applicability to a nonlinear Cauchy singular integral equation arising from a free surface seepage problem are discussed.

1. Introduction

In [10] two of the authors considered the modified Newton iteration method based on a collocation method w.r.t. Chebyshev nodes of second kind for Cauchy singular integral equations (CSIEs) of the form

$$F(x,u(x)) + \frac{1}{\pi}\int_{-1}^{1} \frac{u(y)}{y-x}\,dy + \delta = 0\,, \quad -1 < x < 1\,.$$

Using some new results from [11] on the stability of collocation methods w.r.t. Chebyshev nodes of first and second kind, in the present paper we investigate respective collocation methods for a wider class of nonlinear CSIEs, namely

$$F(x,u(x)) + \frac{1}{\pi}\int_{-1}^{1} \frac{G(y,u(y))}{y-x}\,dy + \sum_{k=0}^{p-1} \delta_k x^k = 0\,, \quad -1 < x < 1\,, \tag{1}$$

where $F : [-1,1]\times\mathbb{R} \longrightarrow \mathbb{R}$ and $G : [-1,1]\times\mathbb{R} \longrightarrow \mathbb{R}$ are given. The function $u : [-1,1] \longrightarrow \mathbb{R}$ satisfying

$$u(-1) = u(1) = 0 \tag{2}$$

and the coefficients $\delta_k \in \mathbb{C}$ are sought. Concerning the Newton method, which is used for solving the resulting discrete system, we show the convergence not only for the modified but also for the classical Newton method.

In Section 2 we prove a theorem on the convergence of the classical Newton method for operators whose Fréchet derivatives are Hölder continuous. In Section 3 we present some auxiliary material from [11] concerning the stability of collocation methods w.r.t. Chebyshev nodes for linear CSIEs and give a generalization to equations involving a Cauchy singular integral operator (CSIO) with negative Fredholm index. In Section 4 the convergence of the Newton iteration method applied to the discrete collocation equations for (1) is proved and error estimates for the collocation solutions are presented. Finally, in Section 5 some computational aspects of the considered Newton collocation methods are discussed.

2. On the classical Newton iteration method

For Banach spaces $\mathbf{X}$ and $\mathbf{Y}$, let $\mathcal{L}(\mathbf{X},\mathbf{Y})$ denote the Banach space of all linear and bounded operators from $\mathbf{X}$ into $\mathbf{Y}$. For a given mapping $P:\mathbf{D}\subseteq\mathbf{X}\longrightarrow\mathbf{Y}$ we study Newton's method

$$u_{n+1}=u_n-[P'(u_n)]^{-1}P(u_n),\quad n=0,1,\dots, \tag{3}$$

and its modified version

$$u_{n+1}=u_n-[P'(u_0)]^{-1}P(u_n),\quad n=0,1,\dots, \tag{4}$$

as methods of solving the operator equation

$$P(u)=0\,. \tag{5}$$

At first let us recall a theorem on the convergence of the modified Newton method.

Theorem 2.1 ([6], Theor. 4.2). *Let $\mathbf{X}$ and $\mathbf{Y}$ be Banach spaces and $P:\mathbf{X}\to\mathbf{Y}$ an operator whose Fréchet derivative $P'(u)\in\mathcal{L}(\mathbf{X},\mathbf{Y})$ exists for all $u\in\mathbf{X}$, and suppose $P'(u_0)$ is invertible for some $u_0\in\mathbf{X}$. Further, assume that there are constants M_0, N_0, C_0 and $\eta\in(0,1]$ such that*

(a) $\|[P'(u_0)]^{-1}\|_{\mathcal{L}(\mathbf{Y},\mathbf{X})}\le M_0\,,$

(b) $\|[P'(u_0)]^{-1}P(u_0)\|_{\mathbf{X}}\le N_0\,,$

(c) $\|P'(u)-P'(u_0)\|_{\mathcal{L}(\mathbf{X},\mathbf{Y})}\le C_0\|u-u_0\|_{\mathbf{X}}^{\eta}\quad\forall u\in\mathbf{X}\,,$

and $h_0:=M_0N_0^{\eta}C_0\le 1/4$. Then, there exists a neighbourhood $\mathbf{U}(u_0,\rho)$ of u_0 such that $\rho\le 2N_0$ and the equation

$$P(u)=0$$

has a unique solution u^* in $\mathbf{U}(u_0,\rho)$, where the sequence $\{u_k\}_{k=1}^{\infty}$ defined by (4) converges to u^*.

The following lemma is well known.

Lemma 2.2. *If $A,B\in\mathcal{L}(\mathbf{X},\mathbf{Y})$, $A^{-1}\in\mathcal{L}(\mathbf{Y},\mathbf{X})$, and $k:=\|A-B\|\,\|A^{-1}\|<1$, then $B^{-1}\in\mathcal{L}(\mathbf{Y},\mathbf{X})$, $\|B^{-1}\|\le(1-k)^{-1}\|A^{-1}\|$, and*

$$\|A^{-1}-B^{-1}\|\le(1-k)^{-1}\|A^{-1}\|^2\|A-B\|\,.$$

The following lemma generalizes the assertion of [21, Probl. 4.1c].

Lemma 2.3. *Let $P : \mathbf{D} \subseteq \mathbf{X} \to \mathbf{Y}$ be a mapping from a nonempty open convex subset $\mathbf{D}$ of a Banach space $\mathbf{X}$ into a Banach space $\mathbf{Y}$. Let P be Fréchet differentiable on $\mathbf{D}$. If P' is Hölder continuous on $\mathbf{D}$, i.e.,*

$$\|P'(y) - P'(x)\| \le H\|y - x\|^\eta \quad \forall x, y \in \mathbf{D},$$

where $H > 0$ and $\eta \in (0, 1]$ are some constants, then

$$\|P(y) - P(x) - P'(x)(y - x)\| \le \frac{H\|y - x\|^{1+\eta}}{1 + \eta} \quad \forall x, y \in \mathbf{D}.$$

Proof. For $t \in [0, 1]$ and $x, y \in \mathbf{D}$, set $\psi(t) = P(x + t(y - x)) - tP'(x)(y - x)$. Then

$$\psi'(t) = [P'(x + t(y - x)) - P'(x)](y - x)$$

and therefore

$$\|\psi'(t)\| \le \|P'(x + t(y - x)) - P'(x)\| \, \|y - x\| \le Ht^\eta \|y - x\|^{1+\eta}.$$

That means

$$\begin{aligned} \|P(y) - P(x) - P'(x)(y - x)\| &= \|\psi(1) - \psi(0)\| = \left\| \int_0^1 \psi'(t)\, dt \right\| \\ &\le \int_0^1 Ht^\eta \|y - x\|^{1+\eta}\, dt = \frac{H\|y - x\|^{1+\eta}}{1 + \eta}, \end{aligned}$$

and the lemma is proved. □

The following theorem generalizes [21, Prop. 5.1].

Theorem 2.4. *Suppose that*

(a) *equation (5) has a solution u^*,*
(b) *the Fréchet-derivative $P'(u)$ exists for all u in an open neighbourhood $\mathbf{U}(u^*, \varepsilon_0) = \{u \in \mathbf{X} : \|u - u^*\| < \varepsilon_0\}$ of u^* and is Hölder continuous there, i.e.,*

$$\|P'(u) - P'(v)\| \le H\|u - v\|^\eta \quad \forall u, v \in \mathbf{U}(u^*, \varepsilon_0), \tag{6}$$

where $H > 0$ and $\eta \in (0, 1]$ are some constants,
(c) *$[P'(u^*)]^{-1}$ exists as a bounded linear operator from $\mathbf{Y}$ into $\mathbf{X}$.*

Then there is a $\delta > 0$ such that the sequence $\{u_n\}$ defined by (3) converges to the solution u^* for every initial point u_0 satisfying $\|u_0 - u^*\| \le \delta$. Furthermore, there exists a constant M (independent of n and $u_0 \in \mathbf{U}(u^*, \delta)$) such that

$$\|u_n - u^*\| \le M\|u_{n-1} - u^*\|^{1+\eta} \quad \text{and} \quad \|u_n - u^*\| \le M^{\frac{(1+\eta)^n - 1}{\eta}} \delta^{(1+\eta)^n}. \tag{7}$$

Proof. For $u \in \mathbf{U}(u*, \varepsilon_0)$, let $k := \|P'(u) - P'(u^*)\| \, \left\|[P'(u^*)]^{-1}\right\|$. Using (6) we get

$$k \le H\|u - u^*\|^\eta \left\|[P'(u^*)]^{-1}\right\| \le H\varepsilon^\eta \left\|[P'(u^*)]^{-1}\right\| < \frac{1}{2}$$

for a sufficiently small $\varepsilon > 0$ and for all $u \in \mathbf{U}(u^*, \varepsilon)$. Now by Lemma 2.2 the inverse operator $[P'(u)]^{-1} \in \mathcal{L}(\mathbf{Y}, \mathbf{X})$ exists for all $u \in \mathbf{U}(u^*, \varepsilon)$ and

$$\left\| [P'(u)]^{-1} \right\| \le (1-k)^{-1} \left\| [P'(u^*)]^{-1} \right\| < 2 \left\| [P'(u^*)]^{-1} \right\| \quad \forall u \in \mathbf{U}(u^*, \varepsilon).$$

So we can define

$$\alpha := \sup \left\{ \left\| [P'(u)]^{-1} \right\| : u \in \mathbf{U}(u^*, \varepsilon) \right\} < \infty. \tag{8}$$

With the notation $K(u) := u - [P'(u)]^{-1} P(u)$ Newton's method (3) becomes the iterative method $u_{n+1} = K(u_n)$. For $u \in \mathbf{U}(u^*, \varepsilon)$, we write $K(u) - u^* = K(u) - K(u^*)$ in the form

$$[P'(u)]^{-1} [P'(u^*)(u-u^*) - P(u) + P(u^*)] + [P'(u)]^{-1} [P'(u) - P'(u^*)](u-u^*).$$

Using (6), (8), and Lemma 2.3 this gives

$$\begin{aligned} \|K(u) - u^*\| &\le \alpha \big[\|P'(u^*)(u-u^*) - P(u) + P(u^*)\| \\ &\qquad + \|P'(u) - P'(u^*)\| \, \|u-u^*\| \big] \\ &\le \alpha \left(\frac{H}{1+\eta} + H \right) \|u-u^*\|^{1+\eta} =: M \|u-u^*\|^{1+\eta}, \end{aligned}$$

so that

$$\|u_{n+1} - u^*\| = \|K(u_n) - K(u^*)\| \le M \, \|u_n - u^*\|^{1+\eta} \quad \text{if} \quad u_n \in \mathbf{U}(u^*, \varepsilon).$$

Consequently,

$$\begin{aligned} \|u_n - u^*\| &\le M \, \|u_{n-1} - u^*\|^{1+\eta} \le M \cdot M^{1+\eta} \|u_{n-2} - u^*\|^{(1+\eta)^2} \\ &\le \cdots \le M^{\frac{(1+\eta)^n - 1}{\eta}} \, \|u_0 - u^*\|^{(1+\eta)^n} \\ &= M^{-1/\eta} \left(M^{1/\eta} \, \|u_0 - u^*\| \right)^{(1+\eta)^n}. \end{aligned}$$

Thus, for sufficiently small $\delta > 0$ and $\|u_0 - u^*\| \le \delta$, u_n belongs to the neighbourhood $\mathbf{U}(u^*, \varepsilon)$ and converges to u^* as n tends to infinity. □

3. Collocation for linear CSIEs

We denote by $\sigma(x) = (1-x^2)^{-\frac{1}{2}}$ and $\varphi(x) = (1-x^2)^{\frac{1}{2}}$ the Chebyshev weights of first and second kind, respectively, and by $\mathbf{L}^2_\sigma := \mathbf{L}^2_\sigma(-1,1)$ the weighted Lebesgue space of all (classes of) measurable functions $u : (-1,1) \longrightarrow \mathbb{C}$ for which

$$\|u\|^2_\sigma := \int_{-1}^{1} |u(x)|^2 \sigma(x) \, dx$$

is finite. $\mathbf{L}^2_\sigma$ is turned into a Hilbert space by defining the inner product

$$\langle u, v \rangle_\sigma := \int_{-1}^{1} u(x) \overline{v(x)} \sigma(x) \, dx.$$

We let $\mathbf{PC} = \mathbf{PC}[-1,1]$ refer to the set of all piecewise continuous functions $a : [-1,1] \longrightarrow \mathbb{C}$ which are continuous from the left and have finite limits from the right at each point $x \in [-1,1)$ and are continuous at $x = -1$. It is well known that the multiplication operator $aI : \mathbf{L}^2_\sigma \longrightarrow \mathbf{L}^2_\sigma$, $u \mapsto a\,u$ ($a \in \mathbf{PC}$) and the CSIO

$$S : \mathbf{L}^2_\sigma \longrightarrow \mathbf{L}^2_\sigma\,, \quad u \mapsto Su\,,\ (Su)(x) = \frac{1}{\pi\mathbf{i}}\int_{-1}^{1}\frac{u(y)}{y-x}\,dy\,,$$

are linear and bounded operators ([2, Theor. I.4.1]). The functions

$$\widetilde{u}_n(x) = \varphi(x)U_n(x)\,, \quad n = 0,1,2\ldots\,,$$

where

$$U_n(\cos s) = \sqrt{\frac{2}{\pi}}\,\frac{\sin(n+1)s}{\sin s}$$

are the Chebyshev polynomials of second kind, form an orthonormal basis in $\mathbf{L}^2_\sigma$. Define the projections $P_n : \mathbf{L}^2_\sigma \longrightarrow \mathbf{L}^2_\sigma$, $n = 1,2,\ldots$, by $P_n u = \sum_{k=0}^{n-1}\langle u, \widetilde{u}_k\rangle_\sigma \widetilde{u}_k$ and consider the C^*-algebra $\mathcal{F}$ of all uniformly bounded sequences of operators $A_n \in \mathcal{L}(\operatorname{im} P_n)$. The norm and the algebraic operations in $\mathcal{F}$ are defined by

$$\|\{A_n\}\|_{\mathcal{F}} := \sup\left\{\|A_nP_n\|_{\mathcal{L}(\mathbf{L}^2_\sigma)} : n = 1,2,\ldots\right\}$$

and

$$\lambda\{A_n\} + \mu\{B_n\} := \{\lambda A_n + \mu B_n\}\,,\ \{A_n\}\{B_n\} := \{A_nB_n\}\,,\ \{A_n\}^* := \{A_n^*\}\,.$$

A sequence $\{A_n\} \in \mathcal{F}$ is called stable if the operators $A_n : \operatorname{im} P_n \longrightarrow \operatorname{im} P_n$ are invertible for all sufficiently large n, say $n \geq n_0$, and if

$$\sup\left\{\|A_n^{-1}P_n\|_{\mathcal{L}(\mathbf{L}^2_\sigma)} : n \geq n_0\right\} < \infty\,.$$

Denote by M_n^ω the "weighted" interpolation operators $M_n^\omega = \varphi L_n^\omega \varphi^{-1} I$, $\omega = \sigma$ or $\omega = \varphi$, where L_n^ω denotes the usual Lagrange interpolation operator, i.e., for a function $f : (-1,1) \longrightarrow \mathbb{C}$, $L_n^\omega f$ is the algebraic polynomial of degree less than n satisfying $(L_n^\omega f)(x_{jn}^\omega) = f(x_{jn}^\omega)$ for $j = 1,\ldots,n$. Let $\mathcal{A}^\omega$ be the smallest C^*-subalgebra of $\mathcal{F}$ containing all sequences of the form $\{M_n^\omega(aI + bS)P_n\}$ and $\{P_nK_1P_n + W_nK_2W_n + C_n\}$ with $a,b \in \mathbf{PC}$, $K_1, K_2 \in \mathcal{L}(\mathbf{L}^2_\sigma)$ compact, and $\lim_{n\to\infty}\|C_nP_n\|_{\mathcal{L}(\mathbf{L}^2_\sigma)} = 0$, where

$$W_n u = \sum_{k=0}^{n-1}\langle u, \widetilde{u}_{n-1-k}\rangle_\sigma \widetilde{u}_k$$

(comp. [11]). Of course, the system $\{T_n\}_{n=0}^\infty$ of the normalized Chebyshev polynomials of first kind,

$$T_0 \equiv \frac{1}{\sqrt{\pi}}\,,\ T_n(\cos s) = \sqrt{\frac{2}{\pi}}\,\cos ns\,,\ n = 1,2,\ldots\,,$$

also forms an orthonormal basis in $\mathbf{L}^2_\sigma$. In the following we need the isomorphism

$$J_\sigma : \mathbf{L}^2_\sigma \longrightarrow \mathbf{L}^2_\sigma , \quad u \mapsto \sqrt{2}\, T_0 + \sum_{n=1}^\infty \langle u, \widetilde{u}_n \rangle_\sigma T_n ,$$

and the shift operator

$$V : \mathbf{L}^2_\sigma \longrightarrow \mathbf{L}^2_\sigma , \quad u \mapsto \sum_{n=0}^\infty \langle u, \widetilde{u}_n \rangle_\sigma \widetilde{u}_{n+1} .$$

For $a, b \in \mathbf{PC}$ and $A_n^\omega = M_n^\omega (aI + bS) P_n$, define (comp. [11])

$$\begin{aligned} W_1^\sigma \{A_n^\sigma\} &:= \text{s-lim}\, A_n^\sigma P_n = aI + bS , \\ W_2^\sigma \{A_n^\sigma\} &:= \text{s-lim}\, W_n A_n^\sigma W_n = J_\sigma^{-1}(aJ_\sigma + \mathrm{i} bV^*) , \end{aligned}$$

and

$$W_1^\varphi \{A_n^\varphi\} := \text{s-lim}\, A_n^\varphi P_n = aI + bS , \quad W_2^\varphi \{A_n^\varphi\} := \text{s-lim}\, W_n A_n^\varphi W_n = aI - bS ,$$

$$W_3^\varphi \{A_n^\varphi\} := a(1)I + b(1)V^* , \qquad W_4^\varphi \{A_n^\varphi\} := a(-1)I + b(-1)V^* .$$

Let $\Theta_\sigma = \{1, 2\}$ and $\Theta_\varphi = \{1, 2, 3, 4\}$.

Theorem 3.1 ([11], Theor. 5.7). *The mappings W_θ^ω, $\theta \in \Theta_\omega$, can be extended to *-homomorphisms $W_\theta^\omega : \mathcal{A}^\omega \longrightarrow \mathcal{L}(\mathbf{L}^2_\sigma)$. A sequence $\{A_n\} \in \mathcal{A}^\omega$ is stable if and only if all operators $W_\theta^\omega \{A_n\} : \mathbf{L}^2_\sigma \longrightarrow \mathbf{L}^2_\sigma$, $\theta \in \Theta_\omega$, are invertible.*

Let $\mathbf{V}$ be a finite dimensional subspace of $\mathbf{L}^2_\sigma$ and look for a solution $(u, v) \in \mathbf{L}^2_\sigma \times \mathbf{V}$ of the equation

$$(aI + Sb)u + v = f , \tag{9}$$

where $a, b \in \mathbf{PC}$ and $f \in \mathbf{L}^2_\sigma$ are given functions.

Obviously, the set $\mathbf{X} := \mathbf{L}^2_\sigma \times \mathbf{V}$ equipped with the norm $\|(u,v)\|_{\mathbf{X}} := \sqrt{\|u\|_\sigma^2 + \|v\|_\sigma^2}$ is a Banach space. Let $\{\widetilde{v}_0, \ldots, \widetilde{v}_{p-1}\}$ be a basis in $\mathbf{V}$. We define the isomorphism $J_p : \mathbf{X} \longrightarrow \mathbf{L}^2_\sigma$ by

$$J_p \left(u, \sum_{k=0}^{p-1} \alpha_k \widetilde{v}_k \right) := \sum_{j=0}^\infty \langle u, \widetilde{u}_j \rangle_\sigma \widetilde{u}_{j+p} + \sum_{k=0}^{p-1} \alpha_k \widetilde{u}_k = V^p u + \sum_{k=0}^{p-1} \alpha_k \widetilde{u}_k .$$

Then the inverse operator $J_p^{-1} : \mathbf{L}^2_\sigma \longrightarrow \mathbf{X}$ is given by

$$J_p^{-1} f = \left(\sum_{j=0}^\infty \langle f, \widetilde{u}_{j+p} \rangle_\sigma \widetilde{u}_j , \sum_{k=0}^{p-1} \langle f, \widetilde{u}_k \rangle_\sigma \widetilde{v}_k \right) = \left((V^*)^p f, \sum_{k=0}^{p-1} \langle f, \widetilde{u}_k \rangle_\sigma \widetilde{v}_k \right) .$$

Thus, setting $w = J_p(u, v)$ equation (9) is equivalent to

$$(B^{(p)} + K)w := (aI + Sb)(V^*)^p w + Kw = f , \quad w \in \mathbf{L}^2_\sigma , \tag{10}$$

where the operator $K : \mathbf{L}^2_\sigma \longrightarrow \mathbf{L}^2_\sigma$ is defined by

$$Kw := \sum_{k=0}^{p-1} \langle w, \widetilde{u}_k \rangle_\sigma \widetilde{v}_k \,.$$

To solve equation (9) approximately we consider the collocation method

$$M_n^\omega (aI + SM_n^\omega b) u_n + M_n^\omega v_n = M_n^\omega f\,, \quad u_n \in \operatorname{im} P_{n-p}\,,\ v_n \in \mathbf{V}\,. \tag{11}$$

Note that if b is not continuous, then $M_n^\omega SbP_n$ is not well defined, so that we use $M_n^\omega SM_n^\omega bP_n$ as an approximation of SbI. Setting $w_n = J_p(u_n, v_n)$ and using the same arguments as above, equation (11) can be written in the form

$$[(M_n^\omega aP_n + M_n^\omega SP_n M_n^\omega bP_n)(M_n^\omega V^* P_n)^p + M_n^\omega KP_n] w_n = M_n^\omega f\,, \tag{12}$$

$w_n \in \operatorname{im} P_n$, where the relation $M_n^\omega (V^*)^p P_n = (V^* P_n)^p$ has been used. Let, for some $\varepsilon \geq 0$, $\mathbf{R}_\varepsilon$ denote the space of all functions $f : (-1,1) \longrightarrow \mathbb{C}$ which are locally bounded and Riemann integrable and satisfy

$$|f(x)| \leq \text{const}\ (1 - x^2)^{-\varepsilon}\,, \quad x \in (-1,1)\,.$$

Equipped with the norm

$$\|f\|_{\mathbf{R}_\varepsilon} = \sup\left\{|f(x)|\,(1-x^2)^\varepsilon : x \in (-1,1)\right\}$$

$\mathbf{R}_\varepsilon$ becomes a Banach space.

Lemma 3.2 ([11], Cor. 3.3, Cor. 3.4). *If $f \in \mathbf{R}_\varepsilon$ for some $\varepsilon < 1/4$, then*

$$\lim_{n\to\infty} \|M_n^\omega f - f\|_\sigma = 0\,.$$

Since $V^* = \psi I + \mathbf{i}\varphi S$, where $\psi(x) = x$ (see [10, 11]), we can conclude the following.

Corollary 3.3. *If $\widetilde{v}_k \in \mathbf{R}_\varepsilon$, $k = 0, \ldots, p-1$, for some $\varepsilon \in [0, 1/4)$ then the sequence $\{B_n^\omega\}$ defined by the left-hand side of (12), i.e.,*

$$B_n^\omega = (M_n^\omega aP_n + M_n^\omega SP_n M_n^\omega bP_n)(M_n^\omega V^* P_n)^p + M_n^\omega KP_n\,, \tag{13}$$

belongs to the algebra $\mathcal{A}^\omega$.

Proof. Since $K : \mathbf{L}^2_\sigma \longrightarrow \mathbf{R}_\varepsilon$ is compact we have

$$\lim_{n\to\infty} \|(M_n^\omega - P_n)KP_n\|_{\mathcal{L}(\mathbf{L}^2_\sigma)} = 0\,,$$

so that $\{M_n^\omega KP_n\} \in \mathcal{A}^\omega$. □

Hence, applying Theorem 3.1 we get the following theorem.

Theorem 3.4. *Let $a, b \in \mathbf{PC}$ and $\widetilde{v}_k \in \mathbf{R}_\varepsilon$ for $k = 0, \ldots, p-1$ and some $\varepsilon \in [0, 1/4)$. Then the sequence $\{B_n^\omega\}$ in (13) is stable in $\mathbf{L}^2_\sigma$ if and only if the operators $(aI + SbI)(V^*)^p + K$ and, in case $\omega = \sigma$, $J_\sigma^{-1}(aJ_\sigma + \mathbf{i}V^* J_\sigma^{-1} bJ_\sigma)V^p$ or, in case $\omega = \varphi$, $(aI - SbI)V^p$ are invertible in $\mathbf{L}^2_\sigma$.*

Proof. We have $W_1^\omega\{M_n^\omega KP_n\} = K$, $W_2^\omega\{M_n^\omega KP_n\} = 0$, $W_1^\omega\{M_n^\omega V^*P_n\} = V^*$, and $W_2^\omega\{M_n^\omega V^*P_n\} = V$. Hence, in case $\omega = \sigma$ the assertion follows immediately from Theorem 3.1. Concerning the case $\omega = \varphi$ we remark that $W_\theta^\varphi\{M^\varphi KP_n\} = 0$ for $\theta = 3,4$ (see [11]), so that $W_{3,4}^\varphi\{B_n^\varphi\} = (\pm 1)^p[a(\pm 1)I + b(\pm 1)V^*]$. If the operators $W_1^\varphi\{B_n^\varphi\}$ and $W_2^\varphi\{B_n^\varphi\}$ are invertible, then $\operatorname{ind}(aI + SbI) = -p$ and $\operatorname{ind}(aI - SbI) = p$. As in the proof of [14, Lemma 6.3] one can show that this implies $|a(\pm 1)| > |b(\pm 1)|$ and hence the invertibility of $W_3^\varphi\{B_n^\varphi\}$ and $W_4^\varphi\{B_n^\varphi\}$. □

4. Newton collocation methods

Let us assume the existence of a continuous function u^* and a vector $d^* = (\delta_0^*, \ldots, \delta_{p-1}^*)$ of complex numbers satisfying (1) and (2). The problem (1), (2) can shortly be written as operator equation

$$H(u,d) = 0\,, \quad (u,d) \in \mathbf{X} := \mathbf{L}_\sigma^2 \times \mathbb{C}^p\,, \tag{14}$$

where H is defined by

$$H(u,d) = F(\,.\,,u) + \mathbf{i}SG(\,.\,,u) + \Pi d\,, \quad \Pi d = \sum_{k=0}^{p-1} \delta_k \widetilde{v}_k\,,$$

and $\widetilde{v}_k(x) = x^k$. To solve this problem numerically we use a collocation method with respect to the Chebyshev nodes x_{kn}^ω and look for an approximate solution $(u_n, d_n) \in \mathbf{X}_n := \operatorname{im} P_{n-p} \times \mathbb{C}^p$ satisfying

$$F(x_{jn}^\omega, u_n(x_{jn}^\omega)) + \frac{1}{\pi}\int_{-1}^{1} \frac{G_n^\omega(y, u_n(y))}{y - x_{jn}^\omega}\,dy + \sum_{k=0}^{p-1} \delta_{kn}(x_{jn}^\omega)^k = 0\,, \tag{15}$$

$j = 1, \ldots, n$, which will briefly be written as

$$H_n^\omega(u_n, d_n) = 0\,,$$

where

$$H_n^\omega(u_n, d_n) = M_n^\omega F(.,u_n) + \mathbf{i}M_n^\omega SG_n^\omega(.,u_n) + M_n^\omega \Pi d_n$$

and $G_n^\omega(x, u_n(x)) := [M_n^\omega G(.,u_n)](x)$. To solve this system of nonlinear equations we consider both the modified and the usual Newton iteration method, which means that (u_n, d_n) is approximated by $\{(u_n^{(m)}, d_n^{(m)})\}_{m=0}^\infty \subset \mathbf{X}_n$, where $u_n^{(m+1)} = u_n^{(m)} + \Delta u_n^{(m)}$, $d_n^{(m+1)} = d_n^{(m)} + \Delta d_n^{(m)}$, and $(\Delta u_n^{(m)}, \Delta d_n^{(m)})$ is the solution of

$$\begin{aligned} &F_u(x_{jn}^\omega, u_n^{(0)}(x_{jn}^\omega))\Delta u_n^{(m)}(x_{jn}^\omega) \\ &\quad + \frac{1}{\pi}\int_{-1}^{1} \frac{G_{u,n}^\omega(y, u_n^{(0)}(y))\Delta u_n^{(m)}(y)}{y - x_{jn}^\omega}\,dy + (\Pi\Delta d_n^{(m)})(x_{jn}^\omega) \\ &\quad = -F(x_{jn}^\omega, u_n^{(m)}(x_{jn}^\omega)) - \frac{1}{\pi}\int_{-1}^{1} \frac{G_n^\omega(y, u_n^{(m)}(y))}{y - x_{jn}^\omega}\,dy - (\Pi d_n^{(m)})(x_{jn}^\omega)\,, \end{aligned} \tag{16}$$

$j=1,\ldots,n$, where $G_{u,n}^{\omega}(x,u_n(x)) := [M_n^{\omega}G_u(.,u_n)](x)$, or

$$F_u(x_{jn}^{\omega},u_n^{(m)}(x_{jn}^{\omega}))\Delta u_n^{(m)}(x_{jn}^{\omega}) + \frac{1}{\pi}\int_{-1}^{1}\frac{G_{u,n}^{\omega}(y,u_n^{(m)}(y))\Delta u_n^{(m)}(y)}{y-x_{jn}^{\omega}}\,dy + (\Pi\Delta d_n^{(m)})(x_{jn}^{\omega}) = -F(x_{jn}^{\omega},u_n^{(m)}(x_{jn}^{\omega})) - \frac{1}{\pi}\int_{-1}^{1}\frac{G_n^{\omega}(y,u_n^{(m)}(y))}{y-x_{jn}^{\omega}}\,dy - (\Pi d_n^{(m)})(x_{jn}^{\omega}), \tag{17}$$

$j=1,\ldots,n$, respectively. Equations (16) and (17) can shortly be written as

$$[(H_n^{\omega})'(u_n^{(0)},d_n^{(0)})]\,(\Delta u_n^{(m)},\Delta d_n^{(m)}) = -H_n^{\omega}(u_n^{(m)},d_n^{(m)}),$$

and

$$[(H_n^{\omega})'(u_n^{(m)},d_n^{(m)})](\Delta u_n^{(m)},\Delta d_n^{(m)}) = -H_n^{\omega}(u_n^{(m)},d_n^{(m)}),$$

respectively.

At first let us consider the modified Newton method (16) and recall that the operator

$$A_n^{\omega} = (H_n^{\omega})'(u_n^{(0)},d_n^{(0)}) : \mathbf{X}_n \to \operatorname{im} P_n$$

is defined by

$$A_n^{\omega}(u_n,d_n) := M_n^{\omega}(a_nI + SM_n^{\omega}b_n)u_n + \Pi d_n \tag{18}$$

with $a_n(x)=F_u(x,u_n^{(0)}(x))$ and $b_n(x)=G_u(x,u_n^{(0)}(x))$. To prove the convergence of the modified Newton method (16) we shall use Theorem 2.1. For checking the conditions (a)–(c) of this theorem w.r.t. the operator $H_n:\mathbf{X}_n\to\operatorname{im}P_n$ we have to choose a suitable initial approximation $u_n^{(0)}\in\operatorname{im}P_{n-1}$ (depending on the level n of discretization).

Remark 4.1. *Since $u^*(x)$ satisfies the boundary conditions (2), one can easily see that there exist algebraic polynomials p_n^* of degree less than n such that*

$$\lim_{n\to\infty}\|\varphi p_n^* - u^*\|_\infty = 0.$$

For all what follows we choose $u_n^{(0)} = \varphi p_n^*$.

Lemma 4.2. *Let, for $a(x) := F_u(x,u^*(x))$, $b(x) := \mathbf{i}G_u(x,u^*(x))$, and $\widetilde{v}_k(x) = x^k$, all conditions of Theorem 3.4 be satisfied. If the functions $F_u:[-1,1]\times\mathbb{R}\longrightarrow\mathbb{R}$ and $G_u:[-1,1]\times\mathbb{R}\longrightarrow\mathbb{R}$ are continuous, then there exists an $n_0\in\mathbb{N}$ such that the operators $A_n^{\omega}:\mathbf{X}_n\longrightarrow\operatorname{im}P_n$ given by (18) are invertible for all $n\ge n_0$ and*

$$M_0 := \sup\left\{\|(A_n^{\omega})^{-1}\|_{\mathcal{L}(\operatorname{im}P_n,\mathbf{X}_n)} : n\ge n_0\right\} < \infty.$$

Proof. We use the equivalent formulation (12) of the collocation method (11) and write

$$\begin{aligned} B_n^\omega &= (M_n^\omega a_n P_n + M_n^\omega S P_n M_n^\omega b_n P_n)(M_n^\omega V^* P_n)^p + M_n^\omega K P_n \\ &= [M_n^\omega (a_n - a) P_n + M_n^\omega S P_n M_n^\omega (b_n - b) P_n](M_n^\omega V^* P_n)^p \\ &\qquad + (M_n^\omega a P_n + M_n^\omega S P_n M_n^\omega b P_n)(M_n^\omega V^* P_n)^p + M_n^\omega K P_n \,. \end{aligned}$$

We show that

$$\lim_{n\to\infty} \|[M_n^\omega (a_n - a) P_n + M_n^\omega S P_n M_n^\omega (b_n - b) P_n](M_n^\omega V^* P_n)^p\|_{\mathcal{L}(\mathbf{L}^2_\sigma, \mathbf{L}^2_\sigma)} = 0 \,. \quad (19)$$

Using the Gaussian rule we can prove (comp. [11, proof of Theor. 3.4]) that, for $g \in \mathbf{PC}$ and $v_n \in \operatorname{im} P_n$,

$$\|M_n^\omega g v_n\|_\sigma \le \text{const}\, \|g\|_\infty \|v_n\|_\sigma \,. \quad (20)$$

It remains to note that $\lim_{n\to\infty} \|a_n - a\|_\infty = 0$ and $\lim_{n\to\infty} \|b_n - b\|_\infty = 0$ in view of Remark 4.1 and the continuity of the functions F_u, $G_u : [-1,1] \times \mathbb{R} \longrightarrow \mathbb{R}$. This proves (19). Since, for

$$(M_n^\omega a P_n + M_n^\omega S P_n M_n^\omega b P_n)(M_n^\omega V^* P_n)^p + M_n^\omega K P_n \,,$$

Theorem 3.4 is in force, we can conclude the stability of B_n^ω. Finally, $A_n^\omega = B_n^\omega J_p P_n$, which proves the lemma. □

For our further considerations we need the following concept of weighted Besov spaces. Let ρ and τ be nonnegative real numbers and use the notation $v^{\rho,\tau}(x) := (1-x)^\rho (1+x)^\tau$. We denote by $\mathbf{C}_{\rho,\tau}$ the Banach space of all continuous functions $u : (-1,1) \longrightarrow \mathbb{C}$ for which the norm

$$\|u\|_{\infty,\rho,\tau} := \|v^{\rho,\tau} u\|_\infty := \sup\{v^{\rho,\tau}(x)\, |u(x)| : x \in (-1,1)\}$$

is finite. Let $E_n^{\rho,\tau}(u)$ be the best weighted uniform approximation of $u \in \mathbf{C}_{\rho,\tau}$ by algebraic polynomials of degree less than n, i.e.,

$$E_n^{\rho,\tau}(u) = \inf\{\|p_n - u\|_{\infty,\rho,\tau} : p_n \in \mathbf{\Pi}_n\}, \quad E_0^{\rho,\tau}(u) := \|u\|_{\infty,\rho,\tau} \,.$$

For real numbers $\gamma > 0$ and $q \ge 0$, we denote by $\mathbf{C}_{\rho,\tau}^{\gamma,q}$ the closed subspace of $\mathbf{C}_{\rho,\tau}$ constituted by the functions $u \in \mathbf{C}_{\rho,\tau}$ for which

$$\|u\|_{\rho,\tau,\gamma,q} = \sup\left\{ \frac{(n+1)^\gamma E_n^{\rho,\tau}(u)}{\log^q(n+2)} : n = 0,1,2,\ldots \right\}$$

is finite (comp., for example, [9]). The proof of the following lemma is taken from [15, Lemma 4.44].

Lemma 4.3. *Let $u \in \mathbf{C}_{\rho,\tau}^{\gamma,q}$, $u_n \in \mathbf{\Pi}_n$, and, for some constants $M > 0$ and $n_0 \ge 2$, $\|u - u_n\|_{\infty,\rho,\tau} \le M n^{-\gamma} \log^q n$, $n \ge n_0$. Then, for $0 < \varepsilon < \gamma$,*

$$\|u - u_n\|_{\rho,\tau,\varepsilon,q} \le c(M + \|u\|_{\rho,\tau,\gamma,q}) n^{\varepsilon-\gamma} \log^q n \,, \quad n \ge n_0 \,,$$

where the constant c does not depend on n, $\{u_n\}$, u, and ε.

Proof. If $m \geq n$ then

$$E_m^{\rho,\tau}(u-u_n) = E_m^{\rho,\tau}(u) \leq \|u\|_{\rho,\tau,\gamma,q}(m+1)^{-\gamma}\log^q(m+2)\,. \tag{21}$$

Clearly,

$$(m+1)^{-\gamma} \leq \text{const}\,(m+1)^{-\varepsilon}n^{\varepsilon-\gamma}\log^q n$$

(since $x^{\varepsilon-\gamma}\log^q(x+1)$ is decreasing for $x \geq x_0(\varepsilon)$). For $m < n$ we have

$$E_m^{\rho,\tau}(u-u_n) \leq \|u-u_n\|_{\infty,\rho,\tau} \leq Mn^{-\gamma}\log^q n \leq Mn^{\varepsilon-\gamma}(m+1)^{-\varepsilon}\log^q n\,. \tag{22}$$

The estimates (21) and (22) lead to

$$\frac{(m+1)^{\varepsilon}E_m^{\rho,\tau}(u-u_n)}{\log^q(m+2)} \leq \text{const}\left\{\begin{array}{ccc} \|u\|_{\rho,\tau,\gamma,q}\,n^{\varepsilon-\gamma}\log^q n & , & m \geq n\,, \\ \log^{-q}(m+2)Mn^{\varepsilon-\gamma}\log^q n & , & m < n\,, \end{array}\right.$$

and the lemma is proved. □

We say that a function $g : [-1,1]\times\mathbb{R} \longrightarrow \mathbb{R}$ is locally bounded if, for each $r > 0$,

$$\sup\left\{|g(x,u)| : (x,u) \in [-1,1]\times[-r,r]\right\} < \infty\,.$$

Lemma 4.4. *Assume that* $\varphi^{-1}u^* \in \mathbf{C}^{\gamma,q}_{\frac{1}{2},\frac{1}{2}}$ *and* $\varphi^{-1}G(\,.\,,u^*) \in \mathbf{C}^{\gamma,q-1}_{\frac{1}{2},\frac{1}{2}}$ *and that*

$$\|\varphi p_n^* - u^*\|_\infty \leq c_* E_{n-1}^{\frac{1}{2},\frac{1}{2}}(\varphi^{-1}u^*) \tag{23}$$

with a constant $c_* > 0$. *If the functions* $F_u\,,\ G_u : [-1,1]\times\mathbb{R} \longrightarrow \mathbb{R}$ *are locally bounded then there exist positive constants* $\widetilde{N}_0$ *and* ε *with* $\varepsilon < \gamma$ *such that the estimate*

$$\left\|H_n^\omega(u_n^{(0)},d^*)\right\|_\sigma \leq \widetilde{N}_0 n^{\varepsilon-\gamma}\log^q(n+1)$$

holds.

Proof. Obviously, $M_n^\omega H(u^*,d^*) = 0$ and, consequently,

$$\begin{aligned}
\left\|H_n^\omega(u_n^{(0)},d^*)\right\|_\sigma &\leq \|M_n^\omega[F(\,.\,,\varphi p_n^*) - F(\,.\,,u^*)]\|_\sigma \\
&\qquad + \|M_n^\omega S M_n^\omega G(\,.\,,\varphi p_n^*) - M_n^\omega S G(\,.\,,u^*)\|_\sigma \\
&\leq \|M_n^\omega[F(\,.\,,\varphi p_n^*) - F(\,.\,,u^*)]\|_\sigma \\
&\qquad + \|M_n^\omega S P_n\|_{\mathbf{L}^2_\sigma\to\mathbf{L}^2_\sigma}\,\|M_n^\omega[G(\,.\,,\varphi p_n^*) - G(\,.\,,u^*)]\|_\sigma \\
&\qquad + \left\|M_n^\omega S\varphi[L_n^\omega\varphi^{-1}G(\,.\,,u^*) - \varphi^{-1}G(\,.\,,u^*)]\right\|_\sigma\,.
\end{aligned}$$

Now we remark that in view of Lemma 3.2 we have

$$c_1 := \sup\left\{\|M_n^\omega\|_{\mathcal{L}(\mathbf{R}_0,\mathbf{L}^2_\sigma)} : n = 1,2,\ldots\right\} < \infty\,, \tag{24}$$

and therefore, taking into account the uniform boundedness of the operators $M_n^\omega S P_n : \mathbf{L}^2_\sigma \longrightarrow \mathbf{L}^2_\sigma$ (comp. [11, 20]), we get

$$\left\| H_n^\omega(u_n^{(0)}, d^*) \right\|_\sigma \le c_1 \left\| F(\,.\,, \varphi p_n^*) - F(\,.\,, u^*) \right\|_\infty + c_2 c_1 \left\| G(\,.\,, \varphi p_n^*) - G(\,.\,, u^*) \right\|_\infty$$

$$+ c_1 \left\| S\varphi [L_n^\omega \varphi^{-1} G(\,.\,, u^*) - \varphi^{-1} G(\,.\,, u^*)] \right\|_\infty . \qquad (25)$$

Since $\|\varphi p_n^*\|_\infty$ is bounded, say by c_0, and because of the local boundedness of F_u and G_u we get

$$\left\| F(\,.\,, \varphi p_n^*) - F(\,.\,, u^*) \right\|_\infty \le c_3 \left\| \varphi p_n^* - u^* \right\|_\infty \qquad (26)$$

and

$$\left\| G(\,.\,, \varphi p_n^*) - G(\,.\,, u^*) \right\|_\infty \le c_4 \left\| \varphi p_n^* - u^* \right\|_\infty , \qquad (27)$$

where

$$c_3 := \sup\{ |F_u(x, u)| : (x, u) \in [-1, 1] \times [-c_0, c_0] \} < \infty$$

and

$$c_4 := \sup\{ |G_u(x, u)| : (x, u) \in [-1, 1] \times [-c_0, c_0] \} < \infty .$$

To estimate the last term on the right-hand side of (25) we use the fact that $S\varphi I : \mathbf{C}^{\varepsilon,q}_{\frac{1}{2},\frac{1}{2}} \longrightarrow \mathbf{C}^{\varepsilon,q+1}_{0,0}$ is a continuous operator ([9, Remark 4.9]). It follows

$$\begin{aligned} &\left\| S\varphi [L_n^\omega \varphi^{-1} G(\,.\,, u^*) - \varphi^{-1} G(\,.\,, u^*)] \right\|_\infty \\ &\quad\le \left\| S\varphi [L_n^\omega \varphi^{-1} G(\,.\,, u^*) - \varphi^{-1} G(\,.\,, u^*)] \right\|_{0,0,\varepsilon,q+1} \\ &\quad\le c_5 \left\| L_n^\omega \varphi^{-1} G(\,.\,, u^*) - \varphi^{-1} G(\,.\,, u^*) \right\|_{\frac{1}{2},\frac{1}{2},\varepsilon,q} . \end{aligned}$$

Using [1, Theor. 4.1] we get

$$\left\| L_n^\omega \varphi^{-1} G(\,.\,, u^*) - \varphi^{-1} G(\,.\,, u^*) \right\|_{\infty,\frac{1}{2},\frac{1}{2}} \le c_6\, n^{-\gamma} \log^q n \left\| \varphi^{-1} G(\,.\,, u^*) \right\|_{\frac{1}{2},\frac{1}{2},\gamma,q-1} .$$

Consequently, the assumptions of Lemma 4.3 are fulfilled with

$$M = c_6 \left\| \varphi^{-1} G(\,.\,, u^*) \right\|_{\frac{1}{2},\frac{1}{2},\gamma,q-1} ,$$

and we get, for $0 < \varepsilon < \gamma$,

$$\left\| L_n^\omega \varphi^{-1} G(\,.\,, u^*) - \varphi^{-1} G(\,.\,, u^*) \right\|_{\frac{1}{2},\frac{1}{2},\varepsilon,q} \le c_7\, n^{\varepsilon-\gamma} \log^q (n+1) . \qquad (28)$$

Together with (26) and (27) and assumption (23) we conclude

$$\begin{aligned} \left\| H_n^\omega(u_n^{(0)}, d^*) \right\|_\sigma &\le c_8 \left\| \varphi p_n^* - u^* \right\|_\infty + c_7\, n^{\varepsilon-\gamma} \log^q (n+1) \\ &\le \widetilde{N}_0\, n^{\varepsilon-\gamma} \log^q (n+1) . \end{aligned}$$

□

As an immediate consequence of Lemma 4.2 and Lemma 4.4 we get the following corollary.

Corollary 4.5. *Under the assumptions of Lemmas 4.2 and 4.4 there exists an n_0 such that, for $n \geq n_0$,*

$$\left\| [(H_n^\omega)'(u_n^{(0)}, d^*)]^{-1} H_n^\omega(u_n^{(0)}, d^*) \right\|_{\mathbf{X}_n} \leq N_0 ,$$

where $N_0 = M_0 \widetilde{N}_0 n^{\varepsilon-\gamma} \log^q(n+1)$.

Now, let us make the following observation. Since $\|\widetilde{u}_n\|_\infty \leq \sqrt{\frac{2}{\pi}}$, for $u_n \in \operatorname{im} P_n$ and $x \in [-1,1]$, we can estimate

$$|u_n(x)| \leq \sqrt{\sum_{k=0}^{n-1} |\langle u_n, \widetilde{u}_n \rangle_\sigma|^2} \sqrt{\sum_{k=0}^{n-1} |\widetilde{u}_n(x)|^2} \leq \|u_n\|_\sigma \sqrt{\frac{2n}{\pi}} .$$

Consequently,

$$\|u_n\|_\infty \leq \sqrt{\frac{2n}{\pi}} \|u_n\|_\sigma \quad \text{for all} \quad u_n \in \operatorname{im} P_n . \tag{29}$$

With the help of this estimate we can prove the following lemma.

Lemma 4.6. *Assume that F_u and G_u satisfy the Hölder conditions*

$$|F_u(x,u^1) - F_u(x,u^2)| \leq c_F |u^1 - u^2|^\eta , \quad x \in [-1,1] , \; u^1, u^2 \in \mathbb{R} , \tag{30}$$

and

$$|G_u(x,u^1) - G_u(x,u^2)| \leq c_G |u^1 - u^2|^\eta , \quad x \in [-1,1] , \; u^1, u^2 \in \mathbb{R} , \tag{31}$$

for some constants $c_F, c_G > 0$ and $\eta \in (0,1]$. Then there exists a constant $\widetilde{C}_0$ such that, for all $u_n^1, u_n^2, u_n \in \operatorname{im} P_{n-1}$ and $d_1, d_2, d \in \mathbb{C}^p$,

$$\left\| \left[(H_n^\omega)'(u_n^1, d_1) - (H_n^\omega)'(u_n^2, d_2) \right] (u_n, d) \right\|_\sigma \leq \widetilde{C}_0 n^{\frac{\eta}{2}} \left\| u_n^1 - u_n^2 \right\|_\sigma^\eta \|u_n\|_\sigma .$$

Proof. Using (20), (30), (31), (29) and the uniform boundedness of the operators $M_n^\omega S P_n : \mathbf{L}_\sigma^2 \to \mathbf{L}_\sigma^2$ we can estimate

$$\begin{aligned}
&\left\| [(H_n^\omega)'(u_n^1, d_1) - (H_n^\omega)'(u_n^2, d_2)](u_n, d) \right\|_\sigma \\
&= \left\| M_n^\omega [F_u(.,u_n^1) - F_u(.,u_n^2) + S P_n M_n^\omega (G_u(.,u_n^1) - G_u(.,u_n^2))] u_n \right\|_\sigma \\
&\leq \left\| M_n^\omega [F_u(.,u_n^1) - F_u(.,u_n^2)] u_n \right\|_\sigma + \left\| M_n^\omega S P_n M_n^\omega [G_u(.,u_n^1) - G_u(.,u_n^2)] u_n \right\|_\sigma \\
&\leq c(c_F + c_G) \left\| u_n^1 - u_n^2 \right\|_\infty^\eta \|u_n\|_\sigma \\
&\leq \widetilde{C}_0 n^{\frac{\eta}{2}} \left\| u_n^1 - u_n^2 \right\| \sigma^\eta \|u_n\|_\sigma ,
\end{aligned}$$

and the lemma is proved. □

Corollary 4.7. *Under the assumptions of Lemma 4.6 we have the estimate*

$$\left\|(H_n^\omega)'(u_n,d_n)-(H_n^\omega)'(u_n^{(0)},d_n^{(0)})\right\|_{\mathbf{X}_n\to\mathbf{L}^2_\sigma}\le C_0\left\|(u_n,d_n)-(u_n^{(0)},d_n^{(0)})\right\|_{\mathbf{X}_n}^\eta$$

for all $(u_n,d_n),(u_n^{(0)},d_n^{(0)})\in\mathbf{X}_n$ *and* $C_0=\widetilde{C}_0 n^{\frac{\eta}{2}}$.

Theorem 4.8. *Let* (u^*,d^*) *be a solution of the equation* $H(u,d)=0$ *satisfying* $\varphi^{-1}u^*\in\mathbf{C}^{\gamma,q}_{\frac12,\frac12}$ *and* $\varphi^{-1}G(\,.\,,u^*)\in\mathbf{C}^{\gamma,q-1}_{\frac12,\frac12}$ *for some* $\gamma>\frac12$. *Assume that the functions* $a(x):=F_u(x,u^*(x))$ *and* $b(x):=\mathbf{i}G_u(x,u^*(x))$ *satisfy all conditions of Theorem 3.4 and that the functions* F_u, $G_u:[-1,1]\times\mathbb{R}\longrightarrow\mathbb{R}$ *are continuous and satisfy the Hölder conditions* (30) *and* (31). *Then, for all sufficiently large* n, *there exists an element* $(u_n^{(0)},d_n^{(0)})\in\mathbf{X}_n$ *such that equation* (17) *possesses a unique solution* $(\Delta u_n^{(m)},\Delta d_n^{(m)})$ *for all* $m=0,1,2,\ldots$ *The sequence* $\{(u_n^{(m)},d_n^{(m)})\}_{m=0}^\infty$ *converges in the norm of* $\mathbf{X}$ *to a solution* (u_n^*,d_n^*) *of equation* (15) *and*

$$\|(u_n^*,d_n^*)-(u^*,d^*)\|_{\mathbf{X}}\le\mathrm{const}\,n^{\varepsilon-\gamma}\log^q n\,,\tag{32}$$

where $0<\varepsilon<\gamma-\frac12$ *and the constant does not depend on* n.

Proof. Choose $u_n^{(0)}=\varphi p_n^*$ satisfying (23), $d_n^{(0)}=d^*$, and apply Lemma 4.2, Corollary 4.5, and Corollary 4.7 as well as Theorem 2.1 with $\mathbf{X}=\mathbf{X}_n$, $\mathbf{Y}=\operatorname{im}P_n$, and $P=H_n^\omega$. We get $h_0=\mathrm{const}\,n^{\eta(\frac12+\varepsilon-\gamma)}\log^{q\eta}(n+1)$, which is small enough if n is sufficiently large. Besides the existence of a solution (u_n^*,d_n^*) of equation (15) this gives the estimate

$$\left\|(u_n^*,d_n^*)-(u_n^{(0)},d_n^{(0)})\right\|_{\mathbf{X}}\le 2\,N_0=2\,M_0\widetilde{N}_0 n^{\varepsilon-\gamma}\log^q(n+1)\,,$$

from which the estimate (32) follows by taking into account that

$$\left\|u_n^{(0)}-u^*\right\|\sigma\le\mathrm{const}\left\|u_n^{(0)}-u^*\right\|_{\infty,\frac12,\frac12}.$$

The theorem is proved. □

To prove the convergence of the classical Newton method (17), we use Theorem 2.4. Since we have already proved the existence of a solution (u_n^*,d_n^*) of (15), it remains to check conditions (b) and (c) of Theorem 2.4. But (b) is already proved by Corollary 4.7 and (c) by Lemma 4.2. Hence we arrive at the following theorem.

Theorem 4.9. *Let* (u^*,d^*) *be a solution of the equation* $H(u,d)=0$ *satisfying* $\varphi^{-1}u^*\in\mathbf{C}^{\gamma,q}_{\frac12,\frac12}$ *and* $\varphi^{-1}G(\,.\,,u^*)\in\mathbf{C}^{\gamma,q-1}_{\frac12,\frac12}$ *for some* $\gamma>\frac12$. *Assume that the functions* $a(x):=F_u(x,u^*(x))$ *and* $b(x):=\mathbf{i}G_u(x,u^*(x))$ *satisfy all conditions of Theorem 3.4 and that the functions* F_u, $G_u:[-1,1]\times\mathbb{R}\longrightarrow\mathbb{R}$ *satisfy the Hölder conditions* (30) *and* (31). *Then, for all sufficiently large* n, *there exists a solution* (u_n^*,d_n^*) *of the discrete equation* (15) *converging to* (u^*,d^*), *where*

$$\|(u_n^*,d_n^*)-(u^*,d^*)\|_{\mathbf{X}}\le\mathrm{const}\,n^{\varepsilon-\gamma}\log^q(n+1)\,.$$

Moreover, for all sufficiently large n, *there exists a* $\delta>0$ *such that for each initial point* $(u_n^{(0)},d_n^{(0)})\in\mathbf{X}_n$ *with* $\|(u_n^{(0)},d_n^{(0)})-(u_n^*,d_n^*)\|\le\delta$ *the sequence*

$\{(u_n^{(m)}, d_n^{(m)})\}_{m=0}^{\infty}$ converges to (u_n^*, d_n^*) in the norm of $\mathbf{X}$. Furthermore, there exists a constant M such that

$$\|(u_n^{(m)}, d_n^{(m)}) - (u_n^*, d_n^*)\| \le M \|(u_n^{(m-1)}, d_n^{(m-1)}) - (u_n^*, d_n^*)\|^{1+\eta}$$

and

$$\|(u_n^{(m)}, d_n^{(m)}) - (u_n^*, d_n^*)\| \le M \frac{(1+\eta)^m - 1}{\eta} \delta^{(1+\eta)^m} .$$

5. Implementation

In this section we present some ideas for the effective solution of the linear equations in the Newton collocation methods investigated in the previous section. We will see that the solution of the (linear) collocation equations (11) is equivalent to the solution of a structured system of linear algebraic equations which can be realized with $O(n^2)$ complexity due to a principle presented in [3]. Moreover, using the afore-mentioned structure and fast trigonometric transforms it seems to be possible to reduce the complexity to an $O(n \log n)$ level.

Let us write (11) as the equivalent system

$$(M_n^\omega a P_n + M_n^\omega S P_n M_n^\omega b P_n) u_n + M_n^\omega v_n = M_n^\omega f , \tag{33}$$

$$\langle u_n, \widetilde{u}_j \rangle_\sigma = 0 , \quad j = n-p, \ldots, n-1 , \tag{34}$$

and search for an approximate solution $(u_n, v_n) \in \operatorname{im} P_n \times \mathbf{V}$ in the form

$$u_n = \sum_{k=1}^{n} \xi_{kn} \widetilde{\ell}_{kn}^\omega , \quad v_n = \sum_{k=1}^{p} \xi_{n+k,n} \widetilde{v}_{k-1} ,$$

where

$$\widetilde{\ell}_{kn}^\omega(x) = \frac{\varphi(x)}{\varphi(x_{kn}^\omega)} \ell_{kn}^\omega(x)$$

and

$$\ell_{kn}^\sigma(x) = \frac{T_n(x)}{(x - x_{kn}^\sigma) T_n'(x_{kn}^\sigma)} , \quad \ell_{kn}^\varphi(x) = \frac{U_n(x)}{(x - x_{kn}^\varphi) U_n'(x_{kn}^\varphi)} .$$

We start with computing the matrix representations $\mathbf{a}_n^\omega$ and $\mathbf{S}_n^\omega$ of $M_n^\omega a P_n : \operatorname{im} P_n \longrightarrow \operatorname{im} P_n$ and $M_n^\omega S P_n : \operatorname{im} P_n \longrightarrow \operatorname{im} P_n$, respectively, w.r.t. the basis $\left\{\widetilde{\ell}_{kn}^\omega\right\}_{k=1}^{n}$ of $\operatorname{im} P_n$.

Obviously, $\mathbf{a}_n^\omega = \operatorname{diag}[a(x_{1n}^\omega), \ldots, a(x_{nn}^\omega)]$. With the help of the relations

$$T_n'(x_{kn}^\sigma) = \sqrt{\frac{2}{\pi}} \frac{n(-1)^{k+1}}{\varphi(x_{kn}^\sigma)} \quad \text{and} \quad U_n'(x_{kn}^\varphi) = \sqrt{\frac{2}{\pi}} \frac{(n+1)(-1)^{k+1}}{[\varphi(x_{kn}^\varphi)]^2}$$

we get

$$\begin{aligned}\left(S\widetilde{\ell}^{\sigma}_{kn}\right)(x) &= \sqrt{\frac{\pi}{2}}\,\frac{(-1)^{k+1}}{n}\,\frac{1}{\pi\mathbf{i}}\int_{-1}^{1}\frac{\varphi(y)T_n(y)}{(y-x^{\sigma}_{kn})(y-x)}\,dy\\ &= \sqrt{\frac{\pi}{2}}\,\frac{(-1)^{k+1}}{n(x^{\sigma}_{kn}-x)}\,\frac{1}{\pi\mathbf{i}}\int_{-1}^{1}\left(\frac{1}{y-x^{\sigma}_{kn}}-\frac{1}{y-x}\right)\varphi(y)T_n(y)\,dy\end{aligned}$$

and, analogously,

$$\left(S\widetilde{\ell}^{\varphi}_{kn}\right)(x) = \sqrt{\frac{\pi}{2}}\,\frac{(-1)^{k+1}\varphi(x^{\varphi}_{kn})}{(n+1)(x^{\varphi}_{kn}-x)}\,\frac{1}{\pi\mathbf{i}}\int_{-1}^{1}\left(\frac{1}{y-x^{\varphi}_{kn}}-\frac{1}{y-x}\right)\varphi(y)U_n(y)\,dy\,.$$

Furthermore, due to the well-known relations

$$S\sigma T_{n+1} = -\mathbf{i}U_n\,,\quad S\varphi U_n = \mathbf{i}T_{n+1}\,,\quad n=0,1,2,\dots\,,$$

it follows that

$$\begin{aligned}\frac{1}{\pi\mathbf{i}}\int_{-1}^{1}\frac{\varphi(y)T_n(y)}{y-x}\,dy &= \frac{1}{\pi\mathbf{i}}\int_{-1}^{1}\frac{(1-y^2)\sigma(y)T_n(y)}{y-x}\,dy\\ &= -\mathbf{i}(1-x^2)U_{n-1}(x)-\frac{1}{\pi\mathbf{i}}\int_{-1}^{1}(y+x)\sigma(y)T_n(y)\,dy\\ &= -\mathbf{i}[\varphi(x)]^2U_{n-1}(x)\end{aligned}$$

and, for $j\neq k$,

$$\begin{aligned}\left(S\widetilde{\ell}^{\sigma}_{kn}\right)(x^{\sigma}_{jn}) &= \frac{1}{n\mathbf{i}}\,\frac{\varphi(x^{\sigma}_{kn})-(-1)^{j+k}\varphi(x^{\sigma}_{jn})}{x^{\sigma}_{kn}-x^{\sigma}_{jn}}\,,\\ \left(S\widetilde{\ell}^{\varphi}_{kn}\right)(x^{\varphi}_{jn}) &= \frac{1}{(n+1)\mathbf{i}}\,\frac{\varphi(x^{\varphi}_{kn})[1-(-1)^{j+k}]}{x^{\varphi}_{kn}-x^{\varphi}_{jn}}\,.\end{aligned}$$

Using

$$\frac{d}{dx}[(1-x^2)U_{n-1}(x)] = (1-x^2)U'_{n-1}(x)-2xU_{n-1}(x) = -nT_n(x)-xU_{n-1}(x)$$

and

$$T'_{n+1}(x) = (n+1)U_n(x)$$

we conclude

$$\left(S\widetilde{\ell}^{\sigma}_{kn}\right)(x^{\sigma}_{kn}) = -\frac{1}{n\mathbf{i}}\,\frac{x^{\sigma}_{kn}}{\varphi(x^{\sigma}_{kn})}\,,\quad \left(S\widetilde{\ell}^{\varphi}_{kn}\right)(x^{\varphi}_{kn}) = 0\,.$$

Thus, the entries of $\mathbf{S}^{\omega}_n$ outside the main diagonal are of the form

$$\frac{\alpha^{\omega}_k-\alpha^{\omega}_j}{x^{\omega}_{kn}-x^{\omega}_{jn}}\,\beta^{\omega}_k\,,$$

where

$$\alpha_k^\sigma = (-1)^k \varphi(x_{kn}^\sigma)\,,\ \beta_k^\sigma = \frac{(-1)^k}{n\mathbf{i}}\,,\ \alpha_k^\varphi = (-1)^k\,,\ \beta_k^\varphi = \frac{(-1)^k \varphi(x_{kn}^\varphi)}{(n+1)\mathbf{i}}\,,$$

which means that $\mathbf{S}_n^\omega$ is the product of a Löwner matrix (comp. [3]) and a diagonal matrix. Of course, the same remains true also for the matrix $\mathbf{a}_n^\omega + \mathbf{S}_n^\omega \mathbf{b}_n^\omega$ in (33). The columns $[\widetilde{v}_k(x_{jn}^\omega)]_{j=1}^n$, $k = 1, \ldots, p$, concerning the unknowns $\xi_{n+k,n}$, $k = 1, \ldots, p$, in (33) and the rows

$$\left[\langle \widetilde{\ell}_{1n}^\omega, \widetilde{u}_j \rangle_\sigma \ \cdots \ \langle \widetilde{\ell}_{nn}^\omega, \widetilde{u}_j \rangle_\sigma \ 0 \ \cdots \ 0\right]\,,\quad j = n-p, \ldots, n-1\,,$$

representing the p equations in (34) can be considered as a small rank perturbation to the structure. Note that, by virtue of the respective Gaussian rules,

$$\langle \widetilde{\ell}_{kn}^\sigma, \widetilde{u}_j \rangle_\sigma = \frac{1}{\varphi(x_{kn}^\sigma)} \langle \ell_{kn}^\sigma, \varphi^2 U_j \rangle_\sigma = \frac{\pi}{n}\, \varphi(x_{kn}^\sigma) U_j(x_{kn}^\sigma)$$

for $j < n-1$ and, due to $(1-x^2)U_{n-1}(x) = \frac{1}{2}\left[T_{n-1}(x) - T_{n+1}(x)\right]$,

$$\langle \widetilde{\ell}_{kn}^\sigma, \widetilde{u}_{n-1} \rangle_\sigma = \frac{T_{n-1}(x_{kn}^\sigma)}{2\varphi(x_{kn}^\sigma)}\,.$$

Moreover,

$$\langle \widetilde{\ell}_{kn}^\varphi, \widetilde{u}_j \rangle_\sigma = \frac{1}{\varphi(x_{kn}^\varphi)} \langle \ell_{kn}^\varphi, U_j \rangle_\varphi = \frac{\pi}{n+1}\, \varphi(x_{kn}^\varphi) U_j(x_{kn}^\varphi)\,,\quad j \le n-1\,.$$

Let us denote the $n \times n$ matrix $\mathbf{a}_n^\omega + \mathbf{S}_n^\omega \mathbf{b}_n^\omega$ by $\mathbf{A} = [a_{jk}]_{j,k=1}^n$. For $m = 1, 2, \ldots, n$, set $\mathbf{A}_m = [a_{jk}]_{j,k=1}^m$. The matrix $\mathbf{A}$ admits a so-called small rank diagonal reduction (cf. [3, Part II, Sect. 3.4]). Indeed,

$$\mathbf{A}\, \mathrm{diag}\, [x_{jn}^\omega]_{j=1}^n - \mathrm{diag}\, [x_{jn}^\omega]_{j=1}^n \mathbf{A} = \mathbf{g}_1 \mathbf{f}_1^T + \mathbf{g}_2 \mathbf{f}_2^T \tag{35}$$

with

$$\mathbf{g}_1 = [1]_{j=1}^n\,,\ \mathbf{f}_1 = [\alpha_j^\omega \beta_j^\omega b(x_{jn}^\omega)]_{j=1}^n\,,\ \mathbf{g}_2 = [-\alpha_j^\omega]_{j=1}^n\,,\ \mathbf{f}_2 = [\beta_j^\omega b(x_{jn}^\omega)]_{j=1}^n\,.$$

Let us consider a more general situation. Assume that $\mathbf{A} \in \mathbb{C}^{n\times n}$ permits a small rank diagonal reduction of the form

$$\mathbf{A}\, \mathrm{diag}\, [\gamma_j]_{j=1}^n - \mathrm{diag}\, [\delta_j]_{j=1}^n \mathbf{A} = \varepsilon \alpha^T - \beta \zeta^T \tag{36}$$

with $\alpha = [\alpha_j]_{j=1}^n$, $\beta = [\beta_j]_{j=1}^n$, $\varepsilon = [\varepsilon_j]_{j=1}^n$, $\zeta = [\zeta_j]_{j=1}^n$. For example, in case $\gamma_j \ne \delta_k$ for all $j, k = 1, \ldots, n$, we have

$$a_{jk} = \frac{\varepsilon_j \alpha_k - \beta_j \zeta_k}{\gamma_k - \delta_j}\,,\quad j, k = 1, \ldots, n\,.$$

In the case we are interested in here (cf. (35)) this structure is disturbed along the diagonal. But, the basis for the following considerations is only the small rank reduction (36). We use the ideas of [3, Part II, Sect. 3.4] to construct an $O(n^2)$ complexity algorithm for solving the system

$$\mathbf{A}\mathbf{x} = \mathbf{y} \tag{37}$$

for given $\mathbf{y} = [y_j]_{j=1}^n \in \mathbb{C}^n$. We assume that $\mathbf{A}_m$ is nonsingular for all $m = 1, 2, \ldots, n$. The following lemma is well known and is even true independently of the structure of the matrix $\mathbf{A}$.

Lemma 5.1. *If* $\mathbf{x}_0^{m-1}$ *and* $\widehat{\mathbf{x}}^{m-1}$ *are the solutions of*

$$\mathbf{A}_{m-1}\mathbf{x}_0^{m-1} = [a_{jm}]_{j=1}^{m-1} \quad \text{and} \quad \mathbf{A}_{m-1}\widehat{\mathbf{x}}^{m-1} = [y_j]_{j=1}^{m-1},$$

respectively, then $\widehat{\mathbf{x}}^m$ *is given by the formula*

$$\widehat{\mathbf{x}}^m = \begin{bmatrix} \widehat{\mathbf{x}}^{m-1} \\ 0 \end{bmatrix} + \frac{\eta_m}{\chi_m} \begin{bmatrix} \mathbf{x}_0^{m-1} \\ -1 \end{bmatrix},$$

where $\eta_m = y_m - \sum_{k=1}^{m-1} a_{mk}\widehat{x}_k^{m-1}$ *and* $\chi_m = \sum_{k=1}^{m-1} a_{mk} x_{0k}^{m-1} - a_{mm}$.

In the following we show that the vectors $\mathbf{x}_0^m$, $m = 1, \ldots, n-1$, can be found with the help of the solutions $\mathbf{x}_1^m = [x_{1j}^m]_{j=1}^m$ and $\mathbf{x}_2^m = [x_{2j}^m]_{j=1}^m$ of the equations

$$\mathbf{A}_m\mathbf{x}_1^m = [\varepsilon_j]_{j=1}^m \quad \text{and} \quad \mathbf{A}_m\mathbf{x}_2^m = [-\beta_j]_{j=1}^m, \tag{38}$$

respectively, which can be recursively determined with the help of Lemma 5.1. For this, let $\mathbf{D}_m(\gamma) := \operatorname{diag}[\gamma_j]_{j=1}^m$,

$$\mathbf{F}_m := \begin{bmatrix} \alpha_1 & \zeta_1 \\ \vdots & \vdots \\ \alpha_m & \zeta_m \end{bmatrix}, \quad \text{and} \quad \mathbf{G}_m := \begin{bmatrix} \varepsilon_1 & -\beta_1 \\ \vdots & \vdots \\ \varepsilon_m & -\beta_m \end{bmatrix}.$$

Then (36) leads to

$$\mathbf{A}_m\mathbf{D}_m(\gamma) - \mathbf{D}_m(\delta)\mathbf{A}_m = \mathbf{G}_m\mathbf{F}_m^T, \quad m = 1, \ldots, n. \tag{39}$$

We denote by $\mathbf{I}_m$ the identity matrix of order m. Let

$$\mathbf{X}_m := [x_{kj}^m]_{j=1,k=1}^{m\ \ 2} \quad \text{and} \quad \mathbf{K}_m = \gamma_{m+1}\mathbf{I}_m - \mathbf{D}_m(\gamma) + \mathbf{X}_m\mathbf{F}_m^T.$$

From (39) and (38) we obtain that

$$\mathbf{A}_m\mathbf{K}_m^{-1} = [\gamma_{m+1}\mathbf{I}_m - \mathbf{D}_m(\delta)]^{-1}\mathbf{A}_m$$

and hence

$$\mathbf{A}_m\mathbf{K}_m^{-1}\mathbf{X}_m\mathbf{F}_{m+1}^T\mathbf{e}_{m+1} = [a_{j,m+1}]_{j=1}^m, \quad \text{i.e.} \quad \mathbf{x}_0^m = \mathbf{K}_m^{-1}\mathbf{X}_m\mathbf{F}_{m+1}^T\mathbf{e}_{m+1},$$

where $\mathbf{e}_{m+1} = [\delta_{j,m+1}]_{j=1}^{m+1}$. The inverse of $\mathbf{K}_m$ admits the representation

$$\mathbf{K}_m^{-1} = [\gamma_{m+1}\mathbf{I}_m - \mathbf{D}_m(\gamma)]^{-1}\left(\mathbf{I}_m - \mathbf{X}_m\mathbf{R}_m\mathbf{F}_m^T[\gamma_{m+1}\mathbf{I}_m - \mathbf{D}_m(\gamma)]^{-1}\right)$$

with $\mathbf{R}_m = \mathbf{I}_2 + \mathbf{F}_m^T[\gamma_{m+1}\mathbf{I}_m - \mathbf{D}_m(\gamma)]^{-1}\mathbf{X}_m$. Putting all these things together we arrive, for $m < n$, at the formula

$$x_{0k}^m = \frac{\rho_{m2}x_{1k}^m + \rho_{m1}x_{2k}^m}{\rho_m(\gamma_{m+1} - \gamma_k)}, \quad k = 1, \ldots, m, \tag{40}$$

where

$$\rho_m = \alpha_{m1}\zeta_{m2} - \alpha_{m2}\zeta_{m1},$$

$$\rho_{m1} = \zeta_{m+1}\alpha_{m1} - \alpha_{m+1}\zeta_{m1}\,, \quad \rho_{m2} = \alpha_{m+1}\zeta_{m2} - \zeta_{m+1}\alpha_{m2}\,,$$

$$\alpha_{m1} = 1 + \sum_{k=1}^{m} \frac{\alpha_k x_{1k}^m}{\gamma_{m+1} - \gamma_k}\,, \qquad \alpha_{m2} = \sum_{k=1}^{m} \frac{\alpha_k x_{2k}^m}{\gamma_{m+1} - \gamma_k}\,,$$

$$\zeta_{m1} = \sum_{k=1}^{m} \frac{\zeta_k x_{1k}^m}{\gamma_{m+1} - \gamma_k}\,, \qquad \zeta_{m2} = 1 + \sum_{k=1}^{m} \frac{\zeta_k x_{2k}^m}{\gamma_{m+1} - \gamma_k}\,.$$

Thus, since in the **case** $p = 0$ the matrix of the system (33), (34) admits a reduction of the form (36) (cf. (35)), one can apply Lemma 5.1 and (40) alternately to find the solution $\mathbf{x} = \widehat{\mathbf{x}}^n = [\xi_{jn}]_{j=1}^n$ of (33) with $O(n^2)$ complexity.

In the **case** $p = 1$ the matrix of (33), (34) is of the form

$$\begin{bmatrix} & & & v_{01} \\ & \mathbf{A}_n & & \vdots \\ & & & v_{0n} \\ s_{1,n-1} & \cdots & s_{n,n-1} & 0 \end{bmatrix},$$

where $v_{kj} = \widetilde{v}_k(x_{jn}^\omega)$ and $s_{kj} = \langle \widetilde{\ell_{kn}^\omega}, \widetilde{u}_j \rangle_\sigma$. Here we can proceed as follows:

1. Solve $\mathbf{A}_n \widehat{\mathbf{x}}^n = [f(x_{jn}^\omega)]_{j=1}^n$ and $\mathbf{A}_n \widetilde{\mathbf{x}}^n = [\widetilde{v}_0(x_{jn}^\omega)]_{j=1}^n$ with the help of Lemma 5.1 and (40).
2. Use Lemma 5.1 to compute

$$\mathbf{x} = [\xi_{jn}]_{j=1}^{n+1} = \begin{bmatrix} \widehat{\mathbf{x}}^n \\ 0 \end{bmatrix} + \frac{\eta_{n+1}}{\chi_{n+1}} \begin{bmatrix} \widetilde{\mathbf{x}}^n \\ -1 \end{bmatrix},$$

where $\eta_{n+1} = -\sum_{k=1}^{n} \langle \widetilde{\ell_{kn}^\omega}, \widetilde{u}_{n-1} \rangle_\sigma \widehat{x}_k^n$ and $\chi_{n+1} = \sum_{k=1}^{n} \langle \widetilde{\ell_{kn}^\omega}, \widetilde{u}_{n-1} \rangle_\sigma \widetilde{x}_k^n$.

In the **case** $p > 1$ the effective solution of

$$\begin{bmatrix} & & & v_{01} & \cdots & v_{p-1,1} \\ & \mathbf{A}_n & & \vdots & & \vdots \\ & & & v_{0,n} & \cdots & v_{p-1,n} \\ s_{1,n-1} & \cdots & s_{n,n-1} & 0 & \cdots & 0 \\ \vdots & & \vdots & \vdots & & \vdots \\ s_{1,n-p} & \cdots & s_{n,n-p} & 0 & \cdots & 0 \end{bmatrix} \begin{bmatrix} \xi_{1n} \\ \vdots \\ \xi_{nn} \\ \xi_{n+1,n} \\ \vdots \\ \xi_{n+p,n} \end{bmatrix} = \begin{bmatrix} f(x_{1n}^\omega) \\ \vdots \\ f(x_{nn}^\omega) \\ 0 \\ \vdots \\ 0 \end{bmatrix}$$

can be performed recursively. For this, set $\mathbf{y}_p = \begin{bmatrix} \mathbf{y}_{p-1} \\ 0 \end{bmatrix}$ with $\mathbf{y}_0 = [f(x_{jn}^\omega)]_{j=1}^n$ and $\mathbf{v}_p = [v_{p-1,j}]_{j=1}^n$.

1. Solve

$$\mathbf{A}_{n+p-1} \widehat{\mathbf{x}}^{n+p-1} = \mathbf{y}_{p-1} \quad \text{and} \quad \mathbf{A}_{n+p-1} \widetilde{\mathbf{x}}^{n+p-1} = \begin{bmatrix} \mathbf{v}_p \\ \mathbf{0} \end{bmatrix} \in \mathbb{C}^{n+p-1}.$$

2. Compute

$$\mathbf{x} = [\xi_{jn}]_{j=1}^{n+p} = \begin{bmatrix} \widehat{\mathbf{x}}^{n+p-1} \\ 0 \end{bmatrix} + \frac{\eta_{n+p}}{\chi_{n+p}} \begin{bmatrix} \widetilde{\mathbf{x}}^{n+p-1} \\ -1 \end{bmatrix},$$

where

$$\eta_{n+p} = -\sum_{k=1}^{n} \langle \widetilde{\ell}_{kn}^{\omega}, \widetilde{u}_{n-p} \rangle_\sigma \widehat{x}_k^{n+p-1} \qquad \text{and} \qquad \eta_{n+p} = \sum_{k=1}^{n} \langle \widetilde{\ell}_{kn}^{\omega}, \widetilde{u}_{n-p} \rangle_\sigma \widetilde{x}_k^{n+p-1}.$$

Consequently, since p is a fixed number independent of n, we are able to save the $O(n^2)$ complexity for solving one linear system in a Newton step also for $p > 0$.

Now, let us present an idea for a further reduction of this complexity. For the sake of simplicity we restrict ourselves to the case $p = 0$ and $\omega = \varphi$. We recall that the matrix $\mathbf{A} = \mathbf{A}_n$ can be written in the form

$$\mathbf{A}_n = \mathbf{a}_n^{\varphi} + \mathbf{S}_n^{\varphi} \mathbf{b}_n^{\varphi} = \mathbf{a}_n^{\varphi} + \mathbf{S}_n^0 \mathbf{D}_n(\beta) \mathbf{b}_n^{\varphi},$$

where

$$\mathbf{S}_n^0 = \left[\frac{\alpha_k^{\varphi} - \alpha_j^{\varphi}}{x_{kn}^{\varphi} - x_{jn}^{\varphi}} \right]_{j,k=1}^{n} = \mathbf{C}_n \mathbf{D}_n(\alpha) - \mathbf{D}_n(\alpha) \mathbf{C}_n$$

and where $\mathbf{C}_n$ denotes the Cauchy matrix

$$\mathbf{C}_n = \left[\frac{1}{x_{kn}^{\varphi} - x_{jn}^{\varphi}} \right]_{j,k=1}^{n}$$

with zero entries along the main diagonal. If we define the discrete sine transform

$$\mathbf{S}_n^1 = \left[\sin \frac{jk\pi}{n+1} \right]_{j,k=1}^{n}$$

and if we set $\Gamma_n = \operatorname{tridiag}[-\mathbf{c}, \mathbf{0}, \mathbf{c}]$ with $\mathbf{c} = \frac{1}{2}[3, 5, \ldots, 2n-1]$ and $\mathbf{D}_n = \operatorname{diag}\left[(-1)^k \sin \frac{k\pi}{n+1}\right]_{k=1}^{n}$ then, due to [4, Theorem 5.5] (see also [5]), $\mathbf{C}_n$ admits the representation

$$\mathbf{C}_n = \frac{1}{n+1} \mathbf{D}_n^{-1} \mathbf{S}_n^1 \Gamma_n \mathbf{S}_n^1 \mathbf{D}_n.$$

Thus, since the discrete sine transform can be realized with $O(n \log n)$ complexity (see, for example, [18, 19]), the matrix-vector multiplication $\mathbf{A}_n \mathbf{x}$ can also be done with $O(n \log n)$ complexity. So, it seems to be of interest to apply a suitable iterative method for solving (37), for example the GMRES-method. In [17] it is shown that k steps of GMRES require $k(k+2)n + k\,n_{\mathbf{A}}$ multiplications, where $n_{\mathbf{A}}$ denotes the number of multiplications needed for the multipication $\mathbf{A}_n \mathbf{x}$, which can be replaced by $O(n \log n)$ in the situation under consideration here. It is an advantage of GMRES that the residual norm can be computed at each step with no extra cost (see [17, Prop. 1]), so that the number k of GMRES-steps can be chosen minimal depending on the wanted accuracy without increasing the complexity.

6. Application to a free surface seepage problem

As an example we consider a two-dimensional steady seepage problem from a channel through a homogeneous and isotropic porous medium underlain by a drain at a finite depth (see the picture).

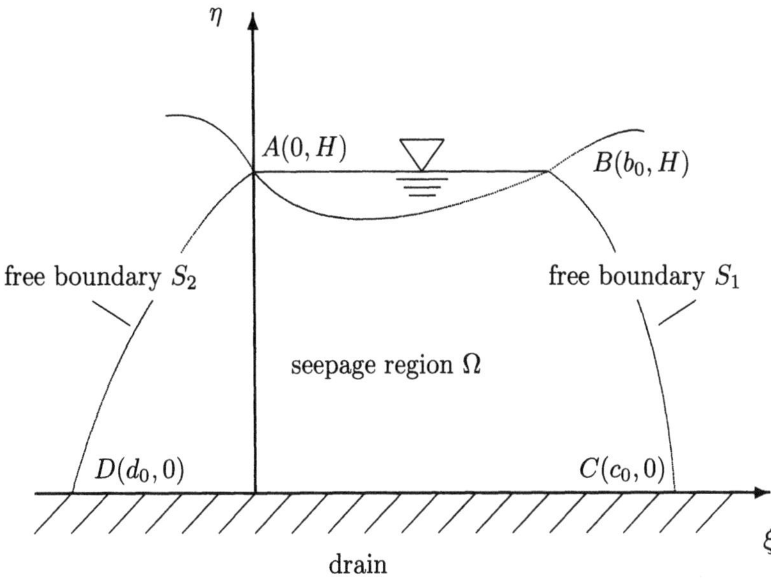

In [16] and [8] the problem is transformed into a singular integral equation problem which can be written in the form (comp. [8, (1.5)–(1.8)])

$$u(x)+\frac{1}{\pi}\int_{-1}^{1}\frac{G(y,u(y))}{y-x}\,dy+\delta_0+\delta_1 x=f(x)\,,\quad -1<x<1\,, \tag{41}$$

where

$$G(x,u)=H-g\left(\frac{b_0(1-x)}{2}+u\right),\quad f(x)=R(x)-\frac{b_0(1-x)}{2}+\frac{H}{\pi}\ln\frac{1-x}{1+x},$$

$$R(x)=\frac{1}{\pi}\int_1^{\frac{1}{\kappa}}r(\kappa,t)\left[\frac{1}{t-x}-\frac{1}{t+x}\right]dt\,,\quad 0<\kappa<1\,,$$

$$r(\kappa,t)=H-\frac{H}{\mathcal{K}'}\int_1^t\frac{ds}{\sqrt{(s^2-1)(1-\kappa^2s^2)}}\,,\quad 1<t<\frac{1}{\kappa}\,,$$

$$\mathcal{K}'=\mathcal{E}(\sqrt{1-\kappa^2},1)\,,\quad \mathcal{E}(\kappa,t)=\int_0^t\frac{ds}{\sqrt{(1-s^2)(1-\kappa^2s^2)}}\,.$$

The function $g : [0, b_0] \longrightarrow [0, H]$ with $g(0) = g(b_0) = H$ describes the shape of the channel, is assumed to be differentiable with

$$|g'(\xi_1) - g'(\xi_2)| \le c_g |\xi_1 - \xi_2|^\eta, \quad \xi_1, \xi_2 \in [0, b_0], \tag{42}$$

for some positive constants c_g and $\eta \in (0, 1]$, and is defined outside the interval $[0, b_0]$ by

$$g(\xi) = \begin{cases} g'(0)\xi + H & , \quad \xi < 0, \\ g'(b_0)(\xi - b_0) + H & , \quad \xi > b_0. \end{cases}$$

The unknown function $u : [-1, 1] \longrightarrow \mathbb{R}$ has to satisfy the boundary conditions

$$u(-1) = u(1) = 0. \tag{43}$$

Beside the unknown constants $\delta_0, \delta_1 \in \mathbb{R}$ the parameter $\kappa \in (0, 1)$ is also looked for, where u, δ_0, δ_1 have to be considered as functions of κ, and κ is determined by the condition

$$\delta_1(\kappa) = 0. \tag{44}$$

Let $\ell_1 = -\min\{g'(\xi) : \xi \in [0, b_0]\}$, $\ell_2 = \max\{g'(\xi) : \xi \in [0, b_0]\}$, and assume that

$$\ell_1 \ell_2 < 1. \tag{45}$$

In [8, Sect. 1] it is shown that, for each fixed $\kappa \in (0, 1)$, the problem (41), (43) has a solution

$$u^*(x) = \int_{-1}^{x} v(y)\, dy \quad \text{with} \quad v \in \bigcap_{1<p<p_0} \mathbf{L}^p(-1, 1) \tag{46}$$

for some $p_0 > 1$. From [13, Theorem 4.1] we infer that $p_0 \ge 2$. In [8] numerical results are presented which are based on the application of the classical collocation method for Cauchy singular integral equations and on the results of [7]. The advantage of the method presented there is that no further conditions occur to guarantee the stability of the operator sequences of the linearized discrete equations. But as a disadvantage we have an expensive preprocessing since the parameters of Gaussian rules w.r.t. generalized Jacobi weights are needed. In what follows we discuss the applicability of the Newton collocation methods proposed here, which are very cheap concerning the preprocessing. The main conditions, which ensure the applicability of these collocation methods based on Chebyshev nodes, are the conditions of Theorem 3.4 for the stable sovability of the sequences of the linearized discrete equations (15) (see Lemma 4.2).

Thus, let us consider the conditions of Theorem 3.4 related to the functions

$$a(x) = F_u(x, u^*(x)) \equiv 1$$

and

$$b(x) = \mathbf{i} G_u(x, u^*(x)) = -\mathbf{i} g'\left(\frac{b_0(1 - x)}{2} + u^*(x)\right).$$

For this define

$$c(x) = \frac{1 + b(x)}{1 - b(x)}, \quad -1 \le x \le 1,$$

and

$$\mathbf{c}(x,\mu)=\begin{cases} c(x) & , \quad x\in(-1,1)\,,\\ c(1)[1-r(\mu)]+r(\mu) & , \quad \mu\in[0,1], x=1\,,\\ 1-r(\mu)+c(-1)r(\mu) & , \quad \mu\in[0,1], x=-1\,,\end{cases}$$

where $r(\mu)=\sin\frac{\pi\mu}{2}\exp\left(\frac{\mathrm{i}\pi(\mu-1)}{2}\right)$. Note that $\{z_1[1-r(\mu)]+z_2r(\mu):\mu\in[0,1]\}$ describes the half-circle line from z_1 to z_2 that lies to the right of the straight line from z_1 to z_2. Let $\operatorname{wind}\mathbf{c}(x,\mu)$ denote the winding number of the closed curve $\{\mathbf{c}(x,\mu):x\in[-1,1],\mu\in[0,1]\}$ (with its natural orientation) with respect to the origin 0. Then, due to [2, Theor. IX.4.1], the operator $A=I+SbI:\mathbf{L}^2_\sigma\longrightarrow\mathbf{L}^2_\sigma$ is Fredholm if and only if $\mathbf{c}(x,\mu)\neq 0$ for all $x\in[-1,1]$ and $\mu\in[0,1]$, in which case $\operatorname{ind}A=-\operatorname{wind}\mathbf{c}(x,\mu)$ and A is one-sided invertible.

Since $b:[-1,1]\longrightarrow\mathrm{i}\mathbb{R}$ is continuous, the function values $c(x)$ cannot meet the point -1, so that the index of the operator $(I+SbI)(V^*)^2+K$ is equal to zero if and only if

$$\operatorname{Re}c(\pm1)<0 \quad\text{and}\quad \operatorname{Im}c(1)>0,\ \operatorname{Im}c(-1)<0, \tag{47}$$

which means

$$g'(0)<-1 \quad\text{and}\quad g'(b_0)>1 \tag{48}$$

in contradiction to (45). Since (45) was only needed to ensure the existence of a solution with property (46) we can refer to [12], where the existence of a solution u of (41), (43) belonging to all $\mathbf{L}^p$, $p\in(1,\infty)$, is proved without condition (45). But if condition (47) is fulfilled then, as already mentioned in the proof of Theorem 3.4, the operator $(I-SbI)V^2$ cannot be invertible in $\mathbf{L}^2_\sigma$. Thus, in case $\omega=\varphi$ (and $a(x)\equiv 1$, $b:[-1,1]\longrightarrow\mathrm{i}\mathbb{R}$ continuous) the conditions of Theorem 3.4 cannot be fulfilled.

For this reason we restrict ourselves to the case $\omega=\sigma$ and assume (48). The operator $J_\sigma^{-1}(J_\sigma+\mathrm{i}V^*J_\sigma^{-1}bJ_\sigma)V^2$ has index 0 if the operator $bI-\mathrm{i}J_\sigma V$ has index 1. Using $V=\psi I-\mathrm{i}\varphi S$, $J_\sigma=\varphi I-\mathrm{i}\psi S$, and $S\varphi S=\varphi I-K_0$, where $K_0u=\frac{1}{\sqrt{2\pi}}\langle u,\widetilde{u}_0\rangle_\sigma$ (see [10, 11]) is a compact operator, we get

$$bI-\mathrm{i}J_\sigma V=bI-S-\mathrm{i}\psi K_0-\psi K_1$$

with $K_1=S\psi-\psi S:\mathbf{L}^2_\sigma\longrightarrow\mathbf{L}^2_\sigma$ compact. The corresponding symbol function $c(x)$ of the operator $bI-S$ is equal to $c(x)=\dfrac{b(x)-1}{b(x)+1}$ and cannot meet the point 1. Moreover, $\operatorname{Re}c(\pm1)>0$, $\operatorname{Im}c(-1)<0$, and $\operatorname{Im}c(1)>0$, so that $\operatorname{wind}\mathbf{c}(x,\mu)=-1$. Consequently, the operator $J_\sigma^{-1}(J_\sigma+\mathrm{i}V^*J_\sigma^{-1}bJ_\sigma)V^2:\mathbf{L}^2_\sigma\longrightarrow\mathbf{L}^2_\sigma$ is Fredholm with index 0. What remains to be shown is that the null spaces of the operators $(I+SbI)(V^*)^2+K:\mathbf{L}^2_\sigma\longrightarrow\mathbf{L}^2_\sigma$ and $J_\sigma^{-1}(J_\sigma+\mathrm{i}V^*J_\sigma^{-1}bJ_\sigma)V^2$ are trivial. For this, in concrete examples, one can use the method presented in [10, Sect. 4].

Summarizing we can state that, for the example considered here, Theorems 4.8 and 4.9 can only be employed for collocation w.r.t. Chebyshev nodes of first kind and under the assumption that (48) is fulfilled. The reason for this seems to be that the weights in the Hilbert space $\mathbf{L}^2_\sigma$ and in the ansatz functions $\widetilde{u}_k(x)$

are not chosen carefully enough with respect to the data in the integral equation (41). A way to overcome these difficulties will be discussed in a second part to the present paper.

References

[1] M.R. Capobianco, P. Junghanns, U. Luther and G. Mastroianni, *Weighted uniform convergence of the quadrature method for Cauchy singular integral wquations,* in: Singular Integral Operators and Related Topics, ed. by A. Böttcher and I. Gohberg, Operator Theory, Advances and Applications, Vol. 90, Birkhäuser Verlag, **1996**, 153–181.

[2] I. Gohberg and N. Krupnik, *One-dimensional Linear Singular Integral Equations,* Birkhäuser Verlag, Basel, Boston, Berlin, **1992**.

[3] G. Heinig and K. Rost, *Algebraic Methods for Toeplitz-like Matrices and Operators,* Akademie Verlag, Berlin, and Birkhäuser Verlag, Boston, **1984**.

[4] G. Heinig and K. Rost, *Representations of Cauchy matrices with Chebyshev nodes using trigonometric transforms,* in: Structured Matrices, Recent Advances and Applications, ed. by D.A. Bini, E. Tyrtyshnikov, P. Yalamov, Advances in the Theory of Computational Mathematics, Vol. 4, Nova Science Publishers, **2001**, 135–147.

[5] T. Huckle, *Cauchy matrices and iterative methods for Toeplitz matrices*, Proceedings of SPIE-95, **2563** (1995), 281–292.

[6] P. Junghanns, *Numerical analysis of Newton projection methods for nonlinear singular integral equations,* J. Comp. Appl. Math., **55** (1994), 145–163.

[7] P. Junghanns, *On the numerical solution of nonlinear singular integral equations,* ZAMM, **76** (1996), 152–166.

[8] P. Junghanns, *Numerical solution of a free surface seepage problem from nonlinear channel*, Applicable Analysis, **63** (1996), 87–110.

[9] P. Junghanns and U. Luther, *Cauchy singular integral equations in spaces of continuous functions and methods for their numerical solution,* J. Comp. Appl. Math., **77** (1997), 201–237.

[10] P. Junghanns and K. Müller, *A collocation method for nonlinear Cauchy singular integral equations,* J. Comp. Appl. Math., **115** (2000), 283–300.

[11] P. Junghanns, S. Roch and B. Silbermann, *Collocation methods for systems of Cauchy singular integral equations on an interval,* Comp. Techn., **6** (2001), 88–124.

[12] P. Junghanns, G. Semmler, U. Weber and E. Wegert, *Non-linear singular integral equations on a finite interval*, Math. Meth. Appl. Sci., **24** (2001).

[13] P. Junghanns and U. Weber, *On the solvability of nonlinear singular integral equations*, Zeitschr. Anal. Anw., **12** (1993), 683–698.

[14] P. Junghanns and U. Weber, *Local theory of projection methods for Cauchy singular integral equations on an interval,* in: Boundary Integral Methods – Numerical and Mathematical Aspects, Comp. Mech. Publ., Southampton, **1999**, 219–256.

[15] U. Luther, *Generalized Besov Spaces and Cauchy Singular Integral Equations,* PhD-Thesis, TU Chemnitz, **1998**.

[16] D. Oestreich, *Singular integral equations applied to free surface seepage from nonlinear channels*, Math. Meth. Appl. Sci., **12** (1990), 209–219.

[17] Y. Saad and M.H. Schultz, *GMRES: A generalized minimal residual algorithm for solving nonsymmetric linear systems*, SIAM J. Sci. Stat. Comput., **7** (1986), 856–869.

[18] G. Steidl and M. Tasche, *A polynomial approach to fast algorithms for discrete Fourier-cosine and Fourier-sine transforms*, Math. Comp., **56** (1991), 281–296.

[19] M. Tasche, *Fast algorithms for discrete Chebyshev-Vandermonde transforms and applications*, Numerical Algorithms, **5** (1993), 453–464.

[20] U. Weber, *Weighted Polynomial Approximation Methods for Cauchy Singular Integral Equations in the Non-Periodic Case*, PhD-Thesis, Logos Verlag, Berlin, **2000**.

[21] E. Zeidler, *Nonlinear Functional Analysis and its Applications, I: Fixed-Point Theorems*, Springer Verlag, New York, **1986**.

P. Junghanns, K. Müller and K. Rost
Department of Mathematics
Chemnitz University of Technology
D-09107 Chemnitz, Germany
e-mail: `peter.junghanns@mathematik.tu-chemnitz.de`
e-mail: `karla.rost@mathematik.tu-chemnitz.de`

Received: 20 December 2001

Operator Theory:
Advances and Applications, Vol. 135, 235–247

Local Method for Nonlocal Operators on Banach Spaces

Yuri Karlovich and Bernd Silbermann

Abstract. The paper is devoted to a general local method for studying the Fredholmness in Banach algebras of bounded linear operators with shifts. This method is applicable to Banach algebras of singular integral operators with piecewise continuous coefficients and discrete subexponential groups of piecewise smooth shifts acting topologically free on composed contours.

1. Introduction

The paper is devoted to the presentation of a general method of studying the Fredholmness and $n(d)$-normality of operators in Banach algebras of bounded linear operators with discrete groups of shifts. As is well known (see, e.g., [3], [5]), a bounded linear operator N is called *n-normal* (*d-normal*) if its image is closed and $\dim \operatorname{Ker} N < \infty$ (resp. $\dim \operatorname{Coker} N < \infty$); N is called *Fredholm* if N is n-normal and d-normal simultaneously. The method developed here has emerged from the investigation of Banach algebras $\mathfrak{B}$ of singular integral operators with shifts acting on Lebesgue spaces $L^p(\Gamma)$, $1 < p < \infty$. The algebras $\mathfrak{B}$ are generated by the operators of multiplication by piecewise continuous functions and the operators S_Γ and U_g $(g \in G)$ where

$$(S_\Gamma\varphi)(t) = \frac{1}{\pi i}\,\text{v.p.}\int_\Gamma \frac{\varphi(\tau)}{\tau - t}\,d\tau, \quad (U_g\varphi)(t) = |g'(t)|^{1/p}\varphi[g(t)], \quad t \in \Gamma,$$

and G is a discrete group of piecewise smooth homeomorphisms of a piecewise smooth composed contour Γ without cusps onto itself.

In Sections 2 and 3 we describe the afore-mentioned general method. It is applied to Banach algebras $\mathcal{B}$ of operators of the form

$$N = \sum_{k=1}^{m} A_k P_k + H_0$$

where all operators A_k belong to a unital Banach algebra $\mathcal{A}$, $\sum P_k = I$, and H_0 belongs to a closed two-sided ideal $\mathcal{H} \subset \mathcal{B}$ containing all compact operators,

---

Yu.I. Karlovich is partially supported by CONACYT project 32726-E, México. B. Silbermann is partially supported by Praxis XXI/BCC120247/99, Portugal.

the operators P_kP_j $(k \neq j)$ and all the commutators $AP_k - P_kA$ $(A \in \mathcal{A},\ k = 1,\dots,m)$.

The method consists of two steps of a successive localization. On the first step (Section 2) we reduce the study of the Fredholmness of N to the study of the invertibility for all operators A_k and investigating the so-called H-Fredholmness of N for a finite set of operators $H \in \mathcal{H}$. For singular integral operators with shifts, such operators $H \in \mathcal{H}$ correspond to the G-orbits generated by the nodes of the contour and the points of discontinuity of the coefficients and the derivatives of the shifts, while A_k are functional operators in the Banach subalgebra $\mathfrak{A} \subset \mathfrak{B}$ generated by the operators of multiplication by piecewise continuous functions and the (isometric) shift operators U_g $(g \in G)$. Finding conditions which guarantee the invertibility of every operator A_k for a Fredholm operator N is based on an idea going back to V.G. Kravchenko's paper [16] and being stipulated by the qualitative difference between singular integral operators with shifts and singular integral operators without shifts: as a rule, the subset of Fredholm singular integral operators with shifts is not dense in the set of all singular integral operators with shifts.

On the second step (Section 3) we deal with the local investigation of the H-Fredholmness on the basis of a lifting theorem being a modification of that in [6, Theorem 1.8]. This lifting theorem has its roots in Banach algebra based numerical analysis and is widely used there. The point is that the occurring algebras have a trivial center, so that familiar localization procedures cannot be applied. The interested reader is also addressed to the books [4], [7] and the literature cited there.

In the end, the problem is reduced to the study of the local representatives of N. In the realization for singular integral operators with a discrete group G of shifts, those local representatives act on the space $l^p(G(t))$ and correspond to the G-orbits $G(t)$ of contour points.

For functional operators $A \in \mathfrak{A}$ it is also possible to construct a local-trajectory invertibility theory by using the concept of the lower norm of operators (see [10]). In the case of a discrete subexponential group G acting topologically free on Γ, the local representatives for functional operators $A \in \mathfrak{A}$ act on the space $l^p(G)$ in contrast to the local representatives for singular integral operators with shifts, which act on the spaces l^p over the G-orbits.

The first version of such a method was developed by V.G. Kravchenko and the first author (see [12], [13]). It was applied by them to the investigation of Fredholmness in Banach algebras of singular integral operators with piecewise continuous coefficients and a cyclic group of piecewise smooth shifts having finite sets of fixed points on the contour (see [13], [17] and the references therein). But that version was not applicable to Banach algebras of singular integral operators with ergodic shifts or non-cyclic groups of shifts in the case of discontinuous coefficients. The approaches developed in parallel by A.P. Soldatov (see [22], [23]), A.G. Myasnikov and L.I. Sazonov [19], [20], V.N. Semenyuta [21], and A.B. Khevelev [15] also do not provide such a possibility.

To study Banach algebras of singular integral operators with piecewise continuous coefficients and subexponential groups of shifts (including arbitrary cyclic groups and a wide class of non-cyclic and non-commutative groups), the first author developed a new version of that method and applied it to the investigation of the Fredholmness of singular integral operators with discrete subexponential groups of shifts on Lebesgue spaces (see [10], [11] and also [14]). That version was based on a property of separation for arbitrary finite subsets lying on different G-orbits and a property called approximate localizability [10] for operators $H \in \mathcal{H}$. The method presented here is a combination of the approach of [10] and the lifting theorem borrowed from [6]. It allows us to simplify and generalize the results announced in [10]. In particular, we do not use a two-parameter system of localizing classes (assumption (Π 3) in [10]) and remove the conditions of the uniform boundedness for the homomorphisms Ψ_λ and the inverse operators N_λ^{-1} (see (Π 7) and Theorem 2 of [10]).

Another approach suitable for studying C^*-algebras of singular integral operators with discrete amenable groups of shifts, which are associated with C^*-dynamical systems, was developed by A.B. Antonevich and A.V. Lebedev [1], [2] and the first author [8], [9].

Applications of the method presented in Sections 2 and 3 to Banach algebras $\mathfrak{B}$ of singular integral operators with discrete groups of shifts will appear in a forthcoming paper.

2. First step of localization

2.1. Fredholm criterion

Let $\mathcal{L} = \mathcal{L}(X)$ be the Banach algebra of bounded linear operators acting on a Banach space X, $\mathcal{K} = \mathcal{K}(X)$ the ideal of compact operators in $\mathcal{L}$, $\mathcal{A}$ a subalgebra of $\mathcal{L}$ with the identity operator $I \in \mathcal{L}$, and let $\mathcal{P} = \{P_1, \dots, P_m\}$ be a fixed set of operators in $\mathcal{L}$ such that $P_1 + P_2 + \dots + P_m = I$. Consider the Banach subalgebra $\mathcal{B} = \operatorname{alg}(\mathcal{A}, \mathcal{P}, \mathcal{K}) \subset \mathcal{L}$ that is generated by the sets $\mathcal{A}, \mathcal{P}$, $\mathcal{K}$. Let $\mathcal{H}$ be the smallest closed two-sided ideal of $\mathcal{B}$ containing the ideal $\mathcal{K}$, the set $\mathcal{P}_0 = \{P_k P_j : k, j = 1, 2, \dots, m;\ k \neq j\}$ and all the commutators $AP - PA$ for $A \in \mathcal{A}$ and $P \in \mathcal{P}$.

For $A, B \in \mathcal{L}$ we write $A \simeq B$ if $A - B \in \mathcal{K}$, and denote by

$$|A| := \inf\{\|A + K\| : K \in \mathcal{K}\}$$

the essential norm of $A \in \mathcal{L}$, i.e., the norm in the Calkin algebra $\mathcal{L}/\mathcal{K}$.

We consider operators of the form

$$N = \sum_{k=1}^{m} A_k P_k + H_0, \tag{1}$$

where $A_k \in \mathcal{A}$ $(k = 1, 2, \dots, m)$ and $H_0 \in \mathcal{H}$. The set of such operators N is dense in $\mathcal{B}$ and can coincide with $\mathcal{B}$. First we will find conditions under which the Fredholmness (n-normality, d-normality) of N implies the invertibility (left invertibility, right invertibility) of every operator A_k, $k = 1, 2, \ldots, m$.

The invertibility of the operators A_k follows from the next assumption.

(A1) For any noninvertible operator $A \in \mathcal{A}$ on X, any $H_0 \in \mathcal{H}$, and any $\varepsilon > 0$, there exist operators $A_\varepsilon \in \mathcal{A}$ and $H_\varepsilon \in \mathcal{H}$ such that $|A - A_\varepsilon| < \varepsilon$, $|H_0 - H_\varepsilon| < \varepsilon$, and one of the following conditions holds:

(i) for every $P \in \mathcal{P}$ there exists a sequence of elements $f_n \in X$ with norm $\|f_n\|_X = 1$ such that for all $H \in \mathcal{H}_\varepsilon \setminus \mathcal{K}$ the sequences $\{A_\varepsilon f_n\}$ and $\{Hf_n\}$ converge in X, but the sequence $\{Pf_n\}$ does not contain subsequences convergent in X;

(ii) for every $P \in \mathcal{P}$ there exists a sequence of elements $f_n \in X^*$ with norm $\|f_n\|_{X^*} = 1$ such that for all $H \in \mathcal{H}_\varepsilon \setminus \mathcal{K}$ the sequences $\{A_\varepsilon^* f_n\}$ and $\{H^* f_n\}$ converge in X^*, but the sequence $\{P^* f_n\}$ does not contain subsequences convergent in X^*.

Here we used the notation

$$\mathcal{H}_\varepsilon := \{A_\varepsilon P - PA_\varepsilon : P \in \mathcal{P}\} \cup \mathcal{P}_0 \cup \{H_\varepsilon P,\ PH_\varepsilon : P \in \mathcal{P}\} \subset \mathcal{H}.$$

The corresponding one-sided invertibility of A_k follows from the $n(d)$-normality of N under the following stronger assumption.

(A1)^0 For any operator $A \in \mathcal{A}$ which is not left-invertible (right-invertible), any $H_0 \in \mathcal{H}$, and any $\varepsilon > 0$, there are operators $A_\varepsilon \in \mathcal{A}$ and $H_\varepsilon \in \mathcal{H}$ such that $|A - A_\varepsilon| < \varepsilon$, $|H_0 - H_\varepsilon| < \varepsilon$, and condition (i) (respectively, condition (ii)) in (A1) holds.

In particular, one can derive from (A1) (respectively, from (A1)^0) that an operator $A \in \mathcal{A}$ is Fredholm (n-normal, d-normal) if and only if A is invertible (left-invertible, right- invertible) (see Theorems 2.1 and 2.2 below). Note also that conditions (i) and (ii) follow, respectively, from the following conditions (cf. [13]):

(i)$'$ for each $P \in \mathcal{P}$ there is an $F \in \mathcal{L}$ such that for every $H \in \mathcal{H}_\varepsilon$,

$$PF \not\simeq 0,\ A_\varepsilon F \simeq 0,\ HF \simeq 0;$$

(ii)$'$ for each $P \in \mathcal{P}$ there is an $F \in \mathcal{L}$ such that for every $H \in \mathcal{H}_\varepsilon$,

$$FP \not\simeq 0,\ FA_\varepsilon \simeq 0,\ FH \simeq 0.$$

Following [13], we call an operator $N_H \in \mathcal{L}$ a *right (left) H-regularizer* of an operator $N \in \mathcal{L}$ if $NN_H \simeq H$ ($N_H N \simeq H$), and we say that N is *H-Fredholm* if it has right and left H-regularizers.

As is known (see, e.g., [18]), an operator $N \in \mathcal{L}$ is Fredholm if and only if it has right and left regularizers. Thus the notion of H-Fredholmness generalizes the usual notion of Fredholmness: it is sufficient to take $H = I$. Obviously, a Fredholm operator $N \in \mathcal{L}$ is H-Fredholm for every $H \in \mathcal{L}$, and every operator $N \in \mathcal{L}$ is H-Fredholm if H is a compact operator.

We set

$$\mathcal{G}(N) := \bigcup_{k=1}^{m} \Big(\mathcal{H}(A_k) \cup \mathcal{H}_k(A_k) \Big) \cup \mathcal{P}_0 \cup \{H_0\} \tag{2}$$

where

$$\mathcal{H}(A) := \{AP - PA : P \in \mathcal{P}\}, \quad \mathcal{H}_k(A) := \{DP_k - P_k D : D \in \mathcal{H}(A)\},$$

and $k = 1, 2, \ldots, m$. Clearly, $\mathcal{G}(N) \subset \mathcal{H}$.

Theorem 2.1. *If* (A1) *is satisfied, then the operator N given by* (1) *is Fredholm if and only if*

(a) *the operators A_k are invertible for all $k = 1, 2, \ldots, m$,*
(b) *for every $H \in \mathcal{G}(N) \setminus \mathcal{K}$ the operator N is H-Fredholm.*

Proof. Sufficiency. For $A, B \in \mathcal{L}$, let $[A, B] := AB - BA$. If

$$N_0 = \sum_{k=1}^{m} P_k A_k^{-1}, \tag{3}$$

then the equalities

$$AP_jP_k = [A, P_j]P_k + [A, P_k]P_j + [P_j, [A, P_k]] + P_jP_kA \qquad (A \in \mathcal{A})$$

imply that

$$NN_0 \simeq I + \sum_{H \in \mathcal{G}(N)} HB_H$$

with some $B_H \in \mathcal{L}$. Then a right regularizer of N has the form

$$N^{(r)} = N_0 - \sum_{H \in \mathcal{G}(N) \setminus \mathcal{K}} N_H^{(r)} B_H$$

where the operators $N_H^{(r)}$ are right H-regularizers of N. A left regularizer of N is constructed similarly.

Necessity. It is clear that condition (b) is satisfied. Suppose that N is Fredholm but condition (a) is violated. To be specific, we suppose that the operator A_1 is not invertible and that in ε-neighborhoods of A_1 and H_0 there exist operators $A_{1,\varepsilon}$ and H_ε satisfying (i). Then for sufficiently small $\varepsilon > 0$ the operator

$$N_\varepsilon = A_{1,\varepsilon}P_1 + \sum_{k=2}^{m} A_k P_k + H_\varepsilon \tag{4}$$

is Fredholm and hence n-normal. By virtue of (i), choosing a sequence $\{f_n\}$ for P_1, we infer from the equality

$$N_\varepsilon P_1 f_n = P_1 A_{1,\varepsilon} f_n + [A_{1,\varepsilon}, P_1] f_n + \sum_{k=2}^{m} (A_k - A_{1,\varepsilon}) P_k P_1 f_n + H_\varepsilon P_1 f_n \tag{5}$$

that the sequence $\{N_\varepsilon P_1 f_n\}$ converges in X. Then, according to [18, Theorem 6.1], the n-normality of N_ε implies that from the sequence $\{P_1 f_n\}$ we can extract a subsequence convergent in X, which contradicts (i). Hence, (a) also holds. □

Analogously, taking into account the stability of $n(d)$-normality [18], one can prove the following assertion.

Theorem 2.2. *If condition* $(\mathrm{A1})^0$ *is satisfied, then the n-normality (d-normality) of operator* (1) *implies the left-invertibility (right-invertibility) of every operator* A_k, $k = 1, 2, \ldots, m$.

Corollary 2.3. *If* $(\mathrm{A1})^0$ *holds and either* $\mathcal{G}(N) \subset \mathcal{K}$, *or* $A_1 = \ldots = A_m$ *and* $H_0 = 0$, *then the n-normality (d-normality) of operator* (1) *is equivalent to the left-invertibility (right-invertibility) of all operators* A_k.

2.2. Cosets versus H-Fredholmness

We now rewrite the notion of H-Fredholmness in terms of the invertibility of cosets in some quotient algebra. Together with the closed two-sided ideal $\mathcal{H} \subset \mathcal{B}$, we consider another closed two-sided ideal in $\mathcal{B}$, namely,

$$\mathcal{I}_{\mathcal{H}} = \{B \in \mathcal{B} : \ BH \simeq HB \simeq 0 \quad \text{for all} \quad H \in \mathcal{H}\}.$$

Clearly, $\mathcal{K} \subset \mathcal{I}_{\mathcal{H}}$.

Proposition 2.4. *If* $B \in \mathcal{B}$ *and* $B + \mathcal{I}_{\mathcal{H}}$ *is invertible in the quotient algebra* $\mathcal{B}/\mathcal{I}_{\mathcal{H}}$, *then* B *is* H*-Fredholm for every* $H \in \mathcal{H}$.

Proof. If $B + \mathcal{I}_{\mathcal{H}}$ is invertible in $\mathcal{B}/\mathcal{I}_{\mathcal{H}}$, then there exists an operator $C \in \mathcal{B}$ such that

$$(C + \mathcal{I}_{\mathcal{H}})(B + \mathcal{I}_{\mathcal{H}}) = (B + \mathcal{I}_{\mathcal{H}})(C + \mathcal{I}_{\mathcal{H}}) = I + \mathcal{I}_{\mathcal{H}}. \tag{6}$$

Since

$$\mathcal{I}_{\mathcal{H}} H, \ H\mathcal{I}_{\mathcal{H}} \subset \mathcal{K} \quad \text{for every} \quad H \in \mathcal{H},$$

from (6) it follows that

$$HCB \simeq H, \quad BCH \simeq H,$$

which implies that $B_H^{(l)} = HC$ and $B_H^{(r)} = CH$ are left and right H-regularizers of B, respectively. □

Lemma 2.5. *If* (A1) *holds and the algebra* $\mathcal{A}$ *is inverse closed in* $\mathcal{L}$, *then the invertibility of all the operators* A_k $(k = 1, 2, \ldots, m)$ *is equivalent to the invertibility of the coset* $N + \mathcal{H}$ *in* $\mathcal{B}/\mathcal{H}$.

Proof. If all A_k are invertible in $\mathcal{L}$ then their inverses A_k^{-1} belong to $\mathcal{B}$ due to the inverse closedness of $\mathcal{A}$ in $\mathcal{L}$. Hence, according to the proof of sufficiency in Theorem 2.1,

$$(N + \mathcal{H})^{-1} = \sum_{k=1}^{m} P_k A_k^{-1} + \mathcal{H}.$$

Conversely, let the coset $N + \mathcal{H}$ be invertible in $\mathcal{B}/\mathcal{H}$ but suppose, for example, A_1 is not invertible in $\mathcal{L}$. Then by (A1), in any ε-neighborhoods of A_1 and H_0 there are operators $A_{1,\varepsilon}$ and H_ε satisfying conditions (i) or (ii). Assume (i) holds.

For every sufficiently small $\varepsilon > 0$, the coset $N_\varepsilon + \mathcal{H}$, with N_ε given by (4), is invertible in $\mathcal{B}/\mathcal{H}$ together with $N + \mathcal{H}$. Hence there exist operators $B \in \mathcal{B}$ and $H \in \mathcal{H}$ such that

$$BN_\varepsilon = I + H.$$

Choosing a sequence $\{f_n\} \subset X$ for P_1 according to (i), we conclude from (5) that

$$\begin{aligned} P_1 f_n &= (BN_\varepsilon - H)P_1 f_n = B\Big(P_1 A_{1,\varepsilon} f_n + [A_{1,\varepsilon}, P_1] f_n \\ &+ \sum_{k=2}^{m} (A_k - A_{1,\varepsilon}) P_k P_1 f_n + H_\varepsilon P_1 f_n\Big) - H P_1 f_n, \end{aligned}$$

which contradicts (i) as in the proof of necessity in Theorem 2.1. □

Making use of Proposition 2.4, we obtain the following simple criterion.

Theorem 2.6. *An operator $N \in \mathcal{B}$ is Fredholm if*

(a) *the coset $N + \mathcal{H}$ is invertible in the quotient algebra $\mathcal{B}/\mathcal{H}$,*
(b) *the coset $N + \mathcal{I}_\mathcal{H}$ is invertible in the quotient algebra $\mathcal{B}/\mathcal{I}_\mathcal{H}$.*

Conversely, if the quotient algebra $\mathcal{B}/\mathcal{K}$ is inverse closed in $\mathcal{L}/\mathcal{K}$ and N is Fredholm, then (a) *and* (b) *hold.*

Proof. By (a), there exist operators $B_1 \in \mathcal{B}$ and $H_1 \in \mathcal{H}$ such that

$$NB_1 = I + H_1. \tag{7}$$

According to (b) and the proof of Proposition 2.4, there exists an operator $B_2 \in \mathcal{B}$ such that

$$NB_2 H_1 \simeq H_1. \tag{8}$$

From (7) and (8) it follows that $N(B_1 - B_2 H_1) \simeq I$, i.e., the operator $B_1 - B_2 H_1$ is a right regularizer for N. Analogously one can prove the existence of a left regularizer for N. Thus N is Fredholm.

Obviously, the Fredholmness of N implies (a) and (b) due to the inverse closedness of $\mathcal{B}/\mathcal{K}$ in $\mathcal{L}/\mathcal{K}$. □

From Theorem 2.1 and Proposition 2.4 we immediately get the following modification of Theorems 2.1 and 2.6.

Theorem 2.7. *If* (A1) *is satisfied and the quotient algebra $\mathcal{B}/\mathcal{K}$ is inverse closed in $\mathcal{L}/\mathcal{K}$, then the operator N given by* (1) *is Fredholm if and only if*

(a) *the operators A_k are invertible for all $k = 1, 2, \ldots, m$,*
(b) *the coset $N + \mathcal{I}_\mathcal{H}$ is invertible in the quotient algebra $\mathcal{B}/\mathcal{I}_\mathcal{H}$.*

Proof. If N is Fredholm, then (a) holds by Theorem 2.1. Furthermore, if $N^{(-1)}$ is a regularizer of N, then $N^{(-1)} \in \mathcal{B}$. Hence $N^{(-1)} + \mathcal{I}_\mathcal{H}$ is the inverse to the coset $N + \mathcal{I}_\mathcal{H}$ in the quotient algebra $\mathcal{B}/\mathcal{I}_\mathcal{H}$, which gives (b).

Conversely, if (b) holds, then N is H-Fredholm for every $H \in \mathcal{H}$ due to Proposition 2.4. Hence, by (a) and Theorem 2.1, N is Fredholm. □

3. Second step of localization

3.1. Local H-Fredholmness

Under additional assumptions we can replace the condition of H-Fredholmness in Theorem 2.1 by its local analogue.

Theorem 3.1. *Let the algebra $\mathcal{A}$ be inverse closed in $\mathcal{L}$, let $\mathcal{H}$ be the smallest closed two-sided ideal of $\mathcal{B}$ containing the closed two-sided ideals $\mathcal{H}_\lambda \subset \mathcal{B}$ $(\lambda \in \Lambda)$ and $\mathcal{K} \subset \mathcal{H}_\lambda$ for every $\lambda \in \Lambda$, where Λ is an arbitrary index set. If* (A1) *is satisfied, then the operator N given by* (1) *is Fredholm if and only if*

(a) *all the operators A_k $(k = 1, \dots, m)$ are invertible,*

(b) *for every $H \in \bigcup_{\lambda \in \Lambda} \mathcal{H}_\lambda$, the operator N is H-Fredholm.*

Proof. Necessity. Part (a) was proved in Theorem 2.1. Obviously, N is H-Fredholm for every $H \in \mathcal{H}$.

Sufficiency. If (a) holds, then, in view of the inverse closedness of $\mathcal{A}$ in $\mathcal{L}$, the operator N_0 given by (3) belongs to the algebra $\mathcal{B}$. Then, following the proof of sufficiency in Theorem 2.1, we infer that $NN_0 \simeq I + H$ where $H \in \mathcal{H}$. Since $\mathcal{H}$ is the smallest closed two-sided ideal generated by all the closed two-sided ideals $\mathcal{H}_\lambda$ $(\lambda \in \Lambda)$, there exist a finite set $\Lambda_0 \subset \Lambda$ and operators $H_\lambda \in \mathcal{H}_\lambda$ $(\lambda \in \Lambda_0)$ such that

$$\Big| H - \sum_{\lambda \in \Lambda_0} H_\lambda \Big| < 1. \tag{9}$$

According to (b), for every $\lambda \in \Lambda_0$ there exists an operator $N_{H_\lambda} \in \mathcal{L}$ such that $NN_{H_\lambda} \simeq H_\lambda$. Hence

$$N\Big(N_0 - \sum_{\lambda \in \Lambda_0} N_{H_\lambda}\Big) \simeq I + \Big(H - \sum_{\lambda \in \Lambda_0} H_\lambda\Big).$$

By (9), there is an operator $\widetilde{H} \in \mathcal{H}$ such that

$$\Big[I + \Big(H - \sum_{\lambda \in \Lambda_0} H_\lambda\Big)\Big](I + \widetilde{H}) \simeq I,$$

which implies that the operator

$$N^{(r)} := \Big(N_0 - \sum_{\lambda \in \Lambda_0} N_{H_\lambda}\Big)(I + \widetilde{H})$$

is a right regularizer of N. A left regularizer of N is constructed analogously. Hence the operator N is Fredholm. □

3.2. Lifting theorem and its application

To obtain a more suitable Fredholm criterion for operators $B \in \mathcal{B}$ and, in particular, more easily verifiable conditions for the local H-Fredholmness of B, we will make use of a lifting theorem (being some modification of [6, Theorem 1.8]) instead of the H-Fredholmness.

Definition 3.2. *Let $\mathcal{B}$ be a Banach subalgebra of a Banach algebra $\mathcal{L}$ and let $\mathcal{H}$ be a closed two-sided ideal of $\mathcal{B}$. We call $\mathcal{B}$ inverse closed in $\mathcal{L}$ with respect to $\mathcal{H}$ if the invertibility of any operator $B \in \mathcal{B}$ in $\mathcal{L}$ and the invertibility of the coset $B + \mathcal{H}$ in the quotient algebra $\mathcal{B}/\mathcal{H}$ imply the invertibility of B in $\mathcal{B}$.*

Theorem 3.3. *Assume the quotient algebra $\mathcal{B}/\mathcal{K}$ is inverse closed in $\mathcal{L}/\mathcal{K}$. Let Λ be an index set and suppose that, for each $\lambda \in \Lambda$, we are given a unital Banach algebra $\mathcal{L}_\lambda$, a unital homomorphism $\Psi_\lambda : \mathcal{B} \to \mathcal{L}_\lambda$, and a closed two-sided ideal $\mathcal{H}_\lambda \subset \mathcal{B}$ such that*

(i) *$\mathcal{K} \subset \mathcal{H}_\lambda \cap \operatorname{Ker} \Psi_\lambda$ and $\mathcal{H}_\mu \subset \operatorname{Ker} \Psi_\lambda$ for all $\mu \in \Lambda \setminus \{\lambda\}$;*

(ii) *the restriction of the quotient homomorphism*

$$\mathcal{B}/\mathcal{K} \to \mathcal{L}_\lambda, \quad B + \mathcal{K} \mapsto \Psi_\lambda(B)$$

(being correctly defined by (i) and denoted by Ψ_λ again) onto the ideal $\mathcal{H}_\lambda/\mathcal{K}$ is an isomorphism between $\mathcal{H}_\lambda/\mathcal{K}$ and the closed two-sided ideal $\mathcal{R}_\lambda := \Psi_\lambda(\mathcal{H}_\lambda)$ of the Banach algebra $\mathcal{B}_\lambda$ which is defined as the closure of the set $\Psi_\lambda(\mathcal{B})$ in $\mathcal{L}_\lambda$;

(iii) *there exists a Banach subalgebra $\mathcal{C}_\lambda \subset \mathcal{L}_\lambda$ which contains $\mathcal{B}_\lambda$, has two closed two-sided ideals, the ideal $\mathcal{R}_\lambda$ and an ideal $\widetilde{\mathcal{R}}_\lambda \supset \mathcal{R}_\lambda$, and which is inverse closed in $\mathcal{L}_\lambda$ with respect to the ideal $\widetilde{\mathcal{R}}_\lambda$;*

(iv) *$\mathcal{H}$ is the smallest closed two-sided ideal of $\mathcal{B}$ containing all the ideals $\mathcal{H}_\lambda$ $(\lambda \in \Lambda)$.*

Then an operator $B \in \mathcal{B}$ is Fredholm if and only if the coset $B + \mathcal{H}$ is invertible in $\mathcal{B}/\mathcal{H}$ and the element $\Psi_\lambda(B)$ is invertible in $\mathcal{L}_\lambda$ for every $\lambda \in \Lambda$.

Proof. *Necessity.* Let $B \in \mathcal{B}$ be Fredholm. Since the quotient algebra $\mathcal{B}/\mathcal{K}$ is inverse closed in $\mathcal{L}/\mathcal{K}$, we conclude that the coset $B + \mathcal{H}$ is invertible in $\mathcal{B}/\mathcal{H}$ and that the elements $\Psi_\lambda(B)$ are invertible in $\mathcal{L}_\lambda$ for all $\lambda \in \Lambda$.

Sufficiency. Let the coset $B + \mathcal{H}$ be invertible in the quotient algebra $\mathcal{B}/\mathcal{H}$. Then there are elements $D \in \mathcal{B}$ and $H \in \mathcal{H}$ such that $DB = I + H$. By the definition of the ideal $\mathcal{H}$ (see (iv)), one can choose a set $\{\lambda_1, \lambda_2, \ldots, \lambda_n\} \subset \Lambda$, elements $H_{\lambda_i} \in \mathcal{H}_{\lambda_i}$ $(i = 1, 2, \ldots, n)$, and $H' \in \mathcal{H}$ such that

$$H' \simeq H_{\lambda_1} + \ldots + H_{\lambda_n} \quad \text{and} \quad |H - H'| < 1.$$

In addition, for each $\lambda \in \Lambda$, since $\Psi_\lambda(H_\mu) = 0$ for all $\mu \in \Lambda \setminus \{\lambda\}$ and all $H_\mu \in \mathcal{H}_\mu$, the invertibility of the coset $B + \mathcal{H}$ in $\mathcal{B}/\mathcal{H}$ implies the invertibility of the coset $\Psi_\lambda(B) + \mathcal{R}_\lambda$ in the quotient algebra $\mathcal{B}_\lambda/\mathcal{R}_\lambda$ and thus in the quotient algebra $\mathcal{C}_\lambda/\mathcal{R}_\lambda$. In its turn, this implies the invertibility of the coset $\Psi_\lambda(B) + \widetilde{\mathcal{R}}_\lambda$ in the quotient algebra $\mathcal{C}_\lambda/\widetilde{\mathcal{R}}_\lambda$.

Let $\Psi_{\lambda_i}^{-1}(B)$ stand for the inverse of $\Psi_{\lambda_i}(B)$ in $\mathcal{L}_{\lambda_i}$, $i = 1, 2, \ldots, n$. Then, according to (iii), $\Psi_{\lambda_i}^{-1}(B) \in \mathcal{C}_{\lambda_i}$, which shows that $\Psi_{\lambda_i}(H_{\lambda_i})\Psi_{\lambda_i}^{-1}(B)$ belongs to the ideal $\mathcal{R}_{\lambda_i}$. Since $\mathcal{R}_{\lambda_i}$ is a closed two-sided ideal of both the algebras $\mathcal{B}_{\lambda_i}$ and $\mathcal{C}_{\lambda_i}$, we deduce from (ii) that there exists an element $D_{\lambda_i} \in \mathcal{H}_{\lambda_i}$ such that

$$\Psi_{\lambda_i}(D_{\lambda_i}) = \Psi_{\lambda_i}(H_{\lambda_i})\Psi_{\lambda_i}^{-1}(B).$$

Put

$$D' = D - D_{\lambda_1} - \ldots - D_{\lambda_n}.$$

Then

$$\begin{aligned} D'B &= I + H - D_{\lambda_1}B - \ldots - D_{\lambda_n}B \\ &= I + (H - H') + (H_{\lambda_1} - D_{\lambda_1}B) + \ldots + (H_{\lambda_n} - D_{\lambda_n}B). \end{aligned} \quad (10)$$

Since, by definition, $\Psi_{\lambda_i}(H_{\lambda_i} - D_{\lambda_i}B) = 0$ and $H_{\lambda_i} - D_{\lambda_i}B \in \mathcal{H}_{\lambda_i}$, we conclude from (i) and (ii) that $H_{\lambda_i} - D_{\lambda_i}B \in \mathcal{K}$. Thus we infer from (10) that

$$D'B \simeq I + H - H'.$$

In view of the inequality $|H - H'| < 1$, the coset $I + H - H' + \mathcal{K}$ is invertible in $\mathcal{B}/\mathcal{K}$, and $(I + H - H' + \mathcal{K})^{-1} = B_0 + \mathcal{K}$ with $B_0 \in \mathcal{B}$. Then $B_0 D'B \simeq I$, which implies that the operator $B_0 D' \in \mathcal{B}$ is a left regularizer of B.

Analogously one can prove the existence of a right regularizer of B, which together with the previous part gives the Fredholmness of B. □

From the proof of sufficiency in Theorem 3.3 we obtain the following important result.

Corollary 3.4. *Let all the conditions of Theorem* 3.3 *except for the inverse closedness of $\mathcal{B}/\mathcal{K}$ be fulfilled. If, for an operator $B \in \mathcal{B}$, the coset $B + \mathcal{H}$ is invertible in $\mathcal{B}/\mathcal{H}$ and for every $\lambda \in \Lambda$ the element $\Psi_\lambda(B)$ is invertible in $\mathcal{L}_\lambda$, then the operator B is Fredholm and all its regularizers belong to $\mathcal{B}$.*

In Lemma 2.5 we have described the invertibility of the coset $N + \mathcal{H}$ in the quotient algebra $\mathcal{B}/\mathcal{H}$. Thus, from Theorems 2.1 and 3.3 we get the following criterion.

Theorem 3.5. *If* (A1) *and all the assumptions of Theorem* 3.3 *are fulfilled, then an operator $N \in \mathcal{B}$ of the form* (1) *is Fredholm if and only if every operator A_k $(k = 1, 2, \ldots, m)$ is invertible in $\mathcal{L}$ and every element $\Psi_\lambda(N)$ $(\lambda \in \Lambda)$ is invertible in $\mathcal{L}_\lambda$.*

Proof. If N is Fredholm, then by Theorem 2.1 all the operators A_k are invertible, and by Theorem 3.3 all the elements $\Psi_\lambda(N)$ are invertible.

Conversely, if all the operators A_k are invertible, then, analogously to Lemma 2.5, the coset $N + \mathcal{H}$ is invertible in the quotient algebra $\mathcal{B}/\mathcal{H}$. Thus, taking into account the invertibility of all the elements $\Psi_\lambda(N)$ in $\mathcal{L}_\lambda$, we get the Fredholmness of N by Theorem 3.3. □

Note that it is sufficient to consider the invertibility of $\Psi_\lambda(N)$ only for λ in some subset $\Lambda(N) \subset \Lambda$. Indeed, let

(A2) $\mathcal{H}_\lambda \mathcal{H}_\mu = \mathcal{K}$ for every $\lambda, \mu \in \Lambda$, $\lambda \neq \mu$.

Definition 3.6. *For $H \in \mathcal{H}$, the set*

$$s(H) = \Lambda \setminus \{\lambda \in \Lambda : HH_\lambda \simeq 0 \quad \text{for every} \quad H_\lambda \in \mathcal{H}_\lambda\}$$

is called the support of the operator H.

By (A2), $s(H_\lambda) \subset \{\lambda\}$ for every $H_\lambda \in \mathcal{H}_\lambda$, and if $H_\lambda^2 \not\simeq 0$ then $s(H_\lambda) = \{\lambda\}$. Thus, if condition (iv) of Theorem 3.3 holds too, then, for every $H \in \mathcal{H}$, the set $s(H) \subset \Lambda$ is at most countable. Obviously, $s(H) = \emptyset$, if $H \in \mathcal{K}$. Suppose

(A3) for every $H \in \mathcal{H} \setminus \mathcal{K}$ there is a nonempty and at most countable set $\{H_\lambda \in \mathcal{H}_\lambda : \lambda \in s(H)\}$ such that $s(H - H_\lambda) = s(H) \setminus \{\lambda\}$ for every $\lambda \in s(H)$ and

$$|H| = \sup_{\lambda \in s(H)} |H_\lambda|.$$

From (A3) it follows that for $H \in \mathcal{H}$, $s(H) = \emptyset$ if and only if $H \in \mathcal{K}$. If (A3) and condition (iv) of Theorem 3.3 are satisfied, then for every operator $H \in \mathcal{H}$ there is a sequence of finite sums

$$\sum_{\lambda \in \Lambda_n} H_\lambda^{(n)}$$

of operators $H_\lambda^{(n)} \in \mathcal{H}_\lambda$ such that $\Lambda_n \subset s(H)$ and

$$\Big| H - \sum_{\lambda \in \Lambda_n} H_\lambda^{(n)} \Big| < n^{-1}. \tag{11}$$

Indeed, in view of condition (iv), the inequalities (11) are fulfilled for finite sets $\Lambda_n \subset \Lambda$. Then by (A3),

$$\Big| H - \sum_{\lambda \in \Lambda_n \cap s(H)} H_\lambda^{(n)} \Big| \le \Big| H - \sum_{\lambda \in \Lambda_n} H_\lambda^{(n)} \Big| < n^{-1}.$$

Let

$$\Lambda(N) := \bigcup_{H \in \mathcal{G}(N)} s(H),$$

where $\mathcal{G}(N)$ is defined by (2). Then, under assumption (A2) and condition (iv) of Theorem 3.3, the set $\Lambda(N) \subset \Lambda$ is at most countable for every $N \in \mathcal{B}$.

Remark 3.7. *If* (A1)–(A3) *and all the assumptions of Theorem* 3.3 *are fulfilled, then in Theorem* 3.5 *we may replace the invertibility of* $\Psi_\lambda(N)$ *for every* $\lambda \in \Lambda$ *by the invertibility of* $\Psi_\lambda(N)$ *for every* $\lambda \in \Lambda(N)$.

Indeed, from the proof of sufficiency in Theorem 2.1 it follows that

$$NN_0 \simeq I + H \quad \text{where} \quad H \in \mathcal{H}, \quad s(H) \subset \Lambda(N).$$

Further, from the proof of Theorem 3.3 we see that the invertibility of $\Psi_\lambda(N)$ in $\mathcal{L}_\lambda$ for all $\lambda \in \Lambda(N)$ implies the H-Fredholmness of N for every

$$H \in \bigcup_{\lambda \in \Lambda(N)} \mathcal{H}_\lambda.$$

Since due to (A3) we can take $\Lambda_0 \subset s(H)$ in (9), the H-Fredholmness of N for every

$$H \in \bigcup_{\lambda \in \Lambda(N)} \mathcal{H}_\lambda$$

and the invertibility of all A_k $(k = 1, \dots, m)$ imply the Fredholmness of N by analogy with Theorem 3.1.

References

[1] A.B. Antonevich, *Linear Functional Equations. Operator Approach*, Operator Theory: Advances and Applications **83**, Birkhäuser Verlag, Basel, 1996.

[2] A. Antonevich and A. Lebedev, *Functional Differential Equations: I. C^*-Theory*, Pitman Monographs and Surveys in Pure and Applied Mathematics **70**, Longman Scientific & Technical, Harlow, 1994.

[3] A. Böttcher and B. Silbermann, *Analysis of Toeplitz Operators*, Akademie-Verlag, Berlin, 1989.

[4] A. Böttcher and B. Silbermann, *Introduction to Large Truncated Toeplitz Matrices*, Springer-Verlag, New York, 1998.

[5] I. Gohberg and N. Krupnik, *One-Dimensional Linear Singular Integral Equations*, Vols. I and II, Birkhäuser Verlag, Basel, Boston, Berlin, 1992.

[6] R. Hagen, S. Roch and B. Silbermann, *Spectral Theory of Approximation Methods for Convolution Equations*, Birkhäuser Verlag, Basel, Boston, Berlin, 1995.

[7] R. Hagen, S. Roch and B. Silbermann, *C^*-Algebras and Numerical Analysis*, Marcel Dekker, Inc., New York-Basel, 2001.

[8] Yu.I. Karlovich, *The local-trajectory method of studying invertibility in C^*-algebras of operators with discrete groups of shifts*, Soviet Math. Dokl., **37** (1988), 407–412.

[9] Yu.I. Karlovich, *C^*-algebras of operators of convolution type with discrete groups of shifts and oscillating coefficients*, Soviet Math. Dokl., **38** (1989), 301–307.

[10] Yu.I. Karlovich, *On algebras of singular integral operators with discrete groups of shifts in L_p-spaces*, Soviet Math. Dokl., **39** (1989), 48–53.

[11] Yu.I. Karlovich, *Algebras of Convolution Type Operators with Discrete Groups of Shifts and Oscillating Coefficients*, Doctoral dissertation, Math. Inst. Georgian Acad. Sci., Tiflis, 1991 [Russian].

[12] Yu.I. Karlovich and V.G. Kravchenko, *On a general method of investigating operators of singular integral type with non-Carleman shift*, Soviet Math. Dokl., **22** (1981), 10–14.

[13] Yu.I. Karlovich and V.G. Kravchenko, *An algebra of singular integral operators with piecewise-continuous coefficients and piecewise-smooth shift on a composite contour*, Math. USSR Izvestiya, **23** (1984), 307–352.

[14] Yu.I. Karlovich, V.G. Kravchenko and G.S. Litvinchuk, *On Noethericity and Mikhlin symbols of operators of the type of singular integral operators with shift* Zeitschrift für Analysis und ihre Anwendungen, **9** (1990), 15–32 [Russian].

[15] A.B. Khevelev, *On Invertibility and Noethericity of Some Classes of Operators with Non-Carleman Shifts*, Cand. dissertation, Rostov-on-Don 1979 [Russian].

[16] V.G. Kravchenko, *On a singular integral operator with a shift*, Soviet. Math. Dokl., **15** (1974), 690–692.

[17] V.G. Kravchenko and G.S. Litvinchuk, *Introduction to the Theory of Singular Integral Operators with Shift*, Kluwer Academic Publishers, Dordrecht, Boston, London, 1994.

[18] S.G. Krein, *Linear Equations in Banach Spaces*, Birkhäuser, Boston, Basel, Stuttgart, 1982.

[19] A.G. Myasnikov and L.I. Sazonov, *Singular integral operators with non-Carleman shift*, Iz. VUZ **24**, (1980), no. 3, 22–31.

[20] A.G. Myasnikov and L.I. Sazonov, *On singular operators with a non-Carleman shift and their symbols*, Soviet Math. Dokl., **22** (1980), 531–535.

[21] V.N. Semenyuta, *On singular operator equations with shift on circle*, Soviet Math. Dokl., **18** (1977), 1572–1574.

[22] A.P. Soldatov, *On the theory of singular operators with a shift*, Differential Equations, **15** (1979), 80–91.

[23] A.P. Soldatov, *Singular integral operators on the line*, Differential Equations, **16** (1980), 98–105.

Yu. Karlovich
Departamento de Matemáticas
CINVESTAV del I.P.N.
Apartado Postal 14–740
07000 México, D.F., México
e-mail: `karlovic@math.cinvestav.mx`

B. Silbermann
Fakultät für Mathematik
Technische Universität Chemnitz
09107 Chemnitz, FRG
e-mail: `silbermn@mathematik.tu-chemnitz.de`

Received: 20 December 2001

Operator Theory:
Advances and Applications, Vol. 135, 249–260

Numerical Solution of Mellin Type Equations via Wiener-Hopf Equations

G. Mastroianni and G. Monegato

To Bernd Silbermann, with admiration for his work

Abstract. Recently the authors have proposed to replace the classical Gauss-Laguerre quadrature formula and its associated rules of product type by truncated versions of them, obtained by ignoring the last part of the nodes. This has the effect of getting optimal order of convergence using a significantly less number of nodes. These rules have then been used to define stable Nyström type interpolants for second kind integral equations on the half-line. Unfortunately, in the case of some well-known Wiener-Hopf equations, the enforcement of stability required a modification of the interpolants near infinity.

Here we present new conditions for the stability and propose to replace a class of Mellin type integral equations by corresponding Wiener-Hopf equations and to solve these by using Nyström interpolants associated with truncated Gauss-Laguerre rules. In this case we have again to modify the interpolants in a neighborhood of infinity in order to prove their stability. Some numerical results which show the efficiency of this approach are presented.

1. Introduction

In [7] and [8], we associated with the classical Gauss-Laguerre quadrature formula

$$\int_0^\infty w(x)f(x)dx = \sum_{i=1}^{m} \lambda_{mi} f(x_{mi}) + R_m(f), \tag{1}$$

where $w(x) = e^{-x}$ and $x_{m1} < x_{m2} < \cdots < x_{mm}$, the "truncated" version

$$\int_0^\infty w(x)f(x)dx = \sum_{0<x_{mi}\leq 4\theta m} \lambda_{mi} f(x_{mi}) + R_m^\theta(f), \tag{2}$$

This work was supported by the Ministero dell'Università e della Ricerca Scientifica e Tecnologica of Italy and by the Italian Research Council under contract n.98.00648.CT11.

where $0 < \theta < 1$ is arbitrarily chosen. Then, in [8], we considered the corresponding product formulas, which are of the type

$$\int_0^\infty w(x)k(x,y)f(x)dx = \int_0^\infty w(x)k(x,y)L_m(f;x)dx \qquad (3)$$
$$= \sum_{i=1}^m w_{mi}(y)f(x_{mi}) + R_m(k_y;f),$$

where $L_m(f;x)$ is the Lagrange interpolation polynomial associated with the function $f(x)$ and the nodes $\{x_{mi}\}$, $k_y(x) := k(x,y)$, and

$$\int_0^\infty w(x)k(x,y)f(x)dx = \sum_{0<x_i\leq 4\theta m} w_{mi}(y)f(x_{mi}) + R_m^\theta(k_y;f). \qquad (4)$$

To examine the behavior of the above truncated rules, the following space setting was introduced. We denote by L^2 the set of all real-valued measurable functions on $R^+ = (0,\infty)$ that are square integrable and let

$$\|f\|_{L^2} = \left(\int_0^\infty f^2(x)dx\right)^{1/2}$$

be the usual norm in L^2. We write $f \in L^2_w$ if $f\sqrt{w} \in L^2$ and set $\|f\|_{L^2_w} = \|f\sqrt{w}\|_{L^2}$. For $f \in L^2_w$,

$$c_k(w,f) = \int_0^\infty f(x)p_k(w,x)w(x)dx$$

is the k-th Fourier coefficient of f with respect to the orthonormal system of Laguerre polynomials $\{p_k(w)\}$. We also need the scale of subspaces [6]

$$L^2_{w,s} = \{f \in L^2_w : \|f\|_{L^2_{w,s}} < \infty\}, \quad s \geq 0,$$

where $\|f\|_{L^2_{w,s}} = \left(\sum_{i=0}^\infty (i+1)^s c_i^2(w,f)\right)^{1/2}$. With the modulus of continuity defined by

$$\Omega_\varphi^r(f,t)_{L^2_w} = \sup_{0<h\leq t} \|\Delta_{h\varphi}^r f\|_{L^2_w(I_h)},$$

where

$$\varphi(x) = \sqrt{x}, \ I_h = [4r^2h^2, 1/h^2],$$

and

$$\Delta_{h\varphi}^r f(x) = \sum_{k=0}^r (-1)^k \binom{r}{k} f\left(x + \frac{h\sqrt{x}}{2}(r-2k)\right),$$

the equivalence

$$\|f\|_{L^2_{w,s}} \sim \|f\|_{L^2_w} + \left(\int_0^1 \left[\frac{\Omega_\varphi^r(f,t)_{L^2_w}}{t^{s+1/2}}\right]^2 dt\right)^{1/2} \qquad (5)$$

is true for $r \geq s \in R^+$ (see [2]).

Finally, we recall that (see [6]) for every positive integer s we have

$$L^2_{w,s} = W^2_s := \{f \in L^2_w : f^{(s-1)} \in AC \text{ and } \|f^{(s)}\varphi^s\|_{L^2_w} < \infty\}$$

and

$$\|f\|_{L^2_{w,s}} \sim \|f\|_{W^2_s} = \|f\|_{L^2_w} + \|f^{(s)}\varphi^s\|_{L^2_w},$$

where AC denotes the set of all functions which are absolutely continuous on any bounded subinterval of $(0,\infty)$, and the equality between the spaces of functions is in the sense of norm equivalence.

Among the many estimates derived in [7] and [8], we recall those which are most important for obtaining the results we will present in the next two sections.

Theorem 1.1. *If $f^{(r-1)} \in AC$ and $\|f^{(r)}\varphi^r w\|_{L^1} < \infty$ for some integer $r \geq 1$, then*

$$|R^\theta_m(f)| \leq c\left(\frac{\|f^{(r)}\varphi^r w\|_{L^1}}{m^{r/2}} + e^{-am}\|fw\|_{L^1}\right), \tag{6}$$

where the constants c and a are independent of m and f. If the f in (2) belongs to $L^2_{w,s}$ with $s > \frac{1}{2}$, then we have

$$|R^\theta_m(f)| \leq \frac{c}{m^{s/2}}\|f\|_{L^2_{w,s}}, \tag{7}$$

where the constant c is independent of m and f.

Theorem 1.2. *If the k_y in (4) belongs to L^2_w and if f is in $L^2_{w,s}$ with $s > \frac{1}{2}$, then*

$$|R^\theta_m(k_y; f)| \leq \frac{c\|k_y\|_{L^2_w}}{m^{s/2}}\|f\|_{L^2_{w,s}}, \tag{8}$$

where the constant c is independent of m, f and k_y.

Remark 1.3. Incidentally, we notice that to obtain an error bound similar to that of Theorem 1.2 for the complete rule (3), in [6] we had to assume that $f \in L^2_{w^q}$ with $w^q = e^{-qx}$ for some $q \in (0,1)$. These results were actually proved for the more general weight function $w(x) = w_\alpha(x) := x^\alpha e^{-x}, \alpha > -1$. The truncation trick employed here allows to take $q = 1$ as well.

In [8], the optimal behavior of rule (4) suggested the construction of Nyström type interpolants for the numerical solution of second kind integral equations on the half-line of the form

$$u(y) - \mu\int_0^\infty k_0(x,y)u(x)dx = f(y), \tag{9}$$

that is,

$$(I - \mu K)u = f \tag{10}$$

which have solutions $u(x)$ that decay exponentially at ∞, at least like $e^{-(\frac{1}{2}+\epsilon)x}$ with $\epsilon > 0$. In particular, equation (9) is preliminarily rewritten in the form

$$v(y) - \mu\int_0^\infty e^{-x}k(x,y)v(x)dx = g(y). \tag{11}$$

This can be accomplished directly by setting either $k(x,y) = e^x k_0(x,y)$ or

$$v(x) = e^x[u(x) - u(\infty)], \qquad k(x,y) = e^y k_0(x,y) \tag{12}$$

and

$$g(y) = e^y \left(f(y) - u(\infty) \left[1 - \mu \int_0^\infty k(x,y)dx \right] \right).$$

Then the integral in (11) is approximated by the product rule (4), which here takes the form

$$K_m v(y) := \sum_{i=1}^{j} w_{mi}(y) v(x_{mi}), \tag{13}$$

where j denotes the largest integer such that $x_{mj} \leq 4\theta m$. Thus we have the approximating equation

$$(I - \mu K_m) v_m = g \tag{14}$$

or, equivalently, the Nyström interpolant

$$v_m(y) = g(y) + \mu K_m v_m(y).$$

The values $\{v_m(x_{mi})\}$ needed to define $v_m(y)$ are obtained by solving the linear system

$$(I - \mu K_m) v_m(x_{ml}) = g(x_{ml}), \quad l = 1, \ldots, j.$$

The truncation of the quadrature rule also significantly reduces the condition number of the above linear system.

In [8], the setting (12) was introduced and the behavior of method (14) was examined in the weighted space

$$X := \{ v \in C[0,\infty) : \|v\|_X := \|\rho(x) v(x)\|_\infty < \infty \}, \tag{15}$$

where $\rho(x) = e^{(-\frac{1}{2}+\epsilon)x}, 0 < \epsilon < 1/2$. For the above Nyström interpolant the following main result was proved.

Theorem 1.3. *If*

$$\|K\|_{X \to X} = \sup_{y \geq 0} \rho(y) \|k_y \rho^{-1}\|_{L^1} < \infty \tag{16}$$

and

$$\sup_{y \geq 0} \rho(y) \|k_y\|_{L^2_w} < \infty, \tag{17}$$

then

$$\lim_{m \to \infty} \|K_m\|_{X \to X} \leq \|K\|_{X \to X}. \tag{18}$$

Moreover, if

$$\|K\|_{X \to X} < |\mu|^{-1}, \tag{19}$$

then for all m sufficiently large the operator $I - \mu K_m$ is invertible and

$$\|(I - \mu K_m)^{-1}\|_{X \to X} \leq c.$$

Under the above assumptions we finally have

$$\|v - v_m\|_X \leq c \frac{\|v\|_{L^2_{w,s}}}{m^{\frac{s}{2}}}$$

whenever $v \in L^2_{w,s}$ *for some* $s > \frac{1}{2}$.

To show the efficiency of the truncated rules (2) and (4), in particular when the second is used to solve an integral equation as described above, several numerical results were presented in [7] and [8].

In the next section we give another sufficient condition for the stability of the method and hence for the above convergence estimate. Unfortunately, this new condition is not satisfied for classical kernels like $k_0 = e^{-|x-y|}$ and $k_0(x, y) = E_1(|x - y|)$ with $E_1(z) = \displaystyle\int_z^\infty \frac{e^{-t}}{t} dt$, although practical computations seem to confirm that, at least in these two cases, our interpolants are indeed stable.

To be able to prove stability, in [8] we had to modify the interpolant $K_m v(y)$ near ∞, that is, for $y > c$ with an arbitrarily large constant $c > 0$. This has to be done also in the case of the new condition introduced here. However, since the solution $u(x)$ of our equations decay exponentially at ∞, in practice this modification is never used.

In the last section we consider a special class of Mellin type equations. After having transformed these into equivalent Wiener-Hopf equations, we propose to approximate them by using Nyström interpolants based on our truncated Gauss-Laguerre rules. Also for these interpolants, to be able to prove their stability we have to modify them around infinity. Some numerical examples are presented.

2. An alternative sufficient condition for stability

Here we consider the weight function $\rho(x) := \sqrt{w(x)}(1 + x)^b$, $b > \frac{1}{2}$, to which we associate the weighted (Banach) space

$$X := \{f \in C[0, \infty) : \|f\|_X := \|\rho f\|_\infty < \infty\}.$$

We also assume that our operator K is bounded on X and that equation (11) has a unique solution $v \in X$. Notice that this space X is different from the corresponding space (15), which was considered in [8] and was needed there only in order to make the calculation of some integrals trivial.

It is well known (see [1]) that the stability of a Nyström interpolant is also guaranteed by the condition

$$\|(K - K_m)K_m\|_{X \to X} \to 0, \ m \to \infty,$$

which can be satisfied even if $\|K\|_{X \to X} \geq |\mu|^{-1}$. The following theorem assures that this alternative stability condition is valid under certain assumptions on the kernel $k(x, y)$.

Theorem 2.1. *If the two assumptions*

$$(i) \quad \sup_{y\geq 0} \rho(y)\|k_y\|_{L^2_w} = A < \infty$$
$$(ii) \quad \sup_{x\geq 0} \rho(x)\|k_x\|_{L^2_{w,s}} = B < \infty \text{ for some } s > 1/2$$

are satisfied, then

$$\sup_{m\geq 1} m^{s/2}\|(K-K_m)K_m\|_{X\to X} \leq cAB$$

with c independent of m and k.

Proof. Recalling Theorem 1.2 we get, for $s > \frac{1}{2}$,

$$\|\rho[(K-K_m)K_m u]\|_\infty \leq \frac{cA}{m^{s/2}}\|K_m u\|_{L^2_{w,s}},$$

where

$$K_m u(y) = \int_0^\infty L_m(w,u_j,x)k(x,y)w(x)dx.$$

Now

$$\|K_m u\|^2_{L^2_{w,s}} = \sum_{k=0}^\infty (1+k)^s c_k^2(w,K_m u)$$

and

$$c_k(w,K_m u) \quad = \quad \int_0^\infty L_m(w,u_j,x)w(x)c_k(w,k_x)dx.$$

Thus, by Minkowski's inequality ([4], p. 148, (6.13.8)),

$$\begin{aligned}\|K_m u\|_{L^2_{w,s}} &\leq \int_0^\infty |L_m(w,u_j,x)|w(x)\left(\sum_{k=0}^\infty (1+k)^s c_k^2(w,k_x)\right)^{1/2} dx \\ &\leq B\|L_m(w,u_j)w\rho^{-1}\|_{L^1} \leq cB\|L_m(w,u_j)\sqrt{w}\|_{L^2}.\end{aligned}$$

Since (see [9])

$$\lambda_{mi} \sim e^{-x_{mi}}(x_{mi}-x_{m,i-1}), \qquad x_{m0}=0, \tag{20}$$

and, consequently,

$$\begin{aligned}\|L_m(w,u_j)\sqrt{w}\|_2^2 &= \sum_{i=1}^j \lambda_{mi}u^2(x_{mi}) \\ &\sim \sum_{i=1}^j [\sqrt{w(x_{mi})}(1+x_{mi})^b u(x_{mi})]^2 \frac{\Delta x_{mi}}{(1+x_{mi})^{2b}} \\ &\leq c\|u\|_X^2,\end{aligned}$$

Theorem 2.1 follows.

Notice that condition (i) of Theorem 2.1 coincides with (16). For the kernels considered here it is certainly satisfied if we take, for example, $\rho(x) = e^{-x/2}$, i.e., $b = 0$, but we had to assume $b > 1/2$ to obtain the final bound in the proof. Condition (ii) is even more severe because of the constraint $s > 1/2$. As in [8], to satisfy the assumptions of the theorem we should modify the interpolant for $y > c$, $c > 0$ being an arbitrarily large constant.

3. Mellin type equations

Here we consider equations of the form

$$u(s) - \mu \int_0^1 k_0\left(\frac{t}{s}\right) \frac{u(t)}{t} dt = f(s), \quad 0 < s \leq 1 \tag{21}$$

with a solution $u(t)$ that is smooth everywhere except at the origin, where it has a singular behavior of the type t^σ (or $t^\sigma \log t$) with $\sigma > 0$. Equations of this form have been analyzed in [3], where results useful for detecting the singular behavior of the solution at the origin were established.

In many applications the kernel $k(\tau)$ is smooth, even analytic. For example, a case of this type is given by the equation

$$u(s) + \frac{4}{\pi} \int_0^1 \frac{s^2 t}{(s^2 + t^2)^2} u(t) dt = s, \quad 0 < s \leq 1, \tag{22}$$

which describes the distribution of stress in a thin elastic plate in the vicinity of a cruciform crack. This equation has been examined in [11]. Its solution $u(t)$ is C^∞ in $(0, 1]$ and has a singular behavior at the origin, bounded by t^σ with $\sigma < 2 - \epsilon$ as close to 2 as we like. The kernel

$$k_0(\tau) = \frac{\tau^2}{(1 + \tau^2)^2}$$

is analytic and has two poles at $\tau = \pm i$. Another example is provided by equation (21) with $\mu = \frac{-\sin \lambda\pi}{\pi}$ and

$$k_0(\tau) = \frac{\tau}{1 + 2\tau \cos \lambda\pi + \tau^2}, \tag{23}$$

which occurs when the double layer potential is used to solve the Dirichlet problem in a domain which contains a corner with interior angle $(1 - \lambda)\pi$, $\lambda \in (-1, 1)$. In this case, the kernel $k_0(\tau)$ is also analytic, having two complex conjugate poles.

Several numerical methods have been proposed to solve equation (21) (see [3], [12]). These include projection methods as well as Nyström type methods. To prove their stability and optimal order of convergence, proper modifications of the approximants near the origin appear to be needed.

Here we propose to convert (21) into an equivalent Wiener-Hopf equation. This is a simple and well-known procedure (see [13]): by setting $t = e^{-x}$ and

$s = e^{-y}$ in (21) we obtain

$$v(y) - \mu \int_0^\infty k_0(e^{-(x-y)})v(x)dx = f(e^{-y}) \tag{24}$$

with

$$v(x) = u(e^{-x}).$$

Notice that in (24) the kernel is smooth, often analytic, and the solution $v(x)$ is smooth and bounded in $[0, \infty)$. The singular behavior of the original solution $u(t)$ at the origin is in this way transformed into an exponential decay of the new solution $v(x)$ at infinity. But this is precisely the behavior we like to have.

Now we can proceed as suggested in the introduction, that is, we rewrite the equation in the form (10) by taking

$$k(x, y) = e^x k_0(x, y),$$

and then we approximate the integral directly by the truncated Gauss-Laguerre formula (2), that is, in (12) we define

$$K_m v(y) := \sum_{i=1}^{j} \lambda_{mi} k(x_{mi}, y) v(x_{mi}).$$

This very simple approach appears more convenient, for example, when the kernel $k_0(\tau)$ is analytic and has (complex conjugate) poles which are not too close to the interval of integration. For instance, this happens in the case of the kernels occurring in (22) and (23).

Under certain assumptions on the kernel $k(x, y)$, we are able to prove stability and optimal convergence order in the Banach space $X = \{f \in C[0, \infty), \|f\|_\infty < \infty\}$ for our interpolant. Our assumptions include that the operator K is bounded and the corresponding equation has a unique solution.

Theorem 3.1. *If $k(x, y)$, $\frac{\partial k(x,y)}{\partial x}$ and $\frac{\partial k(x,y)}{\partial y}$ are piecewise monotonic with respect to x for each $y \geq 0$ and if, furthermore,*

$$\sup_{y \geq 0} \int_0^\infty \left(|k_0(x, y)| + \left| \frac{\partial k_0(x, y)}{\partial y} \right| \right) dx < \infty, \tag{25}$$

$$\sup_{y \geq 0} \int_0^\infty \sqrt{x} \left(|k_0(x, y)| + \left| \frac{\partial k_0(x, y)}{\partial x} \right| \right) dx < \infty, \tag{26}$$

then

$$\|(K - K_m)K_m\|_\infty \leq \frac{c}{\sqrt{m}}.$$

Proof. Estimate (6) with $r = 1$ gives

$$
\begin{aligned}
&|(K - K_m)K_m u(y)| \\
&\le \frac{c}{\sqrt{m}} \left(\int_0^\infty \left| \frac{\partial}{\partial t}[k(t,y)K_m u(t)] \right| \sqrt{t} e^{-t} dt + \int_0^\infty |k(t,y)K_m u(t)| e^{-t} dt \right) \\
&\le \frac{c}{\sqrt{m}} \left(\int_0^\infty \left| \frac{\partial}{\partial t} k(t,y) K_m u(t) \right| \sqrt{t} e^{-t} dt + \int_0^\infty \left| k(t,y) \frac{\partial}{\partial t} K_m u(t) \right| \sqrt{t} e^{-t} dt \right. \\
&\qquad \left. + \int_0^\infty |k(t,y) K_m u(t)| e^{-t} dt \right). \qquad (27)
\end{aligned}
$$

Due to (20) and since, moreover, we have assumed that $k(x,y)$, $\frac{\partial k(x,y)}{\partial x}$, and $\frac{\partial k(x,y)}{\partial y}$ are piecewise monotonic in x, we also have

$$
|K_m u(t)| \le \sum_{i=1}^{j} \lambda_{mi} |k(x_{mi}, t) u(x_{mi})| \le c \|u\|_\infty \int_0^\infty |k(x,t)| e^{-x} dx
$$

and

$$
\left| \frac{\partial}{\partial t} K_m u(t) \right| \le \sum_{i=1}^{j} \lambda_{mi} \left| \frac{\partial}{\partial t} k(x_{mi}, t) u(x_{mi}) \right| \le c \|u\|_\infty \int_0^\infty \left| \frac{\partial}{\partial t} k(x,t) \right| e^{-x} dx.
$$

These estimates follow because the above sums turn out to be bounded by Riemann sums. By inserting these two estimates in (27), we obtain

$$
\begin{aligned}
&|(K - K_m)K_m u(y)| \\
&\le \frac{c\|u\|_\infty}{\sqrt{m}} \left[\int_0^\infty \left| k_0(t,y) + \frac{\partial}{\partial t} k_0(t,y) \right| \sqrt{t} \left(\int_0^\infty |k_0(x,t)| dx \right) dt \right. \\
&\quad + \int_0^\infty |k_0(t,y)| \sqrt{t} \left(\int_0^\infty \left| \frac{\partial}{\partial t} k_0(x,t) \right| dx \right) dt \\
&\quad \left. + \int_0^\infty |k_0(t,y)| \left(\int_0^\infty |k_0(x,t)| dx \right) dt \right].
\end{aligned}
$$

Recalling the bounds stated in (25) and (26), the theorem follows.

Unfortunately, condition (26) is not satisfied by the two Mellin kernels we have considered in this section. This happens also for Picard's kernel $k_0(x,y) = e^{-|x-y|}$. Indeed, a direct calculation gives

$$
|(K - K_m)K_m u(y)| \le c \sqrt{\frac{y}{m}} \|u\|_\infty. \qquad (28)
$$

In this case we have

$$
|(K - K_m)K_m u(y)| = o(1) \|u\|_\infty, \qquad m \to \infty,
$$

uniformly with respect to y, in any interval of the form $[0, c\frac{m}{\log m}]$ with $c > 0$ arbitrarily large, but not on $[0, \infty)$. This means that we have to modify our interpolant $K_m u(y)$ for $y > c\frac{m}{\log m}$.

Thus, for simplicity we define the new interpolant

$$K_m^* u(y) = \begin{cases} K_m u(y), & 0 \le y \le c\frac{m}{\log m}, \\ Ku(y), & y > c\frac{m}{\log m}, \end{cases} \tag{29}$$

so that now the condition

$$\|(K - K_m^*)K_m^*\|_\infty \to 0 \quad (m \to \infty)$$

is guaranteed and, because of Theorem 1.1, the order of convergence is $O(m^{-s/2})$ whenever $f \in L^2_{w,s}$ with $s > 1/2$ or $f^{(s-1)} \in AC$ and $\|f^{(s)}\varphi^s w\|_{L^1} < \infty$ with $s \ge 1$.

To be sure that all the nodes of the truncated Gauss-Laguerre formula (2) belong to the interval $[0, c\frac{m}{\log m}]$ we can proceed in two alternative directions. In (2) we could consider all nodes $0 < x_{mi} \le 4\theta\frac{m}{\log m}$. In this case it is possible to verify that the proofs of Theorems 1.1 and 1.2 lead to similar estimates with m replaced by $\lfloor\frac{m}{\log m}\rfloor$. But we could also recall that a sufficient condition for the stability of our interpolant is

$$\|(K - K_m^*)K_m^*\|_\infty < \frac{1}{\|(I - \mu K)^{-1}\|_\infty},$$

(see [1]) once we have assumed that $\|(I - \mu K)^{-1}\|_\infty < c$. Therefore by modifying our interpolant for $y > 4\theta m$ and taking $0 < \theta \le 1$ sufficiently small, the above stability condition follows from (28). In this case the convergence order is still optimal.

In the following tables we report some relative errors we obtained by solving equation (22) by the unmodified version of the method we have proposed. The reference values $u(x)$ were computed by our Nyström method taking $m = 512$ and $j = 161$ $(\theta = 1/16)$. All computations were performed on a PC using MATLAB.

Table 1

x	$u(x)$
0.01	4.998606531015860e-3
0.05	2.489714170026475e-2
0.1	5.017241798296036e-2
0.2	1.062334280080870e-1
0.4	2.521118292992208e-1
0.6	4.378859011179868e-1
0.8	6.462622497816191e-1
1.0	8.635420987843082e-1

Table 2 θ=1/8 $\lvert u(x) - u_n(x)\rvert$							
x	n=4	n=8	n=16	n=32	n=64	n=128	n=256
0.01	4.15e-3	3.76e-3	2.72e-4	1.13e-5	8.11e-7	1.41e-8	3.01e-12
0.05	6.4 e-3	2.21e-3	4.39e-4	3.7 e-5	3.08e-7	3.18e-10	3.36e-13
0.1	3.3 e-3	1.19e-3	1.83e-4	5.87e-6	7.49e-8	9.05e-10	8.69e-14
0.2	1.15e-3	2.52e-3	2.57e-4	3.52e-6	1.21e-8	1.36e-11	3.68e-15
0.4	1.05e-2	8.49e-4	1.52e-5	1.81e-7	3.38e-9	1.01e-13	3.22e-15
0.6	1.69e-3	2.96e-4	5.38e-5	6.76e-7	4.64e-11	1.79e-12	1.44e-15
0.8	3.92e-3	8.5 e-4	1.54e-6	2.11e-8	1.19e-10	1.32e-12	6.66e-16
1.0	4.83e-3	6.05e-4	1.5 e-5	7.49e-8	1.51e-10	9.16e-13	4.44e-16
	j=2	j=3	j=7	j=14	j=28	j=56	j=113

Table 3 θ=1/16 $\lvert u(x) - u_n(x)\rvert$						
x	n=8	n=16	n=32	n=64	n=128	n=256
0.01	4.74e-3	1.68e-3	1.13e-5	8.11e-7	1.41e-8	3.01e-12
0.05	1.87e-2	4.08e-4	3.7 e-5	3.08e-7	3.19e-10	3.36e-13
0.1	6.6 e-2	1.82e-4	5.87e-6	7.49e-8	9.05e-10	8.69e-14
0.2	2.02e-2	2.56e-4	3.52e-6	1.21e-8	1.36e-11	3.68e-15
0.4	4.18e-3	1.54e-5	1.81e-7	3.38e-9	1.01e-13	3.22e-15
0.6	2.04e-3	5.38e-5	6.76e-7	4.64e-11	1.79e-12	1.44e-15
0.8	1.57e-3	1.56e-6	2.11e-8	1.19e-10	1.32e-12	6.66e-16
1.0	9.4 e-4	1.5 e-5	7.49e-8	1.51e-10	9.16e-13	4.44e-16
	j=2	j=5	j=10	j=20	j=40	j=80

Table 4 θ=1/32 $\lvert u(x) - u_n(x)\rvert$						
x	n=8	n=16	n=32	n=64	n=128	n=256
0.01	4.74e-3	4.69e-3	1.98e-3	8.18e-7	1.41e-8	3.01e-12
0.05	1.87e-2	1.77e-2	1.79e-4	3.08e-7	3.18e-10	3.36e-13
0.1	6.6 e-2	2.29e-2	1.98e-5	7.49e-8	9.06e-10	8.69e-14
0.2	2.02e-2	1.62e-2	3.73e-6	1.21e-8	1.36e-11	3.68e-15
0.4	4.18e-3	4.6 e-3	5.38e-7	3.38e-9	1.01e-13	3.22e-15
0.6	2.04e-3	1.69e-3	5.23e-7	4.65e-11	1.79e-12	1.44e-15
0.8	1.57e-3	6.97e-4	9.47e-8	1.19e-10	1.32e-12	6.66e-16
1.0	9.41e-4	3.25e-4	1.15e-7	1.51e-10	9.16e-13	4.44e-16
	j=2	j=3	j=7	j=14	j=28	j=57

As one can see, the results obtained taking $\theta = 1/32$, that is, by considering less than 1/4 of the nodes, are very satisfactory.

In the case of equation (21) with kernel (23) the corresponding results are very similar.

Acknowledgements. The authors are grateful to Dr. C. Frammartino for performing all the numerical computations reported in Section 4.

References

[1] K.E. Atkinson, *The Numerical Solution of Integral Equations of the Second Kind*, Cambridge University Press, 1997.

[2] M.C. De Bonis, G. Mastroianni and M. Viggiani, *K-functionals, moduli of smoothness and weighted best approximation on the semi-axis*, to appear in the Proceedings of the Conference in Honor of G. Alextis.

[3] J. Elschner, *On spline approximation for a class of non-compact integral equations*, Math. Nachr. 146 (1990), 271–321.

[4] G.H. Hardy, J. E. Littlewood and G. Pólya, *Inequalities*, Cambridge Univ. Press, 1983.

[5] G. Mastroianni and G. Monegato, *Nyström interpolants based on the zeros of Legendre polynomials*, IMA J. Numer. Anal. 14 (1993), 81–95.

[6] G. Mastroianni and G. Monegato, *Nyström interpolants based on zeros of Laguerre polynomials for some Wiener-Hopf equations*, IMA J. Numer. Anal. 17 (1997), 621–642.

[7] G. Mastroianni and G. Monegato, *Truncated Gauss-Laguerre rules*, Recent trends in Numerical Analysis (D.Trigiante, ed.), Nova Science, 2000, pp. 213–221.

[8] G. Mastroianni and G. Monegato, *Truncated quadrature rules over* $(0, \infty)$ *and Nyström type methods*, submitted.

[9] G. Mastroianni and D. Occorsio, *Lagrange interpolation at Laguerre zeros in some weighted uniform spaces*, Acta Math. Hungar. 91 (1-2) (2001), 27–52.

[10] G. Mastroianni, *Polynomial inequalities, functional spaces and best approximation on the real semiaxis with Laguerre weights*, to appear in ETNA.

[11] G. Monegato and S. Prössdorf, *On the numerical treatment of an integral equation arising from a cruciform crack problem*, Math. Meth. Appl. Sci. 12 (1990), 489–502.

[12] G. Monegato, *A stable Nyström interpolant for some Mellin convolution equations*, Numer. Algorithms 11 (1996), 271–283.

[13] S. Prössdorf and B. Silbermann, *Numerical Analysis for Integral and Related Operator Equations*, Birkhäuser Verlag, Basel 1991.

G. Mastroianni
Dipartimento di Matematica
Università della Basilicata
Potenza, Italy
e-mail: `mastroianni@unibas.it`

G. Monegato
Dipartimento di Matematica
Politecnico di Torino
Torino, Italy
e-mail: `monegato@polito.it`

Received: 28 September 2001

Operator Theory:
Advances and Applications, Vol. 135, 261–265

A 1-parameter Scale of Closed Ideals Formed by Strictly Singular Operators

Albrecht Pietsch

To Bernd Silbermann on the occasion of his sixtieth birthday

Abstract. Let $\mathfrak{K}$ and $\mathfrak{S}$ denote the closed ideals consisting of all compact and all strictly singular operators, respectively. Kato [10] observed that $\mathfrak{K} \subseteq \mathfrak{S}$, and Goldberg/Thorp [7] showed that this inclusion is proper. At first glance, the difference $\mathfrak{S} \setminus \mathfrak{K}$ appears to be relatively small. This is not true (in the following sense): there exists a continuum of closed ideals $\mathfrak{X}_p$ with $2 \le p < \infty$ which are located between $\mathfrak{K}$ and $\mathfrak{S}$.

The crucial point of this note is the discovery of ideals having the desired property, while the proofs are elementary.

1. The operator ideals

Let $\mathfrak{L}(X,Y)$ denote the Banach space of all bounded linear operators $T : X \to Y$, where X and Y are (real or complex) Banach spaces. Fix an exponent $1 \le p < \infty$ and $n = 1, 2, \dots$. Then there exists a constant $c \ge 0$ such that

$$\left(\sum_{k=1}^{n} \|Tx_k\|^p\right)^{1/p} \le c \sup\left\{\left(\sum_{k=1}^{n} |\langle x_k, x^*\rangle|^p\right)^{1/p} : \|x^*\| \le 1\right\}$$

for any choice of $x_1, \dots, x_n \in X$. The infimum of all possible c's will be denoted $\pi_p^{(n)}(T)$. This quantity is referred to as the p-summing norm of T *computed with n vectors.*

Here is a list of inequalities which hold for $S, T \in \mathfrak{L}(X,Y)$ as well as for $A \in \mathfrak{L}(X_o, X)$ and $B \in \mathfrak{L}(Y, Y_o)$:

(1) $\pi_p^{(n)}(S+T) \le \pi_p^{(n)}(S) + \pi_p^{(n)}(T)$,

(2) $\pi_p^{(n)}(BTA) \le \|B\| \pi_p^{(n)}(T) \|A\|$,

(3) $\|T\| = \pi_p^{(1)}(T) \le \dots \le \pi_p^{(n)}(T) \le \dots$,

(4) $\pi_p^{(n)}(T) \le n^{1/p} \|T\|$,

(5) $\pi_q^{(n)}(T) \le \pi_p^{(n)}(T) \le n^{1/p-1/q} \pi_q^{(n)}(T)$ whenever $1 \le p \le q < \infty$.

For a proof of the left-hand inequality in **(5)** the reader is referred to [13, pp. 335–336]. The rest is trivial.

As stated in **(4)**, the optimal growth of $(\pi_p^{(n)}(T))$ is like $n^{1/p}$. So we may hope that operators with suboptimal growth form an interesting subclass: $\mathfrak{X}_p$. Suboptimality means that $(\pi_p^{(n)}(T))$ is a 'little oh' of $n^{1/p}$:

$$\lim_{n\to\infty} n^{-1/p}\pi_p^{(n)}(T) = 0.$$

The elementary proof of the following result can be found in [14, p. 31].

Proposition 1. *$\mathfrak{X}_p$ is a closed operator ideal.*

Next, we state a consequence of the right-hand inequality in **(5)**.

Proposition 2. *$\mathfrak{X}_p \supseteq \mathfrak{X}_q$ whenever $1 \le p \le q < \infty$.*

The main result of this paper consists in showing that the scale $(\mathfrak{X}_p)$ is *strictly* decreasing.

The subsequent theorem from [9, p. 310] illustrates why the identity map of l_{p^*} into l_∞ plays an important role as an example and counterexample:

An operator $T : X \to Y$ is a member of $\mathfrak{X}_p$ if and only if it fails to factor the identity maps $l_{p^*}^n \to l_\infty^n$ uniformly.

2. The examples

An operator $T : X \to Y$ is $(p, 1)$-*summing* if there exists a constant $c \ge 0$ such that

$$\Big(\sum_{k=1}^n \|Tx_k\|^p\Big)^{1/p} \le c \sup\Big\{\sum_{k=1}^n |\langle x_k, x^*\rangle| : \|x^*\| \le 1\Big\}$$

for any choice of $x_1, \ldots, x_n \in X$ and $n = 1, 2, \ldots$. The infimum of all possible c's will be denoted $\pi_{(p,1)}(T)$.

Defining the dual exponent p^* by $1/p + 1/p^* = 1$, we conclude from

$$\sum_{k=1}^n \|Tx_k\| \le n^{1/p^*}\Big(\sum_{k=1}^n \|Tx_k\|^p\Big)^{1/p}$$

that

(6) $\pi_1^{(n)}(T) \le n^{1/p^*}\pi_{(p,1)}(T)$.

The crucial tool for the following considerations is the famous Orlicz-Bennett-Carl theorem [11, 1, 3]:

Let $1/u + 1/v \le 1$ and $2 \le v \le \infty$. Then the identity map from l_{u^*} into l_v is $(u^*, 1)$-summing.

If (α_n) and (β_n) are sequences of positive numbers, then the symbol $\alpha_n \asymp \beta_n$ means that $\alpha_n \le c\beta_n$ and $\beta_n \le c\alpha_n$ for $n = 1, 2, \ldots$ and some constant $c \ge 1$.

Lemma 1. *If $1 \le p \le u \le \infty$, $1/u + 1/v \le 1$ and $2 \le v \le \infty$, then*

$$\pi_p^{(n)}(Id : l_{u^*}^n \to l_v^n) \asymp \pi_p^{(n)}(Id : l_{u^*} \to l_v) \asymp n^{1/u}.$$

Proof. The upper estimate follows from

$$\pi_p^{(n)}(Id : l_{u^*}^n \to l_v^n) \le \pi_p^{(n)}(Id : l_{u^*} \to l_v)$$

as well as inequalities **(5)** and **(6)**:

$$\pi_p^{(n)}(Id : l_{u^*} \to l_v) \le \pi_1^{(n)}(Id : l_{u^*} \to l_v) \le n^{1/u}\pi_{(u^*,1)}(Id : l_{u^*} \to l_v).$$

On the other hand, substituting the identity map $Id : l_{u^*}^n \to l_v^n$ and the unit vectors $e_1, \dots, e_n \in l_{u^*}^n$ into

$$\left(\sum_{k=1}^n \|Tx_k\|^p\right)^{1/p} \le \pi_p^{(n)}(T) \sup\left\{\left(\sum_{k=1}^n |\langle x_k, x^*\rangle|^p\right)^{1/p} : \|x^*\| \le 1\right\}$$

yields

$$\begin{aligned} n^{1/p} &\le \pi_p^{(n)}(Id : l_{u^*}^n \to l_v^n) \sup\left\{\|x^*|l_p^n\| : \|x^*|l_u^n\| \le 1\right\} \\ &\le \pi_p^{(n)}(Id : l_{u^*}^n \to l_v^n) n^{1/p-1/u}. \end{aligned}$$

Example. *If $1 \le p < q < \infty$, then the identity map from l_{q^*} into l_∞ is contained in $\mathfrak{X}_p \setminus \mathfrak{X}_q$.*

Proof. Note that

$$\pi_p^{(n)}(Id : l_{q^*} \to l_\infty) \asymp n^{1/q} = o(n^{1/p})$$

and

$$\pi_q^{(n)}(Id : l_{q^*} \to l_\infty) \asymp n^{1/q} \ne o(n^{1/q}).$$

Lemma 2. $\pi_p^{(n)}(Id : l_2^n \to l_2^n) = n^{1/p}$ *if* $2 \le p < \infty$.

Proof. The upper estimate follows from inequality **(4)**.

On the other hand, substituting the identity map $Id : l_2^n \to l_2^n$ and the unit vectors $e_1, \dots, e_n \in l_2^n$ into

$$\left(\sum_{k=1}^n \|Tx_k\|^p\right)^{1/p} \le \pi_p^{(n)}(T) \sup\left\{\left(\sum_{k=1}^n |\langle x_k, x^*\rangle|^p\right)^{1/p} : \|x^*\| \le 1\right\}$$

yields

$$n^{1/p} \le \pi_p^{(n)}(Id : l_2^n \to l_2^n) \sup\left\{\|x^*|l_p^n\| : \|x^*|l_2^n\| \le 1\right\} \le \pi_p^{(n)}(Id : l_2^n \to l_2^n).$$

3. Strictly singular operators

Let Y_o be a superspace of a Banach space Y, and denote the embedding map from Y into Y_o by J. Given any ideal $\mathfrak{A}$, then

$$T \in \mathfrak{A}(X, Y) \;\Rightarrow\; JT \in \mathfrak{A}(X, Y_o).$$

The ideal $\mathfrak{A}$ is called *injective* [12, pp. 70–72] if we even have

$$T \in \mathfrak{A}(X, Y) \;\Leftrightarrow\; JT \in \mathfrak{A}(X, Y_o).$$

In other words, the question whether an operator $T : X \to Y$ belongs to $\mathfrak{A}$ does not depend on the size of its target space Y.

An ideal $\mathfrak{A}$ is said to be *proper* if it contains the identity map I_X of a Banach space X only when X is finite-dimensional.

Following Kato [10, p. 284], an operator $T : X \to Y$ is called *strictly singular* if

$$\|Tx\| \geq c\|x\| \quad \text{for all members } x \text{ of a closed subspace } M \text{ and some } c > 0$$

implies that M is finite-dimensional; see also [6, pp. 76–93] and [12, pp. 48–50]. The collection of these operators will be denoted by $\mathfrak{S}$. It easily turns out that *$\mathfrak{S}$ is the largest proper, injective and closed ideal.*

As a counterpart, we state that

$\mathfrak{K}$ is the smallest proper, injective and closed ideal.

Letting $J : Y \to Y_o$ as above, we have $\pi_p^{(n)}(T) = \pi_p^{(n)}(JT)$. Therefore all ideals $\mathfrak{X}_p$ are injective.

The next result is a little bit surprising.

Proposition 3. *The ideal $\mathfrak{X}_p$ is proper if and only if $2 \leq p < \infty$.*

Proof. In the case $1 \leq p < 2$, we infer from $\pi_p^{(n)}(Id : l_2 \to l_2) \asymp n^{1/2}$ that the identity map $Id : l_2 \to l_2$ is a member of $\mathfrak{X}_p$.

Next, we let $2 \leq p < \infty$. By Dvoretzky's theorem [4], every infinite-dimensional Banach space X contains n-dimensional subspaces E_n which are close to the Euclidian space l_2^n:

$$E_n \underset{U_n^{-1}}{\overset{U_n}{\rightleftarrows}} l_2^n \quad \text{and} \quad \|U_n\|\|U_n^{-1}\| \leq 2.$$

Hence

$$2\pi_p^{(n)}(Id : X \to X) \geq \|U_n\|\pi_p^{(n)}(Id : E_n \to E_n)\|U_n^{-1}\| \geq \pi_p^{(n)}(Id : l_2^n \to l_2^n) = n^{1/p}.$$

This implies that $I_X \notin \mathfrak{X}_p$. So $\mathfrak{X}_p$ is proper.

Finally, we use the fact that every proper, injective and closed ideals is located between $\mathfrak{K}$ (the smallest) and $\mathfrak{S}$ (the largest).

Theorem. $\mathfrak{K} \subset \mathfrak{X}_q \subset \mathfrak{X}_p \subset \mathfrak{X}_2 \subset \mathfrak{S}$ *for* $2 < p < q < \infty$.

The preceding inclusions are strict.

4. Banach spaces with many closed ideals

The starting point of the theory of operator ideals was Calkin's theorem [2]. It says that the ring $\mathfrak{L}(l_2)$ only contains one non-trivial closed ideal, namely $\mathfrak{K}(l_2)$. The same result was proved by Gohberg/Markus/Feldman [5] for the Banach spaces l_p with $1 \leq p < \infty$ and c_o. It seems to be unknown whether there exist further spaces having this property.

On the other hand, Gramsch [8] showed that, for a non-separable Hilbert spaces H, the ring $\mathfrak{L}(H)$ contains a chain of closed ideals whose length is determined by the dimension of H.

Therefore, one may use the ideal structure of $\mathfrak{L}(X)$ as a measure of complexity of the underlying space X. In this sense, it is interesting to state that there are *separable* (and reflexive) Banach spaces X such that $\mathfrak{L}(X)$ contains a continuum of different closed ideals. The examples are easy: take any sequence of exponents $1 < p_n < \infty$ which is dense in $(1, \infty)$ and form the direct l_2-sum of the corresponding l_{p_n}'s.

References

[1] G. Bennett, *Inclusion mappings between l^p spaces*, J. Funct. Anal. **13** (1973), 20–27.

[2] J.W. Calkin, *Two-sided ideals and congruences in the ring of bounded operators in Hilbert space*, Ann. of Math. **42** (1941), 839–873.

[3] B. Carl, *Absolut-$(p,1)$-summierende Operatoren von l_u in l_v*, Math. Nachr. **63** (1974), 353–360.

[4] A. Dvoretzky, *Some results on convex bodies and Banach spaces*, in: Proc. Symp. on Linear Spaces, Jerusalem 1961, 123–160.

[5] I.C. Gohberg, A.S. Markus and I.A. Feldman, *Normally solvable operators and ideals associated with them*, Izvestiya Moldavskogo Filiala Akad. Nauk SSSR **10** (1960), 51–70 [Russian].

[6] S. Goldberg, *Unbounded Linear Operators*, MacGraw–Hill, New York, 1966.

[7] S. Goldberg and E.O. Thorp, *On some open questions concerning strictly singular operators*, Proc. Amer. Math. Soc. **14** (1963), 334–336.

[8] B. Gramsch, *Eine Idealstruktur Banachscher Operatoralgebren*, J. Reine Angew. Math. **225** (1967), 97–115.

[9] A. Hinrichs and A. Pietsch, *The closed ideal of (c_o, p, q)-summing operators*, Integral Equations Operator Theory **38** (2000), 302–316.

[10] T. Kato, *Perturbation theory for nullity, deficiency and other quantities of linear operators*, J. Analyse Math. **6** (1958), 261–322.

[11] W. Orlicz, *Über unbedingte Konvergenz in Funktionenräumen* I–II, Studia Math. **1** (1933), 33–37; 41–47.

[12] A. Pietsch, *Operator Ideals*, Deutscher Verlag der Wissenschaften, Berlin, 1978; North-Holland, Amsterdam, London, New York, Tokyo, 1980.

[13] A. Pietsch, *Absolut p-summierende Operatoren in normierten Räumen*, Studia Math. **28** (1967), 333–353.

[14] A. Pietsch and J. Wenzel, *Orthonormal Systems and Banach Space Geometry*, Cambridge Univ. Press, 1998.

A. Pietsch
Mathematisches Institut
Friedrich–Schiller–Universität
D–07740 Jena
e-mail: `pietsch@minet.uni-jena.de`

Received: 29 August 2001

Operator Theory:
Advances and Applications, Vol. 135, 267–291

Local Theory of the Fredholmness of Band-Dominated Operators with Slowly Oscillating Coefficients

Vladimir S. Rabinovich and Steffen Roch

Dedicated to Bernd Silbermann on his sixtieth birthday

Abstract. A band-dominated operator on an l^p-space of vector-valued functions is known to be a Fredholm operator (in a generalized sense) if and only if all of its limit operators are invertible and if their inverses are uniformly bounded. We show that the limit operators approach is also compatible with the local Fredholmness of band-dominated operators with respect to localization over the maximal ideal space of the algebra of the slowly oscillating scalar-valued functions. A corollary of this result is that the uniform boundedness condition is redundant for band-dominated operators with slowly oscillating operator-valued coefficients.

1. Introduction

Let X be a complex Banach space. For $p \in (1, \infty)$ and N a positive integer, consider the Banach spaces $l^p(\mathbb{Z}^N, X)$ and $l^\infty(\mathbb{Z}^N, X)$ of all functions f which are defined on $\mathbb{Z}^N$ and take values in X such that

$$\|f\|_p^p := \sum_{x \in \mathbb{Z}^N} \|f(x)\|_X^p < \infty \quad \text{and} \quad \|f\|_\infty := \sup_{x \in \mathbb{Z}^N} \|f(x)\|_X < \infty,$$

respectively. Further, $c_0(\mathbb{Z}^N, X)$ refers to the closed subspace of $l^\infty(\mathbb{Z}^N, X)$ consisting of all functions f with

$$\lim_{x \to \infty} \|f(x)\|_X = 0.$$

In case $X = \mathbb{C}$, we will simply write $l^p(\mathbb{Z}^N)$ and $c_0(\mathbb{Z}^N)$, and we let E stand for one of the spaces $l^p(\mathbb{Z}^N, X)$ with $p \in (1, \infty)$.

Every function $a \in l^\infty_{L(X)} := l^\infty(\mathbb{Z}^N, L(X))$ gives rise to a multiplication operator on E:

$$(af)(x) = a(x)f(x), \quad x \in \mathbb{Z}^N.$$

We denote this operator by aI. Evidently, $aI \in L(E)$ and $\|aI\|_{L(E)} = \|a\|_\infty$. Finally, for $\alpha \in \mathbb{Z}^N$, let V_α refer to the shift operator

$$(V_\alpha f)(x) = f(x - \alpha), \quad x \in \mathbb{Z}^N,$$

which also belongs to $L(E)$ and has norm 1.

Definition 1.1. *A* band operator *is a finite sum of the form* $\sum_\alpha a_\alpha V_\alpha$ *where* $\alpha \in \mathbb{Z}^N$ *and* $a_\alpha \in l^\infty(\mathbb{Z}^N, L(X))$. *A* band-dominated operator *is the uniform limit of a sequence of band operators.*

The band-dominated operators on E form a closed subalgebra of $L(E)$ which we denote by $\mathcal{A}_E$. (For this and the following facts we refer to the papers [5, 6].)

Given $m \in \mathbb{Z}^N$, let s_m stand for the function on $\mathbb{Z}^N$ which is $I \in L(X)$ at m and 0 at all other points. The operator of multiplication by s_m will be denoted by S_m. For $n \geq 0$, define $P_n := \sum_{|m| \leq n} S_m$ and $Q_n := I - P_n$, and let $\mathcal{P}$ refer to the family (P_n).

Definition 1.2. *An operator* $K \in L(E)$ *is* $\mathcal{P}$-compact *if*

$$\|KQ_n\| \to 0 \quad \textit{and} \quad \|Q_n K\| \to 0 \quad \textit{as } n \to \infty.$$

By $K(E, \mathcal{P})$ *we denote the set of all* $\mathcal{P}$*-compact operators on* E*, and by* $L(E, \mathcal{P})$ *the set of all operators* $A \in L(E)$ *for which both* AK *and* KA *are* $\mathcal{P}$*-compact whenever* K *is* $\mathcal{P}$*-compact.*

It turns out that $L(E, \mathcal{P})$ is a closed subalgebra of $L(E)$, $K(E, \mathcal{P})$ is a closed two-sided ideal of $L(E, \mathcal{P})$, and $K(E, \mathcal{P}) \subset \mathcal{A}_E \subset L(E, \mathcal{P})$. Operators $A \in L(E, \mathcal{P})$ for which the coset $A + K(E, \mathcal{P})$ is invertible in the quotient algebra $L(E, \mathcal{P})/K(E, \mathcal{P})$ are called $\mathcal{P}$-Fredholm. If X is a finite-dimensional space, then $L(E, \mathcal{P}) = L(E)$, $K(E, \mathcal{P})$ is the ideal of the compact operators on E, and the $\mathcal{P}$-Fredholm operators are just the Fredholm operators in the common sense. Let further $\mathcal{H}$ stand for the set of all sequences $h = (h(m))_{m=0}^\infty \subset \mathbb{Z}^N$ which tend to infinity.

Definition 1.3. *Let* $A \in L(E, \mathcal{P})$ *and* $h \in \mathcal{H}$. *The operator* $A_h \in L(E)$ *is called* limit operator of A with respect to h *if*

$$\lim_{n \to \infty} \|(V_{-h(n)} A V_{h(n)} - A_h) P_m\| = \lim_{n \to \infty} \|P_m (V_{-h(n)} A V_{h(n)} - A_h)\| = 0 \qquad (1)$$

for every $P_m \in \mathcal{P}$. *The set* $\sigma_{op}(A)$ *of all limit operators of* A *is called the* operator spectrum *of* A.

An operator can possess only one limit operator with respect to a given sequence which justifies the notation A_h.

We let finally $\mathcal{A}_E^{rich}$ refer to the set of all operators $A \in \mathcal{A}_E$ the operator spectrum of which contains sufficiently many operators (or is *rich*) in the following sense: every sequence h tending to infinity possesses a subsequence g for which the limit operator A_g exists. Then the main result of [6] can be stated as follows.

Theorem 1.4. *An operator $A \in \mathcal{A}_E^{rich}$ is $\mathcal{P}$-Fredholm if and only if all of its limit operators are invertible and if*

$$\sup\{\|(A_h)^{-1}\| : A_h \in \sigma_{op}(A)\} < \infty. \tag{2}$$

It is the main goal of this paper to discuss and weaken the uniform invertibility condition (2). To reach this goal, we examine several local theories of $\mathcal{P}$-Fredholmness. To describe some typical ideas and results we have to introduce some more notations. Let S^{N-1} denote the unit sphere $\{\eta \in \mathbb{R}^N : |\eta|_2 = 1\}$ where $|\eta|_2$ stands for the Euclidean norm of η. Given a 'radius' $R > 0$, a 'direction' $\eta \in S^{N-1}$, and a neighborhood $U \subseteq S^{N-1}$ of η, define

$$W_{R,U} := \{z \in \mathbb{Z}^N : |z| > R \text{ and } z/|z| \in U\}. \tag{3}$$

We will call $W_{R,U}$ a *neighborhood at infinity of* η. If h is a sequence which tends to infinity, then we say that h tends into the direction of $\eta \in S^{N-1}$ if, for every neighborhood at infinity $W_{R,U}$ of η, there is an m_0 such that

$$h(m) \in W_{R,U} \quad \text{for all } m \geq m_0.$$

Definition 1.5. *Let $\eta \in S^{N-1}$ and $A \in L(E)$.*

(a) *The* local operator spectrum $\sigma_\eta(A)$ of A at η *is the set of all limit operators A_h of A with respect to sequences h tending into the direction of η.*

(b) *The operator A is* locally invertible at η *if there are operators $B, C \in L(E)$ and a neighborhood at infinity W of η such that*

$$BA\hat{\chi}_W I = \hat{\chi}_W AC = \hat{\chi}_W I$$

where $\hat{\chi}_W$ refers to the characteristic function of W.

The following theorem and its corollary (which is also partially based on Theorem 6.8 below) have been shown in [5, 6].

Theorem 1.6. *Let $A \in \mathcal{A}_E^{rich}$ and $\eta \in S^{N-1}$. Then the operator A is locally invertible at η if and only if all limit operators in $\sigma_\eta(A)$ are invertible and if*

$$\sup\{\|(A_h)^{-1}\| : A_h \in \sigma_\eta(A)\} < \infty.$$

Corollary 1.7. *An operator $A \in \mathcal{A}_E^{rich}$ is $\mathcal{P}$-Fredholm if and only if all of its limit operators are invertible and if*

$$\sup\{\|(A_h)^{-1}\| : A_h \in \sigma_\eta(A)\} < \infty \quad \textit{for all } \eta \in S^{N-1}.$$

Observe that this is a true generalization of Theorem 1.4 since it is not required in the corollary that the suprema are uniformly bounded with respect to η.

In the present paper we will show that an analogous result holds if the sphere S^{N-1} is replaced by the fiber $M^\infty(SO)$ at infinity of the maximal ideal space of the algebra of the slowly oscillating functions on $\mathbb{Z}^N$. This fiber is much larger than S^{N-1}, hence, the resulting localization is much finer, and this localization will provide a further essential improvement of Theorem 1.4. It should be also noted that the localization over $M^\infty(SO)$ is, in some sense, the finest possible.

It is due to the topological properties of the maximal ideal space of the algebra of the slowly oscillating functions that we have to replace sequences tending to infinity by general nets tending to infinity. This requires some additional work which is done in Sections 2–5. In particular, we will derive a version of Cantor's diagonalization procedure for nets in place of sequences. Sections 6 and 7 are devoted to the proof of the local Fredholm criterion and of one of its consequences, which states that a band-dominated operator with rich slowly oscillating coefficients is $\mathcal{P}$-Fredholm if and only if all of its limit operators are invertible (Theorem 7.2). Thus, for these operators, the uniform invertibility of the inverses of the limit operators is not needed to guarantee their $\mathcal{P}$-Fredholmness, which is a second main result of the present paper. In the course of the proof we will also see that the method of limit operators is compatible with another local theory, the so-called local principle by Allan (Theorems 6.8 and 6.10 below). The final section contains an alternative proof of Theorem 7.2 which borrows some arguments from the symbol calculus for pseudodifferential operators, and which sheds new light upon the properties of band-dominated operators with slowly oscillating coefficients.

The authors are grateful for the support by the CONACYT project 32424-E and by the DFG grant 436 RUS 17/24/01.

2. Slowly oscillating functions

A function $a \in l^\infty_{L(X)}$ is *slowly oscillating* if

$$\lim_{x \to \infty} (a(x+k) - a(x)) = 0 \quad \text{for all } k \in \mathbb{Z}^n. \tag{4}$$

We denote the class of all slowly oscillating functions in $l^\infty_{L(X)}$ by $SO_{L(X)}$ and write SO instead of $SO_{L(\mathbb{C})}$ for brevity. Trivial examples of slowly oscillating functions are provided by the continuous functions on $\mathbb{Z}^N$, which possess a limit at infinity, whereas $\mathbb{Z} \to \mathbb{C} : x \mapsto \sin \sqrt{|x|}$ is an example of a slowly oscillating function which does not have this property.

It follows essentially from the definition of the class SO that a function a is slowly oscillating if and only if the operator $V_{-k}aV_k - aI$ is $\mathcal{P}$-compact for every $k \in \mathbb{Z}^N$ or, equivalently, if and only if the commutator $aV_k - V_kaI = V_k(V_{-k}aV_k - aI)$ is $\mathcal{P}$-compact for every k. Since $K(E, \mathcal{P})$ is a closed ideal of $L(E, \mathcal{P})$, we conclude that $SO_{L(X)}$ is a closed subalgebra of $l^\infty_{L(X)}$. If, moreover, the slowly oscillating function a is scalar-valued, then the operator of multiplication by a also commutes with every multiplication operator. Summarizing we get the following.

Proposition 2.1. *If $f \in SO$ and $A \in \mathcal{A}_E$, then the operator $fA - AfI$ is $\mathcal{P}$-compact on E. If, conversely, $f \in l^\infty_{L(X)}$ is a function for which $fA - AfI$ is $\mathcal{P}$-compact for every $A \in \mathcal{A}_E$, then $f \in SO$.*

Thus, the algebra SO (more precisely, the image of SO in $L(E, \mathcal{P})/K(E, \mathcal{P})$ under the canonical embedding) is *the* natural candidate for localizing the algebra

$\mathcal{A}_E/K(E,\mathcal{P})$ by means of the local principle by Allan. We will pursue this idea in Section 6.

Another special feature of slowly oscillating functions concerns the limit operators of their multiplication operators.

Proposition 2.2. *Let $a \in SO_{L(X)}$. Then every limit operator of aI is a multiplication operator in $\mathbb{C}_{L(X)}$, i.e. an operator of multiplication by a constant function with values in $L(X)$.*

Proof. Let $a \in SO_{L(X)}$. From (4) we conclude that

$$\lim_{k\to\infty} (a(x' + h(k)) - a(x'' + h(k))) = 0$$

for all sequences h tending to infinity and for all x', $x'' \in \mathbb{Z}^n$. Hence, if h is a sequence such that the limit operator $(aI)_h$ exists, then $\lim_{k\to\infty} a(x + h_k)$ is independent of $x \in \mathbb{Z}^n$, i.e. $(aI)_h = AI$ with an operator $A \in L(X)$. □

3. Inadequacy of sequences

Let $M(SO)$ denote the maximal ideal space of the commutative C^*-algebra SO, and write $M^\infty(SO)$ for the fiber of $M(SO)$ consisting of all characters $\eta \in M(SO)$ such that $\eta(a) = 0$ whenever $a \in c_0$. Every $m \in \mathbb{Z}^N$ defines a character of SO by $f \mapsto f(m)$. In this sense, $\mathbb{Z}^N$ is embedded into $M(SO)$, and $M(SO)$ is the union of its disjoint subsets $\mathbb{Z}^N$ and $M^\infty(SO)$.

Theorem 3.1. *$\mathbb{Z}^N$ is densely and homeomorphically embedded into $M(SO)$ with respect to the Gelfand topology.*

This is a special case of a general result on compactifications of topological spaces, see [3], Chapter I, Theorem 8.2.

We will run into a lot of trouble when trying to realize the simple and natural idea of localizing the algebra $\mathcal{A}_E/K(E,\mathcal{P})$ over SO. The main reason for this is the following observation.

Proposition 3.2. *Let $\eta \in M^\infty(SO)$. Then $\eta \in \mathrm{clos}_{M(SO)}\mathbb{Z}^N$, but there is* no sequence *in $\mathbb{Z}^N$ which tends to η with respect to the Gelfand topology of $M(SO)$.*

Proof. We know from Theorem 3.1 that η is in $\mathrm{clos}_{M(SO)}\mathbb{Z}^N$ and that, hence, there is a *net* with values in $\mathbb{Z}^N$ which converges to η. Assume there is a *sequence* h with values in $\mathbb{Z}^N$ and with limit η in the Gelfand topology. Since every subsequence of h also converges to η, we can assume without loss that

$$|h(n+1)| \geq |h(n)| + 2^{n+2} \quad \text{for all } n.$$

Let $\varphi_0 : \mathbb{R}^N \to [0, 1]$ be a continuous function with support in $\{t \in \mathbb{R}^N : |t| \leq 1\}$ and with $\varphi_0(0) = 1$, and set $\varphi_n(t) := \varphi(t/2^n)$ for $n \geq 1$. Then the function

$$\varphi(t) := \sum_{n\geq 0}^{\infty} \hat{\varphi}_{2n}(t - h(2n))$$

is slowly oscillating, and $\varphi(h(2n)) = 1$ and $\varphi(h(2n+1)) = 0$ for all n. The assumed convergence of h to η implies that both sequences $(\varphi(h(2n)))$ and $(\varphi(h(2n+1)))$ converge to $\varphi(\eta)$. Contradiction. □

Thus, the topological nature of $M(SO)$ requires the use of nets rather than sequences. In the following two sections, we are going to recall and provide some facts about nets and about limit operators with respect to nets.

4. Preliminaries on nets

Nets and subnets. A set T is *directed* if there is a binary relation $\geq$ on T such that

$$\begin{array}{lll} \forall t \in T: & t \geq t & \text{(reflexivity)}, \\ \forall r, s, t \in T: & r \geq s,\ s \geq t \Rightarrow s \geq t & \text{(transitivity)}, \\ \forall r, s \in T\ \exists t \in T: & t \geq r \text{ and } t \geq s & \text{(inductivity)}, \end{array}$$

A mapping x from a directed set T into a topological space X is called a *net*, and this net *converges to a point* $x^* \in X$ if, for every neighborhood U of x^*, there is a $t_0 \in T$ such that $x(t) \in U$ for all $t \geq t_0$. The net $x : T \to X$ is sometimes also denoted by $(x_t)_{t \in T}$ where $x_t = x(t)$. Accordingly, if $x : T \to X$ converges to x^*, we will write

$$\lim_{t \in T} x_t = x^* \quad \text{or} \quad x_t \to x^* \text{ with respect to } T.$$

A net $(y_s)_{s \in S}$ is a *subnet* of the net $(x_t)_{t \in T}$ if there is a mapping $F : S \to T$ such that

$$\begin{array}{ll} \forall s \in S: & y_s = x_{F(s)}, \\ \forall t \in T\ \exists s_0 \in S: & F(s) \geq t \quad \text{for all } s \geq s_0. \end{array}$$

A subset S of a directed set T is called *cofinal* if

$$\forall t \in T\ \exists s \in S: \quad s \geq t.$$

Every cofinal subset S of a directed set T is again a directed set with respect to the restriction of the order relation $\geq$ onto S. If S is a cofinal subset of T, and if $(x_t)_{t \in T}$ is a net, then the restriction of $(x_t)_{t \in T}$ onto S is a subnet of $(x_t)_{t \in T}$. We will be mainly interested in subnets which do *not* arise in this simple manner.

Nets tending to infinity. In what follows we will only be concerned with nets in $\mathbb{Z}^N$. A net $(x_t)_{t \in T}$ with values in $\mathbb{Z}^N$ is said to *converge to infinity* if

$$\forall k \in \mathbb{N}\ \exists t_0 \in T: \quad |x_t| \geq k \quad \text{for all } t \geq t_0.$$

Let $\mathcal{N}$ denote the set of all nets in $\mathbb{Z}^N$ which converge to infinity.

Lemma 4.1. (a) *For every net $(x_t)_{t \in T} \in \mathcal{N}$, the set $\{x_t : t \in T\}$ of its values is countably infinite.*

(b) *If $h : \mathbb{N} \to \mathbb{Z}^N$ is injective, then the sequence h belongs to $\mathcal{N}$.*

Proof. (*a*) Since $\mathbb{Z}^N$ is countable, $(x_t)_{t\in T} \subseteq \mathbb{Z}^N$ is an at most countable set, and since $(x_t)_{t\in T}$ tends to infinity, this set cannot be finite.
(*b*) Suppose the sequence h does not converge to infinity. Then

$$\exists\, k \in \mathbb{N}\ \forall n_0 \in \mathbb{N}\ \exists\, n \geq n_0 : |h(n)| \leq k.$$

Repeating this argument we get an infinite sequence $n_0 < n_1 < n_2 < \cdots$ such that $|h(n_r)| \leq k$ for all r. But h is injective. Thus, $h(n_r) \neq h(n_s)$ whenever $r \neq s$. So we have infinitely many points in $\{z \in \mathbb{Z}^N : |z| \leq k\}$, which is nonsense. □

Lemma 4.2. *Let $x \in \mathcal{N}$ be a net, and let h be a bijection from $\mathbb{N}$ onto the set of the values of x. Then x is a subnet of the sequence h. In particular, every net $x \in \mathcal{N}$ is a subnet of a sequence $h \in \mathcal{H}$.*

Proof. Let $x = (x_t)_{t\in T} \in \mathcal{N}$, and let $h : \mathbb{N} \to \{x_t : t \in T\}$ be a bijection. Such bijections exist by Lemma 4.1.

To show that x is a subnet of h, define $F : T \to \mathbb{N}$ by $F(t) := h^{-1}(x_t)$. Then, clearly, $x_t = h_{F(t)}$ for every $t \in T$, and it remains to check whether

$$\forall n \in \mathbb{N}\ \exists\, t_0 \in T : \quad F(t) \geq n \quad \text{for all } t \geq t_0. \tag{5}$$

Given $n \in \mathbb{N}$, set $k := \max\{|h_1|, \ldots, |h_n|\}$. Since $(x_t)_{t\in T}$ belongs to $\mathcal{N}$, there is a $t_0 \in T$ such that

$$|x_t| \geq k + 1 \quad \text{for all } t \geq t_0.$$

By the definition of F, this implies $F(t) \geq n$ for all $t \geq t_0$ which gives (5). Hence, x is a subnet of h, and this sequence belongs to $\mathcal{H}$ due to Lemma 4.1 (*b*). □

A version of Cantor's diagonalization procedure. The following result can be regarded as a substitute for the well-known diagonalization argument for sequences due to Cantor.

Theorem 4.3. *Let Z be a set, and let $(f_n)_{n\geq 1}$ be a sequence of functions $f_n : Z \to \mathbb{R}^+$ which converges uniformly on Z to a function $f : Z \to \mathbb{R}^+$. Assume further that $(x^0_{t_0})_{t_0\in T_0}$ is a net with values in Z and with the property that, for every $n \geq 1$, there is a subnet $(x^n_{t_n})_{t_n\in T_n}$ of $(x^{n-1}_{t_{n-1}})_{t_{n-1}\in T_{n-1}}$ such that*

$$\lim_{t_n\in T_n} f_n(x^n_{t_n}) = 0. \tag{6}$$

Then there is a subnet $(y_w)_{w\in W}$ of $(x^0_{t_0})_{t_0\in T_0}$ with $\lim_{w\in W} f(y_w) = 0$.

Proof. We split the proof into several steps and emphasize some partial results as lemmas. Our starting point is a net $(x^0_{t_0})_{t_0\in T_0}$ in Z and, for every $n \geq 1$, a subnet $(x^n_{t_n})_{t_n\in T_n}$ of $(x^{n-1}_{t_{n-1}})_{t_{n-1}\in T_{n-1}}$ with (6). In particular, we have mappings $F_n : T_n \to T_{n-1}$ with $x^n_{t_n} = x^{n-1}_{F_n(t_n)}$ for all $t_n \in T_n$ and such that

$$\forall t_{n-1} \in T_{n-1}\ \exists\, t^0_n \in T_n : \quad F(t_n) \geq t_{n-1} \quad \text{for all } t_n \geq t^0_n. \tag{7}$$

Step 1. We show that the directed sets $T_0, T_1, \ldots$ can be replaced be one and the same directed set S.

Indeed, set $S := T_0 \times T_1 \times T_2 \times \ldots$ and provide S with the order

$$(s_0, s_1, s_2, \ldots) \geq (s_0', s_1', s_2', \ldots) \iff s_k \geq s_k' \text{ for all } k$$

which makes S to a directed set. Further, there are canonical mappings

$$G_n : S \to T_n, \quad (s_0, s_1, s_2, \ldots) \mapsto s_n.$$

For every $n \in \mathbb{N}$, define a net $(y_s^n)_{s \in S}$ by $y_s^n := x_{G_n(s)}^n$.

Lemma 4.4. *(a) For all $n \geq 0$, $(y_s^n)_{s\in S}$ is a subnet of $(x_{t_n}^n)_{t_n \in T_n}$.*
(b) For all $n \geq 1$, $(y_s^n)_{s\in S}$ is a subnet of $(y_s^{n-1})_{s \in S}$.

Proof of Lemma 4.4. (a) By the definition of y_s^n, what we have to check is whether

$$\forall t_n \in T_n \, \exists \, s^0 \in S : \quad G_n(s) \geq t_n \quad \text{for all } s \geq s^0.$$

But this is obvious: Set $s^0 := (t_0, t_1, t_2, \ldots) \in S$. Then, for $s \geq s^0$, one indeed has $G_n(s) \geq t_n$.
(b) For $n \geq 1$, define

$$H_n : S \to S, \ (s_0, s_1, s_2, \ldots) \mapsto (s_0, \ldots, s_{n-2}, F_n(s_n), s_n, s_{n+1}, \ldots)$$

with $F_n(s_n)$ standing at the $n-1$ th position. Then, for all $s = (s_0, s_1, s_2, \ldots) \in S$ and all $n \geq 1$,

$$y_s^n = x_{G_n(s)}^n = x_{s_n}^n = x_{F_n(s_n)}^{n-1} = x_{G_{n-1}(H_n(s))}^{n-1} = y_{H_n(s)}^{n-1}, \tag{8}$$

and it remains to show that

$$\forall \hat{s} \in S \, \exists s^0 \in S : \quad H_n(s) \geq \hat{s} \quad \text{for all } s \geq s^0. \tag{9}$$

Let $\hat{s} = (\hat{s}_0, \hat{s}_1, \hat{s}_2, \ldots) \in S$. For $k \neq n$, set $s_k^0 := \hat{s}_k$. In case $k = n$, we first choose $s_n^{00} \in T_n$ such that

$$\forall s_n \geq s_n^{00} : \quad F_n(s_n) \geq \hat{s}_{n-1} \tag{10}$$

(which is possible due to (7)), and then we choose $s_n^0 \in T_n$ such that both $s_n^0 \geq s_n^{00}$ and $s_n^0 \geq \hat{s}_n$. Define $s^0 := (s_0^0, s_1^0, s_2^0, \ldots) \in S$. Then, for all $s = (s_0, s_1, s_2, \ldots) \geq s^0$, we have

$$\begin{array}{lllll} s_k & \geq & s_0^k & = & \hat{s}_k \quad \text{for all } 0 \leq k \leq n-2, \\ s_n & \geq & s_n^0 & \geq & s_n^{00}, \quad \text{whence } F_n(s_n) \geq \hat{s}_{n-1} \text{ due to (10)}, \\ s_n & \geq & s_n^0 & \geq & \hat{s}_n, \\ s_k & \geq & s_k^0 & = & \hat{s}_k \quad \text{for all } k \geq n+1. \end{array}$$

Consequently,

$$\begin{aligned} H_n(s_0, s_1, s_2, \ldots) &= (s_0, \ldots, s_{n-2}, F_n(s_n), s_n, s_{n+1}, \ldots) \\ &\geq (\hat{s}_0, \ldots, \hat{s}_{n-2}, \hat{s}_{n-1}, \hat{s}_n, \hat{s}_{n+1}, \ldots) = \hat{s}. \end{aligned}$$

This proves (9) and the lemma. $\square$

Step 2. Choice of the diagonal net.

Let $\Omega := S \times \mathbb{N}$. This set becomes directed by the order relation

$$(s, n) \geq (s', n') \iff s \geq s' \text{ and } n \geq n'.$$

Consider the net

$$y : \Omega \to \mathbb{Z}^N, \quad y_{(s,n)} := y_s^n. \tag{11}$$

Of course (and as in the standard diagonalization procedure for sequences) one cannot expect that $(y_{(s,n)})_{(s,n)\in\Omega}$ is a subnet of $(y_s^n)_{s\in S}$. But (also as for standard diagonalization) one has the following result where we write $\Omega_{n_0} := \{(s, n) \in \Omega : n > n_0\}$ for brevity. Clearly, Ω_{n_0} is a cofinal subset of Ω for every $n_0 \in \mathbb{N}$.

Lemma 4.5. *For all $n_0 \in \mathbb{N}$, $(y_{(s,n)})_{(s,n)\in\Omega_{n_0}}$ is a subnet of $(y_s^{n_0})_{s\in S}$.*

Proof of Lemma 4.5. For all $s \in S$ and all $n > n_0$, we have

$$y_{(s,n)} = y_s^n = y^{n-1}_{H_n(s)} = y^{n-2}_{H_{n-1}(H_n(s))} = \ldots = y^{n_0}_{(H_{n_0+1}\circ H_{n_0+2}\circ\ldots\circ H_n)(s)}$$

(compare (8)). This equality suggests to define

$$K_{n_0} : \Omega_{n_0} \to S, \quad (s, n) \mapsto (H_{n_0+1} \circ H_{n_0+2} \circ \ldots \circ H_n)(s).$$

Then, obviously,

$$y_{(s,n)} = y^{n_0}_{K_{n_0}(s,n)} \quad \text{for all } (s, n) \in \Omega_{n_0},$$

and what remains to be verified is

$$\forall \hat{s} \in S\ \exists (\tilde{s}, \tilde{n}) \in \Omega_{n_0} : \quad K_{n_0}(s, n) \geq \hat{s} \quad \text{for all } (s, n) \geq (\tilde{s}, \tilde{n}).$$

Set $\tilde{n} := n_0 + 1$ and construct $\tilde{s} := (\tilde{s}_0, \tilde{s}_1, \ldots)$ successively as follows. Let $\hat{s} = (\hat{s}_0, \hat{s}_1, \ldots) \in S$. We set $\tilde{s}_k := \hat{s}_k$ for $k \leq n_0$. Further, by (7), given $\hat{s}_{n_0} \in T_{n_0}$,

$$\exists\, \overline{s}_{n_0+1} \in T_{n_0+1} : \quad F_{n_0+1}(s) \geq \hat{s}_{n_0} \quad \forall s \geq \overline{s}_{n_0+1}.$$

Then choose $\tilde{s}_{n_0+1}$ larger than both $\overline{s}_{n_0+1}$ and $\hat{s}_{n_0+1}$.

For $\tilde{s}_{n_0+1} \in T_{n_0+1}$, we choose $\overline{s}_{n_0+2} \in T_{n_0+2}$ such that

$$\forall s \geq \overline{s}_{n_0+2} : \quad F_{n_0+2}(s) \geq \tilde{s}_{n_0+1}\ (\geq \hat{s}_{n_0+1})$$

and, hence,

$$F_{n_0+1}(F_{n_0+2}(s)) \geq \hat{s}_{n_0}.$$

Then choose $\tilde{s}_{n_0+2}$ larger than both $\overline{s}_{n_0+2}$ and $\hat{s}_{n_0+2}$.

We proceed in this way, i.e. we choose $\overline{s}_{n_0+3} \in T_{n_0+3}$ such that

$$\forall s \geq \overline{s}_{n_0+3} : \quad F_{n_0+3}(s) \geq \tilde{s}_{n_0+2}\ (\geq \hat{s}_{n_0+2})$$

which implies that

$$F_{n_0+2}(F_{n_0+3}(s)) \geq \hat{s}_{n_0+1}$$

and, hence,

$$F_{n_0+1}(F_{n_0+2}(F_{n_0+3}(s))) \geq \hat{s}_{n_0}.$$

Then choose $\tilde{s}_{n_0+3}$ larger than $\overline{s}_{n_0+3}$ and $\hat{s}_{n_0+3}$.

Thus we have fixed $\tilde{s}$. Let now $s = (s_0, s_1, \ldots) \geq \tilde{s}$. Then, due to our construction,

$$\begin{array}{rcll} s_k & \geq & \hat{s}_k & \text{for all } k \leq n_0 - 1, \\ (F_{n_0+1} \circ F_{n_0+2} \circ \ldots \circ F_n)(s_n) & \geq & \hat{s}_{n_0}, & \\ (F_{n_0+2} \circ F_{n_0+3} \circ \ldots \circ F_n)(s_n) & \geq & \hat{s}_{n_0+1}, & \\ & \vdots & & \\ F_n(s_n) & \geq & \hat{s}_{n-1}, & \\ s_k \geq \tilde{s}_k & \geq & \hat{s}_k & \text{for all } k \geq n. \end{array}$$

This shows that

$$K_{n_0}(s, n) = (H_{n_0+1} \circ \ldots \circ H_n)(s) \geq \hat{s}$$

since

$$H_n(s) = (s_0, \ldots, s_{n-2}, F_n(s_n), s_n, s_{n+1}, \ldots),$$
$$(H_{n-1} \circ H_n)(s) = (s_0, \ldots, s_{n-3}, F_{n-1}(F_n(s_n)), F_n(s_n), s_n, s_{n+1}, \ldots),$$
$$(H_{n-2} \circ H_{n-1} \circ H_n)(s) =$$
$$(s_0, \ldots, s_{n-4}, F_{n-2}(F_{n-1}(F_n(s_n))), F_{n-1}(F_n(s_n)), F_n(s_n), s_n, s_{n+1}, \ldots),$$

and so on. This finishes the proof of Lemma 4.5. □

Step 3. Let $W := \Omega_0$. Then $(y_w)_{w \in W}$ is the net we are looking for.

It is obvious from the above construction that $(y_w)_{w \in W}$ is a subnet of $(x^0_{t_0})_{t_0 \in T_0}$. So we are left with verifying that $\lim_{w \in W} f(y_w) = 0$.

Given $\varepsilon > 0$, choose and fix $n \geq 1$ such that $\|f - f_n\| < \varepsilon/2$. Then, by hypothesis,

$$\lim_{t_n \in T_n} f_n(x^n_{t_n}) = 0.$$

Since $(y_w)_{w \in \Omega_n}$ is a subnet of $(x^n_{t_n})_{t_n \in T_n}$, we also have $\lim_{w \in \Omega_n} f_n(y_w) = 0$, which implies the existence of a $w_n \in \Omega_n$ with

$$|f_n(y_w)| < \varepsilon/2 \quad \text{for all } w \geq w_n. \tag{12}$$

Let now $w \in W$ with $w \geq w_n$. Then, evidently, $w \in \Omega_n$, and from (12) we conclude

$$|f(y_w)| \leq |f(y_w) - f_n(y_w)| + |f_n(y_w)| \leq \|f - f_n\|_\infty + |f_n(y_w)| < \varepsilon.$$

Hence, $\lim_{w \in W} f(y_w) = 0$, which finishes the proof of Theorem 4.3. □

5. Limit operators with respect to nets

Now we return to band-dominated operators on one of the sequence spaces E. If $y := (y_w)_{w \in W}$ is a net in $\mathcal{N}$, then we call the operator A_y the *limit operator* of the operator $A \in L(E)$ with respect to y if

$$\lim_{n \to \infty} \|(V_{-y_w} A V_{y_w} - A_y) P_m\| = \lim_{n \to \infty} \|P_m (V_{-y_w} A V_{y_w} - A_y)\| = 0$$

for every $P_m \in \mathcal{P}$. Roughly speaking, the properties of limit operators with respect to sequences (as derived in [5, 6]), remain valid without changes also for limit operators with respect to nets. We will illustrate this fact by two results for which the Cantor diagonalization procedure for nets is employed.

Theorem 5.1. *Let $A = aI \in L(E)$ be a rich multiplication operator. Then every net $(x_t)_{t\in T} \in \mathcal{N}$ possesses a subnet $y := (y_w)_{w\in W}$ such that the limit operator A_y exists.*

Proof. Recall that A_y is the limit operator of A with respect to the net y if and only if

$$\lim_{w\in W} \|(V_{-y_w} A V_{y_w} - A_y)S_k\| = 0 \quad \text{for every } k \in \mathbb{Z}^N$$

where, as before, S_k refers to the operator of multiplication by the function which is I at $k \in \mathbb{Z}^N$ and 0 at all other points.

Set $(x^0_{t_0})_{t_0\in T_0} := (x_t)_{t\in T}$ and choose a bijection $m : \mathbb{N} \to \mathbb{Z}^N$. Since A is rich we find, for every $n \geq 1$, a subnet $(x^n_{t_n})_{t_n\in T_n}$ of $(x^{n-1}_{t_{n-1}})_{t_{n-1}} \in T_{n-1}$ as well as an operator $B_n \in L(\operatorname{Im} S_{m(n)})$ such that

$$\|(V_{-x^n_{t_n}} A V_{x^n_{t_n}} - B_n)S_{m(n)}\| \to 0. \tag{13}$$

Let B stand for the operator of multiplication by the function

$$\mathbb{Z}^N \to L(X), \quad k \mapsto B_{m^{-1}(k)}.$$

We claim that B is the limit operator of A with respect to the net y. For, we reify Cantor's scheme (= Theorem 4.3) as follows. Set $Z := \mathbb{Z}^N$. For $n \geq 1$ and $z \in \mathbb{Z}^N$, define

$$f_n(z) := \sum_{k=1}^{n} 2^{-k} \|(V_{-z} A V_z - B)S_{m(k)}\|,$$

and let

$$f(z) := \sum_{k=1}^{\infty} 2^{-k} \|(V_{-z} A V_z - B)S_{m(k)}\|.$$

Then, obviously, $\|f_n - f\|_\infty \to 0$. Further, by (13), we have $\lim_{t_n\in T_n} f_n(x^n_{t_n}) = 0$. Now we conclude from Theorem 4.3 that there is a subnet $(y_w)_{w\in W}$ of $(x_t)_{t\in T}$ such that $\lim_{w\in W} f(y_w) = 0$. This immediately implies the $\mathcal{P}$-strong convergence of the net $(V_{-y_w} A V_{y_w})_{w\in W}$ to B, whence $B = A_y$. □

Of course, a similar result holds for rich band operators. For another application of Theorem 4.3, consider the set of all operators $A \in L(E)$ having the following property: every net $(x_t)_{t\in T} \in \mathcal{N}$ possesses a subnet $y := (y_w)_{w\in W}$ such that the limit operator A_y exists. We denote this class by $\mathcal{L}_E^{nets}$ for a moment. As we have just remarked, every rich band operator belongs to $\mathcal{L}_E^{nets}$.

Theorem 5.2. *$\mathcal{L}_E^{nets}$ is norm-closed.*

Proof. Let $(A_n)_{n\geq 1} \subseteq \mathcal{L}_E^{nets}$ be a sequence with norm limit $A \in L(E)$, and let $(x^0_{t_0})_{t_0\in T_0} \in \mathcal{N}$. By hypothesis, for every $n \geq 1$, there exists a subnet $x^n := (x^n_{t_n})_{t_n\in T_n}$ of $(x^{n-1}_{t_{n-1}})_{t_{n-1}\in T_{n-1}}$ such that the limit operator A_{n,x^n} of A_n with respect to x^n exists. If $n \geq m$, then $(x^n_{t_n})_{t_n\in T_n}$ is a subnet of $(x^m_{t_m})_{t_m\in T_m}$, thus, the limit operator A_{m,x^n} also exists, and it coincides with A_{m,x^m}. Since $\|A_h\| \leq \|A\|$ for every limit operator A_h of A, we obtain

$$\|A_{n,x^n} - A_{m,x^m}\| = \|A_{n,x^n} - A_{m,x^n}\| = \|(A_n - A_m)_{x^n}\| \leq \|A_n - A_m\|$$

for all $n \geq m$. Hence, the sequence (A_{n,x^n}) converges in the norm, and we let B denote its norm limit.

Now define for all $n \geq 1$ and $z \in \mathbb{Z}^N$ (with the notations S_k and m as in the proof of Theorem 5.1)

$$f_n(z) := \sum_{k=1}^{\infty} 2^{-k} \|(V_{-z}A_nV_z - A_{n,x^n})S_{m(k)}\|$$

and

$$f(z) := \sum_{k=1}^{\infty} 2^{-k} \|(V_{-z}AV_z - B)S_{m(k)}\|.$$

Then again $\|f_n - f\| \to 0$ and $\lim_{t_n\in T_n} f_n(x^n_{t_n}) = 0$, which, via Theorem 4.3, implies the existence of a subnet $y = (y_w)_{w\in W}$ of $(x^0_{t_0})_{t_0\in T_0}$ such that $\lim_{w\in W} f(y_w) = 0$. Thus, $B = A_y$. □

As a consequence we get $\mathcal{A}_E^{rich} \subseteq \mathcal{L}_E^{nets}$. Now one might ask whether one gets something new when considering limit operators with respect to nets instead of sequences. The next theorem says that the answer is *no* in some sense: every limit operator which is defined with respect to a net can also be reached by a sequence! (Nevertheless, limit operators with respect to nets *are* useful as we will point out in the next sections when we will use them to study the local invertibility of band-dominated operators at points in $M^\infty(SO)$.)

Theorem 5.3. *Let $A \in L(E)$, and let $y = (y_w)_{w\in W} \in \mathcal{N}$ be a net for which the limit operator A_y of A exists. Then there is a sequence $z = (z_n)_{n\in\mathbb{N}} \in \mathcal{H}$ for which the limit operator A_z of A exists, and $A_z = A_y$.*

Proof. Let $y = (y_w)_{w\in W}$ be a net for which the limit operator A_y of A exists, and define a function $f : \mathbb{Z}^N \to \mathbb{R}^+$ by

$$f(z) := \sum_{k=1}^{\infty} 2^{-k} \|(V_{-z}AV_z - A_y)S_{m(k)}\|,$$

with the notation as in the proof of Theorem 5.1. Then $\lim_{w\in W} f(y_w) = 0$. For every $n \in \mathbb{N}$, choose $w_n \in W$ such that

$$|y_{w_n}| \geq n \quad \text{and} \quad 0 \leq f(y_w) < 1/n \quad \text{for all } w \geq w_n, \tag{14}$$

and set $z_n := y_{w_n}$. Then the sequence $z := (z_n)$ tends to infinity, and $f(z_n)$ tends to 0 as $n \to \infty$. Hence, A_y is also the limit operator of A with respect to the sequence z. $\square$

6. Local invertibility at points in $M^\infty(SO)$

After these preparations, we can turn over to the formulation and to the proof of the analogue of Theorem 1.6 for localization over $M^\infty(SO)$.

Definition 6.1. *Let $\eta \in M^\infty(SO)$ and $A \in L(E)$. The* local operator spectrum *$\sigma_\eta(A)$ of A at η is the set of all limit operators A_y of A with respect to nets y which tend to η.*

By Theorem 5.3, $\sigma_\eta(A) \subseteq \sigma_{op}(A)$ (recall that $\sigma_{op}(A)$ is the set of all limit operators with respect to *sequences*). Let, conversely, $A_h \in \sigma_{op}(A)$ for some sequence $h \in \mathcal{H}$. Then the intersection $\text{clos}_{M(SO)}\{h(m) : m \in \mathbb{Z}^N\} \cap M^\infty(SO)$ is non-empty by Theorem 3.1. Consequently, there is a subnet y of h which converges to a point $\eta \in M^\infty(SO)$. Clearly, the limit operator A_y exists and is equal to A_h. Hence,

$$\sigma_{op}(A) = \cup_{\eta \in M^\infty(SO)} \sigma_\eta(A) \quad \text{for every } A \in L(E).$$

Let $\eta \in M^\infty(SO)$, and let U be a neighborhood of η in $M(SO)$ with respect to the Gelfand topology. Then we agree upon calling the intersection $U \cap \mathbb{Z}^N$ a *neighborhood at infinity of η*.

Definition 6.2. *Let $\eta \in M^\infty(SO)$ and $A \in L(E)$. The operator A is* locally invertible at η *if there are operators $B, C \in L(E)$ and a neighborhood at infinity W of η such that*

$$BA\hat{\chi}_W I = \hat{\chi}_W AC = \hat{\chi}_W I$$

where $\hat{\chi}_W$ refers to the characteristic function of W.

The following result, which states the analogue of Theorem 1.6 with respect to the much finer localization over points in $M^\infty(SO)$ instead of points in S^{N-1}, is the main outcome of this paper.

Theorem 6.3. *Let $A \in \mathcal{A}_E^{rich}$ and $\eta \in M^\infty(SO)$. Then the operator A is locally invertible at η if and only if all limit operators in $\sigma_\eta(A)$ are invertible and if*

$$\sup\{\|(A_h)^{-1}\| : A_h \in \sigma_\eta(A)\} < \infty.$$

The proof of this result will follow the line of the proof of Theorem 1.6, and we will pay our attention mainly to the differences which are involved by the topology of $M(SO)$ and, hence, by the need of using nets instead of sequences.

A basic step is the specification of Proposition 14 from [5] resp. Proposition 2.17 from [6] to the present context. For, we need some more notations. Let $\varphi : \mathbb{R} \to [0, 1]$ be a continuous function with

$$\varphi(x) \begin{cases} = 1 & \text{for } |x| \le 1/3 \\ > 0 & \text{for } |x| < 2/3 \\ = 0 & \text{for } |x| \ge 2/3. \end{cases} \tag{15}$$

We further suppose that the family $\{\varphi_\alpha^2\}_{\alpha\in\mathbb{Z}}$ with $\varphi_\alpha(x) := \varphi(x-\alpha)$ forms a partition of unity on $\mathbb{R}$ in the sense that

$$\sum_{\alpha\in\mathbb{Z}} \varphi_\alpha(x)^2 = 1 \quad \text{for all } x \in \mathbb{R}.$$

This choice of φ can always be forced as follows: If $f : \mathbb{R} \to [0,\, 1]$ is a continuous function satisfying (15) in place of φ, then the function

$$\varphi(x) := \frac{f(x)^2}{\sum_{\alpha\in\mathbb{Z}} f(x-\alpha)^2}, \quad x \in \mathbb{R},$$

has the desired properties. This definition makes sense since the series $\sum f(x-\alpha)^2$ is strictly positive and has only finitely many non-vanishing terms for each fixed x.

Given $x = (x_1, \ldots, x_N) \in \mathbb{R}^N$, $\alpha \in \mathbb{Z}^N$, and $R > 0$, define $\varphi^{(N)}(x) := \varphi(x_1)\ldots\varphi(x_N)$, $\varphi_\alpha^{(N)}(x) := \varphi^{(N)}(x-\alpha)$ and $\varphi_{\alpha,R}^{(N)}(x) := \varphi_\alpha(x/R)$. Further, let $\psi : \mathbb{R} \to [0,\, 1]$ be a continuous function which also satisfies (15) in place of φ, but with the constants $1/3$ and $2/3$ replaced by $3/4$ and $4/5$, respectively. For this function, we define $\psi_{\alpha,R}^{(N)}$ analogously. Clearly, $\varphi_{\alpha,R}^{(N)}\psi_{\alpha,R}^{(N)} = \varphi_{\alpha,R}^{(N)}$ for all α and R. The family $\{\varphi_{\alpha,R}\}$ is a partition of unity on $\mathbb{R}^N$ for every fixed R (but observe that the family $\{\psi_\alpha\}$ is not required to form a partition of unity. With these notations, the announced analogue of Proposition 14 from [5] reads as follows.

Proposition 6.4. *Let $A \in \mathcal{A}_E$, $\eta \in M^\infty(SO)$. Suppose there is a constant $M > 0$ such that, for all positive integers R, there is a neighborhood at infinity U of η such that, for all $\alpha \in U$, there are operators $B_{\alpha,R}$ and $C_{\alpha,R}$ with $\|B_{\alpha,R}\|_{L(E)} \le M$, $\|C_{\alpha,R}\|_{L(E)} \le M$ and*

$$B_{\alpha,R}\, A\, \hat{\psi}_{\alpha,R} I = \hat{\psi}_{\alpha,R}\, A\, C_{\alpha,R} = \hat{\psi}_{\alpha,R} I.$$

Then the operator A is locally invertible at η, i.e. there are operators $B, C \in \mathcal{A}_E$ and a neighborhood at infinity W of η such that

$$BA\hat{\chi}_W I = \hat{\chi}_W AC = \hat{\chi}_W I. \tag{16}$$

Proof. We follow exactly the proof of Proposition 14 from [5] where we replace the condition $|\alpha| \ge \rho(R)$ by $\alpha \in U$. What results is the existence of a positive integer R such that

$$(I+T_R)^{-1}B_R A = I - (I+T_R)^{-1} \sum_{\alpha\in\mathbb{Z}^N\setminus U} \hat{\varphi}_{\alpha,R} I.$$

The assertion will follow once we have shown that there is a neighborhood at infinity W of η such that $\sum_{\alpha\in\mathbb{Z}^N\setminus U} \hat{\varphi}_{\alpha,R}\, \chi_W = 0$. This will be done in Proposition 6.7 below. □

Filling the gap in the preceding proof requires more precise knowledge on subsets of $M(SO)$. The following definition as well as Theorem 6.6 and its proof are taken from [4].

Definition 6.5. (a) *A subset* $V \subseteq \mathbb{Z}^N$ *is called* growing *if, for every bounded set* $D \subset \mathbb{Z}^N$, *there is an* $x \in \mathbb{Z}^N$ *such that* $x + D \subseteq V$.

(b) *An unbounded subset* V_0 *of a growing set* V *is called a* center *if, for every bounded set* $D \subset \mathbb{Z}^N$, *there is a bounded set* M *such that* $(V_0 \setminus M) + D \subseteq V$.

Theorem 6.6. *Let* W *be an unbounded subset of* $\mathbb{Z}^N$ *and* $\eta \in \overline{W} \cap M^\infty(SO)$ *(where the bar refers to the closure with respect to the Gelfand topology on* $M(SO)$*), and let* $U \subseteq M(SO)$ *be a neighborhood of* η. *Then* $V := U \cap \mathbb{Z}^N$ *is a growing set, and there is a neighborhood* $U_0 \subseteq U$ *of* η *such that* $V_0 := U_0 \cap \mathbb{Z}^N$ *is contained in* W *and a center of* V.

Proof. By Uryson's lemma, there is a continuous function $f : M(SO) \to [0, 1]$ which is 0 at η and 1 on $M(SO) \setminus U$. Since f is continuous on $M(SO)$, the restriction of f onto $\mathbb{Z}^N$ is a slowly oscillating function. Set

$$U_0' := \{x \in M(SO) : f(x) < 1/2\} \quad \text{and} \quad U_0 := U_0' \cap W,$$

and define $V := U \cap \mathbb{Z}^N$ and $V_0 := U_0 \cap \mathbb{Z}^N$. Then $V_0 \subseteq W \cap V$. Moreover, since U_0' is a neighborhood of η, the set V_0 is unbounded. We claim that, for every bounded set M, there is a bounded set D such that $(V_0 \setminus D) + M \subseteq V$. The claim implies that V is growing and that V_0 is a center of V.

Assume the claim is wrong. Then there exists a bounded set M such that $(V_0 \setminus D) + M \not\subseteq V$, hence, $V_1 := (V_0 + M) \setminus V$ is an unbounded set. So it makes sense to consider the limes superior of $|f(x)|$ when $x \in V_1$ tends to infinity. Since $V_1 \subseteq V_0 + M$ and f is slowly oscillating, we get

$$\begin{aligned} \limsup_{x \in V_1, x \to \infty} |f(x)| &\le \limsup_{y \in V_0, y \to \infty} \max_{m \in M} |f(y+m)| \\ &\le \limsup_{y \in V_0, y \to \infty} \max_{m \in M} |f(y+m) - f(y)| + \limsup_{y \in V_0, y \to \infty} |f(y)| \le 0 + 1/2 = 1/2. \end{aligned}$$

This is impossible since V_1 is in the complement of U and, hence, f is 1 on V_1. □

Proposition 6.7. *Let* R *be a positive integer, let* $\eta \in M^\infty(SO)$, *and let* $U \subseteq M(SO)$ *a neighborhood of* η. *Then there exists a neighborhood at infinity* $\tilde{U}$ *of* η *such that* $\sum_{\alpha \in \mathbb{Z}^N \setminus U} \hat{\varphi}_{\alpha,R} \chi_{\tilde{U}} = 0$.

Proof. We apply Theorem 6.6 (with the W in that theorem being $\mathbb{Z}^N$) to obtain: $V := U \cap \mathbb{Z}^N$ is a growing set, and there is a neighborhood $U_0 \subseteq U$ of η such that $V_0 := U_0 \cap \mathbb{Z}^N$ is a center of V.

The support of every function $\hat{\varphi}_{\alpha,R}$ is contained in a smallest ball with center αR and with a radius r which depends on R but not on α. From V, we remove all points z for which the ball with center z and radius r is not completely contained in V. What we get is a set $\tilde{V}$, and we set $\tilde{U} := V_0 \cap \tilde{V}$.

We claim that $\tilde{V}$ is a growing set and that $\tilde{U}$ is one of its centers. Let $D \subset \mathbb{Z}^N$ be bounded, and let B be the ball with center 0 and radius r. Then $D + B$ is a bounded set, and since V_0 is a center of V, there is a bounded set M such that

$$(V_0 \setminus M) + (D + B) \subseteq V.$$

Then, of course, $(V_0 \setminus M) + D \subseteq \tilde{V}$, whence

$$(\tilde{U} \setminus M) + D \subseteq \tilde{V}. \tag{17}$$

Analogously, there is a bounded set N such that $(V_0 \setminus N) + B \subseteq V$. Thus, all points in $V_0 \setminus N$ belong to $\tilde{V}$ and, consequently, also to $\tilde{U}$. This shows that $\tilde{U}$ and V_0 differ by a bounded set only:

$$V_0 \setminus N \subseteq \tilde{U} \subseteq V_0. \tag{18}$$

A first consequence of (18) is that $\tilde{U}$ is an unbounded set. Together with (17) this implies that $\tilde{V}$ is a growing set, and that $\tilde{U}$ is a center of $\tilde{V}$. As another consequence of (18) we observe that, since V_0 is a neighborhood at infinity of η, also $\tilde{U}$ is a neighborhood at infinity of η. This finishes the proof since the support of every function $\hat{\varphi}_{\alpha,R}$ with $\alpha \in \mathbb{Z}^N \setminus U$ is contained in the complement of $\tilde{V}$, hence in the complement of $\tilde{U}$. □

Proof of Theorem 6.3. We will only prove that the uniform invertibility of the operators in $\sigma_\eta(A)$ implies the local invertibility of A at η. Let $A \in \mathcal{A}_E^{rich}$ be an operator with

$$M_A := \sup\{\|A_h^{-1}\| : A_h \in \sigma_\eta(A)\} < \infty,$$

but suppose A is not locally invertible at η. Then, by Proposition 6.4, there is a net $(y_t)_{t \in T}$ with values in $\mathbb{Z}^N$ which converges to η in the topology of $M(SO)$ and which has the property that

$$BA\hat{\psi}_{y_t,R}I \neq \hat{\psi}_{y_t,R}I \tag{19}$$

for all $t \in T$ and all B with $\|B\| \leq M_A$. Since A belongs to $\mathcal{A}_E^{rich} \subseteq \mathcal{L}_E^{nets}$, the net $(y_t)_{t \in T}$ possesses a subnet $x = (x_s)_{s \in S}$ such that the limit operator A_x exists. Clearly, the net $(x_s)_{s \in S}$ still converges to η, and

$$BA\hat{\psi}_{x_s,R}I \neq \hat{\psi}_{x_s,R}I \tag{20}$$

for all $s \in S$ and all B with $\|B\| \leq M_A$. From Theorem 5.3 we conclude that there is a sequence $z = (z_n)_{n \in \mathbb{N}}$ which tends to infinity and for which the limit operator A_z exists and coincides with A_x. Hence, A_z belongs to $\sigma_\eta(A)$. By hypothesis, A_z is invertible, and $\|A_z^{-1}\| \leq M_A$. This yields a contradiction in the very same way as in the proof of Theorem 1.4 by using Proposition 15 from [5]. □

Our next goal is to point out the connections between local invertibility at η and localization by means of the local principle. For the reader's convenience, we state this principle here. Let $\mathcal{B}$ be a unital Banach algebra. By a *central* subalgebra $\mathcal{C}$ of $\mathcal{B}$ we mean a closed subalgebra of the center of $\mathcal{B}$ which contains the identity element. Thus, every element of $\mathcal{C}$ commutes with every element from $\mathcal{B}$, and $\mathcal{C}$ is a commutative Banach algebra with maximal ideal space $M(\mathcal{B})$. With each maximal ideal x of $\mathcal{C}$, we associate the smallest closed two-sided ideal $\mathcal{I}_x$ of $\mathcal{B}$ which contains x, and we let Φ_x refer to the canonical homomorphism from $\mathcal{B}$ onto the quotient algebra $\mathcal{B}/\mathcal{I}_x$. Notice that, in contrast to the commutative setting, the quotient algebras $\mathcal{B}/\mathcal{I}_x$ can differ from each other in dependence on $x \in M(\mathcal{C})$. Moreover, it may happen that $\mathcal{I}_x = \mathcal{B}$ for some points x. In this case we *define* that $\Phi_x(a)$ is invertible in $\mathcal{B}/\mathcal{I}_x$ and that $\|\Phi_x(a)\| = 0$ for each $a \in \mathcal{B}$.

Theorem 6.8. (Allan) *Let $\mathcal{C}$ be a central subalgebra of the unital Banach algebra $\mathcal{B}$ which contains the identity element. Then an element $a \in \mathcal{B}$ is invertible if and only if the cosets $\Phi_x(a)$ are invertible in $\mathcal{B}/\mathcal{I}_x$ for every $x \in M(\mathcal{C})$.*

For a proof see [1] or Theorem 1.34 in [2].

We have seen in Proposition 2.1 that the algebra $\mathcal{C}$ of all cosets $aI + K(E,\mathcal{P})$ with $a \in SO$ lies in the center of the quotient algebra $\mathcal{A}_E/K(E,\mathcal{P})$. From the isomorphy

$$\mathcal{C} \cong (SO \cdot I + K(E,\mathcal{P}))/K(E,\mathcal{P}) \cong SO \cdot I/(SO \cdot I \cap K(E,\mathcal{P})) \cong SO/c_0$$

we conclude that the maximal ideal space of the algebra $\mathcal{C}$ is homeomorphic to the fiber $M^\infty(SO)$. Given $\eta \in M^\infty(SO)$, we denote the local algebra of $\mathcal{A}_E/K(E,\mathcal{P})$ which is associated with η by $\mathcal{A}_{E,\eta}$, and we write π_η for the canonical homomorphism from $\mathcal{A}_E$ onto $\mathcal{A}_{E,\eta}$. Applying Theorem 6.8 to the current situation yields the following.

Theorem 6.9. *An operator $A \in \mathcal{A}_E$ is $\mathcal{P}$-Fredholm if and only if the cosets $\pi_\eta(A)$ are invertible for all $\eta \in M^\infty(SO)$.*

The following theorem relates the invertibility of the coset $\pi_\eta(A)$ to the local invertibility of A at η and can be proved in the very same way as Proposition 23 in [5].

Theorem 6.10. *Let $A \in \mathcal{A}_E$ and $\eta \in M^\infty(SO)$. The coset $\pi_\eta(A)$ is invertible in $\mathcal{A}_{E,\eta}$ if and only if A is locally invertible at η.*

Together with Allan's local principle and with Theorem 6.3, this results implies a further and essential refinement of Theorem 1.4 and Corollary 1.7.

Corollary 6.11. *An operator $A \in \mathcal{A}_E^{rich}$ is $\mathcal{P}$-Fredholm if and only if all of its limit operators are invertible and if*

$$\sup\{\|(A_h)^{-1}\| : A_h \in \sigma_\eta(A)\} < \infty \quad \textit{for every } \eta \in M^\infty(SO).$$

7. Fredholmness of band-dominated operators with slowly oscillating coefficients

We will now specify Corollary 6.11 to band-dominated operators with slowly oscillating coefficients. Let $SO^{rich}_{L(X)}$ refer to the class of all slowly oscillating functions with values in $L(X)$ for which the associated multiplication operator is rich. Further we let $\mathcal{A}_E(SO_{L(X)})$ (resp. $\mathcal{A}_E(SO^{rich}_{L(X)})$) stand for the smallest closed subalgebra of $\mathcal{A}_E$ which contains all band operators $\sum_{|\alpha|\le k} a_\alpha V_\alpha$ with $a_\alpha \in SO_{L(X)}$ (resp. $a_\alpha \in SO^{rich}_{L(X)}$). For the limit operators of operators with slowly oscillating coefficients we have the following.

Proposition 7.1. *(a) If $A \in \mathcal{A}_E(SO_{L(X)})$, then every limit operator of A belongs to $\mathcal{A}_E(\mathbb{C}_{L(X)})$.*
(b) For $A \in \mathcal{A}_E(SO^{rich}_{L(X)})$, every local operator spectrum $\sigma_\eta(A)$ with $\eta \in M^\infty(SO)$ is a singleton.

Proof. (a) Limit operators of shift operators are shift operators and, hence, in $\mathcal{A}_E(\mathbb{C}_{L(X)})$. By Proposition 2.2, the same is true for operators of multiplication by slowly oscillating functions.
(b) If $a \in SO^{rich}_{L(X)}$, then $\sigma_\eta(aI)$ is not empty since $\mathcal{A}^{rich}_E \subseteq \mathcal{L}^{nets}_E$ (see Section 5), and this spectrum is clearly a singleton. With Proposition 1 from [5] we conclude first that every local spectrum of a band operator with coefficients in $SO^{rich}_{L(X)}$ is a singleton, too, and get then the assertion also in the general case. □

Now we can formulate and prove the $\mathcal{P}$-Fredholm criterion for operators with rich slowly oscillating coefficients. It turns out that the uniform boundedness condition is redundant.

Theorem 7.2. *Operators in $\mathcal{A}_E(SO^{rich}_{L(X)})$ are $\mathcal{P}$-Fredholm if and only if all of their limit operators are invertible.*

Since $\sigma_\eta(A)$ is a singleton, the assertion follows immediately from Corollary 6.11.

8. Nets versus sequences

By Proposition 3.2, if $h \in \mathcal{H}$, then the closure $\overline{h}$ of the set $\{h(m) : m \in \mathbb{Z}^N\}$ of the values of h in the Gelfand topology cannot consist of a single point of $M^\infty(SO)$ only. Nevertheless, the sequences in $\mathcal{H}$ separate the points of $M^\infty(SO)$ in the following sense.

Proposition 8.1. *Given η, $\theta \in M^\infty(SO)$, there is a function $h \in \mathcal{H}$ such that $\eta \in \overline{h}$ and $\theta \notin \overline{h}$.*

Proof. Choose disjoint neighborhoods U_η and U_θ of η and θ in $M(SO)$, and let $h \in \mathcal{H}$ be a sequence such that

$$\{h(m) : m \in \mathbb{Z}^N\} = U_\eta \cap \mathbb{Z}^N.$$

(Recall that the intersection $U_\eta \cap \mathbb{Z}^N$ is not empty by Theorem 3.1 and, hence, countable. Thus, h can be even chosen as a bijection from $\mathbb{Z}^N$ onto $U_\eta \cap \mathbb{Z}^N$.) Since $\mathbb{Z}^N$ is dense in $M(SO)$, it is clear that $\eta \in \overline{U_\eta \cap \mathbb{Z}^N} = \overline{h}$, but θ cannot belong to $\overline{h}$ since

$$\theta \in U_\theta \subseteq M(SO) \setminus \overline{U_\eta} = M(SO) \setminus \overline{h},$$

i.e. θ is an interior point of the complement of $\overline{h}$. □

The Proposition 8.1 suggests the following definition.

Definition 8.2. *Let $\eta \in M^\infty(SO)$ and $A \in L(E)$. The* local operator spectrum of A at η with respect to sequences *is the set*

$$\sigma_\eta^{seq}(A) := \{A_h : h \in \mathcal{H}_A \text{ and } \eta \in \overline{h}\}.$$

If h is a sequence with $\eta \in \overline{h}$ and for which the limit operator A_h exists, then there is a subnet y of h which tends to η. Further, if $A \in \mathcal{A}_E$, then there is a subnet x of y for which the limit operator A_x exists. Clearly, x also tends to η and $A_x = A_h$. Thus,

$$\sigma_\eta^{seq}(A) \subseteq \sigma_\eta(A) \quad \text{for every } A \in \mathcal{A}_E. \tag{21}$$

We conjecture that in fact equality holds in (21). Some evidence to this conjecture is given by the following result.

Proposition 8.3. *Let $\eta \in M^\infty(SO)$.*
(a) If $A = aI$ with $a \in SO$, then $\sigma_\eta^{seq}(A) = \{a(\eta)\}$ (where we use the same notation for a function in SO and its Gelfand transform).
(b) If $A = aI$ with $a \in SO_{L(X)}$, then $\sigma_\eta^{seq}(A)$ contains at most one operator.

Proof. (a) Let $h \in \mathcal{H}$ be a sequence such that $\eta \in \overline{h}$ and such that the limit operator $(aI)_h$ exists. By Proposition 2.2, $(aI)_h = \alpha I$ with the complex number $\alpha := \lim a(h(n))$. We claim that $\alpha = a(\eta)$.

Let $\varepsilon > 0$. Since a is continuous at η, there is an open neighborhood U of η such that

$$|a(\eta) - a(\theta)| < \varepsilon/2 \quad \text{for all } \theta \in U.$$

Further, since $\eta \in \overline{h}$, there is an infinite subsequence g of h the values of which are in U. Choose m such that $|a(g(m)) - \alpha| < \varepsilon/2$. Then

$$|a(\eta) - \alpha| \le |a(\eta) - a(g(m))| + |a(g(m)) - \alpha| < \varepsilon.$$

This estimate holds for arbitrary $\varepsilon > 0$; hence, $a(\eta) = \alpha$.

(b) Suppose there are sequences $h_1, h_2 \in \mathcal{H}$ such that $\eta \in \overline{h_1} \cap \overline{h_2}$ and that the limit operators $(aI)_{h_1}$ and $(aI)_{h_2}$ exist, but that $(aI)_{h_1} \neq (aI)_{h_2}$. By Proposition 2.2, $(aI)_{h_1}$ and $(aI)_{h_2}$ are the operators of multiplication by the constant functions $x \mapsto A_1$ and $x \mapsto A_2$ with $A_1, A_2 \in L(X)$. Since $A_1 \neq A_2$, there is a functional $\varphi \in L(X)^*$ such that $\varphi(A_1) \neq \varphi(A_2)$. Consider the function $\hat{a} : \mathbb{Z}^N \to \mathbb{C} : x \mapsto \varphi(a(x))$. This function is in SO:

$$|\hat{a}(x+k) - \hat{a}(x)| \le \|\varphi\| \, \|a(x+k) - a(x)\|_{L(X)} \to 0 \quad \text{as } x \to \infty.$$

From $\|a(h_i(m)) - A_i\| \to 0$ for $i = 1, 2$ we conclude that

$$\|\hat{a}(h_i(m)) - \varphi(A_i)\| \to 0 \quad \text{for } i = 1, 2.$$

Hence, both $\varphi(A_1)I$ and $\varphi(A_2)I$ are limit operators of $\hat{a}I$ at η. This contradicts assertion (a) of this proposition, stating that $\sigma_\eta^{seq}(\hat{a}I)$ is a singleton. □

9. An alternative proof of Theorem 7.2

This section is devoted to an alternative proof of the preceding theorem, which works under more restrictive assumptions only, but which also has its own merits, and which sheds new light upon the properties of band-dominated operators with slowly oscillating coefficients. We let H be a Hilbert space and $E := l^2(\mathbb{Z}^N, H)$. Further, we again write $SO_{L(H)}$ and $SO^{rich}_{L(H)}$ for the algebra of all slowly oscillating functions $\mathbb{Z}^N \to L(H)$ and for the algebra of all slowly oscillating functions $\mathbb{Z}^N \to L(H)$ for which the associated multiplication operator is rich, respectively, and we let $\mathcal{A}_E(SO_{L(H)})$ and $\mathcal{A}_E(SO^{rich}_{L(H)})$ stand for the closures in $L(E)$ of the algebra of the band operators with coefficients in $SO_{L(H)}$ and in $SO^{rich}_{L(H)}$.

Generating functions. The alternative proof is based on the notion of the generating function of a band-dominated operator. This notion is borrowed from the pseudodifferential operator calculus (where the generating function is usually referred to as the symbol of the operator) and adapted for our purposes.

We start with defining the generating function of a band operator. For

$$A = \sum_{|\alpha| \le M} a_\alpha V_\alpha \quad \text{with } a_\alpha \in SO_{L(H)}, \tag{22}$$

let the *generating function* of A be

$$\text{gen}_A : \mathbb{Z}^N \times \mathbb{T}^N \to L(H), \quad (x,t) \mapsto \sum_{|\alpha| \le M} a_\alpha(x) t^\alpha \tag{23}$$

where $t^\alpha := t_1^{\alpha_1} \ldots t_n^{\alpha_n}$. There is a one-to-one correspondence between band operators and their generating functions.

We denote by $C_b(\mathbb{Z}^N \times \mathbb{T}^N, L(H))$ the set of all continuous functions on $\mathbb{Z}^N \times \mathbb{T}^N$ with values in $L(H)$. Provided with pointwisely defined operations and the supremum norm, this set becomes a C^*-algebra, and the set $c_0(\mathbb{Z}^N \times \mathbb{T}^N, L(H))$ of all functions $a \in C_b(\mathbb{Z}^N \times \mathbb{T}^N, L(H))$ with

$$\lim_{x \to \infty} \sup_{t \in \mathbb{T}^N} \|a(x,t)\|_{L(H)} = 0$$

is a closed ideal of $C_b(\mathbb{Z}^N \times \mathbb{T}^N, L(H))$. The quotient algebra C_b/c_0 will be abbreviated to $\widehat{C}_b$, and the coset which contains $a \in C_b(\mathbb{Z}^N \times \mathbb{T}^N, L(H))$ to $\widehat{a}$. Notice that

$$\|\widehat{a}\|_0 := \limsup_{x \to \infty} \sup_{t \in \mathbb{T}^N} \|a(x,t)\|_{L(H)}$$

is just the canonical quotient norm of the coset $\widehat{a}$ in the quotient algebra $\widehat{C}_b$.

Evidently, if A is a band operator of the form (22), then its generating function belongs to $C_b(\mathbb{Z}^N \times \mathbb{T}^N, L(H))$.

Proposition 9.1. *Let A be as in* (22). *Then* $\|\widehat{\text{gen}_A}\|_0 \le \|A\|$.

Proof. Choose a sequence $(x_n) \subset \mathbb{Z}^N$ tending to infinity, a sequence $(t_n) \in \mathbb{T}^N$, and a sequence (v_n) of unit vectors in H, such that

$$\|\widehat{\text{gen}_A}\|_0 = \lim_{n\to\infty} \|\text{gen}_A(x_n, t_n)v_n\|_H.$$

Since $\mathbb{T}^N$ is compact, we can moreover assume that (t_n) is a convergent sequence with limit $t_0 \in \mathbb{T}^N$. The assertion will follow once we have shown that, given $\varepsilon > 0$, there is an n_0 such that

$$\|\text{gen}_A(x_n, t_n)v_n\|_H \le \|A\| + \varepsilon \tag{24}$$

for all $n \ge n_0$.

Given vectors $v \in H$ and $u = (u_k)_{k\in\mathbb{Z}^N} \in l^2$, let $v \otimes u$ denote the sequence $(u_k v)_{k\in\mathbb{Z}^N}$ in $E = l^2(\mathbb{Z}^N, H)$. Let further $A_{n,n}$, A_n and B_n stand for the band operators with the generating functions

$$(x,t) \mapsto \text{gen}_A(x_n, t_n), \quad (x,t) \mapsto \text{gen}_A(x + x_n, t), \quad (x,t) \mapsto \text{gen}_A(x_n, t),$$

respectively. Then we have, for every unit vector $u \in l^2$,

$$\begin{aligned} \|\text{gen}_A(x_n, t_n)v_n\|_H &= \|A_{n,n}(v_n \otimes u)\|_E \\ &\le \|(A_{n,n} - B_n)(v_n \otimes u)\|_E \\ &\quad + \|(B_n - A_n)(v_n \otimes u)\|_E + \|A_n(v_n \otimes u)\|_E. \end{aligned} \tag{25}$$

Since $A_n = V_{-x_n} A V_{x_n}$, we get

$$\|A_n(v_n \otimes u)\|_E = \|V_{-x_n} A V_{x_n}(v_n \otimes u)\|_E \le \|A\|$$

for the last term in (25). The middle term on the right-hand side of (25) is not greater than $\|B_n - A_n\|$, which goes to zero as $n \to \infty$ since the coefficients of A are slowly oscillating. Thus, this middle becomes less than $\varepsilon/2$ uniformly with respect to u and v_n if only n is large enough.

To estimate the first term, choose $\delta > 0$ such that

$$\sup_{x\in\mathbb{Z}^N} \|\text{gen}_A(x,t) - \text{gen}_A(x, t_0)\| \le \varepsilon/4 \quad \text{for all } |t - t_0| < \delta,$$

and choose the unit vector $u = (u_k)_{k\in\mathbb{Z}^N}$ in l^2 such that the u_k are the Fourier coefficients of a continuous function $\hat{u}$ on $\mathbb{T}^N$ with support in $\{t \in \mathbb{T}^N : |t - t_0| < \delta\}$. Since $A_{n,n} - B_n$ is the operator of convolution by the function $\text{gen}_{A_{n,n}} - \text{gen}_{B_n}$, we get

$$\begin{aligned} \|(A_{n,n} - B_n)(v_n \otimes u)\|^2_{l^2(\mathbb{Z}^N,H)} &= \|(\text{gen}_{A_{n,n}} - \text{gen}_{B_n})(\hat{u}v_n)\|^2_{L^2(\mathbb{T}^N,H)} \\ &= \int_{\mathbb{T}^N} \|(\text{gen}_A(x_n, t_n) - \text{gen}_B(x_n, t))\hat{u}(t)v_n\|^2_H \, dt \\ &\le \sup_{|t-t_0|<\delta} \|\text{gen}_A(x_n, t_n) - \text{gen}_A(x_n, t)\|^2_{L(H)} \, \|v_n \otimes u\|^2. \end{aligned}$$

Due to the choice of δ, this term becomes less than $\varepsilon/2$ if n becomes large. □

This proposition allows us to associate with every operator A in $\mathcal{A}_E(SO_{L(H)})$ a uniquely determined coset in $\widehat{C}_b$ which we denote by $\Gamma(A)$.

In what follows, we will make use of the notion of the main diagonal of a band-dominated operator. If A is the band operator $\sum_{|\alpha|\le k} a_\alpha V_\alpha$, then its main diagonal is, by definition, the function $D(A) := a_0$. Since

$$\|D(A)\|_\infty = \sup_k \|a_0(k)\| = \sup_k \|S_k A S_k\| \le \|A\|$$

(where the S_k are as in the introduction), we can extend the mapping D by continuity onto the set of all band-dominated operators. For a band-dominated operator A, we call $D(A)$ its main diagonal and $D(AV_{-\alpha})$ its αth diagonal.

Proposition 9.2. *Γ is a *-homomorphism from $\mathcal{A}_E(SO^{rich}_{L(H)})$ into $\widehat{C}_b$ with kernel $K(E, \mathcal{P})$.*

Proof. It is elementary to check that Γ acts as a *-homomorphism on the algebra of all band operators with slowly oscillating coefficients. Since this algebra is dense in $\mathcal{A}_E(SO_{L(H)})$, and since Γ is continuous on this algebra by the preceding proposition, this proves the first assertion. It is further evident that the ideal $K(E, \mathcal{P})$ lies in the kernel of Γ. Let, finally, A be an operator in $\mathcal{A}_E(SO_{L(H)})$ with $\Gamma(A) = 0$. We have to show that A lies in $K(E, \mathcal{P})$.

Let (A_n) be a sequence of band operators in $\mathcal{A}_E(SO^{rich}_{L(H)})$ which converges to A. Then, trivially, $\|\Gamma(A_n)\| \to 0$. For $\alpha = (\alpha_1, \ldots, \alpha_N) \in \mathbb{Z}^N$, consider the functions $a^{(n)}_\alpha : \mathbb{Z}^N \to L(H)$ which take at $x \in \mathbb{Z}^N$ the value

$$\frac{1}{(2\pi)^N} \int_0^{2\pi} \cdots \int_0^{2\pi} \text{gen}_{A_n}(x, (e^{is_1}, \ldots, e^{is_N}))\, e^{-i\alpha_1 s_1} \ldots e^{-i\alpha_N s_N}\, ds_1 \ldots ds_N. \quad (26)$$

If the band operator A_n is of the form $\sum b^{(n)}_\alpha V_\alpha$, then its αth diagonal $b^{(n)}_\alpha$ just coincides with the function $a^{(n)}_\alpha$ given by (26). From (26) we immediately conclude that

$$\|a^{(n)}_\alpha(x)\|_{L(H)} \le \sup_{t\in\mathbb{T}^N} \|\text{gen}_{A_n}(x, t)\|_\infty$$

whence, in particular,

$$\limsup_{x\to\infty} \|a^{(n)}_\alpha(x)\| \le \limsup_{x\to\infty} \sup_{t\in\mathbb{T}^N} \|\text{gen}_{A_n}(x, t)\|_\infty = \|\Gamma(A_n)\| \to 0$$

as $n \to \infty$. Thus, if a_α denotes the αth diagonal of A, then

$$\begin{aligned} \limsup_{x\to\infty} \|a_\alpha(x)\| &\le \sup_{x\in\mathbb{Z}^N} \|a_\alpha(x) - a^{(n)}_\alpha(x)\| + \limsup_{x\to\infty} \|a^{(n)}_\alpha(x)\| \\ &\le \|A - A_n\| + \|\Gamma(A_n)\|. \end{aligned} \quad (27)$$

This shows that every diagonal of A lies in $c_0(\mathbb{Z}^N, L(H))$, which on its hand implies that all limit operators of A are 0: Indeed, let h be a sequence for which the limit operator A_h exists. Then the operators $S_i V_{-h(n)} A V_{h(n)} S_j$ converge in the norm to $S_i A_h S_j$ for every pair of indices i, $j \in \mathbb{Z}^N$. Since

$$\lim_{n\to\infty} \|S_i V_{-h(n)} A V_{h(n)} S_j\| = \lim_{n\to\infty} \|a_{i-j}(i + h(n))\| = 0$$

due to (27), this shows that $S_i A_h S_j = 0$ for all i and j, whence $A_h = 0$. But a rich band-dominated operator having 0 as its only limit operator lies in $K(E, \mathcal{P})$ due to Theorems 2.24 and 2.24 in [6]. $\square$

Now we can present an alternative proof of Theorem 7.2.

Theorem 9.3. *The following assertions are equivalent for $A \in \mathcal{A}_E(SO^{rich}_{L(H)})$:*

(a) A is $\mathcal{P}$-Fredholm.
(b) $\Gamma(A)$ is invertible in $\widehat{C}_b$.
(c) All limit operators of A are invertible.

Proof. The equivalence of (a) and (b) is quite obvious. Indeed, if the coset $A + K(E, \mathcal{P})$ is invertible in $L(E, \mathcal{P})/K(E, \mathcal{P})$, then it is also invertible in the algebra $\mathcal{A}_E(SO^{rich}_{L(H)})/K(E, \mathcal{P})$ (inverse closedness of C^*-algebras). Hence, there are operators $B \in \mathcal{A}_E(SO^{rich}_{L(H)})$ and K_1, $K_2 \in K(E, \mathcal{P})$ such that $AB = I + K_1$ and $BA = I + K_2$. Applying the homomorphism Γ to these equalities yields the invertibility of $\Gamma(A)$. If, conversely, $\Gamma(A)$ is invertible in $\widehat{C}_b$, then it is also invertible in the image of $\mathcal{A}_E(SO^{rich}_{L(H)})$ under the mapping Γ (again by the inverse closedness of C^*-algebras). Thus, one can find a $B \in \mathcal{A}_E(SO^{rich}_{L(H)})$ with $\Gamma(A)\Gamma(B) = \Gamma(B)\Gamma(A) = 1$, showing that $AB - I$ and $BA - I$ belong to $\ker \Gamma = K(E, \mathcal{P})$.

Since (a) obviously implies (c) (see also the first lines of the proof of Theorem 1 in [5]), we are left with the implication $(c) \Rightarrow (b)$. Assume that all limit operators of $A \in \mathcal{A}_E(SO^{rich}_{L(H)})$ are invertible, but that $\Gamma(A)$ is not invertible in $\widehat{C}_b$. If A is not a band operator, then we let gen_A be any function in the coset $\Gamma(A)$.

Let $\nu(C) := \inf_{x \neq 0} \|Cx\|/\|x\|$ denote the lower norm of the operator $C \in L(H)$. It is well known that C is invertible if both $\nu(C)$ and $\nu(C^*)$ are positive and that, conversely, invertibility of C implies $\nu(C) = \nu(C^*) = 1/\|A^{-1}\|$. Thus, if both

$$\lim_{R \to \infty} \inf_{|x| \geq R,\, t \in \mathbb{T}^N} \nu(\text{gen}_A(x, t)) > 0 \tag{28}$$

and

$$\lim_{R \to \infty} \inf_{|x| \geq R,\, t \in \mathbb{T}^N} \nu(\text{gen}_A(x, t)^*) > 0, \tag{29}$$

then the function gen_A is invertible in C_b modulo functions in c_0. Since $\Gamma(A)$ is non-invertible by assumption, one of the conditions (28) and (29) must be violated, say the first one for definiteness. Then there exist a sequence $x = (x_m)_{m \geq 1} \subset \mathbb{Z}^N$ which tends to infinity, a sequence $(t_m)_{m \geq 1} \subset \mathbb{T}^N$ which we can also suppose to be convergent to a point $t_0 \in \mathbb{T}^N$, as well as a sequence $(v_m)_{m \geq 1}$ of unit vectors in H such that

$$\|\text{gen}_A(x_m, t_m) v_m\| \to 0 \quad \text{as } m \to \infty.$$

We will further suppose without loss that the limit operator A_x of A with respect to the sequence x exists.

Let $\varepsilon < 1/(4\,\|A_x^{-1}\|)$, and let A' be a band operator with coefficients in $SO^{rich}_{L(H)}$ such that $\|A - A'\| < \varepsilon$. Then $\|\Gamma(A) - \Gamma(A')\|_0 < \varepsilon$, which implies that

$$\begin{aligned}
&\limsup_{m\to\infty} \|\text{gen}_{A'}(x_m, t_m)v_m\| \\
&\quad \le \limsup_{m\to\infty} \|\text{gen}_{A'}(x_m, t_m)v_m - \text{gen}_A(x_m, t_m)v_m\| + \lim_{m\to\infty} \|\text{gen}_A(x_m, t_m)v_m\| \\
&\quad \le \limsup_{m\to\infty} \sup_{t\in\mathbb{T}^N} \|\text{gen}_{A'}(x_m, t) - \text{gen}_A(x_m, t)\| \\
&\quad = \|\Gamma(A) - \Gamma(A')\|_0 < \varepsilon.
\end{aligned}$$

Hence, $\|\text{gen}_{A'}(x_m, t_m)v_m\| < \varepsilon$ for all sufficiently large m. We further suppose without loss that the limit operator of A' with respect to the sequence x exists (otherwise we pass to a suitable subsequence of x). As in the proof of Proposition 9.1, we can find a unit vector $u \in l^2$ such that

$$\|V_{-x_m} A' V_{x_m}(v_m \otimes u)\| < 2\varepsilon \quad \text{for all sufficiently large } m$$

and, according to the definition of limit operators, we further have

$$\|(V_{-x_m} A' V_{x_m} - A'_x)(v \otimes u)\| \to 0$$

uniformly with respect to the unit vectors v. Hence, $\|A'_x(v_m \otimes u)\| < 3\varepsilon$ for all sufficiently large m. Since $\|v_m \otimes u\| = 1$, we conclude that

$$\text{either } A'_x \text{ is not invertible or } \|(A'_x)^{-1}\| > 1/(3\varepsilon). \tag{30}$$

On the other hand,

$$\|A_x - A'_x\| \le \|A - A'\| < \varepsilon < 1/(4\,\|A_x^{-1}\|).$$

Thus, by a Neumann series argument, A'_x is invertible, and

$$\|(A'_x)^{-1}\| \le \frac{\|(A_x)^{-1}\|}{1 - \|(A_x)^{-1}\|\,\|A_x - A'_x\|} \le \frac{\|(A_x)^{-1}\|}{1 - \varepsilon\,\|(A_x)^{-1}\|}.$$

Together with (30), this yields

$$\frac{1}{3\varepsilon} < \frac{\|(A_x)^{-1}\|}{1 - \varepsilon\,\|(A_x)^{-1}\|}$$

or, equivalently, $\varepsilon > 1/(4\,\|(A_x)^{-1}\|$. The obtained estimate contradicts the choice of ε. □

In a similar way, the following refinement of the local Fredholm criterion (Theorem 1.6 and its corollary) can be derived.

Theorem 9.4. *The following assertions are equivalent for $A \in \mathcal{A}_E(SO^{rich}_{L(H)})$:*

(a) A is locally invertible at $\eta \in S^{N-1}$.
(b) The local coset $\pi_\eta(A)$ is invertible.
(c) All operators in local operator spectrum $\sigma_\eta(A)$ of A are invertible.

References

[1] G.R. Allan, *Ideals of vector valued functions,* Proc. Lond. Math. Soc. **18** (1968), 3, 193–216.

[2] A. Böttcher and B. Silbermann, *Analysis of Toeplitz Operators,* Akademie-Verlag, Berlin 1989 and Springer-Verlag, Berlin, Heidelberg, New York 1990.

[3] T.W. Gamelin, *Uniform Algebras,* Prentice-Hall, Inc., Englewood Cliffs, N.J., 1969.

[4] B.Ya. Shteinberg, *Compactification of locally compact groups and Fredholmness of convolution operators with coefficients in factor groups,* Tr. St-Peterbg. Mat. Obshch. **6** (1998), 242–260 (Russian).

[5] V.S. Rabinovich, S. Roch and B. Silbermann, *Fredholm theory and finite section method for band-dominated operators,* Integral Equations Oper. Theory **30** (1998), 4, 452–495.

[6] V.S. Rabinovich, S. Roch and B. Silbermann, *Band-dominated operators with operator-valued coefficients, their Fredholm properties and finite sections,* Integral Equations Oper. Theory **40** (2001), 3, 342–381.

V.S. Rabinovich
Instituto Politechnico National
ESIME-Zacatenco, Ed.1, 2-do piso
Av.IPN, Mexico, D.F., 07738
Mexico
e-mail: `rabinov@maya.esimez.ipn.mx`

S. Roch
Technical University Darmstadt
Faculty of Mathematics
Schlossgartenstrasse 7
64289 Darmstadt
Germany
e-mail: `roch@mathematik.tu-darmstadt.de`

Received: 10 September 2001

Operator Theory:
Advances and Applications, Vol. 135, 293–315

More Inequalities and Asymptotics for Matrix Valued Linear Positive Operators: the Noncommutative Case

Stefano Serra-Capizzano

Dedicated to Bernd Silbermann on His 60th Birthday

Abstract. In a recent paper, the author and P. Tilli have analyzed Cauchy-Schwarz type inequalities and associated relations in the case of matrix valued linear positive operators acting on some space $\mathcal{A}$ of scalar valued functions. Here we deal with the noncommutative case where the space $\mathcal{A}$ is constituted by matrix valued functions. Some inequalities are formally the same, but many of them are substantially affected by the noncommutativity of the underlying space. Several applications to Toeplitz sequences with matrix valued symbols are finally derived and discussed.

1. Introduction

Let $\mathcal{M}_n \equiv \mathcal{M}_n(\mathbf{C})$ be the space of complex $n \times n$ matrices. Let $\mathcal{A}$ be a collection of functions with a given domain Ω and range $\mathcal{M}_s$. Assume that $\mathcal{A}$ is a complex vector space which is closed under passage to the Hermitian adjoint and absolute value (i.e. if $f \in \mathcal{A}$ then the functions f^* and $|f| = (f^* f)^{1/2}$ belong to $\mathcal{A}$). We will consider linear maps $T : \mathcal{A} \to \mathcal{M}_n$ that are positive in the following sense: if f is an element of $\mathcal{A}$ such that $f(t)$ is positively semi-definite for all t, then $T(f)$ is positively semi-definite. We then say that T is a *matrix valued linear positive operator* (LPO).

This kind of operators arises quite naturally in many applications, such as multilevel Toeplitz matrices with matrix valued generating functions [22], discretizations (by finite differences or finite elements) of elliptic systems of PDEs [15, 16], and orthogonal polynomials with periodic recurrence coefficients [5].

In this paper we prove some inequalities for unitarily invariant norms of $T(f)$ for such operators. The scalar case $s = 1$ (scalar valued functions) has been considered in [10, 17]. Basic properties of unitarily invariant norms may be found in Chapter 4 of [1].

A matrix norm $||\cdot||$ is called *unitarily invariant* (u.i. for brevity) if $||UAV|| = ||A||$ for arbitrary A whenever U and V are unitary matrices. It is well known that a function $||\cdot||$ defined on the $n \times n$ matrices is a u. i. norm if and only if there exists a symmetric gauge function Φ on $\mathbf{R}^n$ such that

$$||A|| = \Phi\big(\sigma_1(A), \ldots, \sigma_n(A)\big),$$

where $\sigma_j(A)$ denotes the j-th singular value of A. If Φ is a symmetric gauge function, then $\|\cdot\|_\Phi$ will denote the corresponding u.i. norm.

We recall that a function Φ on $\mathbf{R}^n$ is called a *symmetric gauge* function if it satisfies the following properties:

1. Φ is a norm.
2. Φ is absolute, i.e., $\Phi(x_1, \ldots, x_n) = \Phi(|x_1|, \ldots, |x_n|)$.
3. Φ is symmetric, i.e., $\Phi(x_1, \ldots, x_n) = \Phi(x_{\pi_1}, \ldots, x_{\pi_n})$ for every permutation $\pi : \{1, \ldots, n\} \to \{1, \ldots, n\}$.

In what follows, $[x_i]_{i=1}^n$, or simply $[x_i]$ if n is clear from the context, denotes the vector with entries x_i, so that we may write $\Phi\big([x_i]\big)$ instead of $\Phi(x_1, \ldots, x_n)$. Moreover, $\langle x, y\rangle := \sum_{i=1}^n x_i \overline{y_i}$ denotes the Euclidean inner product.

Every symmetric gauge function Φ has the following properties (see [1]), which will be widely used in what follows:

$$\Phi\big(\,[x_i]\,\big) \le \Phi\big(\,[y_i]\,\big) \quad \text{if } |x_i| \le |y_i| \text{ for all } i, \tag{1}$$

$$\Phi\big(\,[\,|x_i\, y_i|\,]\,\big) \le \Phi\big(\,[\,|x_i|^p\,]\,\big)^{1/p}\, \Phi\big(\,[\,|y_i|^q\,]\,\big)^{1/q} \quad \text{if } p > 1 \text{ and } 1/p + 1/q = 1. \tag{2}$$

If $P_1, \ldots, P_k$ are mutually orthogonal projections on $\mathbf{C}^n$ such that $P_1 \oplus \cdots \oplus P_k = I$, then

$$\left\|\sum_{i=1}^{k} P_i A P_i\right\| \le ||A||\,. \tag{3}$$

The last inequality is known as the "pinching inequality" (see [1]) and is the main ingredient for proving the following basic variational characterization of u.i. norms.

Theorem 1.1. [17] *Let Φ be a symmetric gauge function on $\mathbf{R}^n$. Then for every matrix $A \in \mathcal{M}_n$ we have*

$$\|A\|_\Phi = \sup \Phi\Big(\big[\,|\langle Au_i,\, v_i\rangle|\,\big]_{i=1}^n\Big), \tag{4}$$

where the supremum is taken over all pairs of orthonormal bases $\{u_i\}_{i=1}^n$ and $\{v_i\}_{i=1}^n$, i.e.,

$$u_i, v_i \in \mathbf{C}^n \quad \textit{and} \quad \langle u_i,\, u_j\rangle = \langle v_i,\, v_j\rangle = \delta_{i,j}, \quad 1 \le i, j \le n.$$

If, moreover, A is positively semi-definite, then we have

$$\|A\|_\Phi = \sup \Phi\Big(\big[\,\langle Au_i,\, u_i\rangle\,\big]_{i=1}^n\Big), \tag{5}$$

where the supremum is taken over all orthonormal bases $\{u_i\}_{i=1}^n$.

Among u.i. norms, the so-called Schatten p-norms $\|\cdot\|_p$ are of particular interest (see [1]). These are defined as follows:

$$\|A\|_\infty := \max_i\{\sigma_i\} \quad \text{and} \quad \|A\|_p := \left(\sum_{i=1}^n \sigma_i^p\right)^{\frac{1}{p}} \quad \text{if } p \geq 1,$$

where $\sigma_1, \ldots, \sigma_n$ denote the singular values of A.

The paper is organized as follows. Section 2 is devoted to proving a Cauchy-Schwarz inequality when $\mathcal{A} = L^p(\Omega, \mu, \mathcal{M}_s)$ for some $p, s \geq 1$ and several other inequalities for u.i. norms (in particular, Schatten norms) of matrices arising from matrix valued LPOs. In Section 3, we discuss some examples and applications concerning u.i. norms of multilevel block Toeplitz and Hankel matrices, and in Section 4 we finally employ some of these inequalities to derive distributional results on the algebra generated by Toeplitz sequences with matrix valued L^1 generating functions.

2. Some inequalities for u.i. norms of LPOs

2.1. Basic theory

Our first objective is the search for a kind of Cauchy-Schwarz inequality for matrices arising from a linear positive operator T defined over $\mathcal{A}$:

$$|\langle T(f)u, v\rangle|^2 \leq \langle T(|f|)u, u\rangle \langle T(|f^*|)v, v\rangle \quad \text{for all } u, v \in \mathbf{C}^n,\ f \in \mathcal{A} \tag{6}$$

In the following proposition we consider a parametric generalization of inequality (6) and show that it can be rewritten in several equivalent ways.

Proposition 2.1. *Let $T : \mathcal{A} \to \mathcal{M}_n$ be an LPO, where $\mathcal{A}$ is a linear space closed under passage to the Hermitian adjoint and to the absolute value, and let γ be a positive constant. Then the following three statements are equivalent:*

(a) *for all $u, v \in \mathbf{C}^n$ and all $f \in \mathcal{A}$,*

$$|\langle T(f)u, v\rangle|^2 \leq \gamma^2 \langle T(|f|)u, u\rangle \langle T(|f^*|)v, v\rangle; \tag{7}$$

(b) *the (nonlinear) operator $G_\gamma : \mathcal{A} \to \mathcal{M}_{2n}$ defined as*

$$G_\gamma(f) \equiv \begin{pmatrix} \gamma T(|f|) & T(f^*) \\ T(f) & \gamma T(|f^*|) \end{pmatrix} \tag{8}$$

is a positive operator, i.e., $\langle G_\gamma(f)w, w\rangle \geq 0$ for all $w \in \mathbf{C}^{2n}$ and all $f \in \mathcal{A}$;

(c) *for all $u, v \in \mathbf{C}^n$ and all $f \in \mathcal{A}$,*

$$|\langle T(f)u, v\rangle| \leq \frac{\gamma}{2}\langle T(|f|)u, u\rangle + \frac{\gamma}{2}\langle T(|f^*|)v, v\rangle. \tag{9}$$

Proof. Clearly, (c) follows from (a) using the inequality between the arithmetic and geometric means. Concerning (b), we first note that $G_\gamma(f)$ is Hermitian since

$T(f^*) = T(f)^*$. Decomposing $w \in \mathbf{C}^{2n}$ as (u, v), we see that $G_\gamma(f) \geq 0$ is equivalent to

$$\gamma \langle T(|f|)u, u\rangle + \gamma \langle T(|f^*|)v, v\rangle + 2\mathrm{Re}\,\langle T(f)u, v\rangle \geq 0 \text{ for all } u, v \in \mathbf{C}^n, \quad (10)$$

and hence (b) follows from (c) using the inequality $2\mathrm{Re}\, z \geq -2|z|$. On the other hand, given $u, v \in \mathbf{C}^n$ we can find $\omega \in \mathbf{C}$ such that $|\omega| = 1$ and $2\mathrm{Re}\,\omega \langle T(f)u, v\rangle = -2|\langle T(f)u, v\rangle|$. Hence, if $G_\gamma(f) \geq 0$ then rewriting (10) with ωu in place of u we obtain that (b) implies (c). Finally, if (c) holds, then (9) with u/t in place of u and tv in place of v, where $t > 0$ is a parameter, we obtain (a) by minimizing the resulting right-hand side with respect to t. □

Example 2.2. If the linear space $\mathcal{A}$ is not closed under taking absolute values, then relation (6) is generally false even for spaces $\mathcal{A}$ of scalar valued functions. Take $\mathcal{A} = \mathrm{span}\{e^{i\alpha x} : \alpha \in \{0, \pm 1\}\}$ and let $T : \mathcal{A} \to \mathcal{M}_2$ be the linear operator that is defined by

$$T(1) = I_2, \quad T(e^{ix}) = \begin{pmatrix} 0 & 2 \\ 0 & 0 \end{pmatrix}, \quad T(e^{-ix}) = \begin{pmatrix} 0 & 0 \\ 2 & 0 \end{pmatrix}.$$

It is trivial to check that T is an LPO. However, for $f = e^{ix}$, $u = (0,1)^T$, $v = (1,0)^T$, we have $|f| = 1$ and $|\langle T(f)u, v\rangle|^2 = 4$ with $\langle T(|f|)u, u\rangle = \langle T(|f|)v, v\rangle = 1$, so that (6) does not hold.

Example 2.3. If on the right-hand side of inequality (7) we omit the symbol "$*$" and the function space is not scalar valued, then the considered relation is generally false with any choice of a positive γ. Take the identity operator on $\mathcal{M}_2$, which is clearly linear and positive, and consider $v = e_1$, $u = e_2$, $f = vu^T$. Then $|f| = uu^T$, $\langle T(|f|)u, u\rangle = 1$, $\langle T(|f|)v, v\rangle = 0$, but $|\langle T(f)u, v\rangle|^2 = 1$.

We remark that inequality (7) with $\gamma = 1$ is exactly inequality (6). Moreover, it is useful to recall that a linear positive operator T is called 2-positive if the map $T[2] : \mathcal{M}_2(\mathcal{A}) \to \mathcal{M}_2(\mathcal{M}_n)$ is positive, i.e., if

$$T[2](f) = \begin{pmatrix} T(f_{1,1}) & T(f_{1,2}) \\ T(f_{1,2}^*) & T(f_{2,2}) \end{pmatrix}$$

is positively semi-definite for every positively semi-definite

$$f = \begin{pmatrix} f_{1,1} & f_{1,2} \\ f_{1,2}^* & f_{2,2} \end{pmatrix}.$$

Now, in the light of the definition of 2-positivity, which implies the positive semi-definiteness of the operator G_1, we obtain (6) provided we can prove the 2-positivity of T. In the following we will also briefly consider the case where $\gamma = 2$ and $\gamma = \sqrt{2}$. Indeed (7) with $\gamma = 2$ (and therefore its equivalent forms) holds for any linear positive operator acting on scalar valued spaces $\mathcal{A}$ (see [17]).

In the matrix valued case the situation is dramatically different since (7) is generally false with any positive γ as the following example shows.

Example 2.4. Consider the operator T ="transpose", which is positive but not 2-positive from $\mathcal{M}_s$ onto itself provided $s \geq 2$. Then (7) is false with any choice of a positive γ. Take $u = e_1$, $v = e_2$, $f = uv^T$. Then $|f| = uu^T$, $\langle T(|f|)u, u\rangle = 0$, $\langle T(|f^*|)v, v\rangle = 0$, but $|\langle T(f)u, v\rangle|^2 = |\langle v, v\rangle|^2 = 1$.

However if we restrict our attention to Hermitian arguments then (7) with $\gamma = \sqrt{2}$ holds for purely algebraic reasons exactly as in the scalar case (refer to [17]).

Theorem 2.5. *Let $T : \mathcal{A} \to \mathcal{M}_n$ be an LPO and suppose $\mathcal{A}$ is closed under passage to the Hermitian adjoint and to the absolute value. Define $\mathcal{A}_{\rm sym} = \{f \in \mathcal{A} : f^* = f\}$. Then inequality (7) is true with $\gamma = \sqrt{2}$ for all u, v and all $f \in \mathcal{A}_{\rm sym}$.*

Proof. Step 1. First we assume that f is Hermitian valued and positively semi-definite. Therefore $T(f) = T(|f|)$ is positively semi-definite and (6) is true by the classical Cauchy-Schwarz inequality in $\mathbf{C}^n$.

Step 2. If f is Hermitian valued, then by the Schur decomposition we have $f = U\,\mathrm{diag}(d_1, \ldots, d_s)\,U^*$ where U is unitary and the d_j are real valued. Therefore we can always write $f = f^+ - f^-$ where

$$f^+ = U\,\mathrm{diag}(d_1^+, \ldots, d_s^+)\,U^*, \quad f^- = U\,\mathrm{diag}(d_1^-, \ldots, d_s^-)\,U^*$$

are positively semi-definite with $d_j^+ = \max\{0, d_j\}$ and $d_j^- = \max\{0, -d_j\}$. On observing that $|f| = f^+ + f^-$, recalling the first step, and invoking the monotonicity of the operator $T(\cdot)$, we get

$$\begin{aligned}
|\langle T(f)u, v\rangle|^2 &\leq (|\langle T(f^+)u, v\rangle| + |\langle T(f^-)u, v\rangle|)^2 \\
&= (\langle(|\langle T(f^+)u, v\rangle|, |\langle T(f^-)u, v\rangle|), (1,1)\rangle)^2 \\
&\leq_{\text{CS in } \mathbf{C}^2} 2(|\langle T(f^+)u, v\rangle|^2 + |\langle T(f^-)u, v\rangle|^2) \\
&\leq_{\text{by Step 1}} 2(\langle T(f^+)u, u\rangle \langle T(f^+)v, v\rangle + \\
&\qquad + \langle T(f^-)u, u\rangle \langle T(f^-)v, v\rangle) \\
&\leq 2\,\langle T(|f|)u, u\rangle \langle T(|f|)v, v\rangle
\end{aligned}$$

which is (7) with $\gamma = \sqrt{2}$. □

Now we prove (6) in the case where the map is 2-positive and then we show that if $\mathcal{A}$ is an L^p space for a certain $p \in [1, \infty)$ and T is an LPO from $\mathcal{A}$ to $\mathcal{M}_n$ then 2-positivity is automatically satisfied. Actually the arguments of the proof of the first result are purely algebraic while the arguments of the proof of the second result have some essentially analytic flavour as well.

Theorem 2.6. *Let $T : \mathcal{A} \to \mathcal{M}_n$ be an LPO and suppose that $\mathcal{A}$ is closed under passage to the Hermitian adjoint and to the absolute value. If T is 2-positive, then inequality (6) is satisfied.*

Proof. Taking into account the definition of 2-positivity and the equivalence between (7) and (8) proven in Proposition 2.1, the proof is reduced to the claim that, for every matrix $A \in \mathcal{M}_s$, the $2s \times 2s$ matrix

$$\widehat{A} = \begin{pmatrix} |A| & A^* \\ A & |A^*| \end{pmatrix}$$

is positively semi-definite. But this is a straightforward consequence of the singular value decomposition of the matrix $A = U\Sigma V^*$ with unitary U and V and with diagonal Σ such that $(\Sigma)_{i,i} \geq 0$. More precisely we have

$$\widehat{A} = \begin{pmatrix} V & \\ & U \end{pmatrix} \left[\begin{pmatrix} 1 & 1 \\ 1 & 1 \end{pmatrix} \otimes \Sigma \right] \begin{pmatrix} V^* & \\ & U^* \end{pmatrix}$$

and therefore the nonzero eigenvalues of $\widehat{A}$ are $2(\Sigma)_{j,j} \geq 0$, $j = 1, \ldots, s$. □

Theorem 2.7. *Let $p \in [1, \infty)$, let n be a positive integer and let (Ω, μ) be a measure space with a σ-finite positive measure μ. If $T : L^p(\Omega, \mu, \mathcal{M}_s) \to \mathcal{M}_n$ is an LPO, then*

- **(a)** *T is continuous;*
- **(b)** *the operator T is 2-positive;*
- **(c)** *inequality (6) is true for all $u, v \in \mathbf{C}^n$ and all $f \in L^p$.*

Proof of part (a). We first suppose that $n = 1$, i.e. that T is a positive linear functional on $L^p(\Omega, \mu, \mathcal{M}_s)$, so that it can be identified as the sum of s^2 linear functionals on $L^p(\Omega, \mu, \mathbf{C})$. In order to prove that T is bounded, it suffices to prove that for every sequence $\{u_k\}$ which is convergent in $L^p(\Omega, \mu, \mathcal{M}_s)$, there exists a subsequence $\{u_{k_j}\}$ such that $T(u_{k_j})$ is bounded (see [9]). On the other hand, it is well known that for every sequence $\{v_k\}$ convergent in $L^p(\Omega, \mu, \mathbf{C})$, there exists $v \in L^p(\Omega, \mu, \mathbf{C})$ and a subsequence $\{u_{k_j}\}$ such that $|v_{k_j}| \leq v$ μ-a.e. in Ω. Since both the Hermitian $s \times s$ matrices $\operatorname{Re} u_{k_j}$ and $\operatorname{Im} u_{k_j}$ have all the entries bounded by $|v|$, we deduce that $\operatorname{Re} u_{k_j}$, $\operatorname{Im} u_{k_j} \leq s|v|I_s$ in the sense of the partial ordering between matrices. From the positivity of T we therefore obtain

$$|T(u_{k_j})| \leq T(|\operatorname{Re} u_{k_j}|) + T(|\operatorname{Im} u_{k_j}|) \leq 2sT(|v|I_s),$$

and hence T is continuous.

In the general case, after letting $T(f) =: \{T_{ij}(f)\}$ we see that T_{ii} is a positive linear functional on $L^p(\Omega, \mu, \mathcal{M}_s)$, which implies that it is continuous. Finally, the continuity of T_{ij} for $i \neq j$ follows from the inequality $|T_{ij}(u)|^2 \leq 4\, T_{ii}(|u|)T_{jj}(|u|)$, which is an easy consequence of positivity.

Proof of part (b). First observe that $\mathcal{A}$ is closed under passage to the Hermitian adjoint and absolute value so that each term involved in (6) and in its equivalent rewritings is well defined. Each entry $(T(\cdot))_{i,j}$ can be viewed as the sum of s^2 continuous linear functionals from $L^p(\Omega, \mu, \mathbf{C})$ to $\mathbf{C}$ (by part (a)), and therefore by the Riesz representation theorem there exist functions $u_{i,j}^{k,q} \in L^q(\Omega, \mu, \mathbf{C})$, $i, j = 1, \ldots, n$, $k, q = 1, \ldots, s$ such that $1/p + 1/q = 1$ and

$$(T(f))_{i,j} = \int_\Omega \sum_{k,q=1}^{s} u_{i,j}^{k,q}(x) f_{k,q}(x)\, d\mu. \tag{11}$$

Since $T(\cdot)$ is globally positive, it follows that, for every Hermitian positively semi-definite f,

$$\begin{aligned}\langle T(f)z, z\rangle &= \int_\Omega \sum_{i,j=1}^{n} z_i\bar{z}_j \sum_{k,q=1}^{s} u_{i,j}^{k,q}(x) f_{k,q}(x)\, d\mu \\ &= \int_\Omega \sum_{k,q=1}^{s} f_{k,q}(x) \sum_{i,j=1}^{n} z_i\bar{z}_j u_{i,j}^{k,q}(x)\, d\mu \geq 0\end{aligned}$$

for all $z \in \mathbf{C}^n$. Consequently, setting

$$U = (U^{k,q})_{k,q=1}^{s} \in \mathcal{M}_{sn}, \quad U^{k,q} = \left(u_{i,j}^{k,q}\right)_{i,j=1}^{n} \in \mathcal{M}_n, \quad e = (1)_{j=1}^{s} \in \mathbf{R}^s, \tag{12}$$

we arrive at the more compact representation

$$\langle T(f)z, z\rangle = \int_\Omega \langle [(f \otimes I_n) \circ U]\,(e \otimes z), (e \otimes z)\rangle\, d\mu \geq 0, \tag{13}$$

where $\otimes$ is the tensor product and $\circ$ is the Hadamard componentwise product. Therefore, by standard measure theory arguments, it follows that the function

$$\langle [(f \otimes I_n) \circ U]\,(e \otimes z), (e \otimes z)\rangle$$

is nonnegative for any choice of z. Finally, due to the arbitrariness of the positively semi-definite matrix f we deduce that the matrix U must be positively semi-definite almost everywhere.

The 2-positivity of T is reduced to claim that $T_2(f)$ is negative semi-definite for any $f \geq 0$. Let u and v belong to $\mathbf{C}^n$ and let $f \in \mathcal{M}_{2s}$ be positively semi-definite. Then

$$\langle T_2(f)(u,v), (u,v)\rangle = \int_\Omega k\, d\mu,$$

where

$$\begin{aligned} k &= \left\langle \begin{bmatrix} (f_{1,1} \otimes I_n) \circ U & (f_{1,2} \otimes I_n) \circ U \\ (f_{1,2}^* \otimes I_n) \circ U & (f_{2,2} \otimes I_n) \circ U \end{bmatrix} (e \otimes u, e \otimes v), (e \otimes u, e \otimes v) \right\rangle \\ &= \langle (f \otimes I_n) \circ (E_2 \otimes U)\,(e \otimes u, e \otimes v), (e \otimes u, e \otimes v)\rangle, \\ E_2 &= \begin{bmatrix} 1 & 1 \\ 1 & 1 \end{bmatrix}. \end{aligned}$$

Since f, I_n, E_2 and U are positively semi-definite a.e., and since the Hadamard and tensor products preserve positivity, we deduce that k is a nonnegative a.e. on Ω. This implies our claim.

Proof of part (c). This is a direct consequence of (b) and Proposition 2.1. □

Theorem 2.8. *Let $p \in [1, \infty)$, let n be a positive integer and (Ω, μ) be a measure space with a σ-finite measure μ. If $T : L^p(\Omega, \mu, \mathcal{M}_s) \to \mathcal{M}_n$ is an LPO, then*

$$|||T(f)||| \leq \frac{1}{2}|||T(|f|)||| + \frac{1}{2}|||T(|f^*|)||| \tag{14}$$

for every u. i. norm $|||\cdot|||$ and every $f \in L^p$.

Proof. Let $\{u_i\}_{i=1}^n$ and $\{v_i\}_{i=1}^n$ be two systems of unit vectors. From Theorem 2.7 we know that (6) holds. Therefore, by (9) of Proposition 2.1 with $\gamma = 1$, we have, for every i,

$$|\langle T(f)u_i, v_i\rangle| \le \frac{1}{2}\langle T(|f|)\, u_i, u_i\rangle + \frac{1}{2}\langle T(|f^*|)\, v_i, v_i\rangle,$$

where

$$\langle T(f)u_i, v_i\rangle = \int_\Omega \langle [(f\otimes I_n)\circ U](e\otimes u_i), (e\otimes v_i)\rangle \, d\mu$$

by (13) and where the matrix $U(x)$ is defined as in (12). Let $||\cdot||$ be a u.i. norm and let Φ be the associated symmetric gauge function. From the last inequality and the convexity of Φ we obtain

$$\begin{aligned}
\Phi\Big(\Big[|\langle T(f)u_i, v_i\rangle|\Big]_{i=1}^n\Big) &\le \Phi\Big(\frac{1}{2}\Big[\langle T(|f|)\, u_i, u_i\rangle\Big]_{i=1}^n + \frac{1}{2}\Big[\langle T(|f^*|)\, v_i, v_i\rangle\Big]_{i=1}^n\Big) \\
&\le \frac{1}{2}\Phi\Big(\Big[\langle T(|f|)\, u_i, u_i\rangle\Big]_{i=1}^n\Big) + \frac{1}{2}\Phi\Big(\Big[\langle T(|f^*|)\, v_i, v_i\rangle\Big]_{i=1}^n\Big) \\
&\le \frac{1}{2}\|T(|f|)\|_\Phi + \frac{1}{2}\|T(|f^*|)\|_\Phi,
\end{aligned}$$

where the last inequality follows from (4). Finally, from the arbitrariness of $\{u_i\}_{i=1}^n$ and $\{v_i\}_{i=1}^n$ and from one more recourse to (4) we get the inequality

$$\|T(f)\|_\Phi \le \frac{1}{2}\|T(|f|)\|_\Phi + \frac{1}{2}\|T(|f^*|)\|_\Phi,$$

which completes the proof. □

Corollary 2.9. *Under the assumptions of Theorem* 2.8, *suppose that* f, F *are functions in* $L^p(\Omega, \mu, \mathcal{M}_s)$ *such that* F *is Hermitian valued and* $F(x) \ge |f(x)|$, $F(x) \ge |f^*(x)|$ *at* μ*-a.e.* $x \in \Omega$. *Then*

$$|||T(f)||| \le |||T(F)||| \tag{15}$$

for every u. i. norm $|||\cdot|||$.

Proof. From the inequalities $F \ge |f|, |f^*|$ and the positivity of T it follows that $T(F) \ge T(|f|)$ and $T(F) \ge T(|f^*|)$ (in the sense of the partial ordering of Hermitian matrices). From the Fan dominance theorem (see [1]) we obtain $\frac{1}{2}|||T(|f|)||| + \frac{1}{2}|||T(|f^*|)||| \le |||T(F)|||$ for every u.i. norm, and hence (15) follows from (14). □

With the last two results we have, in a sense, reduced the analysis to the positively semi-definite Hermitian valued case. In the following subsection, before proving the main inequalities, we introduce certain Jensen-Hölder inequalities in the positively semi-definite Hermitian valued case that will be used later in Subsection 2.3.

2.2. Jensen-Hölder type inequalities

We start by a simple lemma.

Lemma 2.10. *Let $A \in \mathcal{M}_s$ be a positively semi-definite matrix. Then the following basic inequalities hold:*

(a) *for every unit vector $u \in \mathbf{C}^s$ and every $r \geq 1$,*

$$\langle Au, u \rangle \leq \langle A^r u, u \rangle^{1/r};$$

(b) *for every $B \in \mathcal{M}_t$, every unit vector $u \in \mathbf{C}^{st}$ and every $r, t \geq 1$,*

$$\langle (A \otimes B)u, u \rangle \leq \langle (A^r \otimes B^r)u, u \rangle^{1/r};$$

(c) *for every $B \in \mathcal{M}_s$, every unit vector $u \in \mathbf{C}^s$ and every $r \geq 1$,*

$$\langle (A \circ B)u, u \rangle \leq \langle (A^r \circ B^r)u, u \rangle^{1/r};$$

(d) *for every $z \in \mathbf{C}^s$, $u \in \mathbf{C}^n$ with $\|z \circ u\|_2 = 1$ and every $r \geq 1$,*

$$\langle (A \circ B)u, u \rangle \leq \langle (A^r \circ B)u, u \rangle^{1/r} \quad \text{where} \quad B = zz^*;$$

(e) *for every $B \in \mathcal{M}_s$, $B \geq 0$, $u \in \mathbf{C}^n$ and every $r \geq 1$,*

$$\langle (A \circ B)u, u \rangle \leq \langle (A^r \circ B)u, u \rangle^{1/r} \langle (I_s \circ B)u, u \rangle^{1/\tilde{r}}$$

with $1/r + 1/\tilde{r} = 1$.

Proof of part (a) *and part* (b). By the Schur decomposition, $A = Q\Lambda Q^*$ with unitary Q and diagonal positively semi-definite Λ. Therefore, setting $w = Q^* u$, we get

$$\langle Au, u \rangle = \langle \Lambda w, w \rangle = \sum_{i=1}^{s} |w_i|^2 \Lambda_{i,i}$$

with $\sum_{i=1}^{s} |w_i|^2 = 1$. Now application of the Jensen inequality with the convex function $t \mapsto t^r$ yields

$$\sum_{i=1}^{s} |w_i|^2 \Lambda_{i,i} \leq \left(\sum_{i=1}^{s} |w_i|^2 \Lambda_{i,i}^r \right)^{1/r} = \langle A^r u, u \rangle^{1/r}.$$

Part (b) is a special case of (a) because $(A \otimes B)^r = A^r \otimes B^r$.

Proof of part (c). First we stress that $A \circ B$ is a principal submatrix of $A \otimes B$. Therefore $\langle (A \circ B)u, u \rangle = \langle (A \otimes B)Pu, Pu \rangle$, where P is the $s^2 \times s$ matrix such that $A \circ B = P^T (A \otimes B) P$. It is transparent that Pu is still a unit vector, so that, by part (b),

$$\langle (A \circ B)u, u \rangle \leq \langle (A^r \otimes B^r)Pu, Pu \rangle^{1/r} = \langle (A^r \circ B^r)u, u \rangle^{1/r}.$$

Proof of part (d). By direct inspection, $\langle (A \circ B)u, u \rangle = \langle A(z \circ u), (z \circ u) \rangle$. Since $z \circ u$ is unitary, we infer from part (a) that

$$\langle (A \circ B)u, u \rangle \leq \langle A^r (z \circ u), (z \circ u) \rangle^{1/r} = \langle (A^r \circ B)u, u \rangle^{1/r}.$$

Proof of part (e). From the Schur decomposition we have $B = \sum_{i=1}^{s} \alpha_i z_i z_i^*$, so that

$$\langle (A \circ B)u,\, u\rangle = \sum_{i=1}^{s} \alpha_i \langle A(z_i \circ u),\, (z_i \circ u)\rangle \,.$$

Setting $\theta_i = \|z_i \circ u\|_2$ and $v_i = (z_i \circ u)/\theta_i$ and applying the Hölder inequality we obtain

$$\langle (A \circ B)u,\, u\rangle \le \left(\sum_{i=1}^{s} \alpha_i \theta_i^2 \langle A v_i,\, v_i\rangle^r\right)^{1/r} \left(\sum_{i=1}^{s} \alpha_i \theta_i^2\right)^{1/\tilde{r}}$$

with $1/r + 1/\tilde{r} = 1$. Finally, from (a) we deduce

$$\begin{aligned} \langle (A \circ B)u,\, u\rangle &\le \left(\sum_{i=1}^{s} \alpha_i \theta_i^2 \langle A^r v_i,\, v_i\rangle\right)^{1/r} \left(\sum_{i=1}^{s} \alpha_i \theta_i^2\right)^{1/\tilde{r}} \\ &= \langle (A^r \circ B)u,\, u\rangle^{1/r} \langle (I_s \circ B)u,\, u\rangle^{1/\tilde{r}}, \end{aligned}$$

which completes the proof. □

Theorem 2.11. *Let (Ω, μ) be a measure space with a σ-finite measure μ and let $1 \le r < +\infty$. If $T : L^r(\Omega, \mu, \mathcal{M}_s) \cap L^1(\Omega, \mu, \mathcal{M}_s) \to \mathcal{M}_n$ is an LPO, then for every Hermitian positively semi-definite $f \in L^r \cap L^1$ and for every unit vector u,*

$$\langle T(f)u,\, u\rangle \le \langle T(f^r)u,\, u\rangle^{1/r} \|T(I_s)\|_\infty^{1/\tilde{r}}, \quad 1/r + 1/\tilde{r} = 1. \tag{16}$$

Proof. From the representation given in equation (13) we have

$$\langle T(f)u,\, u\rangle = \int_\Omega \langle [(f \otimes I_n) \circ U]\,(e \otimes u),\, (e \otimes u)\rangle \, d\mu$$

and consequently, by part (e) of the previous lemma,

$$\begin{aligned} \langle T(f)u,\, u\rangle \;\le\; \int_\Omega & \langle [(f^r \otimes I_n) \circ U]\,(e \otimes u),\, (e \otimes u)\rangle^{1/r} \\ & \langle [(I_s \otimes I_n) \circ U]\,(e \otimes u),\, (e \otimes u)\rangle^{1/\tilde{r}} \, d\mu. \end{aligned}$$

Hence, from the Hölder inequality and from (13) we obtain that

$$\begin{aligned} \langle T(f)u,\, u\rangle &\le \left(\int_\Omega \langle [(f^r \otimes I_n) \circ U]\,(e \otimes u),\, (e \otimes u)\rangle \, d\mu\right)^{1/r} \\ &\quad \left(\int_\Omega \langle [(I_s \otimes I_n) \circ U]\,(e \otimes u),\, (e \otimes u)\rangle \, d\mu\right)^{1/\tilde{r}} \\ &= \langle T(f^r)u,\, u\rangle^{1/r} \langle T(I_s)u,\, u\rangle^{1/\tilde{r}}. \end{aligned}$$

Since u has norm 1, it follows that

$$\langle T(I_s)u,\, u\rangle \le \|T(I_s)\|_\infty,$$

and the proof is complete. □

2.3. Main inequalities

Theorem 2.12. *Let (Ω, μ) be a measure space with a σ-finite measure μ and let $1 \le r_1, r_2 < +\infty$. If $T : L^r(\Omega, \mu, \mathcal{M}_s) \cap L^1(\Omega, \mu, \mathcal{M}_s) \to \mathcal{M}_n$ is an LPO with $r = \max\{r_1, r_2\}$, then for every $f \in L^r \cap L^1$ the following estimate holds:*

$$|||T(f)||| \le |||T(|f|^{r_1})^{\alpha/(2r_1)}|||^{\frac{1}{\alpha}} |||T(|f^*|^{r_2})^{\beta/(2r_2)}|||^{\frac{1}{\beta}} \|T(I_s)\|_\infty^{\theta(r_1,r_2)} \tag{17}$$

where $\alpha, \beta \in (1, \infty)$ with $1/\alpha + 1/\beta = 1$, $|||\cdot|||$ is any unitarily invariant norm, and $1/r_1 + 1/\tilde{r}_1 = 1/r_2 + 1/\tilde{r}_2 = 1$ with $\theta(r_1, r_2) = 1/(2\tilde{r}_1) + 1/(2\tilde{r}_2)$.

Proof. Let $\{u_i\}_{i=1}^n$ and $\{v_i\}_{i=1}^n$ be two systems of unit vectors. From Theorem 2.7 we know that (6) is true. Thus, for any gauge function Φ, we infer from (1) and from (2) that

$$\begin{aligned}\Phi(|\langle T(f)u_i, v_i\rangle|) &\le \Phi(\langle T(|f|)u_i, u_i\rangle^{1/2} \langle T(|f^*|)v_i, v_i\rangle^{1/2}) \\ &\le \Phi(\langle T(|f|)u_i, u_i\rangle^{\alpha/2})^{1/\alpha} \Phi(\langle T(|f^*|)v_i, v_i\rangle^{\beta/2})^{1/\beta}\end{aligned}$$

if $\alpha > 1$ and $1/\alpha + 1/\beta = 1$. From Theorem 2.11 we see that, for any positively semi-definite g and any $r \ge 1$,

$$\langle T(g)v_i, v_i\rangle \le \langle T(g^r)v_i, v_i\rangle^{1/r} \|T(I_s)\|_\infty^{1/\tilde{r}}, \quad 1/r + 1/\tilde{r} = 1.$$

Therefore, by invoking the previous argument twice, once with $g = |f|$ and once with $g = |f^*|$, and by using relation (1), we deduce that

$$\begin{aligned}\Phi(|\langle T(f)u_i, v_i\rangle|) &\le \Phi(\langle T(|f|^{r_1})u_i, u_i\rangle^{\alpha/(2r_1)})^{1/\alpha} \\ &\quad \Phi(\langle T(|f^*|^{r_2})v_i, v_i\rangle^{\beta/(2r_2)})^{1/\beta} \|T(I_s)\|_\infty^{1/(2\tilde{r}_1)+1/(2\tilde{r}_2)}.\end{aligned}$$

Finally, from the arbitrariness of the orthonormal systems and the variational characterization of u.i. norms we get

$$|||T(f)||| \le |||T(|f|^{r_1})^{\alpha/(2r_1)}|||^{\frac{1}{\alpha}} |||T(|f^*|^{r_2})^{\beta/(2r_2)}|||^{\frac{1}{\beta}} \|T(I_s)\|_\infty^{1/(2\tilde{r}_1)+1/(2\tilde{r}_2)}.$$

□

Corollary 2.13. *Let (Ω, μ) be a measure space with a σ-finite measure μ and let $1 \le r_1, r_2 < +\infty$, $1 \le p < \infty$. If $T : L^r(\Omega, \mu, \mathcal{M}_s) \cap L^1(\Omega, \mu, \mathcal{M}_s) \to \mathcal{M}_n$ is an LPO with $r = \max\{r_1, r_2\}$, then for every $f \in L^r \cap L^1$ the following estimate for the Schatten p-norm of $T(f)$ holds:*

$$\|T(f)\|_p \le \left\|T(|f|^{r_1})^{\alpha/(2r_1)}\right\|_p^{\frac{1}{\alpha}} \left\|T(|f^*|^{r_2})^{\beta/(2r_2)}\right\|_p^{\frac{1}{\beta}} \|T(I_s)\|_\infty^{\theta(r_1,r_2)} \tag{18}$$

where $\alpha, \beta \in (1, \infty)$ with $1/\alpha + 1/\beta = 1$, $1/r_1 + 1/\tilde{r}_1 = 1/r_2 + 1/\tilde{r}_2 = 1$, and $\theta(r_1, r_2) = 1/(2\tilde{r}_1) + 1/(2\tilde{r}_2)$.

Proof. Since any Schatten p norm is unitarily invariant, it follows that (18) is a special case of (17). □

3. Examples and applications

In what follows we give an idea of the several contexts in which matrix valued linear positive operators arise.

3.1. Multilevel block Toeplitz matrices

We first turn our attention to u.i. norms of (multilevel) Toeplitz matrices.

Let f be an $\mathcal{M}_s$ valued function of k variables, integrable on the k-cube $I_k := (0, 2\pi)^k$. Throughout, the symbol $⨍_{I_k}$ stands for $(2\pi)^{-k} \int_{I_k}$ and the symbol L^p stands for $L^p(I^k, (2\pi)^{-k}dx, \mathcal{M}_s)$. The Fourier coefficients of f, given by

$$\widehat{f}_j := ⨍_{I_k} f(x)e^{-i\,\langle j,\,x\rangle}\, dx, \quad i^2 = -1, \quad j \in \mathbf{Z}^k, \tag{19}$$

are the entries of the k-level Toeplitz matrices generated by f. More precisely, if $n = (n_1, \dots, n_k)$ is a k-index with positive entries, we put $\widehat{n} := \prod_{i=1}^k n_i$ and let $T_n(f)$ denote the matrix of order $s\widehat{n}$ given by

$$T_n(f) = \sum_{|j_1|<n_1} \cdots \sum_{|j_k|<n_k} \left[J_{n_1}^{(j_1)} \otimes \cdots \otimes J_{n_k}^{(j_k)}\right] \otimes \widehat{f}_{(j_1,\dots,j_k)}. \tag{20}$$

In the above equation, $\otimes$ denotes the tensor product, while $J_m^{(l)}$ denotes the matrix of order m whose (i, j) entry equals 1 if $j - i = l$ and equals zero otherwise. The reader is referred to [3, 4, 12, 13, 20] for more details on multilevel block Toeplitz matrices. Here we just recall the following elementary fact (see, e.g., [13, 20]).

Proposition 3.1. *Let $n = (n_1, \dots, n_k)$ be a positive k-index with $k \geq 1$ and define $T_n : L^1 \to \mathcal{M}_{s\widehat{n}}$ by (20). Then T_n is a matrix valued LPO.*

In order to give one specific illustration of the results of Section 2 in the case of multilevel Toeplitz matrices, we identify $\widehat{n}$-vectors with entries in $\mathbf{C}^s$ and $\mathbf{C}^s$ valued polynomials of degree n via a suitable isomorphism. Namely, given $k \geq 1$ and a k-index $n = (n_1, \dots, n_k)$ as above, we define

$$\mathcal{P}_n^k(\mathbf{C}^s) := \left\{u : I_k \mapsto \mathbf{C}^s \mid u(x) = \sum_{j \in \mathcal{I}_n^k} a_j\, e^{i\,\langle j,\,x\rangle}\right\},$$

where

$$\mathcal{I}_n^k := \{(j_1, \dots, j_k) \mid j_i \in \mathbf{N} \text{ and } 0 \leq j_i < n_i \text{ for } i = 1, 2, \dots, k\}.$$

The set $\mathcal{P}_k^n(\mathbf{C}^s)$ is a vector space of trigonometric polynomials of dimension $s\widehat{n}$, and we endow it with the L^2 inner product,

$$\langle u,\, v\rangle_{L^2} := ⨍_{I_k} \langle u(x),\, v(x)\rangle\, dx.$$

Given $u \in \mathcal{P}_k^n(\mathbf{C}^s)$, $u(x) := \sum_{j \in \mathcal{I}_n^k} a_j \, e^{i \langle j, x \rangle}$, we can associate with it the vector $\tilde{u} \in \mathbf{C}^{s\hat{n}}$ given by

$$\tilde{u} := \sum_{i_1=0}^{n_1-1} \cdots \sum_{i_k=1}^{n_k-1} \left[e_{n_1}^{(i_1)} \otimes \cdots \otimes e_{n_k}^{(i_k)} \right] \otimes a_{(i_1,\ldots,i_k)},$$

where $\{e_m^{(i)}\}_{i=0}^{m-1}$ is the canonical basis of $\mathbf{C}^m$. It is clear that the map $u \to \tilde{u}$ is a linear isomorphism. In fact, it is easy to check that

$$\langle u, v \rangle_{L^2} = \langle \tilde{u}, \tilde{v} \rangle \quad \text{for all } u, v \in \mathcal{P}_k^n, \tag{21}$$

and therefore we obtain a linear isometry between $\mathcal{P}_k^n(\mathbf{C}^s)$ and $\mathbf{C}^{s\hat{n}}$. By virtue of this, we can drop the notation $\tilde{u}$ and we can regard u as a polynomial from $\mathcal{P}_k^n(\mathbf{C}^s)$ or as a vector from $\mathbf{C}^{s\hat{n}}$, according to necessity.

Observing that $\left\langle J_m^{(j)} e_m^{(h)}, e_m^{(i)} \right\rangle = \delta_{j,h-i}$ and using elementary properties of the tensor product, we obtain from (20) after straightforward computations that

$$\fint_{I_k} \langle f(x)u(x), v(x) \rangle \, dx = \langle T_n(f) \, u, v \rangle \quad \text{for all } u, v \in \mathcal{P}_k^n(\mathbf{C}^s) \tag{22}$$

(on the right-hand side, u and v are meant as vectors from $\mathbf{C}^{s\hat{n}}$). We remark that (21) can be obtained from (22) by letting $f(x) = I_s$, in which case we have $T_n(f) = I_{s\hat{n}}$.

Remark 3.2. From (22) one can see that if $f(x)$ is Hermitian valued and positively semi-definite, then $T_n(f)$ is positively semi-definite (the converse is also true provided $T_n(f) \geq 0$ for every n). In this case, it is immediate to check that

$$\|T_n(f)\|_1 = \operatorname{tr}\big(T_n(f)\big) = \hat{n} \fint_{I_k} \operatorname{tr}(f(x)) \, dx. \tag{23}$$

Representation (22) is extremely useful. From this formula and from Theorem 1.1 we can deduce a variational characterization for any u.i. norm of any Toeplitz matrix.

Corollary 3.3. *Let $n = (n_1, \ldots, n_k)$ be a positive k-index and let $f \in L^1$. If $T_n(f)$ is the k-level Toeplitz matrix associated with the $\mathcal{M}_s$ valued function f, then for every symmetric gauge function Φ on $\mathbf{R}^{s\hat{n}}$,*

$$\|T_n(f)\|_\Phi = \sup \Phi\left(\left[\left| \fint_{I_k} \langle f(x)u_i(x), v_i(x) \rangle \, dx \right| \right]_{i=1}^{s\hat{n}} \right), \tag{24}$$

where the supremum is taken over all pairs of orthonormal systems $\{u_i\}_{i=1}^{s\hat{n}}$ and $\{v_i\}_{i=1}^{s\hat{n}}$ of trigonometric $\mathbf{C}^s$ valued polynomials such that

$$\fint_{I_k} \langle u_i(x), u_j(x) \rangle \, dx = \fint_{I_k} \langle v_i(x), v_j(x) \rangle \, dx = \delta_{i,j}, \tag{25}$$

$u_i, v_i \in \mathcal{P}_k^n(\mathbf{C}^s)$, $1 \le i, j \le s\widehat{n}$. Moreover, if f is Hermitian and positively semi-definite a.e. then

$$\|T_n(f)\|_\Phi = \sup \Phi\left(\left[\fint_{I_k} \langle f(x)u_i(x),\, u_i(x)\rangle \, dx\right]_{i=1}^{s\widehat{n}}\right), \tag{26}$$

where the supremum is taken over all orthonormal systems $\{u_i\}_{i=1}^{s\widehat{n}}$ of trigonometric polynomials satisfying (25).

Proof. Using (21) and (22) and employing Theorem 1.1 with $A = T_n(f)$, we obtain (24). Similarly, in order to establish (26) it suffices to observe that $T_n(f)$ is positively semi-definite if f is Hermitian and positively semi-definite a.e. (by Proposition 3.1). □

Further consequences of the results of Section 2 are the following estimates, which generalize the estimate contained in Corollary 3.2 of [17] (see also Lemma 4 of [6]).

Corollary 3.4. *Let $r_1, r_2, p \in [1, \infty)$ and let $f \in L^r$ with $r = \max\{r_1, r_2\}$. If $n = (n_1, \dots, n_k)$ is a positive k-index, then the following estimate for the Schatten norms holds:*

$$\|T_n(f)\|_p \le (\widehat{n})^{\frac{1}{p}} \, \|f\|_{L^{r_1}}^{1/2} \, \|f\|_{L^{r_2}}^{1/2} \tag{27}$$

with $1/r_1 + 1/r_2 = 2/p$.

Proof. From Corollary 2.13 with $T(\cdot) = T_n(\cdot)$ we have

$$\|T_n(f)\|_p \le \left\| T_n(|f|^{r_1})^{\alpha/(2r_1)} \right\|_p^{\frac{1}{\alpha}} \left\| T_n(|f^*|^{r_2})^{\beta/(2r_2)} \right\|_p^{\frac{1}{\beta}}$$

since $\|T_n(I_s)\|_\infty = 1$. Choosing α and β such that $\alpha p/(2r_1) = \beta p/(2r_2) = 1$ we deduce that

$$\begin{aligned} \left\| T_n(|f|^{r_1})^{\alpha/(2r_1)} \right\|_p^{\frac{1}{\alpha}} &= \|T_n(|f|^{r_1})\|_1^{\frac{1}{2r_1}}, \\ \left\| T_n(|f^*|^{r_2})^{\beta/(2r_2)} \right\|_p^{\frac{1}{\beta}} &= \|T_n(|f^*|^{r_2})\|_1^{\frac{1}{2r_2}}. \end{aligned}$$

Taking into account that f and f^* have the same singular values and taking into account (23) with $|f|^{r_1}$, $|f^*|^{r_2}$ in place of f, we obtain

$$\begin{aligned} \|T_n(|f|^{r_1})\|_1 &= \operatorname{tr} T_n(|f|^{r_1}) = \widehat{n}\, \|f\|_{L^{r_1}}^{r_1}, \\ \|T_n(|f^*|^{r_2})\|_1 &= \operatorname{tr} T_n(|f^*|^{r_2}) = \widehat{n}\, \|f\|_{L^{r_2}}^{r_2}, \end{aligned}$$

and therefore it follows that

$$\|T_n(f)\|_p \le (\widehat{n})^{\frac{1}{2r_1} + \frac{1}{2r_2}} \, \|f\|_{L^{r_1}}^{1/2} \, \|f\|_{L^{r_2}}^{1/2}$$

where the identity $1/r_1 + 1/r_2 = 2/p$ results from the relations $\alpha p/(2r_1) = \beta p/(2r_2) = 1$ and $1/\alpha + 1/\beta = 1$. □

Corollary 3.5. *Let $f \in L^p$ for some p with $1 \leq p \leq \infty$. If $n = (n_1, \ldots, n_k)$ is a positive k-index, then the following estimate for the Schatten norms holds:*

$$\|T_n(f)\|_p \leq (\widehat{n})^{\frac{1}{p}} \|f\|_{L^p}. \tag{28}$$

Proof. In the case where $p = \infty$, this is a well-known fact concerning the spectral norm of Toeplitz matrices. If $p < \infty$, we obtain (28) from (27) by choosing $r_1 = r_2 = p$. □

Finally we want to mention a link between Toeplitz operators and the general representation formulas of LPOs given in Theorem 2.7. By equation (13), a linear positive operator $T : L^p(\Omega, \mu, \mathcal{M}_s) \to \mathcal{M}_n$ is fully determined by its integral kernel $U(x)$, which is a positively semi-definite $\mathcal{M}_{sn}$ valued function. In the specific case of Toeplitz operators $T_n : L^1(I_k, (2\pi)^{-k}dx, \mathcal{M}_s) \to \mathcal{M}_{s\widehat{n}}$ it turns out that $U(x)$ has a very special structure, and in fact it is a rank-one matrix $\theta(x)[\theta(x)]^*$ where

$$\theta(x) = \left[\sum_{j=1}^{s}(e_j \otimes e_j)\right] \otimes v(x), \quad v(x) = \bigotimes_{j=1}^{s} v_j(x_j)$$

with

$$v_j(x_j) \in \mathbf{C}^{n_j}, \quad e_j \in \mathbf{C}^s, \quad (v_j(x_j))_k = e^{i(k-1)x_j}, \quad (e_j)_k = \delta_{j,k}.$$

It is not surprising that a Toeplitz operator has such a sparse representation. In fact, multilevel block Toeplitz matrices have a low informative content since they are described by just $s^2\widehat{n}$ parameters instead of $s^2(\widehat{n})^2$, which is the usual thing for matrices of size $s\widehat{n}$.

3.2. Multilevel block Hankel matrices

A Hankel matrix is one whose entries are constant along every lower-left to upper-right diagonal. As with Toeplitz matrices, one can consider multilevel Hankel matrices generated by a multivariate symbol f integrable on the k-cube $I_k := (0, 2\pi)^k$. With the same notations as in the previous subsection, let $\{\widehat{f}_j\}$ denote the Fourier coefficients of f given by (19). If $n = (n_1, \ldots, n_k)$ is a k-index with positive entries, then $H_n(f)$ denotes k-level Hankel matrix of order $s\widehat{n}$ generated by f. This matrix is defined as

$$H_n(f) = \sum_{j_1=1}^{2n_1-1} \cdots \sum_{j_k=1}^{2n_k-1} \left[K_{n_1}^{(j_1)} \otimes \cdots \otimes K_{n_k}^{(j_k)}\right] \otimes \widehat{f}_{(j_1,\ldots,j_k)}. \tag{29}$$

Here $K_m^{(l)}$ denotes the matrix of order m whose (i,j) entry equals 1 if $i + j = l + 1$ and equals zero otherwise. As with Toeplitz matrices, the tensor product $\otimes$ stresses the k-level block structure of the matrices we are considering.

It is well known that the operator H_n is not an LPO: to see this, take $k = 1$, $f(x) = 2(1 - \cos x)$, $n = (n_1) = (1)$ and note that $H_1(f) = \widehat{f}_1 = -1$ is negative despite that $f \geq 0$. However, in [6] it was proved that, given $f \in L^1$, for every multiindex n there exist a unitary $\mathcal{M}_s$ valued function g_n and a unitary matrix U_n such that

$$U_n H_n(f) = T_n(g_n f). \tag{30}$$

In view of this fact, we obtain the following.

Corollary 3.6. *Let $f \in L^p$ for some p with $1 \le p \le \infty$. If $n = (n_1, \ldots, n_k)$ is a positive k-index, then the following estimate for the Schatten norms holds:*

$$\|H_n(f)\|_p \le (\widehat{n})^{\frac{1}{p}} \|f\|_{L^p}, \tag{31}$$

and if $p < \infty$ then

$$\|H_n(f)\|_p \le (\widehat{n})^{\frac{1}{p}} \|f\|_{L^{r_1}}^{1/2} \|f\|_{L^{r_2}}^{1/2} \tag{32}$$

with $1/r_1 + 1/r_2 = 2/p$.

Proof. Using (30), we see that $||H_n(f)|| = ||T_n(g_n f)||$ for every u.i. norm, since U_n is unitary. Moreover, we have $\|f\|_{L^p} = \|g_n f\|_{L^p}$, and hence estimates (31) and (32) are a consequence of the corresponding estimates for Toeplitz matrices, stated in Corollary 3.5 and Corollary 3.4. □

We emphasize that this corollary improves and generalizes the estimates obtained in Lemma 5 of [6] and in Corollary 3.3 of [17]. As previously observed in [17] in connection with the scalar case, there exists a substantial difference between relations (28) and (31). Indeed those pertaining Toeplitz sequences are asymptotically tight, while those concerning Hankel sequences are never tight if the multiindex n tends to infinity.

3.3. The nonsquare case

We now consider the case where the generating function f defined on I_k takes values in the space $\mathcal{M}_{(s,t)}$ of the complex $s \times t$ matrices. According to the rule given in (19) we define the Fourier coefficients of f and then the sequence of Toeplitz matrices $\{T_n(f)\}$, $T_n(f) \in \mathcal{M}_{\widehat{n}\cdot(s,t)}$, by

$$T_n(f) = \sum_{|j_1|<n_1} \cdots \sum_{|j_k|<n_k} \left[J_{n_1}^{(j_1)} \otimes \cdots \otimes J_{n_k}^{(j_k)}\right] \otimes \widehat{f}_{(j_1,\ldots,j_k)}. \tag{33}$$

In what follows we briefly show that the nonsquare case can be easily reduced to the square setting. Let $S_{(s,t)} : \mathcal{M}_{(s,t)} \mapsto \mathcal{M}_{\max\{s,t\}}$ be the *minimal completion* operator, which transforms a nonsquare matrix into a square one by adding $\max\{s,t\} - \min\{s,t\}$ zero colums or rows according to necessity. It is evident that, given an $\mathcal{M}_{(s,t)}$ valued function f, a corresponding $\mathcal{M}_{\max\{s,t\}}$ valued function $S(f)$ can be defined. Therefore it easily follows that for all $f \in L^p$ we have

$$\|f\|_{L^p} = \|S(f)\|_{L^p}, \tag{34}$$

since $\|f(x)\|_p = \|S(f)(x)\|_p$ a.e. on I_k. Moreover, from definition (33) we deduce that there exists a permutation matrix $\Pi \in \mathcal{M}_{\max\{s,t\}\widehat{n}}$ such that

$$T_n(S_{(s,t)}(f)) = \Pi S_{(s,t)\widehat{n}}\left(T_n(f)\right)\Pi^*$$

and consequently,

$$\|T_n(S_{(s,t)}(f))\|_p = \|S_{(s,t)\hat{n}}(T_n(f))\|_p = \|T_n(f)\|_p. \tag{35}$$

An immediate consequence of (34) and (35) is that Corollary 3.4 and Corollary 3.5 hold unchanged in the nonsquare setting as well.

3.4. Other examples of matrix valued LPOs

The list of the matrix valued LPOs that arise in the applications is very long. We want just to mention the discretizations by finite elements or finite differences of some boundary value problems [15, 16] and the case of Laplacian graph matrices coming from optimization problems [7]. In all these cases the vector $n = (n_1, \ldots, n_k)$ is related to a finesse parameter of the discretization process, so that we have to deal with sequences of matrices of increasing order. The inequalities proved in the preceding sections can be used in order to analyze the behavior of these structures for large n under some perturbations as it occurs and is of interest in a numerical analysis context.

As an example, let us consider the second order partial differential equation

$$-\sum_{i,j=1}^{k} \frac{\partial}{\partial x_i}\left(a_{i,j}(x)\frac{\partial}{\partial x_j}u\right) = f(x) \quad \text{on a bounded open set } \Omega \subset \mathbf{R}^k$$

with Dirichlet boundary condition. We assume that

$$A(x) = (a_{i,j}(x))_{i,j=1}^{k} \in L^p(\Omega, dx, \mathcal{M}_k) \cap L^1(\Omega, dx, \mathcal{M}_k).$$

Then by using a finite element approach we obtain a sequence of operators $\mathcal{F}_n$: $L^p(\Omega, dx, \mathcal{M}_k) \cap L^1(\Omega, dx, \mathcal{M}_k) \to \mathcal{M}_{d_n}$ which are linear and positive in the sense that the stiffness matrix $\mathcal{F}_n(A)$ is positively semi-definite if A is positively semi-definite. Corollary 2.13 and the arguments used in the proofs of Corollaries 3.4 and 3.5 allow us to compare Schatten norms in the nondefinite case with other norms in the positively semi-definite case. For instance, if A is symmetric and nondefinite, then we find

$$\|\mathcal{F}_n(A)\|_p \le \|\mathcal{F}_n(|A|^{r_1})\|_1^{1/(2r_1)} \|\mathcal{F}_n(|A|^{r_2})\|_1^{1/(2r_2)} \|\mathcal{F}_n(I_k)\|_\infty^{1/q}$$

with $1/r_1 + 1/r_2 = 2/p$ and $1/p + 1/q = 1$, and after setting $r_1 = r_2 = p$ we obtain

$$\|\mathcal{F}_n(A)\|_p \le \|\mathcal{F}_n(|A|^p)\|_1^{1/p} \|\mathcal{F}_n(I_k)\|_\infty^{1/q}.$$

4. Application to asymptotics on the algebra generated by Toeplitz sequences

In this section we use some results of the preceding sections for deriving asymptotics on sequences belonging to the algebra generated by multilevel block Toeplitz sequences. The following theorem concerns the basic case of a single sequence of Toeplitz matrices (see [2, 4, 20, 22] for the history of the topic).

Theorem 4.1 (Tilli [20]). *If f is matrix valued and integrable over I_k, then the singular values of $\{T_n(f)\}$ are distributed as $|f|$, that is, for every $F \in \mathcal{C}_0$ (i.e., for every continuous F with bounded support) we have*

$$\lim_{n\to\infty} \frac{1}{\min\{s,t\}\widehat{n}} \sum_{j=1}^{\min\{s,t\}\widehat{n}} F\Big(\sigma_j\big(T_n(f)\big)\Big) = \int_{I_k} \frac{1}{\min\{s,t\}} \sum_{j=1}^{\min\{s,t\}} F(\sigma_j(|f(s)|))\, ds, \tag{36}$$

where $\sigma_j\big(T_n(f)\big)$, $j = 1, \ldots, \min\{s,t\}\widehat{n}$, are the singular values of $T_n(f)$. In short $\{T_n(f)\}$ is distributed as f, i.e., $\{T_n(f)\} \sim_\sigma f$.

As for notation we recall that when we write $n \to \infty$ with $n = (n_1, \ldots, n_k)$ being a multi-index we mean that each n_j tends to infinity.

Remark 4.2. In [14], it has been proved that if f is complex valued (i.e., $s = t = 1$) and belongs to L^p, then relation (36) holds for any test function $F \in C(p)$ where

$$C(p) = \{F : \mathbf{R} \to \mathbf{R},\ \text{continuous}:\ F(z)/(1+|z|^p) \in L^\infty(\mathbf{R}^+, dx, \mathbf{R})\}.$$

The key point is the scalar version ($s = t = 1$) of inequality (28). Therefore by using the same ideas as in the proof of Theorem 1.2 of [14] and inequality (28), we obtain the limit relation (36) in the case of matrix valued symbols $f \in L^p$ and test functions $F \in C(p)$ as well.

Consider now a finite set $\{f_{\alpha\beta}\}$ of L^1 matrix valued functions and the measurable matrix valued function

$$\theta = \textstyle\sum_{\alpha=1}^{k} [\prod(r)]_{\beta=1}^{q_\alpha} f_{\alpha\beta}$$

where $[\prod(r)]$ is the right multiplication symbol and where the sizes of the matrix valued functions $f_{\alpha\beta}$ agree in such a way that the right products and the sums make sense. We remark that θ may fail to belong to L^1, so that the sequence $\{T_n(\theta)\}$ is not defined according to the rule in (19). However we can consider the sequence of matrices $\{\sum_{\alpha=1}^{k} [\prod(r)]_{\beta=1}^{q_\alpha} T_n(f_{\alpha\beta})\}$.

Theorem 4.3 ([11]). *Let $\{f_{\alpha\beta}\}$ be a finite set of complex valued L^1 functions over I_k, define the measurable function θ by*

$$\theta = \textstyle\sum_{\alpha=1}^{k} \prod_{\beta=1}^{q_\alpha} f_{\alpha\beta},$$

and put

$$A_n = \{\textstyle\sum_{\alpha=1}^{k} \prod_{\beta=1}^{q_\alpha} T_n(f_{\alpha\beta})\}.$$

Then the singular values of $\{A_n\}$ are distributed as $|\theta|$, that is, for every $F \in \mathcal{C}_0$ (i.e., for every continuous F with bounded support) we have

$$\lim_{n\to\infty} \frac{1}{\widehat{n}} \sum_{j=1}^{\widehat{n}} F\Big(\sigma_j\big(A_n\big)\Big) = \int_{I_k} F(|\theta(s)|)\, ds, \tag{37}$$

where $\sigma_j(A_n)$, $j = 1, \ldots, \widehat{n}$ *are the singular values of* A_n. *In short* $\{A_n\}$ *is distributed as* θ, *i.e.*, $\{A_n\} \sim_\sigma \theta$.

The purpose of what follows is to extend (37) to the block case:

$$\{\textstyle\sum_{\alpha=1}^{k}[\prod(r)]_{\beta=1}^{q_\alpha} T_n(f_{\alpha\beta})\} \sim_\sigma \theta. \tag{38}$$

Here $\{A_n\} \sim_\sigma \theta$, with θ being $\mathcal{M}_{(s,t)}$ valued, means that for any $F \in \mathcal{C}_0$ (continuous with bounded support) we have

$$\lim_{n\to\infty} \frac{1}{d_n} \sum_{j=1}^{d_n} F\Big(\sigma_j(A_n)\Big) = \int_{I_k} \frac{1}{\min\{s,t\}} \sum_{j=1}^{\min\{s,t\}} F\big(|\theta(s)|\big)\, ds,$$

where d_n denotes the size of A_n and $d_n < d_{n+1}$ for all n. When $k = 1$ and $q_1 = 1$, equation (38) reduces to Theorem 4.1. If in addition $s = t = 1$, then the result is Tyrtyshnikov's [23]. The case where all of the functions $f_{\alpha\beta}$ belong to L^∞ was analyzed in [21, 19, 18] by using mainly matrix theory techniques and in [4] by employing C^*-algebra techniques. The general case with complex valued symbols was considered and disposed of in [11].

For any real valued function F defined on $\mathbf{R}$ and for any matrix A_n of size d_n, we denote by $\Sigma(F, A_n)$ the mean

$$\frac{1}{d_n} \sum_{j=1}^{d_n} F[\sigma_j(A_n)].$$

Definition 4.4. *Suppose a sequence of matrices* $\{A_n\}$ *of size* d_n *is given. We say that* $\{\{B_{n,m}\}\}_m$, $m \in \mathbf{N}$, *is an approximating class of sequences (a.c.s.) for* $\{A_n\}$ *if, for all sufficiently large* $m \in \mathbf{N}$, *the following splitting holds:*

$$A_n = B_{n,m} + R_{n,m} + N_{n,m} \quad \textit{for all } n > n_m, \tag{39}$$

with

$$\text{rank } R_{n,m} \le d_n\, c(m), \quad \|N_{n,m}\| \le \omega(m), \tag{40}$$

where n_m, $c(m)$ *and* $\omega(m)$ *depend only on* m *and, moreover,*

$$\lim_{m\to\infty} \omega(m) = 0, \quad \lim_{m\to\infty} c(m) = 0. \tag{41}$$

Proposition 4.5. *Let* d_n *be an increasing sequence of natural numbers. Suppose a sequence of matrices* $\{A_n\}$ *of size* d_n *is given such that* $\{\{B_{n,m}\}\}_m$, $m \in \hat{\mathbf{N}} \subset \mathbf{N}$, $\#\hat{\mathbf{N}} = \infty$, *is an a.c.s. for* $\{A_n\}$ *in the sense of Definition 4.4. Suppose that* $\{B_{n,m}\} \sim_\sigma \theta_m$ *and that* θ_m *converges in measure to the measurable function* θ. *Then necessarily*

$$\{A_n\} \sim_\sigma \theta. \tag{42}$$

Proof. This is a special case of Proposition 2.3 of [11]. □

A sequence of matrices $\{A_n\}_n$ is said to be sparsely unbounded (s.u.) if for each $M > 0$, there exists an $\bar{n}_M$ such that for $n \geq \bar{n}_M$ we have

$$\frac{\#\{i : \sigma_i(A_n) > M\}}{d_n} \leq r(M), \quad \lim_{M\to\infty} r(M) = 0. \tag{43}$$

By invoking the singular value decomposition, we get

$$A_n = A^{(1)}_{n,M} + A^{(2)}_{n,M}, \quad \|A^{(1)}_{n,M}\|_\infty \leq M, \quad \text{rank } A^{(2)}_{n,M} \leq r(M)d_n. \tag{44}$$

It is almost trivial to see that if $\{A_n\} \sim_\sigma \theta$ with a measurable θ taking values in $\mathbf{C}\cup\{\infty\}$, then $\{A_n\}$ s.u. if and only if θ is sparsely unbounded, that is, $\lim_{M\to\infty} \mu\{x : |\theta(x)| > M\} = 0$ with $\mu\{\cdot\}$ denoting the usual Lebesgue measure. Furthermore, we observe that any function f belonging to L^1 is sparsely unbounded and that the product $\nu(x)$ of a finite number of s.u. functions is s.u. since the Lebesgue measure of the set where $|\nu(x)| = \infty$ is zero.

Proposition 4.6. [11] *Let $\{A_n\}_n$ and $\{B_n\}_n$, $A_n, B_n \in \mathcal{M}_{d_n}$, be two given s.u. matrix sequences. Suppose that*

$$\{\{Y_{A,n,m}\}_n\}_m \quad \text{and} \quad \{\{Y_{B,n,m}\}_n\}_m,$$

$m \in \hat{\mathbf{N}} \subset \mathbf{N}$, $\#\hat{\mathbf{N}} = \infty$, are two a.c.s. for $\{A_n\}$ and $\{B_n\}$, respectively. Then $\{\{Y_{A,n,m}Y_{B,n,m}\}_n\}_m$ is an a.c.s. for the sequence $\{A_nB_n\}_n$.

Proposition 4.7. *For any (α, β) belonging to a finite set, let $\{A_n^{(\alpha,\beta)}\}_n$ be a s.u. sequence of matrices of increasing dimensions, and let $\{\{Y_{(\alpha,\beta),n,m}\}_n\}_m$ be an a.c.s for $\{A_n^{(\alpha,\beta)}\}_n$. Then*

$$\{\{\textstyle\sum_{\alpha=1}^k [\prod(r)]_{\beta=1}^{q_\alpha} Y_{(\alpha,\beta),n,m}\}_n\}_m$$

is an a.c.s. for

$$\{\textstyle\sum_{\alpha=1}^k [\prod(r)]_{\beta=1}^{q_\alpha} A_n^{(\alpha,\beta)}\}_n$$

under the trivial condition that the dimensions agree.

Proof. Since the sum of two a.c.s for $\{A_n\}$ and $\{B_n\}$ is always an a.c.s for $\{A_n + B_n\}$, it is enough to use an inductive argument and Proposition 4.6 for the inductive step. □

Lemma 4.8. *Let $f \in L^1$ and let $\{p_m\}$ be a sequence of polynomials converging to f in the L^1 norm. Then $\{\{T_n(p_m)\}\}_m$ is an a.c.s. for $\{T_n(f)\}$.*

Proof. We point out that, for any m, the sequence $\{T_n(f)-T_n(p_m)\}$ coincides with $\{T_n(f - p_m)\}$ and therefore it suffices to exploit the singular value decomposition of $T_n(a-p_m)$ for large m and n and the assumption that p_m converges in L^1 norm (and therefore in measure) to f. Indeed, by assumption, by Corollary 3.5, and by what was said in Section 3.3, the estimate

$$\sum_{j=1}^{\min\{s,t\}\hat{n}} \sigma_j(T_n(f - p_m)) = \|T_n(f - p_m)\|_1 \leq \hat{n}\|f - p_m\|_{L^1} \tag{45}$$

holds for all n and m. Setting $k(m) = \|f - p_m\|_{L^1}$, we have $\#\{j : \sigma_j(T_n(f - p_m)) > \sqrt{k(m)}\} \le \hat{n}\sqrt{k(m)}$ and, by the singular value decomposition,

$$T_n(f - p_m)) = R_{n,m} + N_{n,m}$$

with rank $R_{n,m} \le \hat{n}\, c(m)$ and $\|N_{n,m}\|_\infty \le \omega(m)$, where $c(m) = \omega(m) = \sqrt{k(m)}$. Since $k(m)$ converges to zero as m tends to infinity the proof is complete. □

Theorem 4.9. *Let k and q_α, $\alpha = 1, \ldots, k$, be positive natural numbers and let $\{f_{\alpha,\beta} : \alpha = 1, \ldots, k,\ \beta = 1, \ldots, q_\alpha\}$ be a finite set of functions belonging to L^1. Then*

$$\{\textstyle\sum_{\alpha=1}^{k} [\prod(r)]_{\beta=1}^{q_\alpha} T_n(f_{\alpha\beta})\} \sim_\sigma \theta = \sum_{\alpha=1}^{k} [\prod(r)]_{\beta=1}^{q_\alpha} f_{\alpha\beta}.$$

Proof. It suffices to combine Lemma 4.8 and Proposition 4.7 with $A_n^{(\alpha,\beta)} = T_n(f_{\alpha\beta})$ and $Y_{(\alpha,\beta),n,m} = T_n(p_{m,\alpha\beta})$, where $p_{m,\alpha\beta}$ converges in L^1 norm to $f_{\alpha\beta}$. □

Remark 4.10. If $\theta = \sum_{\alpha=1}^{k} [\prod(r)]_{\beta=1}^{q_\alpha} f_{\alpha\beta}$ belongs to the L^1 class then it makes sense to consider the sequence $\{T_n(\theta)\}$. Actually it is easy to see that

$$\{\textstyle\sum_{\alpha=1}^{k} [\prod(r)]_{\beta=1}^{q_\alpha} T_n(f_{\alpha\beta})\} \quad \text{and} \quad \{T_n(\theta)\}$$

are equally distributed since $\{\sum_{\alpha=1}^{k} [\prod(r)]_{\beta=1}^{q_\alpha} T_n(f_{\alpha\beta}) - T_n(\theta)\}$ is clustered at zero.

Remark 4.11. The most classical (and successful) approach to the asymptotics for finite Toeplitz structures consists in using the corresponding infinite-dimensional Toeplitz operators $T(\cdot) = T_\infty(\cdot)$ (see e.g. [2, 3, 4, 24]). However, if the generating functions of the Toeplitz matrices belong to L^1, this approach becomes inconvenient, since the resulting operators are unbounded or even not at all defined. Hence, the "finite-dimensional" approach considered in [11, 22, 23] and in this note seems to be more versatile and flexible at least in this context. We nevertheless wish to point out that, for various much more intricate asymptotics, the C^*-algebra approach developed by Silbermann and his group furnishes, at least at the present moment, the most successful and general technique (see e.g. [2, 4, 8]).

Acknowledgement. I am indebted to Rajendra Bhatia for many suggestions and many discussions concerning the subject of this paper.

References

[1] R. Bhatia, *Matrix Analysis*, Springer-Verlag, New York 1997.

[2] A. Böttcher and S. Grudsky, *Toeplitz matrices, Asymptotic Linear Algebra, and Functional Analysis*, Hindustan Book Agency, New Delhi 2000 and Birkäuser Verlag, Basel 2000.

[3] A. Böttcher and B. Silbermann, *Analysis of Toeplitz Operators*, Springer-Verlag, Berlin, Heidelberg, New York 1990.

[4] A. Böttcher and B. Silbermann, *Introduction to Large Truncated Toeplitz Matrices*, Springer-Verlag, New York 1999.

[5] D. Fasino and S. Serra-Capizzano, *From Toeplitz matrix sequences to zero distribution of orthogonal polynomials, manuscript*, 2001.

[6] D. Fasino and P. Tilli, *Spectral clustering properties of block multilevel Hankel matrices*, Linear Algebra Appl. **306** (2000), 155–163.

[7] A. Frangioni and S. Serra-Capizzano, *Spectral analysis of (sequences of) graph matrices*, SIAM J. Matrix Anal. Appl. **23** (2001), 339–348.

[8] R. Hagen, S. Roch and B. Silbermann, *C^*-Algebras and Numerical Analysis*, Marcel Dekker Inc., New York 2001.

[9] W. Rudin, *Functional Analysis*, Tata Mc Graw-Hill, New Delhi 1974.

[10] S. Serra-Capizzano, *An ergodic theorem for classes of preconditioned matrices*, Linear Algebra Appl. **282** (1998), 161–183.

[11] S. Serra-Capizzano, *Distribution results on the algebra generated by Toeplitz sequences: a finite dimensional approach*, Linear Algebra Appl. **328** (2001), 121–130.

[12] S. Serra-Capizzano, *Spectral and computational analysis of block Toeplitz matrices with nonnegative definite generating functions*, BIT **39** (1999) 152–175.

[13] S. Serra-Capizzano, *Asymptotic results on the spectra of block Toeplitz preconditioned matrices*, SIAM J. Matrix Anal. Appl. **20** (1998), 31–44.

[14] S. Serra-Capizzano, *Test functions, growth conditions and Toeplitz matrices*, Proc. 4th Int. Conf. on Functional Anal. and Approx. Theory, Maratea 2000. Rend. Circolo Mat. Palermo, in press.

[15] S. Serra-Capizzano and C. Tablino-Possio, *Spectral and structural analysis of high precision finite difference matrices for elliptic operators*, Linear Algebra Appl. **293** (1999), 85–131.

[16] S. Serra-Capizzano and C. Tablino-Possio, *Positive representation formulas for second order (elliptic) partial differential equations*, Contemp. Math. **281** (2001), 295–317.

[17] S. Serra-Capizzano and P. Tilli, *On unitarily invariant norms of matrix valued linear positive operators*, J. Inequal. Appl., in press.

[18] S. Serra-Capizzano and P. Tilli, *From partial differential equations to generalized locally Toeplitz sequences*, TR. 12-99, Dept. Informatica, Univ. of Pisa, 1999. (http://www.di.unipi.it).

[19] P. Tilli, *Locally Toeplitz sequences: spectral properties and applications*, Linear Algebra Appl. **278** (1998), 91–120.

[20] P. Tilli, *A note on the spectral distribution of Toeplitz matrices*, Linear Multilinear Algebra **45** (1998), 147–159.

[21] E. Tyrtyshnikov, *Influence of matrix operations on the distribution of eigenvalues and singular values of Toeplitz matrices*, Linear Algebra Appl. **207** (1994), 225–249.

[22] E. Tyrtyshnikov, *A unifying approach to some old and new theorems on distribution and clustering*, Linear Algebra Appl. **232** (1996), 1–43.

[23] E. Tyrtyshnikov and N. Zamarashkin, *Spectra of multilevel Toeplitz matrices: advanced theory via simple matrix relationships*, Linear Algebra Appl. **270** (1998), 15–27.

[24] H. Widom, *Asymptotic behavior of block Toeplitz matrices and determinants II*, Adv. Math. **21** (1976), 1–29.

S. Serra-Capizzano
Dipartimento di Chimica, Fisica e Matematica
University of "Insubria" – Sede di Como
Via Valleggio 11
22100 Como, Italy
e-mail: `serra@mail.dm.unipi.it`
e-mail: `stefano.serrac@uninsubria.it`

Received: 24 September 2001

Operator Theory:
Advances and Applications, Vol. 135, 317–328

Toeplitz Determinants, Random Matrices and Random Permutations

Harold Widom

For Bernd Silbermann on the occasion of his sixtieth birthday

Abstract. In this article we describe how Toeplitz determinants arise in the theory of random matrices (the distribution of eigenvalues of a random unitary matrix) and in a problem concerning random permutations (the length of the longest increasing subsequence in a random permutation). In the first case the connection is direct and in the second case the connection is indirect, via some highly nontrivial combinatorics. In all examples the problem can be recast as one of determining the asymptotics of large Toeplitz determinants with variable symbol.

1. The unitary ensemle

An *ensemble* is a set with a probability measure. In physics, for which the theory of random matrices was invented [21], the most important matrix ensembles consist of Hermitian matrices of one sort or another. However, mathematically simpler, and where Toeplitz determinants arise directly, is the ensemble $\mathcal{U}(N)$ of $N \times N$ unitary matrices with Haar measure. (This is Dyson's *circular ensemble*, so called because all eigenvalues lie on the unit circle.)

The questions we will discuss concern the distribution of the eigenvalues $\lambda_1, \cdots, \lambda_N$ of a random unitary matrix U. Because of this we are interested more in the measure on this set of N-tuples than on $\mathcal{U}(N)$ itself. This is given by the *Weyl integration formula*, which states that the measure is given by the density function (probability density for the eigenvalues)

$$c_N \prod_{j<k} |\lambda_j - \lambda_k|^2.$$

Here c_N is a normalization constant making the total integral over $\mathbb{T}^N$ equal to one. Thus if $F(\lambda_1, \cdots, \lambda_N)$ is a function of the eigenvalues (best taken to be symmetric, since we do not order the λ_j) then

$$\mathrm{E}\left(F(\lambda_1, \cdots, \lambda_N)\right) = c_N \int \cdots \int \prod_{j<k} |\lambda_j - \lambda_k|^2 \, F(\lambda_1, \cdots, \lambda_N) \, d\theta_1 \cdots d\theta_N, \quad (1)$$

where $\lambda_j = e^{i\theta_j}$ and "E" denotes expected value.

When $F(\lambda_1, \cdots, \lambda_N)$ is a product of the form $\prod_{j=1}^N \varphi(\lambda_j)$ this is in fact a Toeplitz determinant. This is a consequence of a general identity expressing a multiple integral whose integrand is a product of two determinants in terms of a determinant whose entries are integrals [1]. Specifically, for a general measure μ

$$\int \cdots \int \det(f_j(x_k)) \det(g_j(x_k)) \, d\mu(x_1) \cdots d\mu(x_N)$$
$$= N! \det \left(\int f_j(x) \, g_k(x) \, d\mu(x) \right).$$

Now in the integrand in (1) we see a product of two Vandermonde determinants, one the complex conjugate of the other. Therefore if F is that product the factors $\varphi(\lambda_j)$ can be combined with the $d\theta_j$ and used as the $d\mu_j$ in the general formula. Carrying this through gives the formula

$$\mathrm{E}\left(\prod \varphi(\lambda_j)\right) = c_N' \det \left(\frac{1}{2\pi} \int e^{-ij\theta} \, e^{ik\theta} \, \varphi(e^{i\theta}) \, d\theta \right)$$

for another constant c_N'. Of course the determinant is just the Toeplitz determinant $D_N(\varphi)$. Also it is clear from taking $\varphi = 1$ that $c_N' = 1$ (and so c_N can also be determined). Thus

$$\mathrm{E}\left(\prod \varphi(\lambda_j)\right) = D_N(\varphi). \tag{2}$$

1.1. Eigenvalue distribution

How are the eigenvalues of a random unitary matrix U distributed? Given a subset A of the unit circle, let

$$\#A = \text{number of } j \text{ such that } \lambda_j \in A.$$

It is an easy consequence of the invariance of Haar measure that the expected value $\mathrm{E}(N^{-1}\#A)$ equals $m(A)$, the normalized Lebesgue measure of A. What we want is something sharper, the actual distribution function for $N^{-1}\#A$.

The random variable $\#A$ is a special case of a *linear statistic*. This is a sum of the form

$$\sum_{j=1}^N f(\lambda_j).$$

Its distribution measure μ is defined by

$$\mu(E) = \Pr\left(\sum f(\lambda_j) \in E\right),$$

and for a general function F we have $\int F(x) \, d\mu(x) = \mathrm{E}\,(F(\sum f(\lambda_j)))$. In particular the Fourier transform $\hat{\mu}(\alpha)$ of $\mu(x)$ equals $\mathrm{E}\,(e^{i\alpha \sum f(\lambda_j)})$, and by (2) this equals the Toeplitz determinant $D_N(e^{i\alpha f})$.

What comes to mind immediately is the Szegő limit theorem

$$D_N(e^{\psi}) = e^{N \int \psi \, dm(\theta) + o(N)}.$$

For the linear statistic $\psi = i\alpha f$ and in the case under consideration $f = N^{-1}\chi_A$. Of course this depends on N, but the larger the N the nicer the ψ and it can be checked by looking at a proof of the limit theorem that in this case the error term $o(N)$ is in fact $o(1)$. Hence

$$\hat{\mu}(\alpha) \to e^{i\,\alpha\, m(A)}$$

pointwise, and it follows from this that $\mu \to \delta_{m(A)}$ weakly. In other words, with high probability $N^{-1}\#A$ is very close to $m(A)$ for large N. Thus the eigenvalues of a random unitary matrix are in the limit uniformly distributed.[1]

1.2. Fluctuations

The quantity $N^{-1}\sum_{j=1}^{N} f(\lambda_j)$ is reminiscent of what one sees in the law of large numbers. If the λ_j were independent then we could say that this has as limit the expected value of f, in other words $\int f dm$. Even though the λ_j are not independent this same conclusion holds, as we saw in the last section. Once again if the λ_j were independent then the fluctuations

$$\sum_{j=1}^{N} f(\lambda_j) - N \int f dm$$

would be of order $\sqrt{N}$. More precisely, by the central limit theorem, the difference would be to a first approximation $\sqrt{N}$ times a Gaussian random variable. But in our case the central limit theorem does not apply and so the question is whether the fluctuations do have a nice limiting behavior. By subtracting from f its mean we may assume that f has mean zero.

Once again Szegő's theorem gives the answer, this time the strong Szegő limit theorem

$$D_N(e^{\psi}) \sim e^{N\int \psi\; dm(\theta) + \sum k\,\psi_k\,\psi_{-k}}.$$

which holds if ψ is well behaved and has mean zero. Hence in this case the distribution measure μ for the linear statistic satisfies

$$\hat{\mu}(\alpha) = D_N(e^{i\,\alpha\,f}) \to e^{-\alpha^2 \sum k\, f_k\, f_{-k}}.$$

Thus $\hat{\mu}(\alpha)$ converges to a Gaussian and it follows that μ itself converges to a Gaussian limit with variance $\sum k\, f_k\, f_{-k}$. Notice that there is no factor $\sqrt{N}$.[2]

What of the fluctuations for $\#A$ assuming, for example, that A is a finite union of intervals? Since χ_A has jump discontinuities the strong Szegő limit theorem as stated does not hold. But there is a limit theorem for symbols with jump discontinuities [3],

$$D_N(e^{\psi}) \sim e^{N\int \psi\, dm(\theta)}\, N^{\sum(\psi(\theta_r^+)-\psi(\theta_r^-))^2/4\pi^2}\, \gamma(\psi), \tag{3}$$

where $\gamma(\psi)$ is a certain constant, a product involving the Barnes G-function and the jumps $\psi(\theta_r^+) - \psi(\theta_r^-)$ of ψ. If we apply this with

$$\psi = i\,\alpha\, f, \quad f = \frac{\chi_A - N\, m(A)}{\sqrt{\log N}}$$

---

[1]This fact was first proved in [9] using representation theory and the moment method.

[2]This argument first appears in [13]. In [8] an independent derivation of the Gaussian limit was given, and this gives yet another proof of the strong Szegő theorem.

we find a limiting distribution for

$$\frac{\#A - N\,m(A)}{\sqrt{\log N}}.$$

Thus the fluctuations are of the order $\sqrt{\log N}$ rather than $O(1)$ as in the smooth case and from the form of the constant in (3) one finds, here also, that the limit is Gaussian.[3]

2. The Gaussian unitary ensemble

This is one of the matrix ensembles of greatest interest to physicists and fundamental contributions were made by Mehta, Dyson and Gaudin among others. (A comprehensive treatment is in Mehta's book [16].) Despite its name the ensemble consists of $N \times N$ Hermitian matrices. The *Gaussian* part of the name derives from the fact that each matrix entry is a Gaussian random variable and the *unitary* part from the fact that for each unitary U the underlying probability measure is invariant under the mapping $H \to UHU^{-1}$. For this ensemble, known for short as GUE, the probability density for the eigenvalues, which are now real, is given by

$$c_N \prod_{j<k} (\lambda_j - \lambda_k)^2 \, e^{-\sum \lambda_j^2} \tag{4}$$

for some normalization constant c_N. The expected value of a product $\prod \varphi(\lambda_j)$ is now equal to a Hankel determinant.

Although not the case in $\mathcal{U}(n)$, in GUE there is a largest eigenvalue $\lambda_{\max}$ and its behavior for large N is known. To a first approximation $\lambda_{\max} \sim \sqrt{2N}$ and the finer result giving the fluctuations [10] is

$$\lim_{N\to\infty} \Pr\left(\lambda_{\max} \le \sqrt{2N} + \frac{s}{\sqrt{2}\,N^{1/6}}\right) = \det\left(I - K_{\mathrm{Airy}}\right)$$

on $L^2(s,\infty)$. Here the *Airy kernel* is given by

$$K_{\mathrm{Airy}}(x,\,y) = \int_0^\infty \mathrm{Ai}(x+z)\,\mathrm{Ai}(z+y)\,dz, \tag{5}$$

where Ai is the Airy function. It satisfies the differential equation $\mathrm{Ai}''(x) = x\,\mathrm{Ai}(x)$ and vanishes rapidly at $+\infty$.

This determinant, as a function of s, is one of the distribution functions of random matrix theory which is expressible in terms of a Painlevé function [18].[4] The distribution function is denoted by $F_2(s)$. (The subscript comes from the first exponent in (4). There are analogous distribution functions F_1 and F_4 representable in terms of the same Painlevé function.)

---

[3]This argument is formal and to carry it through one has to establish uniformity in (3) for the family of symbols $e^{i\alpha f}$. This was carried out in [20]. A completely different proof is in [8].

[4]The solution to the Painlevé II equation $q''(s) = s\,q(s) + 2\,q(s)^3$ which is asymptotically equal to $\mathrm{Ai}(s)$ as $s \to +\infty$.

3. Random permutations

There is a natural probility measure on the symmetric group S_N. Simply give each $\sigma \in S_N$ the propability $1/N!$. An *increasing subsequence* of a permutation $k \to \sigma_k$ is a sequence of the form $\sigma_{k_1} < \sigma_{k_2} < \cdots < \sigma_{k_n}$ with $k_1 < k_2 < \cdots < k_n$. For example, the permutation σ written in the usual way as

$$\sigma = \begin{pmatrix} 1\ 2\ 3\ 4\ 5 \\ 3\ 2\ 4\ 5\ 1 \end{pmatrix}$$

has increasing subsequences 3, 4, 5 and 2, 4, 5 and 2, 5 among others. Denote by $\ell_N(\sigma)$ the length of the longest increasing subsequence of σ. *Ulam's conjecture* states that the expected value of ℓ_N has the asymptotic behavior

$$E(\ell_N) \sim c\sqrt{N}$$

as $N \to \infty$ for some c. The conjecture was proved true in the late 70s [15, 19], the value of c being 2. There the matter stood for some 20 years, with the question remaining of how to get a finer result. Namely, can one find the asymptotic behavior of the fluctuations $\ell_N - 2\sqrt{N}$? The answer is yes [2], and the formula is

$$\lim_{N\to\infty} \Pr\left(\ell_N \le 2\sqrt{N} + s\,N^{1/6}\right) = F_2(s). \tag{6}$$

Thus ℓ_N, after centering and normalization, has the same limiting distribution as the largest eigenvalue in the GUE random matrix ensemble!

How this is established is a long story involving some highly nontrivial combinatorics before getting to the Toeplitz determinant connection. The reader not interested in the abbreviated version of the story we are about to give can go directly to formula (8). A good reference for the combinatorics is Stanley's book [17].

3.1. Combinatorics

3.1.1. Young tableaux Take a rectangular array of boxes with n rows, and for $k = 1, \ldots, n$ remove from the kth row all but the first μ_k boxes, where $\mu_1 \ge \mu_2 \ge \cdots \ge \mu_n$. If $N = \sum \mu_k$ then $\mu = (\mu_1, \mu_2, \ldots, \mu_n)$ is called a *partition of* N (denoted $\mu \vdash N$) of length at most n and the array of remaining boxes is called the *Young diagram* for this partition. The diagram is said to have the *shape* μ. A *standard Young tableau* is such a Young diagram in which the integers 1, 2, ..., N are placed in the boxes in such a way that the integers in each row form an increasing sequence and the integers in each column form an increasing sequence.

The *Robinson-Schensted correspondence* is a one-one correspondence between permutations $\sigma \in S_N$ and pairs of standard Young tableaux with the same shape $\mu \vdash N$. Under this correspondence $\ell_N(\sigma)$ is equal to the length μ_1 of the first row. Hence the number of permutations $\sigma \in S_N$ such that $\ell_N(\sigma) \le n$ equals the square of the number of standard Young tableaux whose first row length μ_1 is at most n. By taking the transposes of the tableaux we see that this equals the square of the number of standard Young tableaux for partitions $\mu \vdash N$ whose length $\ell(\mu)$ is at most n. Thus, if f^μ denotes the number of standard Young tableaux with shape μ, then

$$\Pr(\ell_N \le n) = \frac{1}{N!} \sum_{\substack{\mu \vdash N \\ \ell(\mu) \le n}} (f^\mu)^2 . \tag{7}$$

3.1.2. Schur functions A *semi-standard* Young tableau differs from a standard tableau in that the entries in each row may form a nondecreasing sequence rather than an increasing sequence. (The entries in each column must still be increasing.) The Schur function $s_\mu(x)$, where $x = (x_1, x_2, \cdots)$, equals the sum over all semi-standard Young tableaux of $x_1^{\alpha_1} x_2^{\alpha_2} \cdots$, where α_k denotes the number of times the number k appears in the tableau. Thus one expects sums of the form $\sum_\mu s_\mu(x)\, s_\mu(y)$ to prove useful in counting pairs of Young tableaux with the same shape.

3.1.3. Gessel's theorem Schur function also have representations as determinants. Denote by $h_r(x)$ the rth complete symmetric function of the x_i. These can be defined by

$$\sum_{r=0}^{\infty} h_r(x)\, t^r = \prod_i (1 - x_i\, t)^{-1} .$$

If $\mu = (\mu_1, \mu_2, \cdots, \mu_n)$ is a partition the *Jacobi-Trudi indentity* for s_μ is

$$s_\mu(x) = \det \left(h_{\mu_i + j - i}(x)\right)_{1 \le i,j \le n} .$$

Using this representation Gessel [11] showed that

$$\sum_{\ell(\mu) \le n} s_\mu(x)\, s_\mu(y) = D_n(\varphi),$$

where

$$\varphi(z) = \sum_{i=-\infty}^{\infty} z^i \sum_{\ell=0}^{\infty} h_{\ell+i}(x) h_\ell(y) = \prod_i \left(1 - y_i\, z^{-1}\right)^{-1} \prod_i (1 - x_i\, z)^{-1} .$$

Using this identity and certain homomorphisms on the algebra of polynomials in x and y (*specializations*) he deduced the following representation for the generating function for the probabilities we are interested in:

$$\sum_{N=0}^{\infty} \Pr(\ell_N \le n)\, \frac{t^N}{N!} = D_n(e^{\sqrt{t}\,(z + z^{-1})}). \tag{8}$$

The fact that it is the generating function that is so represented rather than the probabilities themselves is due to the fact that the Schur functions are defined in terms of semi-standard Young tableaux whereas the probabilities in (7) are given in terms of standard tableaux.

3.2. Asymptotics

The main contribution to the sum in (8) comes from where N is approximately equal to t. Thus it seems likely that for large integer t we have the asymptotics $\Pr(\ell_t \le n)\, e^t \sim D_n(e^{\sqrt{t}\,(z+z^{-1})})$ or, replacing t by N,

$$\Pr(\ell_N \le n) \sim e^{-N}\, D_n(e^{\sqrt{N}\,(z+z^{-1})}).$$

It is not difficult to show that this is correct in the range of n which will arise, so the problem becomes the asymptotics of the Toeplitz determinant on the right. In view of the solution to Ulam's problem one expects that taking n approximately $2\sqrt{N}$ is what will lead to the interesting asymptotics. (If n is taken a little too large the limiting probability will be one and if n is taken a little too small the limiting probability will be zero.)

In [2] the asymptotics were derived by an analysis of a matrix Riemann-Hilbert problem. A certain 2×2 piecewise analytic matrix with prescribed discontinuities had encoded in it information about the Toeplitz determinant and an asymptotic analysis of the solution gave the asymptotics of the determinant. The method is very powerful and applies to a large class of asymptotic problems. (For a description of the method see the expository article [7].) But it is also very involved and for problems of the class we are considering there is a simpler method. It begins with a formula of Borodin and Okounkov which allows one to handle the asymptotics by classical steepest descent methods. The method is used, with full details worked out, for a different problem in [12]. The method as applied to the current problem has not appeared in the literature, which is why we go into some detail here.[5] It is an interesting fact that in [2] F_2 appears in its representation in terms of the Painlevé II function whereas here it will arise as the Fredholm determinant of the Airy kernel.

3.2.1. The Borodin-Okounkov identity Suppose φ has a Wiener-Hopf factorization $\varphi = \varphi_+\, \varphi_-$, so that φ_+ (resp. φ_-) extends to a nonzero analytic function in the interior (resp. exterior) of $\mathbb{T}$. Assume that φ has geometric mean one, and the factors are normalized so that $\varphi_+(0) = \varphi_-(\infty) = 1$. The Borodin-Okounkov identity [4] (a very simple proof is in [6]) states that under general conditions

$$D_n(\varphi) = Z\ \det\left(\delta_{jk} - \sum_{\ell=1}^{\infty} \left(\frac{\varphi_-}{\varphi_+}\right)_{n+j+l} \left(\frac{\varphi_+}{\varphi_-}\right)_{-n-l-k}\right)_{j,\,k\ge 0} \tag{9}$$

where $Z = \lim_{n\to\infty} D_n(\varphi)$ and the subscripts on the right side denote Fourier coefficients. Although the formula replaces a finite determinant by an infinite determinant it also replaces a Toeplitz matrix by a product of two Hankel matrices, which is why the right side is easier to handle than the left. In our case $\varphi(z) = e^{\sqrt{N}\,(z+z^{-1})}$ so by Szegő $Z = e^N$. Therefore $\Pr(\ell_N \le n)$ is asymptotically equal to the determinant on the right side.

[5] This is not the first "non-Riemann-Hilbert" proof of the result. See the next footnote.

We see there the product of two Hankel matrices acting on $\ell^2(\mathbb{Z}^+)$ with j, k entries respectively[6]

$$\left(\frac{\varphi_-}{\varphi_+}\right)_{n+j+k+1} = \frac{1}{2\pi i}\int_{\mathbb{T}} e^{\sqrt{N}(z^{-1}-z)}\, z^{-n-j-k-1}\,\frac{dz}{z},$$

$$\left(\frac{\varphi_+}{\varphi_-}\right)_{-n-j-k-1} = \frac{1}{2\pi i}\int_{\mathbb{T}} e^{\sqrt{N}(z-z^{-1})}\, z^{n+j+k+1}\,\frac{dz}{z}.$$

For future convenience we note that the determinant will be unchanged if the entries of both matrices are multiplied by $(-1)^{n+j+k+1}$. Upon making the variable change $z \to -z$ in the two integrals we see that this means that we can replace them by

$$\frac{1}{2\pi i}\int_{\mathbb{T}} e^{\sqrt{N}(z-z^{-1})}\, z^{-n-j-k-1}\,\frac{dz}{z} = M^-(j,k), \tag{10}$$

$$\frac{1}{2\pi i}\int_{\mathbb{T}} e^{\sqrt{N}(z^{-1}-z)}\, z^{n+j+k+1}\,\frac{dz}{z} = M^+(j,k). \tag{11}$$

And we have the asymptotics

$$\Pr(\ell_N \le n) \sim \det(I - M^-\, M^+).$$

It turns out that if we choose n just right then these matrices will *scale*, in a sense to be described, to the same operator kernel, so the product will scale the operator square of this kernel. And that will give the Airy kernel (5).

To explain scaling, let $M(j,k)$ be a matrix depending on a parameter N and thought of as an operator on $\ell^2(\mathbb{Z}^+)$. It has exactly the same Fredholm determinant (operator determinant of I minus the operator) as the kernel $M([x],\,[y])$ acting on $L^2(\mathbb{R}^+)$. This in turn has the same Fredholm determinant as the kernel $N^{1/6}\,M([N^{1/6}x],\,[N^{1/6}y])$. (In general one could use any variable change.) Now suppose that with error having trace norm $o(1)$ we can make the replacements $[N^{1/6}x] \to N^{1/6}x$ and $[N^{1/6}y] \to N^{1/6}y$, and then that as $N \to \infty$

$$N^{1/6}\,M(N^{1/6}x,\,N^{1/6}y) \to K(x,y)$$

in trace norm. Then we say that with the substitutions $j \to N^{1/6}x,\quad y \to N^{1/6}y$ the matrix $M(j,\,k)$ scales in trace norm to $K(x,\,y)$. In this case the Fredholm determinant of the matrix acting on $\ell^2(\mathbb{Z}^+)$ converges as $N \to \infty$ to the Fredholm determinant of $K(x,\,y)$ acting on $L^2(\mathbb{R}^+)$. If our matrix is the product of two matrices which scale (with the same substitutions) to kernels in Hilbert-Schmidt norm then the product of the matrices scales in trace norm to the product of the kernels.

This is what will happen in our case. We shall find that if $n = 2\sqrt{N} + s\,N^{1/6}$ then with the indicated substitutions both matrices $M^{\pm}$ scale in Hilbert-Schmidt

[6]These are Bessel functions and the matrix on the right side of (9), the *discrete Bessel kernel*, is central to the proofs of (6) given in [5, 14].

norm to $\mathrm{Ai}(s+x+y)$ on $(0,\infty)$. Therefore the product scales in trace norm to $K_{\mathrm{Airy}}(s+x,\,s+y)$ on $L^2(\mathbb{R}^+)$, and this gives (6).[7]

3.2.2. THE METHOD OF STEEPEST DESCENT This applies to contour integrals of the form

$$\int_C e^{m\,f(z)}\,g(z)\,dz$$

where f and g are analytic in z and $m \to +\infty$. The method consists of deforming the contour to one on which f has constant imaginary part and the real part of f takes on a maximum. In the simplest case this will occur at a single point z_0, which is necessarily a critical point of f, a point where $f' = 0$. Then the integral is asymptotically equal to what one gets by replacing g by $g(z_0)$, replacing f by the beginning of its Taylor expansion near $z = z_0$, and replacing the contour by the tangent line(s) to it at z_0. One finds in this way that the integral is to a first approximation $e^{m\,f(z_0)}$. Typically $f''(z_0) \neq 0$, in which case the integral is asymptotically

$$e^{m\,f(z_0)}\sqrt{\frac{2\pi}{f''(z_0)\,m}}\,g(z_0).$$

There are many variants but this is the basic idea.

Let us write $n+j+k+1 = n_0 + n'$ in (10) and (11), where n_0 is thought of as the main part of n and it depends only on N. Then in our case we have the two integrals

$$\frac{1}{2\pi i}\int_{\mathbb{T}} \psi(z)^{\pm 1}\, z^{\pm n'}\,\frac{dz}{z},$$

where we have set

$$\psi(z) = e^{\sqrt{N}(z^{-1}-z)}\,z^{n_0}.$$

If we think of either of $\psi(z)^{\pm 1}$ as the $e^{m\,f(z)}$ of the above discussion then in both cases the critical points are

$$\frac{n_0}{2\sqrt{N}} \pm \sqrt{\frac{n_0^2}{4N}-1}.$$

This already suggests that the solution to Ulam's problem should be $E(\ell_N) \sim 2\sqrt{N}$, because that is the most interesting value of n_0.

A little more precisely, suppose we were to choose n_0 to be $2\sqrt{N}$ exactly. Then there would be just one critical point $z = 1$, and to a first approximation the two integrals would be reciprocals of each other and so their product (which is what in the end we are interested in) would be of the order 1. This suggests

---

[7]Since n and N must be integers we should not require s to be fixed. In fact the limit holds uniformly for s in any bounded set. We mention also that since our integrals are Bessel functions, for which uniform asymptotics are known, the discussion which follows is in a sense unnecessary. But it will explain why Ulam's conjecture, with the correct constant, is true, and how the Airy kernel arises naturally. And the method applies equally well to other problems for which the asymptotics are not in the literature.

strongly that we do take $n_0 = 2\sqrt{N}$ (in any case there is nothing that prevents us from doing this) and so define n' by

$$n + j + k + 1 = 2\sqrt{N} + n'.$$

But now, since the two critical points coincide, the behavior of ψ near its critical point is not as described in the first paragraph of this section. In fact, in the neighborhood of $z = 1$

$$\psi(z) \sim e^{-\sqrt{N}(z-1)^3/3}.$$

There will be two steepest descent curves, C^+ on which the real part of ψ has its maximum at $z = 1$ and C^- on which the real part of ψ has its minimum at $z = 1$. The curve C^+ comes into $z = 1$ in the directions $\arg(z-1) = \pm 2\pi/3$ and closes with a cusp at $z = 0$ tangent to the negative real axis while C^- comes into $z = 1$ in the directions $\arg(z-1) = \pm\pi/3$ and its two branches go to infinity parallel to the negative real axis. In (10) the contour of integration $\mathbb{T}$ is replaced by C^- and in (11) it is replaced by C^+.

In preparation for scaling let us make the replacements

$$j \to N^{1/6}x, \quad k \to N^{1/6}y$$

and set $n = 2\sqrt{N} + s\,N^{1/6}$. Then after we multiply by $N^{1/6}$, as scaling says we must, our integrals with their factors become

$$\frac{1}{2\pi i} N^{1/6} \int_{C^\pm} \psi(z)^{\pm 1}\, z^{\pm N^{1/6}(s+x+y)\pm 1} \frac{dz}{z}.$$

In the neighborhood of $z = 1$ the integrands are asymptotically

$$e^{\mp\sqrt{N}(z-1)^3/3 \pm N^{1/6}(s+x+y)(z-1)}.$$

If we replace the integrands by these and integrate only in a small neighborhood of $z = 1$, as we may without changing the asymptotics, we can then without changing the asymptotics integrate over the rays emanating from $z = 1$ at angles $\pm 2\pi/3$ for the integral over C^+ and at angles $\pm\pi/3$ for the integral over C^-. Both integrals are described upward. The resulting integrals are unchanged if the integration is taken over the vertical line through $z = 1$. After making the variable change $z = 1 + iN^{-1/6}\zeta$ in the integrals they become

$$\frac{1}{2\pi} \int_{-\infty}^{\infty} e^{\pm i\zeta^3/3 \pm i(s+x+y)\zeta}\, d\zeta.$$

Both of these equal $\mathrm{Ai}(s + x + y)$.

3.2.3. Conclusion What does this show and what do we need? What we need is that, when $n = 2\sqrt{N} + s\,N^{1/6}$,

$$\lim_{N\to\infty} N^{1/6}\, M^{\pm}([N^{1/6}x],\, [N^{1/6}y]) = \mathrm{Ai}(s + x + y)$$

in Hilbert-Schmidt norm, for then the product $M^- M^+$ would scale in trace norm to $K_{\text{Airy}}(s+x,\, s+y)$ whose Fredholm determinant on $\mathbb{R}^+$ is $F_2(s)$. What the above argument gives (it is easy to fill in the details to make it rigorous) is that

$$\lim_{N\to\infty} N^{1/6}\, M^{\pm}(N^{1/6}x,\, N^{1/6}y) = \text{Ai}(s+x+y)$$

pointwise. There is no problem with the greatest integer function since the last asymptotics hold uniformly for x and y in bounded sets. But in order to prove convergence in Hilbert-Schmidt norm we also need uniform estimates for all x and y. For this the actual steepest descent argument is more complicated than we indicated since the extra factors $z^{\pm n'}$ must be taken into account. Although carrying this out is a little tedious, it is completely straightforward.

References

[1] C. Andréief, *Note sur une relation les intégrales définies des produits des fonctions*, Mém. de la Soc. Sci., Bordeaux **2** (1883), 1–14.

[2] J. Baik, P. Deift and K. Johansson, *On the distribution of the length of the longest increasing subsequence of random permutations*, J. Amer. Math. Soc. **12** (1999), 1119–1178.

[3] E.L. Basor, *A localization theorem for Toeplitz determinants*, Indiana U. Math. J. **28** (1979), 975–983.

[4] A. Borodin and A. Okounkov, *A Fredholm determinant formula for Toeplitz determinants*, Int. Eqs. Oper. Th. **37** (2000), 386–396.

[5] A. Borodin, A. Okounkov and G. Olshanski, *On asymptotics of Plancherel measures for symmetric groups*, J. Amer. Math. Soc. **13** (2000), 481–505.

[6] A. Böttcher, *On the determinant formulas by Borodin, Okounkov, Baik, Deift, and Rains*, This volume.

[7] P. Deift, *Integrable systems and combinatorial theory*, Notices Amer. Math. Soc. **47** (2000), 631–640.

[8] P. Diaconis and S.N. Evans, *Linear functionals of eigenvalues of random matrices*, Trans. Amer. Math. Soc. **353** (2001), 2615–2633.

[9] P. Diaconis and M. Shahshahani, *On the eigenvalues of random matrices*, J. Appl. Prob. **31A** (1994), 49–62.

[10] P.J. Forrester, *The spectrum edge of random matrix ensembles*, Nuclear Phys. **B 402** (1993), 709–728

[11] I.M. Gessel, *Symmetric functions and P-recursiveness*, J. Comb. Theory, Ser. A **53** (1990), 257–285.

[12] J. Gravner, C.A. Tracy and H. Widom, *Limit theorems for height fluctuations in a class of discrete space and time growth models*, J. Stat. Phys. **102** (2001), 1085–1132.

[13] K. Johansson, *On random matrices from the compact classical groups*, Ann. of Math. (2) **145** (1997), 519–545.

[14] K. Johansson, *Discrete orthogonal polynomial ensembles and the Plancherel measure*, arXiv: math.CO/9906120.

[15] B.F. Logan and L.A. Shepp, *A variational problem for random Young tableaux*, Adv. in Math. **26** (1977), 206–222.

[16] M.L. Mehta, *Random Matrices*, 2nd ed., Academic Press, San Diego, 1991.

[17] R.P. Stanley, *Enumerative Combinatorics*, Vol. 2, Cambridge University Press, Cambridge, 1999.

[18] C.A. Tracy and H. Widom, *Level-spacing distributions and the Airy kernel*, Commun. Math. Phys. **159** (1994), 151–174.

[19] A.M. Vershik and S.V. Kerov, *Asymptotics of the Plancherel measure of the symmetric group and the limiting form of Young tables*, Soviet Math. Doklady, **18** (1977), 527–531.

[20] K.L. Wieand, *Eigenvalue distributions of random matrices in the permutation group and compact Lie groups*, Ph. D. Thesis, Harvard University, 1998.

[21] E.P. Wigner, *On the statistical distribution of the widths and spacings of nuclear resonance*, Proc. Cambridge Phil. Soc. **47** (1951), 790–798.

H. Widom
Department of Mathematics
University of California
Santa Cruz, CA 95064, USA
e-mail: `widom@math.ucsc.edu`

Received: 6 September 2001

GPSR Compliance
The European Union's (EU) General Product Safety Regulation (GPSR) is a set of rules that requires consumer products to be safe and our obligations to ensure this.

If you have any concerns about our products, you can contact us on

ProductSafety@springernature.com

In case Publisher is established outside the EU, the EU authorized representative is:

Springer Nature Customer Service Center GmbH
Europaplatz 3
69115 Heidelberg, Germany

www.ingramcontent.com/pod-product-compliance
Ingram Content Group UK Ltd.
Pitfield, Milton Keynes, MK11 3LW, UK
UKHW021447280726
14060UKWH00001BA/282

* 9 7 8 3 7 6 4 3 6 8 7 7 7 *